WEST ACADEMIC PUBLISHING'S
EMERITUS ADVISORY BOARD

JESSE H. CHOPER
Professor of Law and Dean Emeritus
University of California, Berkeley

YALE KAMISAR
Professor of Law Emeritus, University of San Diego
Professor of Law Emeritus, University of Michigan

MARY KAY KANE
Late Professor of Law, Chancellor and Dean Emeritus
University of California, Hastings College of the Law

LARRY D. KRAMER
President, William and Flora Hewlett Foundation

JAMES J. WHITE
Robert A. Sullivan Emeritus Professor of Law
University of Michigan

WEST ACADEMIC PUBLISHING'S
LAW SCHOOL ADVISORY BOARD

MARK C. ALEXANDER
Arthur J. Kania Dean and Professor of Law
Villanova University Charles Widger School of Law

JOSHUA DRESSLER
Distinguished University Professor Emeritus
Michael E. Moritz College of Law, The Ohio State University

MEREDITH J. DUNCAN
Professor of Law
University of Houston Law Center

RENÉE McDONALD HUTCHINS
Dean and Joseph L. Rauh, Jr. Chair of Public Interest Law
University of the District of Columbia David A. Clarke School of Law

RENEE KNAKE JEFFERSON
Joanne and Larry Doherty Chair in Legal Ethics &
Professor of Law, University of Houston Law Center

ORIN S. KERR
William G. Simon Professor of Law
University of California, Berkeley

JONATHAN R. MACEY
Professor of Law,
Yale Law School

DEBORAH JONES MERRITT
Distinguished University Professor,
John Deaver Drinko/Baker & Hostetler Chair in Law Emerita
Michael E. Moritz College of Law, The Ohio State University

ARTHUR R. MILLER
University Professor and Chief Justice Warren E. Burger Professor of
Constitutional Law and the Courts, New York University

GRANT S. NELSON
Professor of Law Emeritus, Pepperdine University
Professor of Law Emeritus, University of California, Los Angeles

A. BENJAMIN SPENCER
Dean & Chancellor Professor of Law
William & Mary Law School

BIOETHICS
HEALTH CARE LAW AND ETHICS

Ninth Edition

■ ■ ■

Brietta R. Clark
Professor of Law and J. Rex Dibble Fellow,
LMU Loyola Law School, Los Angeles

Erin C. Fuse Brown
Catherine C. Henson Professor of Law and Director,
Center for Law, Health & Society
Georgia State University College of Law

Robert Gatter
Professor of Law and Director, Center for Health Law Studies
Saint Louis University School of Law

Elizabeth Y. McCuskey
Professor of Law
University of Massachusetts School of Law

Elizabeth Pendo
Joseph J. Simeone Professor of Law
Saint Louis University School of Law

AMERICAN CASEBOOK SERIES®

WEST
ACADEMIC
PUBLISHING

The publisher is not engaged in rendering legal or other professional advice, and this publication is not a substitute for the advice of an attorney. If you require legal or other expert advice, you should seek the services of a competent attorney or other professional.

American Casebook Series is a trademark registered in the U.S. Patent and Trademark Office.

COPYRIGHT © 1987, 1991 WEST PUBLISHING CO.
© West, a Thomson business, 1997, 2001, 2004, 2008
© 2013 LEG, Inc. d/b/a West Academic Publishing
© 2018 LEG, Inc. d/b/a West Academic
© 2022 LEG, Inc. d/b/a West Academic
      444 Cedar Street, Suite 700
      St. Paul, MN 55101
      1-877-888-1330

West, West Academic Publishing, and West Academic are trademarks of West Publishing Corporation, used under license.

Printed in the United States of America

To our wonderful colleagues—the health law students and teachers who have used this casebook over the last thirty-five years.

PREFACE

This ninth edition of this casebook marks the thirty-fifth anniversary edition of our text, first published in 1987. Since that first edition, the American health care system has undergone striking changes. The transformation in health care and in health law reaches deeply into issues that fall under the umbrella of bioethics. In 1987, the law relating to health care ethics issues was fairly lean, but in the years since, there has been a constant wave of state and federal legislation, judicial opinions, and administrative regulations addressing issues in bioethics.

Since the last edition of this text, the COVID-19 pandemic has inflicted devastating loss of life and health, as well as profound economic and social disruption. It has exposed and exacerbated inequities experienced by racial and ethnic minorities, people with disabilities, and other disadvantaged groups, underscoring the growing view of health law and bioethics as part of a broader framework that encompasses movements addressing social inequities and injustices. While the perspective that we must bring to the legal analysis of issues relating to bioethics is necessarily more extensive now than it was thirty-five years ago, many of the fundamental concerns on which that analysis is brought to bear are surprisingly unchanged. We still want to know what role the law might play in promoting or mediating the values at issue in health care including protection of the human rights and the individual values of those who are provided care within the health care system, individual rights in treatment decision-making, access to necessary health care, competing interests and goals in medical research, and the balance of the individual and community in public health. We want to explore the nature of the relationship between law and health care ethics, both in areas where there are shared values and in areas where there are conflicts. We want to test basic legal and ethical principles in application to specific health care issues. The ninth edition centers the broader goal of a more just and equitable health care system and addresses fundamental concerns of health law and ethics through this equity lens.

The ninth edition of this casebook incorporates issues of equity and justice throughout the broad organization that teachers and students found so helpful in the prior editions. As was the case in previous editions, we employ materials from a variety of sources. This book continues to contain the most significant and useful judicial opinions, state and federal statutes, legislative history, administrative regulations, forms, and a host of other kinds of materials designed to bring the subject to life in the classroom. It also contains many classroom-tested problems that should be helpful in encouraging student engagement with these materials. While many of the problems and other materials have been brought forward from earlier editions of this book, every section of this casebook has been revised to reflect recent changes. Like other areas of health law, the law relating to bioethics is undergoing continuing and rapid change, especially in relation to federal administrative regulations. This book accounts for changes through mid-2021, and teachers using this book can find updates and supplementary material in the Teacher's Manual and on the West website for the *Health Law* casebook. All cases, statutes, regulations, and other materials in the casebook have been edited to enhance their teaching value while assuring that they reflect problems faced by health care professionals, patients, policymakers, and lawyers coping with the health system today and in the future. The notes expose students to a range of the most subtle health law and bioethics inquiries under discussion at the time of publication, including issues raised by the COVID-19 pandemic.

Derived from the comprehensive material contained in our casebook, *Health Law: Cases, Materials, and Problems (West 9th ed.)*, this volume is custom made for courses focusing on issues in bioethics. *Bioethics: Health Care Law and Ethics* is one of three paperback volumes that collect chapters from our *Health Law: Cases, Materials, and Problems* casebook. The other volumes are *The Law of Health Care Organization and Finance* and *Law and Health Care Quality, Patient Safety, and Liability*. Each of these volumes is designed to present materials for courses with a more limited focus than what would be covered in a broader survey. Finally, there is also an abridged version of the comprehensive *Health Law* casebook which includes only limited coverage of bioethics.

Other than the first chapter, the chapters in this paperback volume are the same as they appear in the larger *Health Law* casebook. It is quite easy to teach the course with some students using the complete *Health Law* casebook and others using this volume. Students usually appreciate being able to have that choice.

The first chapter in *Bioethics: Health Care Law and Ethics* introduces students to fundamental theories of health care ethics, including codes of professional ethics. The chapter includes discussion of the relationship between law and ethics, including in matters of conscience, and highlights issues of justice and equity incorporated throughout the remaining chapters. This chapter does not appear in the *Health Law* casebook, so teachers are welcome and have permission to distribute this first chapter to students in their class who may be using the larger casebook for the bioethics class.

This book includes expanded discussion of discrimination and unequal treatment in health care delivery and insurance, as well as fully updated coverage of the current status of laws that regulate human reproduction—abortion, contraception, sterilization, decision-making during pregnancy, and assisted reproductive technology. It provides extensive discussions of the controversies over the definition of death, organ transplantation, the law of health care decision-making, and medically assisted dying. The casebook also includes a review of legal regulation of research involving human subjects, and it concludes with a chapter that addresses developments in public health. Throughout these chapters, we have included materials on genetics where appropriate and have expanded our treatment of justice, equity, and discrimination across virtually every issue covered. All of these materials have been reviewed to assure that a wide range of perspectives leaven the authors' analysis of health law.

This casebook is designed to be a teachable book. We are grateful for the many comments and helpful suggestions that health law teachers across the United States (and from elsewhere, too) have made to help us improve this new edition. We attempt to present all sides of policy issues, not to evangelize for any political, economic or social agenda of our own. This task is made easier, undoubtedly, by the diverse views the several different authors of this casebook bring to this endeavor. A great number of very well-respected health law teachers have contributed to this and previous editions by making suggestions, reviewing problems, or encouraging our more thorough investigation of a wide range of health law subjects. We are especially grateful to Charles Baron, Eugene Basanta, David Bennahum, Robert Berenson, Kathleen Boozang, Kathy Cerminara, Don Chalmers, Ellen Wright Clayton, Judith Daar, Dena Davis, Arthur Derse, Kelly Dineen, Ileana Dominguez-Urban, Stewart Duban, Barbara Evans, Margaret Farrell, Rob Field, David Frankford, Michael Gerhart, Joan McIver Gibson, Susan Goldberg, Jesse Goldner, Andrew Grubb, Sarah Hooper, Art LaFrance, Diane Hoffmann, Jill Horwitz, Amy Jaeger, Eleanor Kinney, Thomasine Kushner, Pam Lambert, Theodore LeBlang, Antoinette Sedillo Lopez, Mary Pareja, Lawrence Singer, Joan Krause, Leslie Mansfield, Thomas Mayo, Maxwell Mehlman, Alan Meisel, Vicki Michel, Frances Miller, John Munich, David Orentlicher, Vernellia Randall, Ben Rich, Arnold Rosoff, Karen Rothenberg, Mark Rothstein, Sallie Sanford, Giles Scofield, Jeff Sconyers, Charity Scott, Ross Silverman, Loane Skene, George Smith, Roy Spece, Jr., Carol Suzuki, Michael Vitiello, Sidney Watson, Lois Weithorn, Ellen Wertheimer, William Winslade and Susan M. Wolf for the benefit of their wisdom and experience.

We wish to thank those remarkable research assistants who provided support for our research and the preparation of the manuscript, Joseph Allen, Debra Au, Elizabeth Bertolino, Elysia Buckley, Laura Hagen, Joshua Hasyniec, Jordan Hobbs, Menqi Rebecca Hsu, Delanie Inman, and Adella Katz. We are also very appreciative of the tremendous support and the publication assistance provided by Greg Olson, Jon Harkness and Cathy Lundeen of West Academic Publishing. We are also indebted to our casebook shepherds—Mary Ann Jauer for the first 25 years of the casebook, and now Cheryl Cooper. We appreciate those who were there for us during the nights and weekends we spent working on this project: Roger Fuse Brown, Maripat Loftus Gatter, Victor Richardson, Ben Walker, and Cary White. Finally, we wish to thank our deans, LaVonda Reed, William Johnson, Eric Mitnick and Michael Waterstone.

It has been a splendid opportunity to work on this casebook. It has been a constant challenge to find a way to teach cutting edge issues influencing our health care system—at times before the courts or legislatures have given us much legal material for our casebook. Each time we have done a new edition, there have been developments that we find difficult to assess as to whether they will become more significant during the lifespan of the edition or are simply blips. It is always difficult to delete materials that required much labor and still remain quite relevant but that have been eclipsed in importance by others, and the length of each succeeding edition attests to our challenge. The good news is that this edition retains the substantially slimmed-down size of the prior edition. We do not write this casebook for our classes alone, but for yours as well. We enjoy teaching, and we hope that comes through to the students and teachers who use this book.

Finally, this edition marks the first edition authored by this group of five authors. We are deeply grateful to the original authors who created this book and sustained it over eight editions, Barry R. Furrow, Thomas L. Greaney, Sandra J. Johnson, Timothy S. Jost, and Robert L. Schwartz, and we thank them for guiding us through the last edition. We know that we stand on your shoulders and are heartened by your trust in us to carry this book forward. We also thank Jaime King, who joined us for the eighth edition, and warmly welcome Liz McCuskey to the fold.

BRIETTA CLARK
LOS ANGELES

ERIN FUSE BROWN
ATLANTA

ROBERT GATTER
ST. LOUIS

ELIZABETH MCCUSKEY
PROVIDENCE

ELIZABETH PENDO
ST. LOUIS

April 2022

ACKNOWLEDGMENTS

American College of Obstetrics and Gynecology Committee on Ethics, The Limits of Conscientious Refusal in Reproductive Medicine, 110 Obstetrics Genecology 1203 (2007, reaffirmed 2016). Reprinted with permission.

American Medical Association, AMA Principles of Medical Ethics VI (adopted June 1957, revised June 1980, revised June 2001). Used with permission of the American Medical Association. All rights reserved.

American Medical Association, CEJA Medical Ethics Opinions 1.1.2, 1.1.7, 5.5, 5.6, 5.7, 5.8, 8.5, 11.1.4 (2016). Used with permission of the American Medical Association.

American Journal of Public Health, Public Health Code of Ethics (2019). Used with permission of the American Public Health Association.

Austin, C.R., Human Embryos: Debate on Assisted Reproduction (1989). Copyright 1989, Oxford University Press. Reprinted by permission of Oxford University Press.

Colorado Medical Orders for Scope of Treatment (2015 form). Reprinted with permission.

Crown copyright is produced with the permission of the Controller of Her Majesty's Stationery Office.

Devers, Kelly and Robert Berenson, Can Accountable Care Organizations Improve the Value of Health Care by Solving the Cost and Quality Quandries?, copyright 2009, The Urban Institute. Reprinted with permission.

Enthoven, Alain, Health Plan: The Only Practical Solution to the Soaring Costs of Health Care 1–12 (1980). Copyright 1980 Alain Enthoven. Reprinted with permission.

Ethics Committee of the American Society for Reproductive Medicine, Informing Offspring of Their Conception by Gamete or Embryo Donation, 109 Fertilization and Sterilization 601 (2018). Reprinted with permission from Elsevier.

Fletcher, Joseph, Indicators of Humanhood, 2 Hastings Center Report (5) 1 (November 1972). Copyright 1972, the Hastings Center. Reprinted with permission of the Hastings Center.

Froedtert Hospital—Medical College of Wisconsin, Futile Medical Care Policy (2020). Reprinted with permission.

Gostin, Lawrence O., Public Health Law: Power, Duty, Restraint. Copyright 2000, University of California Press. Reprinted with permission of the University of California Press.

Hacker, Jacob S., and Theodore R. Marmor, How Not to Think About "Managed Care," 32 University of Michigan Journal of Law Reform 661 (1999). Copyright University of Michigan Journal of Law Reform. Used with permission.

Horney, James R. and Van de Water, Paul N., House-Passed And Senate Health Bills Reduce Deficit, Slow Health Care Costs, and Include Realistic Medicare Savings, Copyright, 2009, Center on Budget and Policy Priorities (used with permission).

Leape, Lucian L., Error in Medicine, 272 JAMA 1851 (1994). Copyright 1994, American Medical Association. Reprinted with permission of the American Medical Association.

National Conference of Commissioners on Uniform State Laws, Uniform Anatomical Gift Act. Copyright 2009, National Conference of Commissioners on Uniform State Laws. Reprinted with permission of National Conference of Commissioners on Uniform State Laws.

National Conference of Commissioners on Uniform State Laws, Uniform Determination Death Act. Copyright 1980, National Conference of Commissioners on Uniform State Laws. Reprinted with permission of National Conference of Commissioners on Uniform State Laws.

National Conference of Commissioners of Uniform State Laws, Uniform Health-Care Decisions Act. Copyright 1994, National Conference of Commissioners on Uniform State Laws. Reprinted with permission of National Conference of Commissioners on Uniform State Laws.

National Conference of Commissioners on Uniform State Laws, Uniform Parentage Act. Copyright 1973, 2000, 2002, and 2017, National Conference of Commissions on Uniform State Laws. Reprinted with permission of National Conference of Commissioners on Uniform State Laws.

National Conference of Commissioners on Uniform State Laws, Uniform Probate Code. Copyright, National Conference of Commissioners on Uniform State Laws. Reprinted with permission of National Conference of Commissioners on Uniform State Laws.

Randall, Vernelia, Trusting the Health Care System Ain't Always Easy! An African-American Perspective on Bioethics, 15 St. Louis U. Public L. Rev. 191 (1996). Reprinted with permission of the St. Louis U. Public Law Review.

Schneider, Eric C., et al., Mirror Mirror 2017: International Comparison Reflects Flaws and Opportunities for Better U.S. Health Care, Commonwealth Fund (2017).

Ulrich, Lawrence P., Reproductive Rights of Genetic Disease, in J. Humber and R. Almeder, eds., Biomedical Ethics and the law. Copyright 1986. Reprinted with permission.

SUMMARY OF CONTENTS

TABLE OF CONTENTS

TABLE OF CASES

The principal cases are in bold type.

BIOETHICS
HEALTH CARE LAW AND ETHICS

Ninth Edition

CHAPTER 1

AN INTRODUCTION TO THE STUDY OF ETHICS AND ETHICAL THEORIES

■ ■ ■

I. THINKING ABOUT THE "ETHICS" IN BIOETHICS

At 4:30 on a Friday afternoon, you, as legal counsel for a 300-bed for-profit hospital, receive a telephone call from a physician, Dr. Rodriguez, asking your advice. For some time, she has been caring for a 37-year-old man, Mr. Jones, who is in the late stages of lung cancer which has metastasized to his bones. Dr. Rodriguez predicts that Mr. Jones may live for another month or so if treatment is continued. Treatment consists of chemotherapy (administration of drugs which slow the growth and spread of the cancer), and pain medication (injection of morphine which reduces, although does not eliminate, the pain Mr. Jones experiences). Mr. Jones also has a pacemaker to regulate his heartbeat.

Dr. Rodriguez has just admitted Mr. Jones to the hospital, and Mr. Jones has made a request that he has made repeatedly in the past. Specifically, Mr. Jones requests that all chemotherapy be stopped, and that the pacemaker be deprogrammed immediately, so that his heart will malfunction and cause his death in a very short time. Dr. Rodriguez has talked at length with Mr. Jones about his request, and she is convinced that the patient's thinking is clear, and that he is determined and consistent in his wishes.

Based on this information, you suggest to Dr. Rodriguez that a meeting with the hospital ethics committee or an ethics consultant at the hospital might be of help. She responds, "How would that help me in this case? This is a legal issue, not an ethical one, right?"

Indeed, just what is meant by "ethics," especially as applied to health care decision-making? Answering this question requires addressing several matters: first, defining and distinguishing certain terms that are used regularly in ethics-related discussions; second, considering some of the processes used in applied ethics and health care decision-making; third, recognizing some strengths and limitations of this enterprise generally;

1

and finally, attempting to figure out whether this is an ethical question or a legal question—and what difference that determination makes.[1]

Definitions

While the terms "values," "morals," and "ethics" are sometimes used interchangeably, in this discussion they will be defined and used as follows. **Value** is, in the broad sense, the worth, goodness, or desirability attributed to something, whether moral or non-moral, such as tidiness, efficiency, honesty, and compassion. **Morals** (or morality) refer to those belief traditions about what is right (or good) in human conduct that develop, are transmitted, and are learned, at least in part, independently of theoretical ethical analysis. Generally, morals are principles that are not amenable to logical proof. **Ethics** (or ethical theory) denotes that branch of philosophy or reasoned inquiry that studies both the nature of, and justification for, general principles governing right or good conduct.

Morality expresses society's basic instructions about what people may and may not do. Religious tenets can also be a source of moral judgments. Ethical theory, on the other hand, describes and justifies such traditions and is often divided into two quite separate areas: scientific and philosophical. The scientific branch of ethics gathers and reports empirical information about existing moral beliefs, without evaluating the worth of moral judgments in any way. Philosophical ethics moves beyond description and either evaluates important moral concepts (for example, freedom, justice, the good) and moral reasoning from a logical perspective, or establishes theoretical justifications for what is right and wrong (or good and bad) in human actions.

Special mention should be made here of one component of formalized moral tradition that is not necessarily the product of formal ethical theory but which is, nevertheless, an important repository of and force for the moral beliefs of certain groups: **codes of ethics**. Sometimes these codes constitute a kind of oath that affirms the highest ideals and standards of the profession and that exhorts individuals always to strive for these standards. Sometimes these codes also provide rules for expected behavior and conduct. They may suggest, usually in general terms, certain ethical principles that professionals ought to consider when determining conduct.

Professional codes are taken seriously and form an important expression of a profession's collective identity and mission at any given time. They are limited, however, in their effectiveness as rational tools for analyzing and resolving ethical issues and dilemmas; there is still a need

---

[1] This portion of the introduction was prepared in part by Dr. Joan Gibson. See Virginia Trotter Betts, Margie Nicks Gale and Joan McIver Gibson, Ethical and Legal Issues, in D. Critchley and J. Mavrin, eds., *The Clinical Specialist in Psychiatric Mental Health Nursing: Theory, Research, and Practice* 453, 454 (1985).

for systematic, comprehensive, and internally consistent analysis of the issues they address.

Ethics and Process

A somewhat different definition of ethics focuses on one aspect of philosophical ethics: establishing theoretical justifications for what is right and wrong in human actions. Why ought/ought not Dr. Rodriguez accede to Mr. Jones's wishes? Should deprogramming the pacemaker be evaluated the same way as stopping the chemotherapy? Indeed, this process asks "Why?" about specific moral value judgments, proffering good reasons in response to that question, and proceeding to the ever-more abstract reasons that characterize critical thinking generally.

For example, a comparison between the giving of good reasons in science, which is called "explanation," and the giving of good reasons in ethics, which is called "moral justification," reveals striking procedural similarities. Consider the following diagram and comparative answers to a series of "Why?" questions:[2]

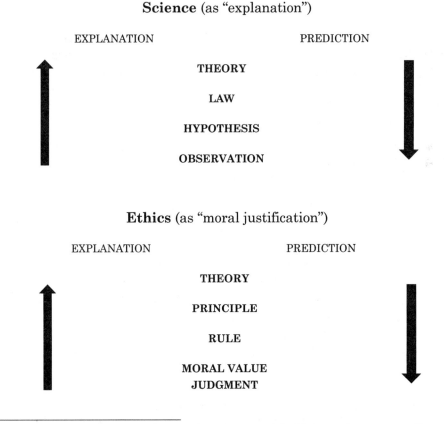

Science (as "explanation")

EXPLANATION   PREDICTION

THEORY

LAW

HYPOTHESIS

OBSERVATION

Ethics (as "moral justification")

EXPLANATION   PREDICTION

THEORY

PRINCIPLE

RULE

MORAL VALUE
JUDGMENT

---

[2] Adapted and expanded from T. Beauchamp and L. Walters, *Contemporary Issues in Bioethics* (8th ed. 2013).

WHY?

. . . did the pencil fall when dropped?

Because of the hypothesis that all objects fall when dropped.

Why?

Because of the law of gravity, i.e., because objects with mass attract each other.

Why?

Because of the theory of Newtonian Mechanics or General Relativity (which we will not try to describe here)

. . . should Mr. Jones's pacemaker not be deprogrammed?

Because of the rule that you should never do an act that will kill another person.

Why?

Because of the principle of non-maleficence, i.e., you should never do something that will cause another harm.

Why?

Because of the theory of (take your pick) Christianity, Judaism, Islam, Hinduism, Aristotelian virtue, Secular Humanism, Utilitarianism (and others which we will not try to describe here)

Of course, the right-hand column could have read:

. . . should Mr. Jones's pacemaker be deprogrammed?

Because of the rule that you should always do what a patient wishes. Why?

Because of the principle of autonomy, i.e., every person ought to control his own body and mind. Why?

Because of the theory of (take your pick) . . . .

Answering the question "Why?" moving up the column under "Science" is known as explanation: the accounting for observed phenomena at levels of increased abstraction, generalization, and simplification. Moving down the column, once the "Why?"s are answered, generates the power of prediction about future similar observations and phenomena.

And the process is similar when giving good reasons for individual moral value judgments, like whether Dr. Rodriguez ought to accede to Mr. Jones's wishes. Answering the question "Why?" moving up the column under "Ethics" is known as moral justification. Moving down the column, once the "Why?"s are answered, yields decisions about similar, future moral value judgments that must be made. Answering "Why?" (and thus moving up the column in either discipline) requires that reasons be elucidated and organized.

"Truth" in science, as well as in ethics, derives not so much from discovering isolated, once-and-for-all answers, as from continually articulating, evaluating, and revising the reasons one gives for the continually modified propositions one asserts and the consistently reevaluated judgments one makes. Extrapolating into the future (predicting and making decisions) is only as sound as the integrity of prior explanations and moral justifications.

Does this sound a great deal like the process of the development of the common law? Is it any surprise that the courts have been at the forefront of our society's debates on bioethics?

Ultimately, how does all of this apply to the case of Mr. Jones and Dr. Rodriguez? When considering whether to stop chemotherapy, Dr. Rodriguez, might say, "As a rule I believe that patient wishes ought to prevail." This first level of abstraction above specific moral value judgments, rule-making, is a level of generalization with which most people who justify moral decisions are familiar. It is the move up to principles, however—the "Why?" asked of rules themselves—that has engaged the energies of bioethicists over the past several decades. Why should Dr. Rodriguez normally respect patients' wishes? Is there a harm that would arise if she were to stop chemotherapy, or deprogram Mr. Jones's pacemaker? Why should Dr. Rodriguez normally avoid harm? What is her professional responsibility as a physician?

II. ETHICAL THEORIES

A. CONSEQUENTIALIST AND DEONTOLOGICAL THEORIES

Most general ethical theories can be categorized as either consequentialist or deontological. Consequentialist theories (also known as teleological theories) judge the quality of an act by the end it achieves. If it maximizes the desired goal, the contemplated act is the appropriate choice. Essentially, the end justifies the means. Deontological theories are those that justify the moral propriety of an act based solely on its compliance with chosen moral principles. A particular act that produces "good" results is still considered immoral in deontology if the act violates identified moral values. Arguments centered on the consequences of a particular action or rule—e.g., what impact a particular rule may have on whether scientific research will flourish—often stand opposed to arguments that essential human characteristics demand certain restrictions on power—e.g., that human dignity requires that consent must precede the use of individuals as subjects in research.

Although consequentialism and deontology are very different theories, they do not necessarily lead to different results in particular situations. So, for example, a consequentialist may argue that requiring consent for participation in research—a rule a deontologist would justify based on its moral value—should be required because it will produce scientific progress through better research design (needed to persuade subjects to participate) and more open sharing of results. In addition, as you will see in the sections that follow, neither approach is entirely absolutist in application. Consequentialist theories can make room for considerations beyond the result produced by the particular, discrete action at issue. Deontological theories likewise generally make room for shades of gray in assessing the moral qualities of actions ordinarily considered inherently immoral by, for example, being open to the analysis of context and intent. Thus, often but not always, consequential analysis and deontological analysis arrive at the same place for different reasons.

Utilitarianism is the most familiar consequentialist theory, and Kantian ethics, the most prominent deontological theory. Both have had great influence in bioethics and in law. Most of us combine consequential and deontological reasoning when we confront ethical dilemmas. Courts and legislatures also tend to look at both the inherent moral qualities of acts and their consequences when they decide which actions should be mandated or prohibited by law.

The "trolley problem" is a classic illustration of an ethical dilemma that tests the boundaries and the opposition of consequentialist and deontological approaches.[3] Suppose you are standing at the switch on a series of train tracks when you see a runaway train about to run through the switch and into a group of people standing on the tracks beyond. If the current switch setting is maintained, the train will barrel through a group of five people gathered on one track, surely killing them all. If you pull the switch, though, the train will be sent off on a siding, where it will kill one person walking on that track. Should you pull the switch, thereby

---

[3] For a full discussion of the trolley problem, see J.J. Thomson, The Trolley Problem, 94 Yale L. J. 1395 (1985).

sacrificing one life to save five? Why? Why not? Before you make your final decision, consider the discussion in the following sections.

Now, suppose instead you are on a bridge overseeing this whole train wreck, and you know that the lives of the five people on the train track can be saved if—and only if—you push the enormous man standing next to you off the bridge and onto the track below just as the train comes by. Should you push your fellow train spotter onto the track, thereby sacrificing his life to save five? Many of us feel comfortable pulling the switch and diverting the train, presumably because we serve the consequentialist end of saving five lives at the cost of one by doing so. But we are far more reluctant to push our neighbor onto the track, sacrificing the life of a known innocent person to save others. It just doesn't seem right, even though the result is equally justifiable in terms of the consequences. It just doesn't seem consistent with the duty we owe another human being.

B. UTILITARIANISM

At its base, utilitarianism provides that the proper act is the one that maximizes the aggregate happiness, or "utility," of society as a whole. If an act tends to do so, it is the right act. If it does not, it is not. Two major branches of utilitarianism form the basis of modern consequentialist analysis. Act utilitarianism asks whether a particular, individual act serves to increase aggregate happiness or utility. Rule utilitarianism focuses on rules as a guide to correct conduct and judges whether a rule is correct by the consequences the rule produces. Rule utilitarianism asks what societal rules, rather than individual actions, are likely to maximize happiness in a society. Rule utilitarianism defines morally appropriate conduct as conduct that is consistent with those rules. Rule utilitarianism allows consequentialists to consider principles (perhaps you would call them moral principles) in judging whether a particular action that produces desirable outcomes in the individual instance is nevertheless unethical. As you can imagine, rule utilitarianism has been an especially powerful tool for legal agencies like courts and legislatures.

This simple description of utilitarianism leaves many questions unanswered, and philosophers have been debating the contours of utilitarianism over the past three centuries. First, it is hard to know just what counts as "happiness," and some utilitarian thinkers would prefer to use some other less hedonistic term to define the ultimate goal of all human conduct. Even if we knew what happiness were, though, we would have to figure out how to measure it (could there be a standard unit of happiness?). And how demanding should we be in requiring that assumptions of what will happen, whether positive or negative, are grounded in empirical evidence? In addition, to judge whether aggregate utility is maximized, we would have to decide how to compare the happiness (or suffering) experienced by one person as a result of the action in question with that experienced by another.

All of these challenges are magnified in the evaluation of the distribution of benefits and burdens (utility and disutility) across society. Most actions and rules produce uneven or unequal distributions of benefits and burdens even when the aggregate effect is to increase utility overall. Utilitarianism also has to confront the question of who counts when we do the math and add up all of the individual levels of happiness so that we can compare the results of different actions and decide which is morally superior. Are children to be treated the same way as adults? How about fetuses? How about unimplanted embryos? How about future generations, children not yet conceived? Those who lack consciousness? Are we to limit the calculus to increasing the utility levels of humankind, or should animals be included too? All animals, from insects to primates, or just some?

Consequentialism is most often criticized for leading to apparently immoral results. It is inconsistent with our notion of morality that an act that would result in the killing of an innocent person, for example, could be justified, even if it maximized happiness. Of course, some utilitarians argue that such results, while theoretically possible, would never govern a society that did the utilitarian calculus properly. Utilitarianism does not mark particular actions—such as murder, rape, lying, and fraud—as inherently unethical. Rather, the acts are to be measured by their aggregated consequences alone. Utilitarianism argues that a consequentialist analysis standing alone could itself reject such actions because of their negative consequences for society as a whole.

Return, then, to the problem at the beginning of these materials. Under utilitarian analysis, should Dr. Rodriguez discontinue the chemotherapy on Mr. Jones? Should she deprogram the pacemaker? Why? Are you applying act utilitarian analysis or rule utilitarian analysis? Are you satisfied with a utilitarian analysis of the potential actions in the trolley dilemma?

C. KANTIAN THEORIES

Immanuel Kant is the most influential deontological philosopher, and his work grounds much bioethics analysis. Kant's work is quite visible, for example, in the ethical framework for the regulation of research with human subjects and in decision-making at the end of life, among other areas in bioethics.

Kant, a leading opponent of utilitarianism, argued that it is not the consequences that determine whether an action is right or wrong. Instead, Kantianism measures moral action primarily by whether it arises from a "good will." This sense of good will is not the equivalent of having a good heart or simply meaning to do well. Instead, good will for Kant refers to the individual's character or moral goodness. One who acts in a morally acceptable way merely because it produces benefits for himself rather than because he has a moral duty to do so is acting in a way that does not have intrinsic moral worth. True moral worth is evidenced by actions based on a sense of duty rather than on personal inclination or the consequences of the act.

For Kant, moral principles (and the duties that proceed from them) are universal rather than subjective or relative. Kant's primary test for right action expresses the universalism of morality. His "Categorical Imperative" states: "I ought never to act except in such a way that I could also will that my maxim should become universal law."

This Categorical Imperative is the core Kantian command, and it is unconditional. Consequently, some argue that, while Kant is regarded as a deontologist, his Categorical Imperative is quite close to the assessment of consequences inherent in rule utilitarianism. There remains a significant distinction, however. Kant strongly asserted that it is not the consequences that determine the goodness or morality of an action but rather the moral principles that inhabit the action.

The Categorical Imperative relies on a critical, but only implied, notion in Kantianism; i.e., the moral equality of all persons. This relational aspect of Kantianism provides a core value in bioethics. It requires respect for all persons as moral beings and holds that persons may never be used merely as means to an end, even where the goal and likely outcome would be a great benefit to other persons. In Kantianism, persons are entirely distinguishable from things, in that all persons have inherent worth and dignity that cannot be denied. The Kantian focus on respect for persons grounds the legal and ethical requirements of informed consent and protection of individuals participating as subjects in medical research.

The Categorical Imperative also appreciates individuals as autonomous beings with free will. Of course, Kant also argued that values are universal, so the emphasis on will in the Categorical Imperative should not be taken to mean that values are subjective and dependent on an individual's personal analysis. Rather, one might expect that freely acting rational individuals will reason by and share universal moral values. The emphasis on individual free will associates Kant with a strong version of autonomy, even though that autonomy is morally bound by freely discerned and accepted principles.

Kant identifies two classes of duties: perfect duties and imperfect duties. Perfect duties are those which support a claim by others to our performance, i.e., rights. Imperfect duties do not support claims of right by others upon us. Examples of perfect duties may include the duty to refrain from unjustly harming another while the related imperfect duty may be the duty to contribute to the well-being of another. The first—the perfect duty to avoid harming another—carries a correlative right on the part of others to demand that this be done. The second—the imperfect duty to benefit others—carries no such rights, and others cannot demand, as a matter of right, our efforts on their behalf. Kant argues that perfect duties cannot conflict, and that perfect duties, those that create rights in others, take priority over imperfect duties, which do not create those rights.

Kant's framework is apparent in bioethics in the U.S., especially in its emphasis on the inherent worth, dignity, and autonomy of individuals. Although Kant's Categorical Imperative can be seen as being concerned about the well-being of the community, the framework, at least as adopted in bioethics, has been criticized as emphasizing the individual to the detriment of the community or society as a whole. Also criticized has been Kant's strong resistance to considering the consequences of actions in making moral judgments.

Under Kantian analysis, should Dr. Rodriguez discontinue the chemotherapy on Mr. Jones? Should she deprogram the pacemaker? Why? Are there perfect or imperfect duties that govern the relationship between Dr. Rodriguez and Mr. Jones? Does Kant's framework contribute to your solution of the trolley dilemma?

D. RELIGIOUSLY BASED ETHICS

Approaches to ethical reasoning in bioethics that are generally considered religious in nature are typically grounded in belief in the existence of a deity. They rely on values that emerge from the guidance of a scripture, respected teachers and disciples, an authoritative central authority, or highly personalized spiritual perceptions, insights, and experiences. Often, some combination of all these influences affects values. Further, as would be expected with the diversity of religious viewpoints in the U.S., religious approaches to bioethics vary considerably. Thus, understanding the core principles—including those relating to the nature and purpose of personhood—and how they apply to particular medical decisions often requires investing significant time and effort in study. Complicating the inquiry is the frequent significant diversity, and perhaps even conflict, within any particular religious tradition about how a specific bioethics issue should be resolved.

The influence of religious belief and tradition upon patient decisions about medical treatment is well known even beyond Jehovah's Witnesses refusing blood transfusions or Christian Scientists treating illness largely in contravention of customary medical standards. In practice, patients often seek pastoral support when considering their treatment options. When a patient, due to a religious perspective, wants to reject the customary or typical treatment approach, legal standards sometimes, although not always, accommodate that individual's health care decision. See, e.g., Chapters 4 and 5.

Scholars with a religious perspective have contributed to the body of work in bioethics, and religious leaders certainly participate in public policy debates on science and medicine, some more vigorously than others. Some religious traditions or doctrines have had a significant influence on the legal framework for certain bioethics issues. In particular, Catholic tenets provided the framework used when the issue of decisions about withdrawal or withholding of treatment first reached the courts. Several early court decisions, such as In re Quinlan, 70 N.J. 10, 355 A.2d 647 (1976), which involved Catholic parents, relied on statements by Pope Pius XII supporting withdrawal of ventilator support and the primacy of the patient's choice. See also Subsection E, below. Similarly, the tenets of some forms of Judaism have influenced the adoption of choice options regarding neurological determination of death. See Chapter 4.

Applying religiously based principles or doctrine when determining public policy in bioethics often engenders considerable conflict in our diverse society, especially when principles based on such perspectives depart from those held by most citizens. This is certainly often the case with the use of contraception and access to abortion, and it appears to be the case with medical assistance in dying. On the public policy level, religious advocates argue that their perspective has a place in the public debate and that it offers principles that will preserve or create a better society. When religious perspectives are enforced by law, however, many argue that religious belief has overstepped its boundaries and violated the tolerance that marks a civil, secular, and pluralistic society.

Conflicts over the role of religion in health policy and medical decision-making have always existed. But they seem to have entered the public consciousness with new vigor as a result of the Affordable Care Act (ACA), especially as it pertains to certain coverage requirements and antidiscrimination protections. For example, contraceptive coverage has been determined to be an "essential health benefit" that must be offered as part of the basic benefits package and has been viewed as essential to gender equity in coverage. Yet, it is also the focus of religious liberty objections by some entities with religious affiliations, including for-profit and non-profit companies that employ and serve diverse constituencies. These entities argue that being compelled to offer any insurance coverage contrary to their religious beliefs violates religious liberty protections under the law. Conversely, employees who do not hold the same religious views argue that allowing employers to deny them the type of comprehensive reproductive health and prescription coverage that is afforded to men not only violates their own liberty regarding intimate decision-making, but also constitutes discrimination on the basis of sex. In the case of contraception, there has been a great deal of regulatory activity attempting to address and resolve this conflict, with the shape of the regulatory compromise offered shifting dramatically from one presidential administration to the next. Throughout this time, courts have been very involved as well; indeed this conflict has resulted in an unusual number of Supreme Court opinions in a very short time period. For a more detailed discussion of the legal and ethical approaches used to address this issue, see Chapters 2 and 3.

Another conflict capturing public attention involves religious objections to the sex antidiscrimination mandate in Section 1557 of the ACA, in particular what this means for access to gender-affirming care. To date, federal courts have tended to interpret Section 1557 as prohibiting gender identity discrimination by insurers and providers. For example, cases have found that the categorical exclusion or denial of gender-affirming care—especially where the ban excludes services that would otherwise be covered or provided—violates the sex antidiscrimination protections of Section 1557, as well as comparable state laws. As with contraceptive coverage, some religious entities have argued that they should be exempted from covering or providing gender-affirming because it violates their religious beliefs respecting the definition of sex and moral judgments connected to traditional gender norms. Although there has not been quite as much regulatory and judicial activity as with contraception, there have been important legal developments indicating that this is an important conflict to watch as well. This particular conflict is discussed in greater detail in Chapter 2. For a general discussion on health care providers asserting religious liberty rights to refuse to provide legally permitted medical interventions more broadly, and the conflict it may create with patients' autonomy and perhaps patients' own religious beliefs, see Section VI.A below.

E. NATURAL LAW

Traditional natural law theories of ethics rely both on the acceptance of a God who guides human beings through divine providence and on the inherent capacity of human beings to be rational actors who know and can choose to act in accord with the precepts of this divine guidance. Under natural law theory, the morally proper act is that to which human beings are naturally drawn by the perfection of human nature. St. Thomas Aquinas, who grounded his theory in Catholic moral theology, is widely viewed as the premier natural law theorist.

While many view the acceptance of a deity as essential to natural law theories, some humanist scholars more recently have argued that a secular natural law theory is possible. They view human beings as transcendent, with an innate knowledge of right principles and values. Rather than identifying the good as coming from God, these theorists argue that the good can emerge from the nature of human beings themselves. The foundational principle in our system of government that human beings are "endowed . . . with certain inalienable rights," for example, may evidence some integration of natural law within our legal system, whether one believes that these rights come from "the Creator" or otherwise.

Natural law theory has had a persistent influence on our legal norms, whether as a remnant of the deity-based natural law approach that informed the early development of our laws or as evidence of the continuing power of religious perspectives in our political system. In either case, natural law, like Kantian ethics, is contrary to consequentialist theories of ethics in maintaining that the good holds inherent value, apart from any of its consequences or effects. Life, knowledge, procreation, friendship—all are aspects of the good under natural law theories without any calculation of benefits to individuals or the group. Also, like Kant, natural law theorists reject subjectivism and relativism that would argue that there are no shared concepts of the good.

Although natural law is criticized for its rigid adherence to a set of divine rules, it still allows significant space for situational analysis and response. Natural law theory, conceiving of humans as essential, rational partners of natural law, accepts that the particulars of distinct situations require an individual to exercise her or his own rational capacity and virtue. Still, the theory steadfastly holds to the notion that some actions are intrinsically and absolutely evil. For example, the killing of an innocent is considered to be intrinsically evil in natural law. Yet, natural law theorists accept that some actions that one knows will result in the death of an innocent person are moral under certain circumstances.

The Doctrine of Double Effect

One tool that natural law theory uses to describe this paradox—that one can view killing as inherently evil but engage in actions that will surely result in death—is the principle of double effect. Double effect has been an influential principle in the development of health care ethics, with application to a range of health care decisions. The New Jersey Supreme Court in *Quinlan*, the first reported appellate case to allow a family to discontinue life-sustaining medical treatment, showcased a 1957 pronouncement by Pope Pius XII that relied on double effect to explain the moral permissibility of withdrawing life-sustaining respirator support from patients in particular circumstances. Its citation by that court gave prominence to the principle of double effect in subsequent cases.

Aquinas originally formulated the doctrine of double effect using the illustration of self-defense. Recognizing self-preservation as the good, and the killing of another as an evil, Aquinas argued that the value or duty of self-preservation is the objective and intent of self-defense, and that the death of the other is the unintended consequence. The moral permissibility of killing in self-defense is entirely conditional and dependent on the proportionality and unavoidability of the response.

The dominant formulation of the doctrine of double effect requires the following four conditions:

1. the act itself must be morally good or at least indifferent;

2. the actor may not intend the bad effect and would avoid it if possible to attain the good effect without the bad;

3. the good effect must be produced directly by the action and not by the bad effect itself; and

4. the good effect must outweigh the bad effect.

The principle of double effect is often credited with having improved care of dying patients by encouraging the treatment of pain even if that treatment could also, as a byproduct, hasten the patient's death. The application of the doctrine of double effect to pain relief was explicitly recognized in Justice Rehnquist's opinion for the majority in Vacco v. Quill, 521 U.S. 793, 117 S. Ct. 2293, 138 L. Ed. 2d 834 (1997), in which the Court rejected a claim of a Constitutional right to medically assisted dying. See Chapter 6.

In the context of pain relief, the principle of double effect holds that killing is intrinsically evil but that providing aggressive pain management, even if it may hasten the death of the patient, is morally acceptable, and some would say mandatory, if certain conditions are met. First, the intention of the caregiver must be to relieve the pain but not to directly kill the patient. In other words, the caregiver must intend to provide the good and avoid the evil, even when the particular action may have both effects— the double effect of good and evil. Second, the means chosen to relieve pain must be reasonably designed to do so. Third, the pain relief must not be produced by the bad effect, i.e., the death of the patient. For example, using pain medication to relieve pain in the dying is acceptable, but using a drug such as potassium chloride that is designed to stop the heart is not, even if it will relieve suffering. Finally, the fourth principle recognizes that the unintended bad effect—the death of the patient—is serious and undesirable, so it requires that the good effect—the relief of pain— outweigh that effect. Thus, the doctrine of double effect as applied in such cases holds that relieving pain is a good and that good may be pursued, despite the foreseeable result of the patient's death, if the actor's intent is right and the actions themselves are reasonably designed to achieve the good and avoid, to the extent possible, the evil.

This application of the principle of double effect has been subject to serious criticism, however. The first criticism is a pragmatic concern. While the application of the principle to pain management at the end of life has had beneficial effects, it may have strengthened a largely mistaken perception that aggressive pain management actually hastens death. The repeated message to provide pain medication "even if it may hasten death" took on the character of a statement of fact (that pain medication does hasten death) rather than a statement of condition. Second, many reject the principle's premise that intentionally shortening the patient's life is an intrinsic evil rather than a legitimate aid in dying. Third, the intent requirement of the analysis is difficult to apply in practice, especially in a legal context. It is quite common that the family, patient, and professional caregivers near the end of a long dying will welcome death in some sense as a release from suffering. This does not mean that they intend to kill the patient, but testimony concerning conversations in which these parties agree that "it is time" or that "it would be better if he were to die" may be construed as evidence of an illegal killing rather than legitimate and necessary pain relief. The intent requirement is also sometimes viewed as a subterfuge that health care professionals can use to disguise their true intentions. Natural law theory itself has been criticized by those who see the need for an elaborate double effect exception to justify humane treatment of some patients as a demonstration of the theory's weakness.

Return one more time to the problem at the beginning of the chapter. Under a natural law analysis, should Dr. Rodriguez discontinue the chemotherapy on Mr. Jones? Should she deprogram the pacemaker? Why? What does natural law require in this case? Is the doctrine of double effect helpful? Does a natural law approach help at all in untying the knot of the dilemma in the trolley problem?

III. PRINCIPLISM

Biomedical ethics has been dominated by a kind of "principlist" orientation over the past 35 years, and this approach is quite evident in judicial opinions addressing legal issues related to bioethics. The principles of this approach have been the hallmarks of bioethics debate ever since the National Commission for the Protection of Human Subjects of Biomedical and Behavioral Research issued its Belmont Report, the *Quinlan* case was decided, and Tom Beauchamp and James Childress published the first edition of their enormously influential *Principles of Biomedical Ethics*, all in the late 1970s.

The principlist approach does not necessarily bother itself with the underlying ethical theories that support the principles, and so it may be inappropriate to include it in a discussion of ethical "theories." Because the principles are untethered by specific theories, they are often referred to as "mid-level principles." Principlism can be criticized for operating without an agreed upon theoretical foundation, but it may be that very characteristic that has led to its great influence in law and public policy.

In principlism, ethical principles "operate primarily as checklists naming issues worth remembering when considering a biomedical moral issue." K. Clouser and B. Gert, A Critique of Principlism, 15 J. Med. & Phil. 200 (1990). The most common checklist of principles includes at least the following:

1. **Autonomy**. The principle that independent actions and choices of the individual should not be constrained by others.

2. **Non-Maleficence (Do No Harm)**. The principle that one has a duty not to inflict evil, harm, or risk of harm on others.

3. **Beneficence**. The principle that one has a duty to help others by doing what is best for them. Obviously, this principle is related to that of non-maleficence.

4. **Confidentiality**. The principle that when information is divulged by one person to another with the promise, implicit or explicit, that it will not be revealed to any other person, that implicit promise should be respected.

5. **Distributive Justice**. The principle that benefits and burdens ought to be distributed equitably, that resources (especially scarce resources) ought to be allocated fairly, and that one ought to act in such a manner that no one person or group holds a disproportionate share of benefits or burdens.

6. **Truth Telling (Honesty, Integrity)**. The principle that one ought to accurately and fully disclose all pertinent information about a person to that person.

There is also one "meta-principle" that is often invoked by those engaged in bioethics discussion, the principle of **Professional Responsibility**. This principle provides that a professional has an obligation to observe the rules, principles, and moral precepts governing relations with patients, colleagues, the profession as a whole, and the community at large. Professional responsibility, as a normative principle, captures the fiduciary duty owed by health care professionals to their patients. It also can imply an obligation to abide by the virtues or codes adopted by the profession itself.

Some argue that principle-oriented bioethics needs major overhauling—an issue of special interest to those in the legal profession because the law still adheres to the principle-oriented approach with such tenacity. Principlism has provided clarity of thought, separation and identification of moral concepts and issues, and a common language of ethics applied to health care decision-making. The approach has its limitations, and challenges to this now "traditional" method of analysis are increasingly frequent. See Section IV, below.

The Principle of Distributive Justice

Although principlism includes distributive justice among the core principles for bioethics and health care decision-making, that principle has largely remained on the sidelines in most of mainstream bioethics. This marginalization is especially apparent in the use of principlism for analyzing medical treatment decision-making "at the bedside," where individual autonomy and beneficence are elevated and distributive issues in health care have been minimized. The classic bioethics paradigm for the discussion of distributive justice in health care is not the social and financial constraints that impact decisions concerning end-of-life care but rather the selection of a recipient for transplantation of an available human organ where demand far outstrips supply. See Chapter 4.

More recently, however, concerns of justice and equity have attracted increased attention in the context of health care disparities and barriers to medical care, especially for racial and ethnic minorities, persons residing in rural areas, women, LGBTQ persons, the poor, and individuals with disabilities. The ACA is seen by many as the culmination of decades of health equity reform efforts, including research and advocacy focused on improving the health of racial minorities and other marginalized populations, as well as persistent calls for a system of health care that would provide affordable health care coverage for all. Although the ACA does not provide universal health care or incorporate every equity reform that advocates have sought, it has come the closest to achieving these goals, by far, when compared to other health reforms. The vigorous discussion and debate around the ACA shined a powerful spotlight on the values and principles underlying the allocation of health care resources and costs within the U.S. And the continuing legal and political challenges to the ACA reflect competing conceptions of how we should allocate resources, as well as the proper role of government with respect to ensuring access to essential goods. Less than ten years later, the COVID-19 pandemic brought the need for health equity reform into devastating relief as it exposed and exacerbated the persistent health and other systemic inequities experienced by racial and ethnic minorities, people with disabilities, and other disadvantaged groups. See Chapters 2 and 8.

Application of general ethical theories to questions of distributive justice in health care typically begins with addressing one issue: whether health care is distinguishable from other goods and services that are generally governed by market transactions. Some argue that health care is unique, or at least special, because the social costs of disease and disability impair the functioning of society generally, or because health is required for human flourishing and identity, or because the relief of suffering is a special imperative, or because the unequal original distribution of health and illness among persons creates unfair competition among citizens for satisfaction of other needs. Others argue that health care is a consumer good like any other, and lack of health care is the result of personal choices and should remain so.

Any of the general theories of ethics discussed in this chapter may be employed in analyzing ethical issues in the distribution of, or access to, health care services. A utilitarian approach, for example, would assess the morality of a particular decision or system distributing health care resources by examining its consequences on society as a whole. A deontological approach might focus on analyzing the source of relevant moral duties and how respect for the moral equality of persons would be observed or violated in particular decisions or systems. Egalitarian and communitarian approaches might both argue that society has an obligation to its members to provide basic minimal care, however that may be defined, the first because it is the only way to assure equal opportunity and the second because respect for persons requires it. Libertarians, on the other hand, might support a market approach to access to health care, arguing that redistributing wealth by force of government power is unjust and that the government controls required to redistribute resources raise costs and limit personal autonomy.

As with other ethical issues in bioethics, those espousing the "same" theory might produce differing results in particular situations. So, for example, utilitarians might argue that providing health care as a matter of right creates disincentives for self-support, inefficiencies because of the added cost of the distribution system, and excessive utilization because of the separation of payment and consumption. On the other hand, another utilitarian might view the harmful societal consequences of denying health care to those unable to pay as outweighing the burdens of providing it.

The forms of discrimination and unequal treatment that have shaped our health care system implicate norms of distributive justice, of course. But they also relate directly to other core principles in the principlist approach. Discriminatory exclusion from health care services, for example, impedes autonomy in health care decision-making and can make that value illusory at best. In addition, discriminatory practices, whether intentional or not, can result in harming individuals by producing poor outcomes as compared to others with identical medical conditions. Systemic or structural barriers to care result in individual harms, as well as producing an unequal distribution of the health and financial effects of illness and disability among particular groups of patients, their families, and their communities. For example, the unequal distribution of health care resources by federal and state governments, housing policies that have permitted and entrenched residential segregation, and the problem of provider and hospital flight out of predominantly minority neighborhoods have combined to create pockets of poorer, predominantly minority neighborhoods without timely access to essential quality care. See Chapter 2.

For many, achieving distributive justice in health care requires going beyond addressing discrimination to deal with underlying health disparities. A "health disparity" is a particular type of population-level health difference that is linked to a history of social, economic, or environmental disadvantage. Related terms, such as health inequity, emphasize that such differences are unfair, unjust, and avoidable. Despite decades of health equity reform efforts, health inequities across race, gender, and disability persist, even after controlling for factors such as ability to pay and medical need. See Chapters 2 and 8. These inequities raise not only ethical concerns about discrimination and social injustice in our health care system, but also pragmatic concerns for that system because they signal a pattern of lapses in quality of care and the creation of excess cost.

IV. CONTEXTUAL THEORIES OF BIOETHICS

Critics of traditional theoretical ethics point out that not all theories can be neatly classified as utilitarian, Kantian, or natural law in origin. In addition, some bioethics scholars have argued that the dominant principle-based ethical analysis is unnecessarily narrow. Not everything, they argue, can be parsed into manifestations of autonomy, beneficence, distributive justice, and a few other secondary principles (including non-maleficence, confidentiality, and truth telling).

A number of ethical approaches offer an alternative to the principlist methodology. Some of these approaches represent a return to earlier ethical frameworks. For example, **virtue bioethics**, grounded in Aristotelian virtue ethics, argues that we can determine what should be done in particular circumstances by asking what a truly good or virtuous person would do. Personal virtues, including honesty and competency for example, will lead to the right actions. Kant's approach is closely associated with virtue ethics: both examine not only the moral principle underlying a decision but also its motivation, requiring that the actor rely on values beyond his or her own self-interest. Virtue ethics dominated medical ethics for many decades and focused on the character of the good doctor as a source of guidance for decision-making. Virtue ethics values the moral agency of health care professionals and argues that other approaches to bioethics mistakenly make professionals the instruments of the patients alone. Critics, however, argue that virtue ethics supports the hegemony of the medical profession.

Casuistry is also a centuries-old approach; it derives principles from the analysis of individual cases. One example of casuistry, in fact, is the common law system, and not surprisingly, judges have been particularly comfortable with this form of practical, case-based reasoning. Casuistry focuses on the analysis and comparison of like cases. For example, this approach might ask whether it is ethical to deprogram a pacemaker by asking whether this action is more like the refusal of life-sustaining medical treatment by a competent adult (which is legally permitted as a general matter) or more like the purposeful commission of homicide (which is legally prohibited as a general matter). Casuistry seeks to discover the proper course of action by closely examining the facts of each case and asking whether it is more like or unlike other cases previously considered. Rather than deriving general principles from theory, casuistry develops principles from case analysis. Just as in the law, marshaling an accurate and adequately rich accounting of the facts to support a principled and generalizable resolution of a case is a challenge in casuistry.

In a related approach, **narrative bioethics** suggests that we can learn a great deal from hearing the stories of those directly involved in episodes that required challenging health care decisions. Hearing such narratives produces a deeper understanding of how people make decisions, how the context of care impacts these individuals and the caregivers, and how complex motivations and experiences affect health care decisions.

Another approach, **new pragmatism**, relies on several different strands of nineteenth century American philosophy and has gained some influence. As applied to bioethics, pragmatism focuses on an empirical approach that examines the context and the consequences of specific actions. This new pragmatism in bioethics is not entirely consequentialist, however: many leading proponents of the approach argue that a vision of justice or sense of the good must animate judgments made when evaluating consequences. Pragmatism in bioethics often focuses on establishing process rather than specific rules to determine right outcomes, arguing that this method is consistent with the respect for democratic decision-making and the diversity of values in America. As you will see in many of the issues you study in this text, statutes and case law often resolve conflicts by prescribing procedures that should be followed in settling the dispute—including who should decide, what procedure should be followed, and what the relevant considerations are—rather than prescribing a particular outcome.

Several more recent and significant contextual approaches to bioethics issues, in contrast, depend on critical analysis of disparities in the provision of health care. These approaches argue that various aspects of the broad social context—including discriminatory practices or effects, the shifting of hidden burdens to familial caregivers, differences in access to care, and power imbalances relating to gender, class, disability, and race, for example—are all relevant to ethical decision-making in health care. They criticize the facially neutral principlist approach as operating from privileged social and professional assumptions and culture. They also emphasize that some morally significant choices for patients may be severely constrained by inadequate financial and social support, especially for minorities, the elderly, the disabled, and others who are marginalized in society.

Among the more influential of these contextual methodologies are Feminist Bioethics, Critical Race Theory, and Disability Theory. They do not operate in separate spheres: they share a central concern about context, particularly the challenges our health care system poses for marginalized persons, and they thus often prioritize the same issues in health care. Furthermore, each of these critical theories explores the impact of intersectionality—the effects of intersecting systems of oppression faced by individuals by virtue of their membership in multiple marginalized categories, such as poor women of color or persons with disabilities who experience poverty.

A. FEMINIST BIOETHICS

The feminist movement over the past five decades has focused on numerous issues relating to health care with special attention to those affecting women, whether as patients or as the caregivers who most often bear the burdens of realizing a patient's choices. Originally focused on women's health care issues relating to reproductive choice, including contraception, abortion, and choice about childbirth assistance, the feminist movement now also reaches into other areas of health care decision-making. Feminist critiques of standard surgical interventions for breast and uterine cancer, as well as revelations about the dearth of research on women's health issues and the approval of new drugs based on male-only research subjects, have all had an impact. In addition, feminist analyses of the principlist approach to health care decisions, such as those at the end of life, have emphasized that the principlist focus on the physician-patient dyad and patient autonomy neglect personal intimate relationships. Feminist theorists also argue that embedded gender-biased power relationships can thwart theoretical autonomy in actual decision-making. For greater exploration of this in the context of reproductive health care, see Chapter 3.

There is a wide diversity of thought within feminist bioethics, however, and the various strands of feminist theory may conflict in the analysis of particular bioethics issues. Liberal feminism, for example, rests on a commitment to formal gender equality and argues that women are entitled to equal rights and equal opportunity. Liberal feminism distrusts the allocation of rights, responsibilities, and opportunities along gender lines, even when gender-specific rights would appear to work to the benefit of women, because female stereotypes have unjustly limited women under the guise of protection.

Cultural feminism highlights gender differences and argues that apparently neutral rules that treat men and women the same can treat women unfairly. This form of feminist theory argues that male paradigms dominate even apparently neutral value systems, including the principlist approach, and thus ignore approaches that are more female in nature, such as an ethic of caring or a more relational approach to decision-making. An ethic of caring fostered by this branch of feminist theory requires that we evaluate how the resolution of bioethical dilemmas will affect important human relationships. This approach stresses the caring, personal, and intimate nature of health care and distinguishes those aspects from medicine's more purely scientific character. The emphasis on relationships also contrasts with the emphasis on bare rights that some argue has characterized legal resolution of conflicts about medical treatment decision-making.

Radical feminism also addresses gender differences. But it argues that these are produced by the oppression of women, and that legal standards ought to protect women more actively from male oppression, especially in areas of sexual activity and male violence against women. See generally, Karen Rothenberg, New Perspectives for Teaching and Scholarship: The Role of Gender in Law and Health Care, 54 Md. L. Rev. 473 (1995). One way to understand the differences among these three types of feminism is to use them to analyze an issue like medically assisted dying. See Chapter 6.

B. CRITICAL RACE THEORY

Critical race theory argues that legal and ethical decisions cannot be analyzed apart from the political, social, and economic systems in which they arise, particularly as these systems reinforce racial hierarchy and subordination. Critical race theory argues, for example, that facially neutral principles that entrench existing resource and power inequities reflect the dominant culture of white privilege and oppression of African Americans and other minorities. Such principles often obscure the discriminatory roots and effects of modern laws, making it difficult, if not impossible, to successfully challenge racially harmful policies. Critical race theory also argues that such principles obstruct the values that have emerged from minority cultures within this country, as well as the values that would be developed and possibly embraced in a more racially and culturally diverse society.

Within health care specifically, critical race theory sheds light on how medical dominance and deference to medical personnel work to the detriment of persons of color where social and economic systems, including health care, are inherently unjust. For example, this dominance has allowed medical professionals to subject pregnant women to forced medical interventions and confinement in the name of fetal well-being. A critical lens helps reveal that pregnant women who are poor and Black, and women suffering from certain kind of medical conditions, are targeted for these liberty infringements, despite the fact that such actions are inconsistent with professional ethical guidelines that highlight how forced medical intervention and confinement can increase the risk of harm for both the mother and fetus. See Chapter 3. A critical race lens also helps to illuminate the intersection between discrimination on the basis of race and national origin, and eligibility restrictions tied to immigration status that discourage immigrants and their families from accessing essential care. See Chapter 2.

At a minimum, critical race theory argues that the significant impact of racism must be acknowledged when developing bioethical principles and processes to assure just outcomes. Studies of the impact of racialized medicine in the U.S. have revealed numerous patterns of abuse and neglect. For example, race-based justifications have undergirded the forcible or deceptive use of African-American individuals as subjects in unethical medical research, while African Americans have been disproportionately denied access to the important medical advances produced by this and similar research. See Chapter 7. Despite insurance expansions and increased attention to racial disparities, health care disparities related to racism persist in terms of access to care, treatment outcomes, and health status. See Chapter 2.

C. DISABILITY THEORY

Disability theory asserts that disability is a social, cultural, and political phenomenon, and not simply a medical one. It embraces a social model of disability rather than a medical model of disability, meaning that disability is a social construct rather than a biological fact and its disadvantages flow from societal attitudes and assumptions rather than simply from physical or mental impairments. The classic example distinguishing the social model from the medical model is the person using a wheelchair who is unable to enter a building because the only means of entry is a staircase. The medical model suggests that the impairment and the need for a wheelchair are the source of disablement. The social model suggests that the socially created barrier—a building design that involves the use of stairs and fails to include a ramp or lift—is the source of disablement.

Disability bioethics brings attention to the ways in which entrenched ableism, including medical dominance and deference to medical authority, shapes health care decisions and experiences involving people with disabilities. It challenges the historical exclusion of disability knowledge and perspectives in bioethics and argues that disability perspectives must be included in the analysis of bioethics issues. Those perspectives are informed by the cultural and legal history of medicalized abuse and neglect of people with disabilities, including the denial of needed services; inadequate or inappropriate treatment; unnecessary institutionalization often under unacceptable conditions; involuntary sterilization; and other harms. Health care disparities related to disability persist in terms of access to care, treatment outcomes, and health status. See Chapter 2. The treatment of individuals with disabilities is discussed in chapters addressing sterilization and reproductive care (Chapter 3), organ transplantation and the determination of death (Chapter 4), life and death decisions (Chapter 5), medically assisted death (Chapter 6), and research (Chapter 7). Disability theory argues that so-called disability is part of the diversity of human experience and rejects the view that life with disability is pitiable or unworthy of living. For example, it calls for examination of assumptions and biases about quality of life that may impact decisions made and advice given by physicians when treating people with disabilities. It also highlights how persistent devaluation of the lives of people with disabilities by the medical community, legislators, researchers, and others have perpetuated inequities in health and access to health care, including life-saving care.

V. CODES OF ETHICS AND PROFESSIONAL OATHS

For the last few millennia, individuals and professional groups have attempted to distinguish themselves from others by adopting and applying principles of ethically appropriate conduct for their professional roles. Sometimes individual professionals accept these principles by taking an oath—a promise—to adhere to them, as in the case of the oath taken by followers of Hippocrates (and many doctors even today). Sometimes these oaths are expressed as contracts, promises, or even prayers, as in the case of the Prayer of Maimonides. Sometimes professional groups, such as the American Medical Association and the American Nurses Association, announce those principles that ought to apply to their membership, as they have in the Principles of Medical Ethics and the Code of Ethics for Nurses.

As you read the oath, prayer, and sets of codes and principles that follow, ask yourself some questions about them. Are oaths, prayers, principles, pledges, and codes truly statements of ethics, or are they statements of professional convention and etiquette? Are they statements of fundamental principles, or parts of a contract among health care providers and between health care providers and their patients? Do they give rise to reasonable expectations among providers? Among patients? For the whole society? Which ethical theories are reflected in each of these documents? Does any one of them have any utilitarian element? Why is utilitarian theory—so popular among philosophers and American courts—so little represented in these statements? Are they consistent with the principles identified in the principlist approach? Do you see imprints of Kantianism or virtue ethics?

Then shift from theory to application. Should any of these statements be enforceable in law? Should they be enforceable in some other way? Does it make sense to approve and accept these statements but maintain them as unenforceable exhortations? As you read the material in the rest of this casebook, ask yourself whether the expectations of these oaths and codes have been accepted by the courts, formally or informally, whether they have been rejected by the courts, or whether they have been ignored.

OATH OF HIPPOCRATES

I swear by Apollo the Healer, by Asclepius, by Hygieia, by Panacea all the gods and goddesses, that, according to my ability and judgment, I will keep this Oath and this stipulation—to reckon him who taught me this Art equally dear to me as my parents, to share my substance with him, and relieve his necessities if required; to look upon his offspring in the same footing as my own brothers, and to teach them this Art, if they shall wish to learn it, without fee or stipulation; and that by precept, lecture and every other mode of instruction, I will impart a knowledge of the Art to my own sons, and those of my teachers, and to disciples bound by a stipulation and oath according to the law of medicine, but to none other. I will follow that system of regimen which, according to my ability and judgment, I consider for the benefit of my patients, and abstain from whatever is deleterious and mischievous. I will give no deadly medicine to anyone if asked, nor suggest any such counsel; and in like manner I will not give to a woman a pessary to produce abortion. With purity and with holiness I will pass my life and practice my Art. I will not cut persons laboring under the stone, but will leave this to be done by men who are practitioners of this work. Into whatever houses I enter, I will go into them for the benefit of the sick, and will abstain from every voluntary act of mischief and corruption; and, further, from the seduction of females, or males, of freemen or slaves. Whatever, in connection with my professional practice, or not in connection with it, I see or hear, in the life of men, which ought not to be spoken of abroad, I will not divulge, as reckoning that all such should be kept secret. While I continue to keep this Oath un-violated, may it be granted to me to enjoy life and the practice of the Art, respected by all men, in all times. But should I trespass and violate this Oath, may the reverse be my lot.

PRAYER OF MAIMONIDES

Almighty God, Thou has created the human body with infinite wisdom. Ten thousand times ten thousand organs hast Thou combined in it that act unceasingly and harmoniously to preserve the whole in all its beauty—the body which is the envelope of the immortal soul. They are ever acting in perfect order, agreement and accord. Yet, when the frailty of matter or the unbridling of passions deranges this order or interrupts this accord, then forces clash and the body crumbles into the primal dust from which it came.

Thou sendest to man diseases as beneficent messengers to foretell approaching danger and to urge him to avert it.

Thou has blest Thine earth, Thy rivers and Thy mountains with healing substances; they enable Thy creatures to alleviate their sufferings and to heal their illnesses. Thou hast endowed man with the wisdom to relieve the suffering of his brother, to recognize his disorders, to extract the healing substances, to discover their powers and to prepare and to apply them to suit every ill. In Thine Eternal Providence Thou hast chosen me to watch over the life and health of Thy creatures. I am now about to apply myself to the duties of my profession. Support me, Almighty God, in these great labors that they may benefit mankind, for without Thy help not even the least thing will succeed.

Inspire me with love for my art and for Thy creatures. Do not allow thirst for profit, ambition for renown and admiration, to interfere with my profession, for these are the enemies of truth and of love for mankind and they can lead astray in the great task of attending to the welfare of Thy creatures. Preserve the strength of my body and of my soul that they ever be ready to cheerfully help and support rich and poor, good and bad, enemy as well as friend. In the sufferer let me see only the human being. Illumine my mind that it recognize what presents itself and that it may comprehend what is absent or hidden. Let it not fail to see what is visible, but do not permit it to arrogate to itself the power to see what cannot be seen, for delicate and indefinite are the bounds of the great art of caring for the lives and health of Thy creatures. Let me never be absent-minded. May no strange thoughts divert my attention at the bedside of the sick, or disturb my mind in its silent labors, for great and sacred are the thoughtful deliberations required to preserve the lives and health of Thy creatures.

Grant that my patients have confidence in me and my art and follow my directions and my counsel. Remove from their midst all charlatans and the whole host of officious relatives and know-all nurses, cruel people who arrogantly frustrate the wisest purposes of our art and often lead Thy creatures to their death.

Should those who are wiser than I wish to improve and instruct me, let my soul gratefully follow their guidance; for vast is the extent of our art. Should conceited fools, however, censure me, then let love for my profession steel me against them, so that I remain steadfast without regard for age, for reputation, or for honor, because surrender would bring to Thy creatures sickness and death.

Imbue my soul with gentleness and calmness when older colleagues, proud of their age, wish to displace me or to scorn me or disdainfully to teach me. May even this be of advantage to me, for they know many things of which I am ignorant, but let not their arrogance give me pain. For they are old and old age is not master of the passions. I also hope to attain old age upon this earth, before Thee, Almighty God!

Let me be contented in everything except in the great science of my profession. Never allow the thought to arise in me that I have attained to sufficient knowledge, but vouchsafe to me the strength, the leisure and the ambition ever to extend my knowledge. For art is great, but the mind of man is ever expanding.

Almighty God! Thou hast chosen me in Thy mercy to watch over the life and death of Thy creatures. I now apply myself to my profession. Support me in this great task so that it may benefit mankind, for without Thy help not even the least thing will succeed.

AMERICAN MEDICAL ASSOCIATION, PRINCIPLES OF MEDICAL ETHICS
(2001).

PREAMBLE

The medical profession has long subscribed to a body of ethical statements developed primarily for the benefit of the patient. As a member of this profession, a physician must recognize responsibility to patients first and foremost, as well as to society, to other health professionals, and to self. The following Principles adopted by the American Medical Association are not laws, but standards of conduct which define the essentials of honorable behavior for the physician.

PRINCIPLES OF MEDICAL ETHICS

I.   A physician shall be dedicated to providing competent medical care, with compassion and respect for human dignity and rights.

II.   A physician shall uphold the standards of professionalism, be honest in all professional interactions, and strive to report physicians deficient in character or competence, or engaging in fraud or deception, to appropriate entities.

III. A physician shall respect the law and also recognize a responsibility to seek changes in those requirements which are contrary to the best interests of the patient.

IV.   A physician shall respect the rights of patients, colleagues, and other health professionals, and shall safeguard patient confidences and privacy within the constraints of the law.

V.   A physician shall continue to study, apply, and advance scientific knowledge, maintain a commitment to medical education, make relevant information available to patients, colleagues, and the public, obtain consultation, and use the talents of other health professionals when indicated.

VI.   A physician shall, in the provision of appropriate patient care, except in emergencies, be free to choose whom to serve, with whom to associate, and the environment in which to provide medical care.

VII.   A physician shall recognize a responsibility to participate in activities contributing to the improvement of the community and the betterment of public health.

VIII. A physician shall, while caring for a patient, regard responsibility to the patient as paramount.

IX.   A physician shall support access to medical care for all people.

American Nurses Association, Provisions of the Code of Ethics for Nurses

(2015).

1.   The nurse practices with compassion and respect for the inherent dignity, worth, and unique attributes of every person.

2.   The nurse's primary commitment is to the patient, whether an individual, family, group, community, or population.

3.   The nurse promotes, advocates for, and protects the rights, health, and safety of the patient.

4.   The nurse has authority, accountability and responsibility for nursing practice; makes decisions; and takes action consistent with the obligation to promote health and to provide optimal patient care.

5.   The nurse owes the same duties to self as to others, including the responsibility to promote health and safety, preserve wholeness of character and integrity, maintain competence, and continue personal and professional growth.

6.   The nurse, through individual and collective effort, establishes, maintains, and improves the ethical environment of the work setting and conditions of employment that are conducive to safe, quality health care.

7.   The nurse, in all roles and settings, advances the profession through research and scholarly inquiry, professional standards development, and the generation of both nursing and health policy.

8.   The nurse collaborates with other health professionals and the public to protect human rights, promote health diplomacy, and reduce health disparities.

9. The profession of nursing, collectively through its professional organizations, must articulate nursing values, maintain the integrity of the profession, and integrate principles of social justice into nursing and health policy.

AMERICAN PUBLIC HEALTH ASSOCIATION, PUBLIC HEALTH CODE OF ETHICS
(2019).

* * *

Section 2.   Public Health Core Values and Related Obligations

The following core ethical values are equally important and are not presented in rank order. These values are multifaceted conceptually and can be realized in practice in different ways. They do not have simple definitions. They require ongoing and explicit reflection and reaffirmation.

A. **Professionalism and Trust**. The effectiveness of public health policies, practices, and actions depends upon public trust gained through decisions based on the highest ethical, scientific, and professional standards. Public health gains public trust in part because its practices are informed by evidence. When the needed evidence is lacking, public health seeks it, and when the evidence reveals faulty or inadequate practices, public health seeks to improve those practices. At times public health practitioners must respond to a situation in the absence of complete scientific information, which highlights the importance of having an ethical framework to drive decision making. Public health practitioners and organizations promote competence, honesty, and accuracy and ensure that their work is not unduly influenced by secondary interests. Public health decision makers need to be transparent and honest about disclosing conflicting interests and influences.

B. **Health and Safety**. Health and safety are essential conditions for human flourishing. Public health practitioners and organizations have an ethical responsibility to prevent, minimize, and mitigate health harms and to promote and protect public safety, health, and well-being.

C. **Health Justice and Equity**. Human flourishing requires the resources and social conditions necessary to secure equal opportunities for the realization of health and other capabilities by individuals and communities. Public health practitioners and organizations have an ethical obligation to use their knowledge, skills, experience, and influence to promote equitable distribution of burdens, benefits, and opportunities for health, regardless of an individual's or a group's relative position in social hierarchies. Health justice and equity also extend to ensuring that public health activities do not exacerbate health inequities. In addition, health justice does not pertain only to the distribution of scarce resources in transactions among individuals; it also involves remediation of structural and institutional forms of domination that arise from inequalities related to voice, power, and wealth. It is difficult for public health to promote health justice at the transactional level if it does not take steps to promote it at the structural and institutional levels as well.

D. **Interdependence and Solidarity.** The health of every individual is linked to the health of every other individual within the human community, to other living creatures, and to the integrity and functioning of environmental ecosystems. Public health practitioners and organizations have an ethical obligation to foster positive—and mitigate negative—relationships among individuals, societies, and environments in ways that protect and promote the flourishing of humans, communities, nonhuman animals, and the ecologies in which they live. Attention to potential intergenerational conflicts over resources can sometimes be essential.

E. **Human Rights and Civil Liberties.** While coercive legal measures limiting behavior can be ethically justified in certain circumstances, overall the effective and ethical practice of public health depends upon social and cultural conditions of respect for personal autonomy, self-determination, privacy, and the absence of domination in its many interpersonal and institutional forms. Contemporary public health respects and helps sustain those social and cultural conditions.

F. **Inclusivity and Engagement.** Preventing adverse health outcomes and protecting and promoting the flourishing of individuals, societies, and ecosystems require informed public decision-making processes that engage affected individuals and communities. Public health practitioners and organizations have an ethical responsibility to be transparent, to be accountable to the public at large, and to include and engage diverse publics, communities, or stakeholders in their decision making.

* * *

PROBLEM: DRAFTING A PROFESSIONAL CODE OF ETHICS

You are legal counsel to several medical professional organizations, including organizations of physicians, nurses, physical therapists, cafeteria workers, and health care administrators. You have been asked to draft codes of ethics for these organizations. Consider: Does this seem like an appropriate endeavor with respect to each of these groups? From whom would you seek advice as you begin such a drafting process? What information will you need? Keeping these questions in mind, draft an appropriate code of ethics for any one of these professions. In addition, draft an oath (or affirmation) professionals should take upon becoming a member of the organization.

VI. THE RELATIONSHIP BETWEEN ETHICS AND LAW

Law and ethics are not identical. In fact, some have viewed the participation of lawyers in the debate over ethical issues in health care as an intrusion into the physician-patient relationship, and some believe that the law is too blunt an instrument to deal with the delicate, intimate, and highly variable situations in which ethical concerns arise. Others argue that the law is essential to protecting patients from medical dominance and overreaching and that it contributes important values to health care decision-making.

It is tempting to assume that all substantial ethical obligations should be supported by the force of law. Paradoxically, however, the "legalization" of a moral obligation, like the "medicalization" of a personal or social problem, can have undesired consequences. For example, the informed consent doctrine discussed in Section VI.B., demonstrates how the law can undermine ethics by "legalizing" a moral obligation. Legal standards, usually based on ethical minimums, have been attacked as stimulating a drive to the bottom and as diminishing commitment to ethical behavior, with the ironic result that more enforcement is needed to achieve less compliance. Law is limited as an enforcer of ethics by the cost or intrusiveness of enforcement, by the variability of context, and even by the lack of consensus on ethical norms. It is usually understood, then, that obligations defined by law do not encompass the complete moral obligations of the health care professional, family, or patient.

Law and ethics do often converge, however, and the two disciplines influence one another. One area of convergence involves some shared methodology. For example, the common law system in the U.S. develops principles on a case-by-case basis, with the courts looking for commonalities and distinctions in diverse fact situations. Casuistry, the development of principles from specific situations or cases, described in Section IV, above, has been a key methodology in bioethics scholarship and education as well. The case method, as you will see in these materials, is a significant teaching and analytical tool in bioethics as well as law.

Another point of convergence between law and ethics is evident in the explicit adoption of ethical norms and ethical modes of reasoning in regulations or case law. For example, the regulation of research with human subjects incorporates the ethical framework established in the Belmont Report, which itself rested on an early version of principlism and is imbued with a Kantian approach to ethical reasoning. In contrast, the debate over stem cell research has implicated primarily natural law and utilitarian theories. Most case law on end-of-life decision-making has explicitly adopted the principlist approach, depending especially upon the principles of autonomy and beneficence. Newer cases on the regulation of assisted reproduction tend to employ utilitarian analysis, while a great deal of the legal regulation of genetic information is explicitly based on Kantian principles.

As you read the legal material in the following chapters, ask yourself which theories are consistent with the judicial opinions in cases concerning bioethics. Consider the *Cruzan* case in Chapter 5, for example. What was the ethical basis of the Cruzan family decision to seek authority to terminate Nancy Cruzan's treatment? What was the ethical basis of the Missouri Supreme Court's majority opinion? How is it different from those ethical frameworks that appear to support the majority and the dissenting opinions in the U.S. Supreme Court? Do any of the U.S. Supreme Court opinions you will study appear to be based in utilitarianism? In Kantian analysis? In natural law? How do the very different ethical theories applied by different parties and judges in *Cruzan* play out in the context of physician-assisted death?

Our law has a number of mechanisms that assure that the law does not directly contradict the demands of ethical behavior, especially on the part of health care professionals. It is common, for example, for courts and legislatures to defer to "professional custom" and "standards of the profession" for the content of legal obligations. Further, various decision-making or consultative functions in health care—including institutional review boards that apply ethical and legal norms to medical research and hospital ethics committees that consider situations of individual patients, offer consultation on difficult cases, and establish policies for the hospital—are dominated by members of the health professions and have acquired official legal status. See Chapters 5 and 7. Accommodating or protecting the ethical norms of individual health care professionals, however, can promote professional domination and conflict directly with the ethical norms and moral decisions of individual patients. Significant concerns are also raised by legislative or institutional policies that allow an individual professional to refuse a patient's choice of treatment on grounds that providing it, although legal, would be inconsistent with the conscience of the provider.

Despite instances of shared methodologies and the adoption of specific ethical norms, the culture of law emphasizes its own particular values. Law focuses on consistency or fairness, for example, as a primary principle. In addition, law tends to respect process over a specific outcome, and legal institutions have resolved some thorny ethical issues in bioethics by specifying a procedure for resolving the conflict rather than by deciding the substantive outcome.

The values in law may thus cause friction with established ethical norms and the culture of the health care professions. In such cases, deference to professional standards and custom is not complete, and legal norms may overtake ethical norms in some circumstances. *Quinlan*, for example, intentionally reached beyond the medical custom and ethics of the time to shift authority, at least partially, from the doctors to the patient and family. In fact, the legal requirement of informed consent was established in contravention of established medical custom, although some argue that medicine was moving in that direction anyway. See Subsection B, below. The greatest culture clash, however, arises with the law's confidence in the adversarial method of advocacy and decision-making.

PROBLEM: ETHICS AND LAW—NEVER THE TWAIN SHALL MEET?

The American Medical Association's "Code of Medical Ethics" consists of both the "Principles of Medical Ethics" and the Opinions of the AMA Council on Ethical and Judicial Affairs (CEJA). The preamble to the Principles of Medical Ethics states:

> The following Principles adopted by the [AMA] are not laws, but standards of conduct which define the essentials of honorable behavior for the physician.

The AMA Code of Medical Ethics contains several provisions relating to access to care. These include the following:

> A physician shall, in the provision of appropriate patient care, except in emergencies, be free to choose whom to serve, with whom to associate, and the environment in which to provide medical care. (AMA Principles of Medical Ethics)

> . . . Physicians must also uphold ethical responsibilities not to discriminate against a prospective patient on the basis of race, gender, sexual orientation or gender identity, or other personal or social characteristics that are not clinically relevant to the individual's care. Nor may a physician decline a patient based solely on the individual's infectious disease status. . . . (Code of Medical Ethics Opinion 1.1.2)

> Individual physicians should: Take steps to promote access to care for individual patients, such as providing pro bono care in their office or through freestanding facilities or government programs that provide health care for the poor, or, when permissible, waiving insurance copayments in individual cases of hardship. . . . (Code of Medical Ethics Opinion 11.1.4)

Assume that you are a member of the state's medical licensure board. Would you recommend that the state adopt some or all of these standards as legal grounds for disciplinary action or conditions for license renewal? Should plaintiffs in private litigation be permitted to submit these ethical principles into evidence to prove the scope of legal obligations or the standard of medical care? Why or why not? See the related discussion in Chapter 2.

Now consider the following:

> [Physicians] should have considerable latitude to practice in accord with well-considered, deeply held beliefs that are central to their self-identities. . . . Physicians have stronger obligations to patients with whom they have a physician-patient relationship, especially one of long standing; when there is imminent risk of foreseeable harm to the patient or delay in access to treatment would significantly adversely affect the patient's physical or emotional well-being; and when the patient is not reasonably able to access needed treatment from another qualified physician. In following conscience, physicians should . . . inform the patient about all relevant options for treatment, including options to which the physician morally objects . . . refer a patient to another physician or institution to provide treatment the physician declines to offer [and when conscience leads a physician to decline to refer], the physician should offer impartial guidance to patients about how to inform themselves regarding access to desired services. (Code of Medical Ethics Opinion 1.1.7)

Should this provision be enforceable in litigation by a patient concerning a physician's refusal to treat the patient or to provide a particular medical intervention the physician is capable of providing? Should a physician be able to rely on this provision to defend against a hospital's requirement that all physicians holding staff privileges must provide legally mandated emergency care when the physician has a moral objection to providing certain interventions? Before you answer these questions, consider the following section.

A. CONFLICTS OF CONSCIENCE AND LAW

The law of bioethics in large part addresses the scope of individual liberties within health care decision-making. For example, case law and statutes create protections for and set limits on abortion access; prohibit or permit medical aid in dying; and regulate the activities of researchers and the participation of individuals as research subjects. See Chapters 3, 6, and 7.

Debates over legal boundaries on individual choice typically contest the appropriate relationship of law and morality. Some argue that the use of the law's power to restrict care believed necessary or desirable by an individual patient (such as abortions or medically assisted dying) enforces a moral, ethical, or religious perspective illegitimately. Others argue that the law must restrict individual choices that they consider immoral because such choices damage the social fabric, violate foundational civic norms, or injure vulnerable persons. See Section II.D above.

Even where a particular medical intervention is permitted by law, patients and health care providers may disagree as to the morality of that intervention. In this context, both parties assert claims to moral agency, personal liberty, and perhaps religious freedom. Individual patients claim a right to make their own moral judgment as to whether to receive treatment, and individual professionals and health care facilities claim a right to refrain from participating in actions they view as immoral even though permitted by law. These conflicts arise in any number of health care treatment decisions but have most prominently concerned abortion, contraception, medical assistance in dying, euthanasia, and most recently gender-affirming care. Beyond these categories, however, these conflicts can arise in other medical treatment decision-making, for example, where health care providers claim that it is unethical to continue treatment that is futile. See Chapter 5.

Some argue that health care professionals and health care organizations surrender some personal liberty when they choose to enter the health care field and should be bound to provide all care allowed by law and chosen by patients as needed. Moreover, some denials of care may be understood as violating nondiscrimination principles, such as in the areas of women's reproductive health and gender-affirming care. If health care professionals or organizations are unwilling to provide treatment permitted or even required by the democratic and social processes that produce legal rules, they should leave the practice or at least confine their field of work. Others argue that health care professionals are themselves moral beings who should never be forced by law to engage in activities that violate their moral principles or subject their own moral agency to the will of another. If the ethics and character of the health professions are compromised, they argue, the moral character of medicine itself is compromised. Most public policy responses to these claims of conscience seek to balance the claims of patients and health care professionals.

Legislative responses to conscience claims over the past four decades have carved out specific areas in which health care professionals can claim a legal right to refuse particular services as a matter of conscience despite the health burdens that refusal may impose on the patient seeking care. Many of these statutes also try to accommodate the interests of health care facilities where particular interventions violate the religious mission of the facility. Some state statutes address how services are to be provided when individual employees or medical staff refuse to participate on moral or religious grounds. All of these efforts attempt to balance the moral agency of the patient and the moral agency of the health care professional within the context of the need for timely medical care. In some situations, these interests are amenable to compromise because the care is quite easily available elsewhere, although some advocates find any compromise unacceptable. In other circumstances, care will be delayed or denied, and no true compromise is available.

Typically, legislation and regulation extending legal protection (from liability, denial of government funding, or adverse employment actions) to health care providers refusing to provide legal health care services set boundaries on such refusals. These boundaries can include delimiting the specific interventions (e.g., abortion) that may be refused; limitations on the categories of professionals who fall within the scope of protection; specification of which bases of objection (e.g., religious but not always moral or psychological) trigger protection; and conditioning protection on the availability of alternative providers or particular preconditions such as specific notice to prospective patients, required referral systems for services, or timely notice to employers by objecting employee caregivers, and so on. Some legislative "conscience clauses" appear as a provision, or clause, within broader legislation. For example, it is quite common for state statutes on advance directives to include a provision recognizing a right of health care professionals to refuse to comply with specific directives under certain circumstances. See Chapter 5.

The most substantial limitations in these conscience clauses concern the scope of services and the character of the objections they address. Early legislative efforts on both the federal and state levels addressed particular identified services only, focusing primarily on abortion and certain forms of end-of-life care (often stated in terms of "euthanasia" or "mercy killing" at the time). Some legislation extends a right of refusal to religious and to moral, non-religiously-based objections alike. These broad terms— "religious" and "moral"—can be difficult to define and to apply. Protecting religious objections may require deciding what counts as a religion or a religious objection. Similarly, expanding the scope of protection to non-religious moral perspectives raises questions as to what counts as a moral objection as opposed to objections made on other grounds. Those statutes extending protection only to religious objections favor religion over secular moral perspectives, raising the question of whether that approach is appropriate in the context of our pluralistic society.

Health care organizations experience the conscientious objection issue in two forms. First, as employers, these organizations are concerned about how protecting individual refusals will affect their ability to deliver care to patients. Although most facilities provide some room for individual conscientious objections, the more leeway that is provided, the more challenging it is for the facility to provide care that is legally permitted or legally demanded as a matter of patient choice or legally enforced as customary medical standards of care. Second, religiously affiliated health care facilities, as institutions, claim protection for their refusal to provide certain forms of treatment when to do so would violate their religious mission and obligations.

Extending conscience protections to health care organizations evokes many conflicts. Legally protecting health care organizations that claim conscience objections, especially if those protections are accorded to religiously based or faith-based facilities that actively and consistently implement their mission, recognizes that the organization carries the religious values of the individuals that created and lead it. It also recognizes the significant contribution that religious health care organizations have made, and continue to make, to serving health care needs. Religious health care organizations are critical to providing adequate health care services to many underserved areas, and they are willing to serve populations that for-profit providers would be happy to avoid.

On the other hand, religiously based health care organizations accept very substantial support from public funds provided by all citizens, including Medicare and Medicaid payments, publicly funded research, and support for residency training. They also can claim exemption from income and property taxes. In addition, in many circumstances, allowing a health care organization to deny treatment to patients if there are no other facilities that can provide timely access to care causes significant harm to those patients.

Further, protecting a health care facility's choice to refuse certain services may compromise the moral agency of the health care professionals working within that facility. A physician may agree with a patient's choice that a particular treatment is medically necessary and ethically appropriate but be forbidden by the hospital from providing that treatment within the hospital because it violates the religious principles claimed by the institution. A hospital, for example, may refuse to allow tubal ligations following the delivery of a child to be performed in the hospital even if the patient explicitly requests it and the physician believes that another pregnancy might be dangerous for the mother. The patient can choose to have the tubal ligation sometime later at another hospital, if one is available, but she will be exposed to the added risk associated with a second surgery. Most conscience legislation, therefore, has permitted facilities to refuse to provide particular services only under certain limiting circumstances.

Given the deeply held beliefs and values animating the various interests at stake, this area will most certainly remain very contentious in the future and one that will continue to command the attention of health care professionals, patients and patient advocates, ethicists, and policy makers. And, of course, law and legal actors will continue to play a crucial role in determining how competing interests will be balanced or whether one interest must give way to another—whether through the kind of longstanding state and federal legislative conscience protections described above, the more dramatic regulatory back and forth that we've seen play out in the context of the ACA's contraceptive mandate, or the more slowly evolving jurisprudence around religious liberty claims which seems increasingly protective with the recent significant shift in the Supreme Court.

B. INFORMED CONSENT: LAW'S IMPERFECT INTERSECTION WITH BIOETHICS

Informed consent exemplifies how the law pursues but often cannot achieve a bioethical ideal. The doctrine imposes two separate duties on physicians: a duty to obtain a patient's consent to treatment, and a duty to disclose material treatment information. Each is enforced through its own cause of action.

First, a physician must obtain a patient's consent before treatment except in emergencies, and this duty is enforced through the intentional tort of battery. Under the battery claim, treatment, in the absence of consent, constitutes "intentional" and "harmful or offensive touching," and even a successful treatment gives rise to damages.

Second, a physician, before seeking consent, must also disclose to a patient all information material to the patient's treatment decision. Breach of this duty is treated as professional negligence. States are divided evenly over the standard for materiality. In some, information is "material" if a reasonable person, in what the physician knows or should know to be the patient's position, would consider it significant to the treatment decision at issue. In the others, information is "material" if a reasonably prudent physician would disclose it to a patient in clinical circumstances similar to those of the patient. The two standards overlap significantly, requiring physicians to disclose a patient's diagnosis and prognosis, viable treatment options, and the risks and benefits of those options. To prevail, a plaintiff must also establish that a breach of the duty to disclose caused the patient to consent to treatment ("decision-causation") and that the breach also caused injury ("injury-causation"). Decision-causation exists when a reasonable person, in what the physician knows or should know to be the patient's position, would have refused treatment if that reasonable person knew about the undisclosed information. Meanwhile, injury-causation requires a plaintiff to show that the undisclosed risk actually materialized in the plaintiff's case and caused harm.

The failure-to-disclose claim in combination with the medical battery claim aligns common-law informed consent with the bioethical principles of truth-telling and respect for individual autonomy. By requiring physicians to disclose to patients information that is objectively relevant to their treatment decisions, the doctrine encourages doctors to be honest with patients about their conditions and treatment options. Additionally, the doctrine forces communication of information, which improves patients' understanding of their treatment choices and their ability to act autonomously.

As much as the informed consent doctrine exemplifies the convergence of law and ethics, though, it also demonstrates their incongruity. For example, under the bioethical ideal, a physician would provide all information necessary to enable a patient to make a treatment choice that reflects the patient's specific values. Yet, the law relies on objective disclosure standards and, thus, does not require physicians to serve the individualized informational needs of each patient. Consequently, the law cannot assure fully autonomous treatment decisions.

Additionally, the informed consent doctrine exemplifies how the law can undermine ethics by "legalizing" a moral obligation. Anxious to avoid liability, a physician may satisfy her common-law duty to disclose material treatment information by delivering a set of rote warnings like a police officer "mirandizes" a criminal suspect. This approach falls well short of the ethical ideal for a physician to help her patient make autonomous treatment decisions, and it may even erode cooperative decision-making by introducing a bureaucratic or adversarial element to the doctor-patient interaction. In fact, informed consent doctrine has been so focused on *what* information should be disclosed, that virtually no meaningful standards have been developed to address *how* providers should communicate to ensure that difficult or complex concepts are being communicated effectively. An essential component of informed decision-making is patient understanding; yet this concept is largely neglected in the common law of tort. Antidiscrimination law provides a bit more guidance on providers' communication obligations in certain contexts, such as for patients with visual or hearing impairments (disability nondiscrimination law) or patients with limited English proficiency (national origin nondiscrimination law). See Chapter 2. Beyond these important but limited antidiscrimination protections, however, a void remains.

There is no question that patients are informed of much more today than they were before informed consent actions became an accepted part of malpractice law in the 1970s. And yet in some areas, such as pregnancy care, contemporary studies show a vast disconnect between provider disclosures and the information that patients say they need for informed decision-making. Such data raise important questions about the impact of the common law duty of informed consent on autonomy and health. How has the law shaped the quality of provider-patient communication and medical decision-making, patient outcomes and experiences of care, and providers' own understanding of what their ethical obligations require? Is the rather blunt instrument of the common law sufficient to achieve autonomy aims, in light of the power imbalance and medical dominance endemic in the health care system? What about other barriers to communication suffered by patients who are disproportionately harmed by existing systemic inequity? In areas where concerns of medical abuse have been very pronounced, as in the case of forced and coerced sterilizations of members of marginalized groups, regulators have created heightened informed consent requirements to better protect patients. Should there be a greater role for regulators in helping to ensure that autonomy goals are better protected in other areas of care where there has not been the same history of abuse and/or the stakes may not seem as high? Or should the law be used more aspirationally—to reward, and create incentives for, innovation that promotes better provider-patient communication. For example, a statute could offer providers some level of protection from informed consent malpractice claims, if the providers go beyond what the common law requires and employ evidence-based, shared decision-making tools that have been developed by relevant medical experts in partnership with other key experts and patient groups. Certainly, law's relationship with bioethics is imperfect. Yet our ever-evolving concept of informed consent reveals how legal and ethical principles are mutually constitutive and together have guided us, however slowly or erratically, toward a more patient-centered understanding of what informed consent demands.

BIBLIOGRAPHICAL NOTE

There is a great deal of literature on the principles underlying bioethics and the ethical theories that give rise to those principles. The foundational text in this area is Tom Beauchamp and James Childress, Principles of Medical Ethics (7th ed. 2013). See also Albert Jonsen, Mark Siegler and William Winslade, Clinical Ethics: A Practical Approach to Ethical Decisions in Clinical Medicine (8th ed. 2015); Bonnie Steinbock, Alex John London, and John D. Arras, Ethical Issues in Modern Medicine (8th ed. 2012). Particularly useful readings can be found in Marsha Garrison and Carl Schneider, The Law of Bioethics: Individual Autonomy and Social Regulation (3rd ed. 2015). The most helpful online reference for those seeking an introduction to philosophy as a discipline is the Stanford Encyclopedia of Philosophy, available at http://plato.stanford.edu/.

CHAPTER 2

DISCRIMINATION AND UNEQUAL TREATMENT IN HEALTH CARE

■ ■ ■

I. INTRODUCTION

Access to health care depends on finding providers who are willing and able to treat you. But some people are refused medical care or given inadequate treatment for economic and social reasons unrelated to medical need. Among the cases you will read in this chapter are cases in which patients were denied care or treated differently because they couldn't pay or had a disfavored form of insurance; because of their race, ethnicity, or national origin; because of their gender; or because they had a particular type of disability.

In addition to individual refusals of care, patients experience systemic or structural barriers to care, directly or indirectly related to the same characteristics that have motivated refusals. People with a lower socioeconomic status, and racial and ethnic minorities, are more likely to live in communities with an inadequate supply of providers. Residential segregation, in combination with public hospital closures and provider "flight" (physicians and private hospitals leaving poorer, predominantly minority neighborhoods in favor of more affluent, predominantly white ones), has deprived entire communities of essential health resources. Private and government insurers have excluded care that uniquely or disproportionately impacts certain genders, including certain types of reproductive care and gender-affirming care, as well as care for certain disabilities, such as HIV treatment and habilitative care (services that help an individual keep, learn, or improve skills and functioning of daily living). And transgender individuals and people with disabilities are at higher risk of not being able to find health professionals or facilities willing to provide the specific care they need. Imagine not being able to get a common diagnostic test, such as an X-ray or MRI, or never being weighed during pregnancy—this is the situation faced by many individuals with mobility impairments.

Although many factors can impact health access and outcomes, this chapter focuses on the role of discrimination and unequal treatment in health care access, quality, financing, and reform. We start, in Section II, by considering the common law—examining whether and to what extent

41

doctors, hospitals and other types of providers may be legally obligated to treat patients needing medical care. The main cases in this section reveal that the common law favors the freedom to contract, typically finding no initial duty to treat regardless of motive for the refusal or seriousness of medical need. Once a provider does undertake care, the common law is less clear about how and when a provider can terminate the relationship for non-medical reasons, such as for concerns about the patient's ability to pay.

This focus on providers' freedom to contract is consistent with a view of health care as an ordinary consumer good whose supply and demand should be subject to free market principles—a view that has dominated the U.S. approach to health care for much of its history. But a countervailing view—that health care is an essential public good to which everyone should have access as a right—has gained increasing support in recent years and was a dominant theme in the push for the Affordable Care Act (ACA).

But a meaningful "right" to health care cannot exist without creating a legal obligation to provide care or to ensure access to an adequate supply of providers. This, in turn, implicates questions about how resources should be allocated and who should bear the cost of care. In many of the cases you will read, access barriers are linked to cost concerns—the risk of being unable to collect payment or being undercompensated for service, as well as concerns about how to distribute limited health care resources, such as hospital services, across different communities. The ACA somewhat ameliorated, but has not eliminated, these concerns. Although socioeconomic and insurance status continue to be important factors influencing access to care, disparities in health care access and outcomes across race, gender, and disability persist even after controlling for factors such as ability to pay and medical need.

Although there is no universal right to health care in the U.S., a patchwork of federal laws imposes certain limits on providers' or payers' right to deny care. The remaining parts of the chapter examine this legal patchwork. Section III focuses on the Emergency Medical Treatment and Labor Act (EMTALA). Enacted to respond to the specific problem of "patient dumping" in hospital emergency departments, EMTALA creates a crucial but very limited duty to treat in certain settings. Section IV reviews a set of statutes that apply broader nondiscrimination principles to health care, with particular focus on discrimination based on race or national origin, sex, and disability. These antidiscrimination principles predate the modern health care system and apply in settings beyond health care delivery and financing. We focus here on how these laws limit providers' right to refuse to treat patients based on particular characteristics and, at times, have been used to limit policies or practices that result in broader systemic inequity.

II.   COMMON LAW APPROACHES

Each of the cases below involves a patient denied care for reasons unrelated to medical need. Consider to what extent, if any, the common law addresses these various forms of discrimination. More specifically, what does the common law say about providers' right to deny care?

CHILDS V. WEIS

Court of Civil Appeals of Texas, 1969.
440 S.W.2d 104.

WILLIAMS, JUSTICE.

On or about November 27, 1966 Daisy Childs, wife of J.C. Childs, a resident of Dallas County, was approximately seven months pregnant. On that date she was visiting in Lone Oak, Texas, and about two o'clock A.M. she presented herself to the Greenville Hospital emergency room. At that time she stated she was bleeding and had labor pains. She was examined by a nurse who identified herself as H. Beckham. According to Mrs. Childs, Nurse Beckham stated that she would call the doctor. She said the nurse returned and stated "that the Dr. said that I would have to go to my doctor in Dallas. I stated to Beckham that I'm not going to make it to Dallas. Beckham replied that yes, I would make it. She stated that I was just starting into labor and that I would make it. The weather was cold that night. About an hour after leaving the Greenville Hospital Authority I had the baby while in a car on the way to medical facilities in Sulphur Springs. The baby lived about 12 hours."

[Dr. Weis] said that he had never examined or treated Daisy Childs and in fact had never seen or spoken to either Daisy Childs or her husband, J.C. Childs, at any time in his life. He further stated that he had never at any time agreed or consented to the examination or treatment of either Daisy Childs or her husband. He said that on a day in November 1966 he recalled a telephone call received by him from a nurse in the emergency room at the Greenville Surgical Hospital; that the nurse told him that there was a negro girl in the emergency room having a "bloody show" and some "labor pains." He said the nurse advised him that this woman had been visiting in Lone Oak, and that her OB doctor lived in Garland, Texas, and that she also resided in Garland. The doctor said, "I told the nurse over the telephone to have the girl call her doctor in Garland and see what he wanted her to do. I knew nothing more about this incident until I was served with the citation and a copy of the petition in this lawsuit."

* * *

Since it is unquestionably the law that the relationship of physician and patient is dependent upon contract, either express or implied, a physician is not to be held liable for arbitrarily refusing to respond to a call of a person even urgently in need of medical or surgical assistance provided that the relation of physician and patient does not exist at the time the call is made or at the time the person presents himself for treatment.

* * *

Applying these principles of law to the factual situation here presented we find an entire absence of evidence of a contract, either express or implied, which would create the relationship of patient and physician as between Dr. Weis and Mrs. Childs. Dr. Weis, under these circumstances, was under no duty whatsoever to examine or treat Mrs. Childs. When advised by telephone that the lady was in the emergency room he did what seems to be a reasonable thing and inquired as to the identity of her doctor who had been treating her. Upon being told that the doctor was in Garland he stated that the patient should call the doctor and find out what should be done. This action on the part of Dr. Weis seems to be not only reasonable but within the bounds of professional ethics.

We cannot agree with appellant that Dr. Weis' statement to the nurse over the telephone amounted to an acceptance of the case and affirmative instructions which she was bound to follow. Rather than give instructions which could be construed to be in the nature of treatment, Dr. Weis told the nurse to have the woman call her physician in Garland and secure instructions from him.

The affidavit of Mrs. Childs would indicate that Nurse Beckham may not have relayed the exact words of Dr. Weis to Mrs. Childs. Instead, it would seem that Nurse Beckham told Mrs. Childs that the doctor said that she would "have to go" to her doctor in Dallas. Assuming this statement was made by Nurse Beckham, and further assuming that it contained the meaning as placed upon it by appellant, yet it is undisputed that such words were uttered by Nurse Beckham, and not by Dr. Weis. . . .

[The court affirmed summary judgment in favor of the defendant.]

LYONS V. GRETHER

Supreme Court of Virginia, 1977.
239 S.E.2d 103.

POFF, JUSTICE.

We awarded a writ of error to a final order entered June 2, 1976 sustaining a demurrer to a motion for judgment filed by Magnolia Lyons (plaintiff) against Dr. Eugene R. Grether (defendant).

A demurrer confesses the truth of the facts alleged and accepts all reasonable inferences therefrom. Plaintiff, a blind person, accompanied by her four year old son and her guide dog, arrived at defendant's "medical office" on the morning of October 18, 1975, a Saturday, to keep an appointment "for a treatment of a vaginal infection". She was told that defendant would not treat her unless the dog was removed from the waiting room. She insisted that the dog remain because she "was not informed of any steps which would be taken to assure the safety of the guide dog, its care, or availability to her after treatment." Defendant "evicted" plaintiff, her son, and her dog, refused to treat her condition, and failed to assist her in finding other medical attention. By reason of defendant's "wrongful conduct", plaintiff was "humiliated" in the presence of other patients and her young son, and "for another two days while she sought medical assistance from other sources", her infection became "aggravated" and she endured "great pain and suffering". Alleging that defendant's waiting room "is a public place and a place to which the general public is invited and where she had a right to have her guide dog with her pursuant to Virginia Code s 63.1–171.2[1]", plaintiff demanded damages resulting from "breach of his duty to treat".

The order sustaining the demurrer was based upon two grounds. Ruling as matters of law, the trial court held that "the defendant had no duty to treat the plaintiff since he had not accepted her as a patient" and that "defendant's waiting room is not a public facility or place contemplated by" the White Cane Act. We address the first ruling in our determination whether the motion for judgment was sufficient to allege the creation of a physician-patient relationship and a duty to treat. If we determine that it was, then the trial court's second ruling bears upon the question whether defendant's withdrawal from the relationship for the reasons and under the circumstances alleged in plaintiff's motion excused non-performance of the duty to treat.

Although there is some conflict of authority, the courts are in substantial accord upon the rules concerning the creation of a physician-patient relationship and the rights and obligations arising therefrom. In the absence of a statute, a physician has no legal obligation to accept as a patient everyone who seeks his services. [ ] A physician's duty arises only upon the creation of a physician-patient relationship; that relationship springs from a consensual transaction, a contract, express or implied, general or special [ ]; and a patient is entitled to damages resulting from a breach of a physician's duty. [ ] Whether a physician-patient relationship

---

[1]   This statute, part of the "White Cane Act" (Acts 1972, c. 156), reads as follows:

"s 63.1–171.2 Rights of blind and physically disabled persons in public places and places of public accommodation. . . .

"(b) The blind, the visually handicapped, and the otherwise physically disabled are entitled to full and equal accommodations, advantages, facilities, and privileges of all . . . places of public accommodation . . . and other places to which the general public is invited, subject only to the conditions and limitations established by law and applicable alike to all persons.

"(c) Every totally or partially blind person shall have the right to be accompanied by a dog guide, especially trained for the purpose, in any of the places listed in subsection (b) without being required to pay an extra charge for the dog guide; provided that he shall be liable for any damage done to the premises or facilities by such dog."

is created is a question of fact, turning upon a determination whether the patient entrusted his treatment to the physician and the physician accepted the case. [ ]

We consider first whether the facts stated in the motion for judgment, and the reasonable inferences deducible therefrom, were sufficient to allege the creation of a physician-patient relationship and a duty to treat. Standing alone, plaintiff's allegation that she "had an appointment with defendant" would be insufficient, for it connotes nothing more than that defendant had agreed to see her. But plaintiff alleged further that the appointment she had been given was "for treatment of a vaginal infection". The unmistakable implication is that plaintiff had sought and defendant had granted an appointment at a designated time and place for the performance of a specific medical service, one within defendant's professional competence, viz., treatment of a particular ailment. It is immaterial that this factual allegation might have been contradicted by evidence at trial. Upon demurrer, the test of the sufficiency of a motion for judgment is whether it states the essential elements of a cause of action, not whether evidence might be adduced to defeat it. [ ]

We are of opinion that the motion for judgment was sufficient to allege a consensual transaction giving rise to a physician-patient relationship and a duty to perform the service contemplated, and that the trial court erred in holding as a matter of law that defendant had not accepted plaintiff as a patient.

We consider next how a physician-patient relationship, once created, may be lawfully terminated.

As a general rule, unless the services to be rendered are conditioned or limited by notice or by the terms of employment, the physician-patient relationship continues until the services are no longer needed [ ]; however, the relationship may be terminated earlier by mutual consent or by the unilateral action of the patient; and under certain circumstances, the physician has a right to withdraw from a case, provided the patient is afforded a reasonable opportunity to acquire the services he needs from another physician. [ ]

Under plaintiff's construction of the White Cane Act, defendant's withdrawal from her case was not justified by the circumstances. She argues that defendant's office was a place "to which the public is invited" within the meaning of Code s 63.1–171.2(b) and that defendant's withdrawal violated the right to which she was entitled under Code s 63.1–171.2(c). Under the trial court's construction, defendant's office was not covered by the Act and plaintiff had no statutory right to take her dog there.

We are persuaded by plaintiff's argument as applied to the facts alleged in this case. It fairly appears from the face of the motion for judgment that defendant's office was a place to which certain members of the public were invited by prior appointment to receive certain treatment at certain scheduled hours. Plaintiff did not allege that defendant's office was a place to which the general public was generally invited to receive general medical services. Accordingly, while we hold that, under the facts alleged here, defendant's office was within the intendment of the White Cane Act and that the trial court erred in ruling otherwise, we believe it would be beyond the issues drawn for us to hold as a matter of law that the Act as presently written covers all physicians' offices under all circumstances.[2]

Even if the trial court had been correct in holding that plaintiff had no statutory right to take her guide dog to defendant's office, the question yet would have remained whether plaintiff's refusal to part with her dog without the assurances she sought constituted a circumstance justifying defendant's withdrawal from her case. Also remaining would have been the other question related to defendant's right to withdraw, viz., whether, as plaintiff expressly alleged, she was denied a reasonable opportunity to acquire the services she needed from another physician. Both questions were questions of fact which, even in the absence of the White Cane Act, were the subjects of proof, and we hold that the trial court erred in sustaining the demurrer.

The judgment is reversed and the case will be remanded with instructions to restore plaintiff's motion for judgment to the docket.

Reversed and remanded.

MUSE V. CHARTER HOSPITAL OF WINSTON-SALEM, INC.

Court of Appeals of North Carolina, 1995.
452 S.E.2d 589.

LEWIS, JUDGE.

This appeal arises from a judgment in favor of plaintiffs in an action for the wrongful death of Delbert Joseph Muse, III (hereinafter "Joe"). Joe was the son of Delbert Joseph Muse, Jr. (hereinafter "Mr. Muse") and Jane K. Muse (hereinafter "Mrs. Muse"), plaintiffs. The jury found that defendant Charter Hospital of Winston-Salem, Inc. (hereinafter "Charter Hospital" or "the hospital") was negligent in that, *inter alia,* it had a policy or practice which required physicians to discharge patients when their insurance expired and that this policy interfered with the exercise of the medical judgment of Joe's treating physician, Dr. L. Jarrett Barnhill, Jr.

---

[2]   Nor is it necessary for purposes of this opinion to decide what effect amendments, adopted since this case arose and addressed to other statutes, may have upon the White Cane Act. We refer to Acts 1976, c. 596, and Acts 1977, c. 608. Under Code ss 35–42.1 and 36–124 as amended by those Acts, "medical and dental offices" are expressly designated as places of public accommodation to which "it shall be lawful for a blind person accompanied by a 'seeing eye' dog to take such dog."

The jury awarded plaintiffs compensatory damages of approximately $1,000,000. The jury found that Mr. and Mrs. Muse were contributorily negligent, but that Charter Hospital's conduct was willful or wanton, and awarded punitive damages of $2,000,000 against Charter Hospital. . . .

The facts on which this case arose may be summarized as follows. On 12 June 1986, Joe, who was sixteen years old at the time, was admitted to Charter Hospital for treatment related to his depression and suicidal thoughts. Joe's treatment team consisted of Dr. Barnhill, as treating physician, Fernando Garzon, as nursing therapist, and Betsey Willard, as social worker. During his hospitalization, Joe experienced auditory hallucinations, suicidal and homicidal thoughts, and major depression. Joe's insurance coverage was set to expire on 12 July 1986. As that date neared, Dr. Barnhill decided that a blood test was needed to determine the proper dosage of a drug he was administering to Joe. The blood test was scheduled for 13 July, the day after Joe's insurance was to expire. Dr. Barnhill requested that the hospital administrator allow Joe to stay at Charter Hospital two more days, until 14 July, with Mr. and Mrs. Muse signing a promissory note to pay for the two extra days. The test results did not come back from the lab until 15 July. Nevertheless, Joe was discharged on 14 July and was referred by Dr. Barnhill to the Guilford County Area Mental Health, Mental Retardation and Substance Abuse Authority (hereinafter "Mental Health Authority") for outpatient treatment. Plaintiffs' evidence tended to show that Joe's condition upon discharge was worse than when he entered the hospital. Defendants' evidence, however, tended to show that while his prognosis remained guarded, Joe's condition at discharge was improved. Upon his discharge, Joe went on a one-week family vacation. On 22 July he began outpatient treatment at the Mental Health Authority, where he was seen by Dr. David Slonaker, a clinical psychologist. Two days later, Joe again met with Dr. Slonaker. Joe failed to show up at his 30 July appointment, and the next day he took a fatal overdose of Desipramine, one of his prescribed drugs.

* * *

II.

Defendants . . . argue that the trial court submitted the case to the jury on an erroneous theory of hospital liability that does not exist under the law of North Carolina. As to the theory in question, the trial court instructed: "[A] hospital is under a duty not to have policies or practices which operate in a way that interferes with the ability of a physician to exercise his medical judgment. A violation of this duty would be negligence." The jury found that there existed "a policy or practice which required physicians to discharge patients when their insurance benefits expire and which interfered with the exercise of Dr. Barnhill's medical judgment." Defendants contend that this theory of liability does not fall within any theories previously accepted by our courts. . . .

Our Supreme Court has recognized that hospitals in this state owe a duty of care to their patients. [ ] [A] hospital has a duty to the patient to obey the instructions of a doctor, absent the instructions being obviously negligent or dangerous. [It also has a] duty to make a reasonable effort to monitor and oversee the treatment prescribed and administered by doctors practicing at the hospital. [ ] In light of these holdings, it seems axiomatic that the hospital has the duty not to institute policies or practices which interfere with the doctor's medical judgment. . . . Charter Hospital had a duty not to institute a policy or practice which required that patients be discharged when their insurance expired and which interfered with the medical judgment of Dr. Barnhill.

III.

* * *

We [next] conclude that in the case at hand, the evidence was sufficient to go to the jury.

Plaintiffs' evidence included the testimony of Charter Hospital employees and outside experts. Fernando Garzon, Joe's nursing therapist at Charter Hospital, testified that the hospital had a policy of discharging patients when their insurance expired. Specifically, when the issue of insurance came up in treatment team meetings, plans were made to discharge the patient. When Dr. Barnhill and the other psychiatrists and therapists spoke of insurance, they seemed to lack autonomy. For example, Garzon testified, they would state, "So and so is to be discharged. We must do this." Finally, Garzon testified that when he returned from a vacation, and Joe was no longer at the hospital, he asked several employees why Joe had been discharged and they all responded that he was discharged because his insurance had expired. Jane Sims, a former staff member at the hospital, testified that several employees expressed alarm about Joe's impending discharge, and that a therapist explained that Joe could no longer stay at the hospital because his insurance had expired. Sims also testified that Dr. Barnhill had misgivings about discharging Joe, and that Dr. Barnhill's frustration was apparent to everyone. One of plaintiffs' experts testified that based on a study regarding the length of patient stays at Charter Hospital, it was his opinion that patients were discharged based on insurance, regardless of their medical condition. Other experts testified that based on Joe's serious condition on the date of discharge, the expiration of insurance coverage must have caused Dr. Barnhill to discharge Joe. The experts further testified as to the relevant standard of care, and concluded that Charter Hospital's practices were below the standard of care and caused Joe's death. We hold that this evidence was sufficient to go to the jury. . . .

IV.

[The court held that the trial court did not err by denying the hospital's motion for judgment notwithstanding the verdict, which had argued that the patient's suicide was a supervening event that interrupted the chain of causation between the hospital's alleged breach and the plaintiffs' damages. The court reasoned that a patient's suicide cannot be deemed a supervening event to prevent liability on part of psychiatric hospital which had assumed care of a suicidal patient.]

ORR, JUDGE, dissenting.

[I] must respectfully dissent from the majority on the submission of the issue on wilful or wanton conduct [which is necessary to justify the award of punitive damages]. . . .

Our Supreme Court [has] defined wilful and wanton conduct as follows:

> An act is done wilfully when it is done purposely and deliberately in violation of law, or when it is done knowingly and of set purpose, or when the mere will has free play, without yielding to reason. 'The true conceptions of wilful negligence involves a deliberate purpose not to discharge some duty necessary to the safety of the person or property of another, which duty the person owing it has assumed by contract, or which is imposed on the person by operation of law.'

> An act is wanton when it is done of wicked purpose, or when done needlessly, manifesting a reckless indifference to the rights of others.

[Citations omitted]. Further,

> While "[o]rdinary negligence has as its basis that a person charged with negligent conduct should have known the probable consequences of his act," we have said "[w]anton and willful negligence rests on the assumption that he knew the probable consequences, but was recklessly, wantonly or intentionally indifferent to the results."

[Citations omitted]. [In this case, the] policy or practice of discharging patients when their insurance ran out . . . was obviously done for a business purpose; however, the evidence reveals that the policy was subject to being overridden on occasion by request of the treating physician or other financial consideration. Although . . . this policy may have affected Dr. Barnhill's decision to discharge the plaintiffs' son, such evidence, while perhaps supporting a negligence theory, does not go beyond that.

. . . No evidence was presented that could lead a jury to conclude that the policy in question involved a deliberate purpose not to discharge some duty necessary to the safety of the person in question. While it can be said that the policy to discharge was deliberate, there is no evidence that the hospital expected, anticipated or intended for the patient to be released in circumstances that put the person's safety in jeopardy. In fact, Joseph Muse, III was discharged into the custody and care of another physician and a community based mental health facility as well as the care of his parents with specific instructions for his care. . . .

. . . A policy to terminate a patient's hospitalization based upon insurance benefits ending in and of itself is not wilful or wanton conduct. . . . If the hospital had simply discharged the patient with no referral to another physician or medical facility, then a cognizable claim for wilful or wanton conduct would have been established. [A]lthough Dr. Barnhill's care in discharging the patient may well have been negligent, there is nothing to suggest that the hospital's policy or its implementation by Dr. Barnhill was done with reckless or deliberate disregard for the patient's safety. . . .

NOTES AND QUESTIONS

1.   *Motive.* Why did the providers refuse to treat Mrs. Childs? Ms. Lyons? The son of Mr. and Mrs. Muse? Was it due to a single motivating factor, or a combination of factors? In determining whether a physician or facility has a legal duty to provide care, should courts be guided by the reason for the refusal, the circumstances of the refusal, or both?

2.   *Ethical vs. Legal Obligations.* Even in the absence of a legal duty to treat, do providers have ethical obligations to provide care? Consider the following ethical prescriptions adopted by the American Medical Association (AMA):

> A physician shall, in the provision of appropriate patient care, except in emergencies, be free to choose whom to serve, with whom to associate, and the environment in which to provide medical care. AMA Principles of Medical Ethics VI (Adopted June 1957; revised June 1980; revised June 2001).

> . . . Differences in treatment that are not directly related to differences in individual patients' clinical needs or preferences constitute inappropriate variations in health care . . . . [P]hysicians . . . ethically are called on to provide the same quality of care to all patients without regard to medically irrelevant personal characteristics. . . . [P]hysicians should [p]rovide care that meets patient needs and respects patient preferences[;] [a]void stereotyping patients[;] [e]xamine their own practices to ensure that inappropriate considerations about race, gender identity, sexual orientation, sociodemographic factors, or other nonclinical factors, do not affect clinical judgment . . . . AMA, Opinion 8.5 Disparities in Health Care, Code of Medical Ethics (2016).

> . . . Individual physicians should . . . promote access to care for individual patients, such as providing pro bono care in their office or through freestanding facilities or government programs that provide health care for the poor, or when permissible, waiving insurance copayments in

individual cases of hardship. . . . AMA, Opinion 11.1.4 Financial Barriers to Health Care Access, Code of Medical Ethics (2009).

Creating an ethical principle that encourages caring for the poor is very different from creating a legal duty to treat that requires physicians to care for patients even if they cannot pay. Should we go this far? Or should the law merely narrow limits on when a provider can refuse treatment? If so, where would you draw the line? Even the AMA seems to draw certain lines in its ethical prescriptions. What factors seem most salient to the AMA's conceptions of physicians' ethical duty of care? What provider interests are implicated by creating a duty to treat?

3.   *Variations in State Common Law.* In *Childs*, the court says that "it is unquestionably the law that the relationship of physician and patient is dependent upon contract, either express or implied." But tort principles have been used by courts to find an implied duty to treat in emergency settings. See, e.g., Wilmington Gen. Hospital v. Manlove, 174 A.2d 135 (Del. 1961), holding that a hospital must provide emergency care to a person who relies on the presence of an emergency department in coming to the hospital. The court analogized this to the negligent termination of gratuitous services because "such a refusal might well result in worsening the condition of the injured person, because of the time lost in a useless attempt to obtain medical aid." The application of tort principles to find a duty to provide emergency care has been inconsistent and has varied by state. See Karen H. Rothenberg, Who Cares? The Evolution of the Legal Duty to Provide Emergency Care, 26 Hous. L. Rev. 21 (1989). Given the variations in state approaches and the often-devastating effects of such denials, the federal government eventually addressed the problem through federal statute. See Section III below.

4.   *Discrimination in Emergency Settings.* The professional and institutional discrimination evident in *Childs* is not an aberration. See, for example, New Biloxi Hospital, Inc. v. Frazier, 245 Miss. 185, 146 So.2d 882 (Miss. 1962), in which the court held a hospital liable for the death of a Black man who remained untreated in the emergency room for over two hours and died twenty-five minutes after transfer to a Veterans Administration hospital. The court based its holding on the hospital's breach of the duty to exercise reasonable care once treatment had been undertaken and it noted the special role of emergency professionals: "A hospital rendering emergency treatment is obligated to do that which is immediately and reasonably necessary for the preservation of the life, limb or health of the patient. It should not discharge a patient in a critical condition without furnishing or procuring suitable medical attention."

Indeed, as recently as 2015, there was a similarly shocking refusal by a hospital to treat a Black woman, who had sought help because she couldn't breathe. After initially being seen in the emergency room, the woman was discharged. She refused to leave, however, because she still felt unwell. The hospital's response was to call law enforcement to have her forcibly removed from the premises. In trying to get her to leave, the officer involved attempted to remove the patient's oxygen mask; when the patient refused to surrender it, hospital staff motioned toward the wall to let the officer know he could disconnect her oxygen hose from there. The officer disconnected the hose, and when he got the patient close to the police car, she collapsed. She lay on the ground—just feet away from the hospital entrance—for nearly 18 minutes before finally being taken back to the hospital, where she was ultimately pronounced dead. The woman had died from a blood clot in her lungs. Throughout this time the patient was begging for help and insisting she could not breathe, but the officer said he thought she was faking it and just being noncompliant, and the hospital treated her as a trespasser. See Michele Goodwin, Black Women Can't Breathe, Online Symposium: Understanding the Role of Race in Health, Bill of Health, Petrie-Flom Center at Harvard Law School, (Oct. 20, 2020). The hospital was ultimately cited for deficiencies in care. Joe Reedy, Florida Hospital Where Woman Died After Being Forcibly Removed Cited for Deficiencies, Orlando Sentinel (Feb. 12, 2016). For more on racial disparities in treatment that persist today, see Section IV.A. below.

5. *When May a Physician Withdraw?* In *Lyons*, the court acknowledged that a physician may withdraw unilaterally from the care of a patient if the patient is "afforded a reasonable opportunity to acquire the services he needs from another physician." In *Muse*, the court focused on the absence of an individualized assessment of continuing need prior to discharge. But both *Lyons* and *Muse* leave unanswered questions about when and how providers can terminate relationships with patients. Once a physician has entered into a health care relationship, must the physician continue to treat the patient until the patient finds someone else to take over the treatment? What if the patient can't find someone else or the alternative facilities are not adequate? This is not merely a theoretical question; questions about the contours of this duty are at the heart of challenges to hospital discharge practices of homeless patients. See, e.g., Joseph Serna, Hospital Agrees to Pay $450,000 to L.A. to Settle Homeless Patient Dumping Lawsuit, L.A. Times (Oct. 25, 2016).

6. *Delineating Common Law Duties. Lyons* illustrates the relevance of statutes in common law questions about provider duties. What role did the state's White Cane Act play in the court's decision? For more information on disability discrimination in health care, see Section IV.C. below.

7. *Intersecting Bases of Discrimination.* In *Muse,* the court affirmed liability against Charter Hospital based on an institutional policy or practice that automatically triggered discharge when insurance coverage ended, effectively disregarding individualized assessments of medical need. This reflects the intersection of two forms of discrimination that have impeded health care access and undermined quality of treatment. One involves discrimination based on insurance status or ability to pay, which has disproportionately impacted groups historically excluded from private insurance, especially racial and ethnic minorities, people with disabilities, and the poor. Even those fortunate enough to qualify for public safety net programs, like Medicaid, may not be able to find providers willing to treat them. The other form of discrimination involves mental health care. Historically, the medical

field and our health financing systems have treated care associated with mental health as less valuable than other kinds of health care. Inadequate provider supply and insurance barriers have made it difficult for patients to get appropriate mental and behavioral care support, as discussed more fully in Section IV.D.2. below.

8.   *State Statutes Requiring Emergency Care.* Some states have statutes or regulations limiting hospitals' ability to refuse care under certain circumstances. For example, regulations promulgated under the hospital licensure statute of North Carolina provide that:

> [a] patient has the right to expect emergency procedures to be implemented without unnecessary delay [and] the right to medical and nursing services without discrimination based upon race, color, religion, sex, sexual orientation, gender identity, national origin, or source of payment. 10A N.C. Admin. Code tit. 13B.3302(6) & (13).

Although the state licensing agency has authority to investigate violations of this provision, at least one court has held that the statute does not create a private right of action. See Williams v. U.S., 242 F.3d 169 (4th Cir. 2001). In contrast, see Thompson v. Sun City Community Hospital, Inc., 141 Ariz. 597, 688 P.2d 605 (1984), where the court relied on state hospital regulations and private accreditation standards to find a duty to provide emergency care enforceable through private litigation.

PROBLEM: CHERYL HANACHEK

Cheryl Hanachek, a law student residing in Boston, learned she was pregnant during an "action" called by the city's obstetricians to protest Medicaid and discounted private insurance payment rates. Ms. Hanachek first called Dr. Cunetto, who had been the obstetrician for the birth of her first child two years earlier. Dr. Cunetto's receptionist told her that Dr. Cunetto wasn't taking any patients covered by her health plan.

About two weeks later, Ms. Hanachek called Dr. Simms, who had been recommended by her sister. Dr. Simms' receptionist told Ms. Hanachek that Dr. Simms was not taking any new patients because his malpractice premiums were so high that he was even considering discontinuing his obstetrical practice. In fact, however, Dr. Simms actually did not accept lawyers as patients. Ms. Hanachek reported to the receptionist that she was having infrequent minor cramping, and the receptionist told her that this was "nothing to worry about at this stage." Later that night Ms. Hanachek was admitted to the hospital on an emergency basis, in shock from blood loss due to a ruptured ectopic pregnancy. As a result of the rupture and other complications, Ms. Hanachek underwent a hysterectomy.

Ms. Hanachek has sued Dr. Cunetto and Dr. Simms. If you were representing Ms. Hanachek, how would you proceed in arguing and proving your case?

III. EMTALA: THE EXCEPTION FOR EMERGENCY CARE

This section and Section IV survey the federal statutory tools used to address certain barriers to health care. For each statutory tool, consider what particular barrier(s) the law is attempting to address, the scope of protection the law provides, and the limits of the statute. How effective is each statute at addressing the problem it was enacted to solve? What about as a tool for protecting health care access generally?

In 1986, the federal Emergency Medical Treatment and Labor Act (EMTALA), 42 U.S.C.A. § 1395dd, was enacted in response to "patient dumping," a practice in which patients are discharged in an unstable condition due to their inability to pay or for other non-therapeutic reasons. This would often take the form of a transfer from one hospital's emergency department to another's—typically from a private hospital to a public hospital that is legally obligated to treat indigent patients.

At the time of EMTALA's enactment, several empirical studies had documented patient dumping as a widespread practice. See, e.g., Robert L. Schiff et al., Transfers to a Public Hospital, 314 NEJM 552 (1986). Researchers determined that transferred patients tended to be uninsured or government-insured patients, with one study revealing that lack of insurance was the reason given for transfer in 87 percent of the cases. The findings suggested a dramatic increase in transfers during the 1980s—coinciding with large cuts in government funding for health care and a growth in uninsured Americans from 25 million in 1977 to 35 million in 1987. Researchers also found that significant percentages of patients transferred were medically unstable at the time of transfer and experienced an average delay of over five hours before receiving treatment. Finally, such transfers were found to disproportionately impact racial minorities. See also Robert L. Schiff & David Ansell, Federal Anti-Patient-Dumping Provisions: The First Decade, 28 Annals Emergency Med. 77 (1996).

EMTALA's focus is narrow: it creates limited duties to treat in the emergency setting. In other ways, however, EMTALA's reach is broad. Although EMTALA applies *only* to hospitals that accept payment from Medicare *and* operate an emergency department, almost all hospitals participate in the Medicare program. And while EMTALA does not require a hospital to offer emergency department services, some state hospital licensure statutes do. In addition, federal tax law encourages tax-exempt hospitals to offer emergency services, and Medicare Conditions of Participation require that all hospitals receiving Medicare payments be capable of providing initial treatment in emergency situations and have an effective procedure for referral or transfer to more comprehensive facilities. Perhaps most importantly, EMTALA applies to *all* patients of such hospitals and not just to Medicare beneficiaries, so its protections apply to the uninsured or those with disfavored forms of insurance.

EMTALA specifically empowers patients to bring civil suits for damages against participating hospitals, but it does not provide a private right of action against a treating physician. The Office of the Inspector General (OIG) of the U.S. Dept. of Health and Human Services (HHS) enforces EMTALA against both hospitals and physicians. Private EMTALA litigation has burgeoned, while government enforcement has been much less active. Administrative enforcement actions under EMTALA are few; monetary penalties are quite small; and exclusion from Medicare is almost unheard of. Despite the passage of EMTALA, and perhaps because of its lax enforcement by the agency, patient dumping continues. See Sara Rosenbaum et al., Case Studies at Denver Health: "Patient Dumping" in the Emergency Department Despite EMTALA, the Law that Banned It, 31 Health Aff. 1749 (2012).

EMERGENCY MEDICAL TREATMENT AND LABOR ACT
42 U.S.C. § 1395dd.

(a) Medical screening requirement. In the case of a hospital that has a hospital emergency department, if any individual . . . comes to the emergency department and a request is made on the individual's behalf for examination or treatment for a medical condition, the hospital must provide for an appropriate medical screening examination within the capability of the hospital's emergency department, including ancillary services routinely available to the emergency department, to determine whether or not an emergency medical condition . . . exists.

(b) Necessary stabilizing treatment for emergency medical conditions and labor.

(1) In general. If any individual . . . comes to a hospital and the hospital determines that the individual has an emergency medical condition, the hospital must provide either—

(A) within the staff and facilities available at the hospital, for such further medical examination and such treatment as may be required to stabilize the medical condition, or

(B) for transfer of the individual to another medical facility in accordance with subsection (c).

* * *

(c) Restricting transfers until individual stabilized.

(1) Rule. If an individual at a hospital has an emergency medical condition which has not been stabilized . . . , the hospital may not transfer the individual unless—

(A)(i)   the individual (or a legally responsible person acting on the individual's behalf) after being informed of the hospital's obligations under this section and of the risk of transfer, in writing requests transfer to another medical facility, [or]

(ii) a physician . . . has signed a certification that[,] based upon the information available at the time of transfer, the medical benefits reasonably expected from the provision of appropriate medical treatment at another medical facility outweigh the increased risks to the individual and, in the case of labor, to the unborn child from effecting the transfer.

* * *

[and]

(B) the transfer is an appropriate transfer . . . to that facility . . . .

* * *

(2) Appropriate transfer. An appropriate transfer to a medical facility is a transfer—

(A) in which the transferring hospital provides the medical treatment within its capacity which minimizes the risks to the individual's health and, in the case of a woman in labor, the health of the unborn child;

(B) in which the receiving facility—

(i) has available space and qualified personnel for the treatment of the individual, and

(ii) has agreed to accept transfer of the individual and to provide appropriate medical treatment;

(C) in which the transferring hospital sends to the receiving facility all medical records . . . related to the emergency condition for which the individual has presented, available at the time of the transfer . . .; [and]

(D) in which the transfer is effected through qualified personnel and transportation equipment. . . .

(d) Enforcement.

(1) Civil monetary penalties. [Fines for negligent violations of the statute by hospitals of up to $50,000 for each violation; and for physicians, up to $50,000 for each negligent violation and exclusion from Medicare and Medicaid for gross and flagrant violations.]

(2) Civil enforcement.

(A) Personal harm. Any individual who suffers personal harm as a direct result of a participating hospital's violation of a requirement of this section may, in a civil action against the participating hospital, obtain those damages available for personal injury under the law of the State in which the hospital is located, and such equitable relief as is appropriate.

(B) Financial loss to other medical facility. Any medical facility that suffers a financial loss as a direct result of a participating hospital's violation of a requirement of this section may, in a civil action against the participating hospital, obtain those damages available for financial loss, under the law of the State in which the hospital is located, and such equitable relief as is appropriate. . . .

* * *

(e) **Definitions.** In this section:

(1) The term "emergency medical condition" means—

(A) a medical condition manifesting itself by acute symptoms of sufficient severity (including severe pain) such that the absence of immediate medical attention could reasonably be expected to result in—

(i) placing the health of the individual (or, with respect to a pregnant woman, the health of the woman or her unborn child) in serious jeopardy,

(ii) serious impairment to bodily functions, or

(iii) serious dysfunction of any bodily organ or part; or

(B) with respect to a pregnant woman who is having contractions—

(i) that there is inadequate time to effect a safe transfer to another hospital before delivery, or

(ii) that transfer may pose a threat to the health or safety of the woman or the unborn child . . . .

* * *

(3)(A) The term "to stabilize" means . . . to provide such medical treatment of the condition as may be necessary to assure, within reasonable medical probability, that no material deterioration of the condition is likely to result from or occur during the transfer of the individual from a facility. . . .

(B) The term "stabilized" means . . . that no material deterioration of the condition is likely, within reasonable medical probability, to result from or occur during the transfer of the individual from a facility, or, with respect to an emergency medical condition described in paragraph (1)(B), that the woman has delivered (including the placenta) . . . .

* * *

(h) No delay in examination or treatment. A participating hospital may not delay provision of an appropriate medical screening examination required under subsection (a) . . . or further medical examination and treatment required under subsection (b) . . . in order to inquire about the individual's method of payment or insurance status.

BABER v. HOSPITAL CORPORATION OF AMERICA

United States Court of Appeals, Fourth Circuit, 1992.
977 F.2d 872.

WILLIAMS, CIRCUIT JUDGE:

Barry Baber, Administrator of the Estate of Brenda Baber, instituted this suit against . . . Raleigh General Hospital (RGH), Beckley Appalachian Regional Hospital (BARH), and the parent corporations of both hospitals. Mr. Baber alleged that the Defendants violated the Emergency Medical Treatment and Active Labor Act (EMTALA)[ ]. The Defendants moved to dismiss the EMTALA claim under Rule 12(b)(6) of the Federal Rules of Civil Procedure. Because the parties submitted affidavits and depositions, the district court treated the motion as one for summary judgment. See Fed.R.Civ.P. 12(b).

* * *

Mr. Baber's complaint charged the various defendants with violating EMTALA in several ways. Specifically, Mr. Baber contends that Dr. Kline, RGH, and its parent corporation violated EMTALA by:

(a) failing to provide his sister with an "appropriate medical screening examination;"

(b) failing to stabilize his sister's "emergency medical condition;" and

(c) transferring his sister to BARH without first providing stabilizing treatment.

* * *

After reviewing the parties' submissions, the district court granted summary judgment for the Defendants. . . . Finding no error, we affirm.

* * *

. . . Brenda Baber, accompanied by her brother, Barry, sought treatment at RGH's emergency department at 10:40 p.m. on August 5,

1987. When she entered the hospital, Ms. Baber was nauseated, agitated, and thought she might be pregnant. She was also tremulous and did not appear to have orderly thought patterns. She had stopped taking her anti-psychosis medications, ... and had been drinking heavily. Dr. Kline, the attending physician, described her behavior and condition in the RGH Encounter Record as follows: Patient refuses to remain on stretcher and cannot be restrained verbally despite repeated requests by staff and by me. Brother has not assisted either verbally or physically in keeping patient from pacing throughout the Emergency Room. Restraints would place patient and staff at risk by increasing her agitation.

In response to Ms. Baber's initial complaints, Dr. Kline examined her central nervous system, lungs, cardiovascular system, and abdomen. He also ordered several laboratory tests, including a pregnancy test.

While awaiting the results of her laboratory tests, Ms. Baber began pacing about the emergency department. In an effort to calm Ms. Baber, Dr. Kline gave her [several medications]. The medication did not immediately control her agitation. Mr. Baber described his sister as becoming restless, "worse and more disoriented after she was given the medication," and wandering around the emergency department.

While roaming in the emergency department around midnight, Ms. Baber ... convulsed and fell, striking her head upon a table and lacerating her scalp. [S]he quickly regained consciousness and emergency department personnel carried her by stretcher to the suturing room, [where] Dr. Kline examined her again. He obtained a blood gas study, which did not reveal any oxygen deprivation or acidosis. Ms. Baber was verbal and could move her head, eyes, and limbs without discomfort. ... Dr. Kline closed the one-inch laceration with a couple of sutures. Although she became calmer and drowsy after the wound was sutured, Ms. Baber was easily arousable and easily disturbed. Ms. Baber experienced some anxiety, disorientation, restlessness, and some speech problems, which Dr. Kline concluded were caused by her pre-existing psychiatric problems of psychosis with paranoia and alcohol withdrawal.

Dr. Kline discussed Ms. Baber's condition with Dr. Whelan, the psychiatrist who had treated Ms. Baber for two years. ... Dr. Whelan concluded that Ms. Baber's hyperactive and uncontrollable behavior during her evening at RGH was compatible with her behavior during a relapse of her serious psychotic and chronic mental illness. Both Dr. Whelan and Dr. Kline were concerned about the seizure she had while at RGH's emergency department because it was the first one she had experienced. ... They also agreed Ms. Baber needed further treatment ... and decided to transfer her to the psychiatric unit at BARH because RGH did not have a psychiatric ward, and both doctors believed it would be beneficial for her to be treated in a familiar setting. The decision to transfer Ms. Baber was further supported by the doctors' belief that any tests to diagnose the cause of her initial seizure, such as a computerized tomography scan (CT scan), could be performed at BARH once her psychiatric condition was under control. The transfer to BARH was discussed with Mr. Baber who neither expressly consented nor objected. His only request was that his sister be x-rayed because of the blow to her head when she fell.

* * *

Because Dr. Kline did not conclude Ms. Baber had a serious head injury, he believed that she could be transferred safely to BARH where she would be under the observation of the BARH psychiatric staff personnel. At 1:35 a.m. on August 6, Ms. Baber was admitted directly to the psychiatric department of BARH upon Dr. Whelan's orders. She was not processed through BARH's emergency department. Although Ms. Baber was restrained and regularly checked every fifteen minutes by the nursing staff while at BARH, no physician gave her an extensive neurological examination upon her arrival. Mr. Baber unsuccessfully repeated his request for an x-ray.

At the 3:45 a.m. check, the nurse found Ms. Baber having a grand mal seizure. At Dr. Whelan's direction, the psychiatric unit staff transported her to BARH's emergency department. Upon arrival in the emergency department, her pupils were unresponsive, and hospital personnel began CPR. The emergency department physician ordered a CT scan, which was performed around 6:30 a.m. The CT report revealed a fractured skull and a right subdural hematoma. BARH personnel immediately transferred Ms. Baber back to RGH because that hospital had a neurosurgeon on staff, and BARH did not have the facility or staff to treat serious neurological problems. When RGH received Ms. Baber for treatment around 7 a.m., she was comatose. She died later that day, apparently as a result of an intracerebrovascular rupture.

* * *

Mr. Baber . . . alleges that RGH, acting through its agent, Dr. Kline, violated several provisions of EMTALA. These allegations can be summarized into two general complaints: (1) RGH failed to provide an appropriate medical screening to discover that Ms. Baber had an emergency medical condition as required by 42 U.S.C.A. § 1395dd(a); and (2) RGH transferred Ms. Baber before her emergency medical condition had been stabilized, and the appropriate paperwork was not completed to transfer a non-stable patient as required by 42 U.S.C.A. § 1395dd(b) & (c). Because we find that RGH did not violate any of these EMTALA provisions, we affirm the district court's grant of summary judgment to RGH.

Mr. Baber first claims that RGH failed to provide his sister with an "appropriate medical screening". He makes two arguments. First, he contends that a medical screening is only "appropriate" if it satisfies a national standard of care. In other words, Mr. Baber urges that we construe EMTALA as a national medical malpractice statute, albeit limited to whether the medical screening was appropriate to identify an emergency medical condition. We conclude instead that EMTALA only requires hospitals to apply their standard screening procedure for identification of an emergency medical condition uniformly to all patients and that Mr. Baber has failed to proffer sufficient evidence showing that RGH did not do so. Second, Mr. Baber contends that EMTALA requires hospitals to provide some medical screening. We agree, but conclude that he has failed to show no screening was provided to his sister.

* * *

While [the Act] requires a hospital's emergency department to provide an "appropriate medical screening examination," it does not define that term other than to state its purpose is to identify an "emergency medical condition."

* * *

[T]he goal of "an appropriate medical screening examination" is to determine whether a patient with acute or severe symptoms has a life threatening or serious medical condition. The plain language of the statute requires a hospital to develop a screening procedure[6] designed to identify such critical conditions that exist in symptomatic patients and to apply that screening procedure uniformly to all patients with similar complaints.

[W]hile EMTALA requires a hospital emergency department to apply its standard screening examination uniformly, it does not guarantee that the emergency personnel will correctly diagnose a patient's condition as a result of this screening.[7] The statutory language clearly indicates that EMTALA does not impose on hospitals a national standard of care in screening patients. The screening requirement only requires a hospital to

---

[6] While a hospital emergency room may develop one general procedure for screening all patients, it may also tailor its screening procedure to the patient's complaints or exhibited symptoms. For example, it may have one screening procedure for a patient with a heart attack and another for women in labor. Under our interpretation of EMTALA, such varying screening procedures would not pose liability under EMTALA as long as all patients complaining of the same problem or exhibiting the same symptoms receive identical screening procedures. We also recognize that the hospital's screening procedure is not limited to personal observation and assessment but may include available ancillary services through departments such as radiology and laboratory.

[7] Some commentators have criticized defining "appropriate" in terms of the hospital's medical screening standard because hospitals could theoretically avoid liability by providing very cursory and substandard screenings to all patients, which might enable the doctor to ignore a medical condition. [ ] Even though we do not believe it is likely that a hospital would endanger all of its patients by establishing such a cursory standard, theoretically it is possible. Our holding, however, does not foreclose the possibility that a future court faced with such a situation may decide that the hospital's standard was so low that it amounted to no "appropriate medical screening." We do not decide that question in this case because Ms. Baber's screening was not so substandard as to amount to no screening at all.

provide a screening examination that is "appropriate" and "within the capability of the hospital's emergency department," including "routinely available" ancillary services. 42 U.S.C.A. § 1395dd(a). This section establishes a standard, which will of necessity be individualized for each hospital, since hospital emergency departments have varying capabilities. Had Congress intended to require hospitals to provide a screening examination which comported with generally-accepted medical standards, it could have clearly specified a national standard. Nor do we believe Congress intended to create a negligence standard based on each hospital's capability. . . . EMTALA is no substitute for state law medical malpractice actions.

* * *

The Sixth Circuit has also held that an appropriate medical screening means "a screening that the hospital would have offered to any paying patient" or at least "not known by the provider to be insufficient or below their own standards."

* * *

Applying our interpretation of section (a) of EMTALA, we must next determine whether there is any genuine issue of material fact regarding whether RGH gave Ms. Baber a medical screening examination that differed from its standard screening procedure. Because Mr. Baber has offered no evidence of disparate treatment, we find that the district court did not err in granting summary judgment.

* * *

Mr. Baber does not allege that RGH's emergency department personnel treated Ms. Baber differently from its other patients. Instead, he merely claims Dr. Kline did not do enough accurately to diagnose her condition or treat her injury.[ ] The critical element of an EMTALA cause of action is not the adequacy of the screening examination but whether the screening examination that was performed deviated from the hospital's evaluation procedures that would have been performed on any patient in a similar condition.

* * *

Dr. Kline testified that he performed a medical screening on Ms. Baber in accordance with standard procedures for examining patients with head injuries. He explained that generally, a patient is not scheduled for advanced tests such as a CT scan or x-rays unless the patient's signs and symptoms so warrant. While Ms. Baber did exhibit some of the signs and symptoms of patients who have severe head injuries, in Dr. Kline's medical judgment these signs were the result of her pre-existing psychiatric condition, not the result of her fall. He, therefore, determined that Ms.

Baber's head injury was not serious and did not indicate the need at that time for a CT scan or x-rays. In his medical judgment, Ms. Baber's condition would be monitored adequately by the usual nursing checks performed every fifteen minutes by the psychiatric unit staff at BARH. Although Dr. Kline's assessment and judgment may have been erroneous and not within acceptable standards of medical care in West Virginia, he did perform a screening examination that was not so substandard as to amount to no examination. No testimony indicated that his procedure deviated from that which RGH would have provided to any other patient in Ms. Baber's condition.

* * *

The essence of Mr. Baber's argument is that the extent of the examination and treatment his sister received while at RGH was deficient. While Mr. Baber's testimony might be sufficient to survive a summary judgment motion in a medical malpractice case, it is clearly insufficient to survive a motion for summary judgment in an EMTALA case because at no point does Mr. Baber present any evidence that RGH deviated from its standard screening procedure in evaluating Ms. Baber's head injury. Therefore, the district court properly granted RGH summary judgment on the medical screening issue.

Mr. Baber also asserts that RGH inappropriately transferred his sister to BARH. EMTALA's transfer requirements do not apply unless the hospital actually determines that the patient suffers from an emergency medical condition. Accordingly, to recover for violations of EMTALA's transfer provisions, the plaintiff must present evidence that (1) the patient had an emergency medical condition; (2) the hospital actually knew of that condition; (3) the patient was not stabilized before being transferred; and (4) prior to transfer of an unstable patient, the transferring hospital did not obtain the proper consent or follow the appropriate certification and transfer procedures.

* * *

Mr. Baber argues that requiring a plaintiff to prove the hospital had actual knowledge of the patient's emergency medical condition would allow hospitals to circumvent the purpose of EMTALA by simply requiring their personnel to state in all hospital records that the patient did not suffer from an emergency medical condition. Because of this concern, Mr. Baber urges us to adopt a standard that would impose liability upon a hospital if it failed to provide stabilizing treatment prior to a transfer when the hospital knew or should have known that the patient suffered from an emergency medical condition.

The statute itself implicitly rejects this proposed standard. Section 1395dd(b)(1) states the stabilization requirement exists if "any individual

. . . comes to a hospital and the hospital determines that the individual has an emergency medical condition." Thus, the plain language of the statute dictates a standard requiring actual knowledge of the emergency medical condition by the hospital staff.

Mr. Baber failed to present any evidence that RGH had actual knowledge that Ms. Baber suffered from an emergency medical condition. Dr. Kline stated in his affidavit that Ms. Baber's condition was stable prior to transfer and that he did not believe she was suffering from an emergency medical condition. While Mr. Baber testified that he believed his sister suffered from an emergency medical condition at transfer, he did not present any evidence beyond his own belief that she actually had an emergency medical condition or that anyone at RGH knew that she suffered from an emergency medical condition. In addition, we note that Mr. Baber's testimony is not competent to prove his sister actually had an emergency medical condition since he is not qualified to diagnose a serious internal brain injury.

. . . [W]e hold that the district court correctly granted RGH summary judgment on Mr. Baber's claim that it transferred Ms. Baber in violation of EMTALA.

* * *

Therefore, the district court's judgment is affirmed.

NOTES AND QUESTIONS

1. *Impact of the ACA.* The ACA leaves EMTALA intact, although it adopted other measures favorable to the provision of emergency care. Most notably, it incentivizes states to expand Medicaid coverage and subsidizes the cost of private coverage for those who cannot afford it. Significantly, the ACA also authorizes states to allow hospitals to make presumptive Medicaid eligibility determinations for individuals, which should reduce the volume of uncompensated care that is provided to individuals eligible for but not enrolled in Medicaid. The ACA also requires insurers to pay for emergency care under a prudent layperson standard, a provision welcomed by hospitals concerned about insurers refusing to pay for emergency services that they deem unnecessary but that hospitals were obligated to provide under EMTALA. 42 U.S.C. § 300gg–19a. The more significant anticipated impact of the ACA, of course, was that it would decrease patients' reliance on emergency department care as a safety net and decrease preventable emergency conditions by providing greater access to primary care. Early evidence suggests, however, that insurance may not significantly reduce emergency department visits in the short term and that other factors, such as primary care wait times, also influence emergency department use. See Scott M. Dresden et al., Increased Emergency Department Use in Illinois After Implementation of the Patient Protection and Affordable Care Act, 69 Ann. Emerg. Med. 172 (2017). See also Kristen M.J. Azar et al., Disparities in Outcomes Among COVID-19 Patients

in a Large Health Care System in California, 39 Health Aff. (May 21, 2020) (discussing research showing that African American patients were more likely to access care later in the acute setting, despite insurance coverage, and pointing to societal factors likely contributing to delayed care, such as structural barriers to timely health care access, unconscious bias of providers, and the distrust generated from patients' prior negative experiences with the health system). Discrimination based on race is discussed further below.

2. *Subjective and Objective Standards.* In contrast to the standard for medical screening, the standard applied to the question of whether the patient was unstable when discharged or transferred is an objective professional standard and not defined by the specific hospital's policy. How should plaintiffs structure discovery to meet each of these two standards? What is the role for expert testimony, if any, in an "unstable transfer or discharge" claim? In an "inappropriate screening" claim?

3. *Motive.* Improper motive is not required for a violation of the EMTALA requirement that the patient be *stabilized*. Roberts v. Galen of Va., Inc., 525 U.S. 249 (1999). The Court expressed no opinion as to whether proof of improper motive is essential for a claim of failure to provide an appropriate screening. The Circuits, except for the Sixth Circuit in Cleland v. Bronson Health Care Group, 917 F.2d 266 (6th Cir. 1990), uniformly have held that EMTALA reaches beyond economically motivated decisions and that proof of motive is not required for either a screening or a stabilization claim. Could proof of improper motive be useful to the plaintiff in distinguishing negligent misdiagnosis from an EMTALA claim? How might such proof assist the plaintiff in making his or her case? How would you go about proving motive once the physician and hospital claim medical judgment as the basis for discharge or transfer?

4. *When Does a Patient "Come to" the Emergency Department?* One of the sticky issues in EMTALA is whether a patient has "come to" the hospital's emergency department. In a rather notorious case, a hospital emergency department refused to aid a teenager who had been shot and lay dying 35 feet from the ER doors. Kristine Marie Meece, The Future of Emergency Department Liability after the Ravenswood Hospital Incident: Redefining the Duty to Treat?, 3 DePaul J. Health Care L. 101 (1999). After years of court opinions with conflicting results, HHS promulgated regulations in 2003 (42 C.F.R. § 489.24(b)) under which a patient is determined to have "come to" an emergency department when the patient:

- "present[s] on hospital property";

- "is in a ground or air ambulance owned and operated by the hospital . . . even if the ambulance is not on hospital grounds [unless the ambulance is directed to another hospital by a communitywide emergency medical service or is operated at the direction of a physician unaffiliated with the hospital]"; or

- "is in a ground or air nonhospital-owned ambulance on hospital property [unless the hospital directs the ambulance elsewhere because the hospital is on 'diversionary status' and the ambulance abides by the hospital's direction]."

5.   *Do EMTALA Obligations End upon Admission?* For many years, courts reached conflicting results on the issue of whether a patient who had been admitted to the hospital would be covered by EMTALA or whether the EMTALA obligations of the hospital ceased upon admission. Finally, in 2003, HHS promulgated regulations providing that EMTALA's obligations end when an emergency department patient is admitted in good faith to the hospital as an inpatient. See 42 C.F.R. 489.24(d)(2). Most courts have deferred to this interpretation of the statute. See, e.g., Thornhill v. Jackson Parish Hospital, 184 F. Supp. 3d 392 (W.D. La. 2016). A lone circuit held, in dicta, that the regulations are contrary to the plain language of the statute, Moses v. Providence Hosp. & Med. Ctrs., 561 F.3d 573 (6th Cir. 2009). Several federal district courts, including the Court in *Thornhill*, have disagreed with that holding. In 2010, the Centers for Medicare and Medicaid Services (CMS) asked for comments on its 2003 regulations and ultimately decided not to change them. 77 Fed. Reg. 5213 (Feb. 2, 2012).

The Fourth Circuit recently ruled that a plaintiff alleging EMTALA violations after the patient was admitted as an inpatient to the hospital bears the burden to prove that that the hospital's admission was not in good faith. "[A] party claiming an admission was not in good faith must present evidence that the hospital admitted the patient solely to satisfy its EMTALA standards with no intent to treat the patient once admitted and then immediately transferred the patient. In other words, the standard requires evidence that the admission was a subterfuge or a ruse." Williams v. Dimensions Health Corp., 952 F.3d 531, 537 (4th Cir. 2020).

6.   *"Appropriate Transfer."* Under EMTALA, an "appropriate transfer" requires that "the receiving facility (i) has available space and qualified personnel for the treatment of the individual, and (ii) has agreed to accept transfer of the individual and to provide appropriate medical treatment." 42 U.S.C. § 1395dd(c)(2)(B). Yet neither the statute nor its associated regulations articulate what the transferring hospital must do to satisfy this requirement. This ambiguity recently arose in Ruloph v. Lammico, No. 2:20-CV-02053, 2021 WL 517044 (W.D. Ark. Feb. 11, 2021) where a patient with a badly injured leg was screened in the defendant hospital's emergency department, and the attending physician determined that the patient required a transfer to a hospital with a vascular surgeon. The attending physician contacted a second hospital, confirmed that it had a vascular surgeon and would accept transfer of the patient, and processed the patient's transfer to that hospital. Only after the patient was in transit did the second hospital discover that it did not have a vascular surgeon on staff who was capable of performing the particular surgery needed by the patient. The resulting delay in surgery caused the patient to lose so much blood that her leg had to be amputated. The patient sued the transferring hospital under EMTALA alleging that, in fact, the

hospital to which the defendant transferred the patient did not have qualified personnel to treat her and thus the defendant's transfer of the patient was not "appropriate." As a matter of first impression, the federal district court held that the determination of whether a transfer is appropriate "must be predicated on a [defendant] hospital's actual knowledge" at the time it initiated the transfer and that the transferring hospital has a right to rely on representations made by the receiving hospital about its ability to treatment the patient.

7. *Problems with On-Call Coverage.* Treatment in the emergency department often will require the services of an on-call specialist. Hospitals generally do not employ on-call specialists, but they may contract with individual physicians to provide on-call services or may require on-call coverage by physicians as a condition of receiving admitting privileges. The division of labor inherent in the emergency department/on-call relationship can be contentious and raise EMTALA risks. Consequently, the management of on-call services remains an intractable problem for hospitals. See Sarah Coyne et al., Using Deferred Compensation to Incent On-Call Coverage, 23 Health Law. 28/No4 (2011) and the Problem below.

8. *EMTALA During the COVID-19 Pandemic.* In response to the COVID-19 pandemic, CMS waived some EMTALA requirements. "Only two aspects of the EMTALA requirements can be waived under 1135 Waiver Authority: 1) Transfer of an individual who has not been stabilized, if the transfer arises out of an emergency or, 2) Redirection to another location (offsite alternate screening location) to receive a medical screening exam under a state emergency preparedness or pandemic plan." CMS, COVID-19 Emergency Declaration Blanket Waivers for Health Care Providers (as revised May 24, 2021). Waivers are retroactive to March 1, 2020. Despite the flexibility offered by these waivers, hospitals subject to EMTALA remain obligated to meet the requirements of EMTALA, and, even when operating under a waiver, such hospitals cannot discriminate between patients who are able to pay and those who are not.

With respect to the "appropriate transfer" requirement, the guidance advises hospitals to rely on the recommendations of public health officials about treating suspected or actual COVID patients when assessing whether they have the capability to treat those patients or to accept transfer of those patients from other hospitals. CMS will "take into account the CDC's recommendations at the time of the event in question in assessing whether a hospital had the requisite capabilities and capacity" when investigating whether a hospital has met the requirement. CMS, Emergency Medical Treatment and Labor Act (EMTALA) Requirements and Implications Related to Coronavirus Disease 2019 (COVID-19) (Revised) (March 30, 2020).

Additionally, a hospital subject to EMTALA can meet its obligation to form a relationship with anyone who comes to its emergency department seeking care and to do so for the purpose of at least providing an appropriate medical screening even if the hospital directs patients to an alternative site for such screening, including a site that is off the hospital's campus. Id. This allows hospitals to meet physical distancing requirements while still treating those who come to their emergency departments: off-site locations may be used for the purpose of establishing a COVID-19 testing program. Id.

9. *"Patient Dumping" Persists.* EMTALA, while helpful, has not ended the problem of "patient dumping," particularly among our most vulnerable sub-populations. Several hospitals in Los Angeles, for example, have settled claims that they improperly discharged homeless patients from their emergency departments back on to the streets. See Maria Castellucci, "Calf. Hospital Pays $1 Million to Settle Patient-Dumping Case," Modern Healthcare (June 24, 2016). At least some of these "skid row dumping" cases involved allegations that patients were discharged while in unstable conditions. See Editorial Board, "No Excuse for 'Patient-Dumping' in L.A.'s Skid Row," L.A. Times (Jan. 20, 2014).

PROBLEMS: EMTALA

Ms. Miller

On May 21, Ms. Nancy Miller, who was eight months pregnant, called her obstetrician, Dr. Jennifer Gibson, at 2:00 a.m. because she was experiencing severe pain which appeared to her to be labor contractions. Dr. Gibson advised Ms. Miller to go to the emergency department of the local hospital and promised to meet her there shortly. Ms. Miller was admitted to the emergency department of General Hospital at 2:30 a.m., and Dr. Gibson joined her there at 3:14 a.m. After examining Ms. Miller, Dr. Gibson concluded that Ms. Miller had begun labor and that, even though the pregnancy had not reached full-term, the labor should be continued to delivery. At that time, Dr. Gibson asked that the on-call anesthesiologist, Dr. Martig, see Ms. Miller to discuss anesthesia during the delivery. At the same time, the procedure to admit Ms. Miller to the hospital's maternity floor was begun. The nurse informed Ms. Miller that there would be a short wait because there was no space available at that point.

Dr. Martig saw Ms. Miller at 4:00 a.m. When asked, Dr. Martig informed Ms. Miller that he was not qualified to and would not be able to perform an epidural (a spinal nerve-block anesthesia, often used in childbirth). Instead, he gave her Demerol and left the emergency department.

At 4:30 a.m., Ms. Miller was admitted to the labor and delivery floor. At 4:45 a.m., the obstetrical nurse observed fetal distress and called Dr. Gibson. At 4:50 a.m., Dr. Gibson concluded that Ms. Miller had a prolapsed umbilical cord and ordered an emergency caesarean section. The OB nurse paged Dr. Martig, but he could not be located. (Dr. Martig later stated that his pager had malfunctioned.) Because Dr. Martig could not be located, Dr. Gibson and a resident performed the C-section without an anesthetic and delivered the child healthy and alive. (These facts are based on Miller v. Martig, 754 N.E.2d 41 (Ind. Ct. App. 2001).)

Assume that Ms. Miller has brought suit against the hospital and Dr. Martig. What federal and state claims might Ms. Miller make? Assume that Dr. Martig and the hospital have filed a motion for summary judgment on all claims. What result?

Mr. Liles

Jesse Liles has no health insurance but went to a local hospital (NMC) complaining of fever and shortness of breath. He was admitted to the hospital with a diagnosis of severe dehydration, bilateral pneumonia, and adult respiratory distress syndrome. According to Mr. Liles, the hospital attempted to transfer him on eighteen separate occasions during the course of his stay between December 28, 2009, and January 24, 2010. Mr. Liles charges that two physicians at NMC falsely certified that he was stable for transfer from the hospital. At 3:35 a.m. on January 1, 2010, emergency medical services (EMS) from the local privately owned ambulance service came at the hospital's request to transfer Mr. Liles to another hospital. Liles went into cardiac arrest in the ambulance where he was resuscitated by EMS personnel and brought back inside NMC. He was placed on a ventilator in the intensive care unit (ICU) and stayed at the hospital until he was discharged to home on January 24.

On January 26, 2010, Liles called the same ambulance company because of his acute respiratory distress. The EMS personnel called NMC en route, but NMC told them to take Mr. Liles to some other hospital because there was no pulmonologist available at NMC. The ambulance was already at the entry to the hospital grounds, and it stayed in contact with NMC to get further instructions. Staff at NMC made numerous calls to potential "receiving hospitals" to identify a facility willing to accept Mr. Liles. Ultimately, he was taken to another hospital where he was admitted to the ICU and underwent surgery for a collapsed lung. (These facts are based on Liles v. TH Healthcare, Ltd., 2012 WL 3930616 (E.D. Tex.).)

Assume that Mr. Liles has sued the hospital for EMTALA violations. You are the hospital's attorney. What arguments do you expect Mr. Liles to make? What are your best defenses to Liles' claims?

IV. FEDERAL ANTIDISCRIMINATION LAW

This section examines federal antidiscrimination laws that apply to a range of programs and activities, including health care, based on certain protected characteristics. It focuses primarily on the key federal statutes addressing health care discrimination based on (1) race, color or national origin; (2) gender; and (3) disability. Racism, sexism, and ableism have long shaped the unequal distribution of benefits and burdens in society and in health care—whether through outright denials of care, unequal treatment in how care is delivered, or inequity in the distribution of health care resources. This has resulted in health care barriers and poorer quality treatment for racial and ethnic minorities, women and LGBTQ individuals, and people with disabilities.

After a brief overview, the section is organized according to the different categories of discrimination that federal laws have been enacted to address. We also include discussions of mental health parity, the use of genetic information, and age discrimination. For each category of discrimination, we discuss relevant federal statutes, explain how preexisting civil rights laws intersect with the newer antidiscrimination prohibitions in Section 1557 of the ACA, and highlight evolving disputes over the scope of antidiscrimination protection playing out among regulators and courts. While antidiscrimination cases are increasingly being brought under Section 1557, many of the cases discussed below are noteworthy decisions under other relevant civil rights laws.

Overview of Antidiscrimination Law

Pre-ACA Civil Rights Laws. The 1964 Civil Rights Act (CRA) bans many forms of discrimination that affect health in profound ways. Title VI of the CRA prohibits discrimination based on race, color, or national origin in programs or activities that receive federal financial assistance, which has been particularly important for combatting race discrimination in hospitals and nursing homes dependent upon Medicare and Medicaid funding. Title VII prohibits employment discrimination based on race and sex, and after it was amended through the Pregnancy Discrimination Act, it became a particularly powerful tool for fighting sex-based discriminatory exclusions in employment-based insurance. Title IX of the Education Amendments of 1972 has extended this antidiscrimination protection to educational settings.

Laws prohibiting disability-based discrimination have been the broadest in scope. Together, Section 504 of the Federal Rehabilitation Act of 1973 (Rehabilitation Act or Section 504) and the Americans with Disabilities Act (ADA), enacted in 1990, require recipients of federal funding, health care entities, employers, educational institutions, state and local governments, and private businesses that serve the public to ensure equal access for people with disabilities.

Section 1557 of the ACA. The ACA, signed into law in 2010, affirmed and expanded the earlier antidiscrimination protections. Section 1557 of the ACA (42 U.S.C. § 18116) provides:

(a) . . . [A]n individual shall not, on the ground prohibited under title VI of the Civil Rights Act of 1964 (42 U.S.C. 2000d et seq.), title IX of the Education Amendments of 1972 (20 U.S.C. 1681 et seq.), the Age Discrimination Act of 1975 (42 U.S.C. 6101 et seq.), or section 504 of the Rehabilitation Act of 1973 (29 U.S.C. 794) . . . be excluded from participation in, be denied the benefits of, or be subjected to discrimination under, any health program or activity, any part of which is receiving Federal financial assistance, including credits, subsidies, or contracts of insurance, or under any program or activity that is administered by an Executive Agency or any entity established under this title (or amendments). The enforcement mechanisms provided for and available under title VI shall apply for purposes of violations of this subsection.

(b) . . . Nothing in this title . . . shall be construed to invalidate or limit the rights, remedies, procedures, or legal standards available to individuals aggrieved under title VI of the Civil Rights Act of 1964 [ ], title IX of the Education Amendments [ ], section 504 of the Rehabilitation Act [ ] or to supersede State laws that provide additional protections against discrimination on any basis described in subsection (a).

Section 1557 has reshaped antidiscrimination law in health care in important ways. First, by incorporating existing civil rights laws that prohibit race (Title VI), sex (Title IX), and disability (Section 504) discrimination, it broadens the application of those prohibitions to more health care entities. Specifically, Section 1557 applies to recipients of "federal financial assistance," which is defined to include grants, loans, credits, subsidies, and insurance contracts. It applies to health insurers that receive premium tax credits or cost-sharing reduction payments for enrollees. Section 1557 also applies to most providers, though there are some limits. For example, HHS maintains that payments for patient care under Medicare Part B (the program that pays for physician services for Medicare beneficiaries) do not count as "Federal financial assistance," and providers paid directly by covered health insurers are not deemed to have received "assistance" for purposes of Section 1557. Nonetheless, most of the physicians in these situations probably receive federal assistance in other forms, and the statute does apply to all physicians who receive Medicaid.

For gender health equity in particular, Section 1557 has been viewed as having transformative potential. The limited applicability of Title VII to employment and Title IX to education meant the problem of gender discrimination in health care was relatively invisible and neglected. Indeed, one of the most anticipated effects of Section 1557's new health-specific prohibition on sex discrimination is its role in combatting discrimination against transgender individuals—a group that has long experienced overt, pervasive, and harmful discrimination by health care providers and insurers. See Subsection B below.

The future of Section 1557 as a vehicle to challenge discrimination in health care is uncertain due to substantial shifts in implementing regulations under different presidential administrations. Initially, an expansive rule implementing Section 1557 was issued under the Obama-

Biden Administration ("2016 Rule"). Nondiscrimination in Health Programs and Activities, 81 Fed. Reg. 31375 (May 18, 2016) (codified at 45 C.F.R. Pt. 92). A few years later, a revised rule issued under the Trump Administration attempted to significantly limit the scope of Section 1557 in terms of covered entities and protected groups ("2020 Rule"). Nondiscrimination in Health and Health Education Programs or Activities, 85 Fed. Reg. 37160–248 (June 19, 2020). Then, in 2021, the Biden-Harris Administration indicated that it would issue a new notice of proposed rulemaking to revise the Section 1557 regulations, and it has taken other steps to reverse Trump Administration policy and regulations that significantly narrowed the reach of Section 1557. Of course, regulators only have so much power to shape Section 1557's reach—implementing regulations must be consistent with the underlying statute, the contours of which are already being adjudicated by courts. These developments are detailed in the notes below, and developments in the interpretation and enforcement of Section 1557 should be watched closely.

Even with the important changes brought about by Section 1557, preexisting civil rights statutes will remain relevant in challenging health discrimination. There are a couple of reasons for this. First, preexisting laws may fill important regulatory gaps where particular entities are not subject to Section 1557. For example, as explained in Section IV.B. below, employment-based plans not subject to Section 1557 under the narrower 2020 Rule are still subject to Title VII's employment protections. Second, courts have often looked beyond the antidiscrimination statute implicated in a particular case to consider how similar antidiscrimination laws have been interpreted. For example, in Section 1557 and Title IX gender discrimination claims, courts have looked to Title VII cases for guidance; for Section 1557 disability claims, the ADA provides useful guidance. Finally, Section 1557's unusual approach to delineating protected categories—specifically, its incorporation by reference of existing federal civil rights statutes—means that the cases interpreting those laws are essential to understanding the scope of nondiscrimination protections under Section 1557. For antidiscrimination challenges brought under Section 1557, courts are looking to the jurisprudence of the incorporated statutes, as well as similar antidiscrimination protections, to help answer a number of questions that we explore in this Section: Who is protected? What kind of conduct is recognized or defined as illegal discrimination? How must plaintiffs prove discrimination claims? What defenses or exceptions exist? What remedies are available?

A. RACE, COLOR, & NATIONAL ORIGIN

U.S. hospitals and other health care facilities were racially segregated by law well into the late 1960s and by custom for some time thereafter. White-only hospitals refused admission to African-American citizens, and those white-dominated hospitals that did admit African Americans segregated them into separate units. Even publicly owned hospitals and hospitals funded by the federal government were segregated by race.

A post-World War II federal health care funding program, known as Hill-Burton, invested millions of federal tax dollars in the construction of hospitals across the country. Yet the legislation specifically institutionalized race discrimination by allowing federally funded hospitals to exclude African Americans if other facilities were available. The segregated facilities available were hardly equal. For example, the ward for African Americans in the community hospital in Wilmington, North Carolina housed only twenty-five beds and two toilets in a building separated from the main building, which meant that surgery patients had to be transported across an open yard. And neither of the two hospitals (one public and one private) in Broward County, Florida admitted patients from the more than 30,000 African Americans residing in the county. Segregation even trumped ability to pay. For example, in 1950s Chicago, African-American union members with generous health insurance plans were steered almost entirely to Cook County Hospital despite closer facilities. See David Barton Smith, Health Care Divided: Race and Healing a Nation (1999). Not until 1962 did a federal Court of Appeals declare the "separate but equal" provision of the Hill-Burton Act unconstitutional. Simkins v. Moses H. Cone Mem'l Hosp., 323 F.2d 959 (4th Cir. 1963), cert. denied, 376 U.S. 938 (1964).

In 1964, Congress passed Title VI of the Civil Rights Act, expressly prohibiting discrimination on the basis of race, color, or national origin by any program receiving federal financial assistance. But it was the creation of Medicare and Medicaid in 1965 that gave the federal government the leverage it needed to desegregate hospitals and fight other forms of race discrimination by hospitals, nursing homes, and other facilities receiving federal funding. Importantly, early regulations clarified that Title VI prohibited the use of criteria or methods of administration that had the *effect* of subjecting individuals to discrimination on the basis of race, color, or national origin. 45 C.F.R. 80.3(b)(2).

For these reasons, some expected Title VI would be a powerful tool for combatting both intentional discrimination in health care and facially neutral health care polices or practices that had discriminatory effects. Indeed, the ability to challenge such policies and practices was understood as vital to combatting more subtle forms of discrimination, especially as antidiscrimination norms made overt race discrimination unpalatable. Today, the policies and practices having discriminatory health effects tend to be facially neutral or the product of more subtle or implicit biases about the affected groups.

The case excerpts below present two examples of Title VI challenges to a facially neutral policy or decision having a racially discriminatory impact. In one case, the challenged policy resulted in the involuntary discharge of nursing home patients. In the other, an underserved community lost its public hospital. As you read the cases, consider why the courts reached different results. What do these cases suggest about the power of Title VI to combat systemic discrimination and other practices that contribute to health inequity?

LINTON V. COMMISSIONER OF HEALTH AND ENVIRONMENT

United States District Court, Middle District, Tennessee, 1990.
779 F. Supp. 925.

JOHN T. NIXON, DISTRICT JUDGE.

Plaintiffs are before the Court seeking to enjoin a Tennessee policy through which only a portion of the beds in Medicaid participating nursing homes are certified to be available for Medicaid patients. Plaintiffs allege that this policy artificially limits the accessibility of nursing home care to indigent Medicaid patients and fosters discrimination against indigent patients by nursing homes. Plaintiffs claim that, as a result of the challenged policy, they and other individuals similarly situated face delay or outright denial of needed nursing home care, as well as displacement from current residency in nursing home facilities. Plaintiffs bring this action under [ ] Title VI of the Civil Rights Act of 1964.

* * *

FINDINGS OF FACT

The present case was initiated on December 1, 1987 on behalf of Mildred Lea Linton. Ms. Linton suffers from rheumatoid arthritis and has been a patient for four years at Green Valley Health Care Center in Dickson, Tennessee [hereinafter "Green Valley"]. A Medicaid patient who had been receiving skilled nursing facility (SNF) level care throughout her stay, the plaintiff received notice from State Medicaid officials that she no longer qualified for such care. The same notice advised her that she would have to move to another nursing home, an intermediate care facility (ICF), to receive the level of care to which the State believed she should be downgraded. Green Valley provides ICF care, and in fact the bed occupied by Ms. Linton was dually certified for Medicaid purposes for provision of both SNF and ICF levels of care. However, Green Valley was unwilling to care for Ms. Linton at an ICF level of reimbursement. The nursing home, which had directed the State to certify only part of its ICF beds as available to Medicaid patients, reserved the right to decertify the plaintiff's bed for Medicaid ICF participation. This decertification would have compelled the plaintiff's involuntary transfer to another facility.

On December 11, 1989, plaintiff Belle Carney, an 89 year old black woman, requested intervention. She had been diagnosed in July 1987 as requiring nursing home treatment due to Alzheimer's disease, however, no nursing home placement was found for her. Plaintiff Carney asserts that the State's limited bed certification policy, which the State refers to as distinct part certification, creates an artificial restriction on the number of available Medicaid beds and that it fosters discrimination against Medicaid patients by nursing homes. Plaintiff Carney's health deteriorated over a period of several months as she was moved from one inadequate placement to another. Finally, Carney's condition declined to the point that she required emergency hospitalization. . . .

* * *

A.   *The Tennessee Medicaid Program*

Tennessee participates in the federal Medicaid program to provide medical assistance to those recipients who are eligible to receive such assistance in accordance with the requirements of Title XIX of the Social Security Act. [ ] . . . The types of medical assistance that are provided under the Tennessee Medicaid program include, among others, services by skilled nursing home facilities [(SNFs)] and intermediate care facilities [(ICFs)]. [ ] [ICF services include institutional, health-related services above the level of room and board, but at a level of care below that of hospital or SNF care. SNF care consists of institutional care above the level of ICF services but below the level of a hospital. The Tennessee Department of Health and Environment (TDHE) is the single state agency responsible for administration of the Medicaid program. The Department is administered under the direction of the defendant Commissioner.] TDHE also acts as the State licensing agency for nursing homes or long term care facilities and certifies such facilities for Medicaid. . . .

* * *

Tennessee has previously had a "Medicaid Bed Management program" which represents an attempt to place a percentage limitation on the number of available Medicaid beds in nursing homes. Federal auditors recommended that this policy be discontinued, and Tennessee abolished this program on October 1, 1985.

Plaintiffs challenge what they refer to as an unwritten limited bed certification policy. Defendants assert that this policy is instead simply Tennessee's version of distinct part certification. [Federal law authorizes State agencies to certify facilities for either SNF or ICF reimbursement. Such certification may be of a "distinct part of an institution." According to federal policy, the term "distinct part" denotes that the unit is organized and operated to give *a distinct type of care within a larger organization which otherwise renders other types or levels of care.* "Distinct" denotes both organizational and physical distinctness. A distinct part SNF must be physically identifiable and be operated distinguishably from the rest of the institution. *It must consist of all the beds within that unit such as a separate building, floor, wing or ward.*] The Court finds, however, that the Tennessee policy varies markedly from federal "distinct part" certification.

Perhaps the most fundamental difference between Tennessee's limited bed certification policy and federal "distinct part" certification is to be found in the different purposes served by the respective policies. Federal distinct part certification is intended to accommodate the delivery of qualitatively different types of health care within the same facility. Tennessee's policy, however, appears to serve the interests of nursing homes who wish to participate in the Medicaid program while also maintaining a separate private pay facility offering the same type of care.

* * *

Federal Medicaid law mandates that states set their Medicaid payments to nursing homes at levels which are "reasonable and adequate to meet the costs which must be incurred by efficiently and economically operated facilities in order to provide care and services in conformity with applicable State and Federal laws, regulations and quality and safety standards . . ." [ ]

By contrast, private pay rates are set by the market, a market in which there are more patients seeking care than there are beds to accommodate them. There are waiting lists to gain admission to nursing homes throughout Tennessee. As a result of this situation, unregulated private pay rates are substantially higher than Medicaid payments, and nursing home operators prefer private pay patients. Tennessee's present certification program allows nursing home operators to give preference to private pay patients by reserving for their exclusive use beds which are, due to lack of certification, unavailable to Medicaid patients.

. . . The [limited bed certification] policy leads to disruption of care and displacement of Medicaid patients after they have been admitted to a nursing home. Such displacement often occurs when a patient exhausts his or her financial resources and attempts transition from private pay to Medicaid. In this situation, a patient who already occupies a bed in a nursing home is told that his or her bed is no longer available to the patient because he or she is dependent upon Medicaid. . . . Involuntary transfers are triggered on other occasions when a patient already on Medicaid at SNF level of reimbursement is reclassified to an ICF level of care [with lower reimbursement].

. . . [T]he Court finds that the limited bed certification policy has caused widespread displacement . . . . The Court is persuaded by the depositions, affidavits and exhibits concerning the severe impact of the limited bed certification policy. Finally, the Court is mindful of the Medicaid eligibility rules which allow eligibility for relatively more affluent patients already residing in nursing homes than those seeking initial admission. This phenomenon combined with the limited bed certification policy often renders the poorest and most medically needy Medicaid applicants unable to obtain the proper nursing home care.

B. *Disparate Impact on Minorities*

The Court finds that the plaintiff has established by a preponderance of the evidence that the Tennessee Medicaid program does have a disparate and adverse impact on minorities. Because of the higher incidence of poverty in the black population, and the concomitant increased dependence on Medicaid, a policy limiting the amount of nursing home beds available to Medicaid patients will disproportionately affect blacks.

Indeed, while blacks comprise 39.4 percent of the Medicaid population, they account for only 15.4 percent of those Medicaid patients who have been able to gain access to Medicaid-covered nursing home services. In addition, testimony indicates that the health status of blacks is generally poorer than that of whites, and their need for nursing home services is correspondingly greater. Finally, such discrimination has caused a "dual system" of long term care for the frail elderly: a statewide system of licensed nursing homes, 70 percent funded by the Medicaid program, serves whites; while blacks are relegated to substandard boarding homes which receive no Medicaid subsidies.

CONCLUSIONS OF LAW

* * *

B. *Plaintiffs' Claims Under Title VI of the Civil Rights Act of 1964*

Title VI of the Civil Rights Act of 1964 provides, in relevant part:

> No person in the United States shall on the ground of race, color or national origin . . . be subjected to discrimination under any program or activity receiving Federal financial assistance.

[ ] Regulations under Title VI provide that a state in its administration of the federally funded program can not:

> . . . directly or through contractual or other arrangements, utilize criteria or methods of administration which have the effect of subjecting individuals to discrimination because of their race, color, or national origin, or have the effect of defeating or substantially impairing accomplishment of the objectives of the program as respect individuals of a particular race, color or national origin.

45 C.F.R. § 80.3(b)(2).

* * *

The plaintiffs have shown that the defendants' limited bed certification policy has a disparate impact on racial minorities in Tennessee. The burden of proof next falls upon the defendants to show that the disparate impact is not unjustifiable. The defendants state that the "self-selection preferences" of the minorities, based upon the minorities' reliance upon the extended family, lack of transportation, and fear of institutional care, adequately explain the disparate impact. [ ] This explanation, however, is not sufficient justification for minority underrepresentation in nursing homes. Therefore, the defendants have failed to meet their burden of proof . . . .

* * *

The Court recognizes that under Title VI, deference is accorded to the Title VI administrative agency to cure the discriminatory effects of the particular program. To be sure, THDE employs Ms. Beverly Bass as a director to monitor Title VI compliance. However, even Bass concedes that under the TDHE certification policy black Medicaid recipients are displaced or denied admission to nursing homes. Further, it is noteworthy that the Bed Management Program that was instituted in 1981, was not repealed as TDHE policy until October 13, 1987. Prior TDHE studies identified the status of minority citizen [sic] in Tennessee's Medicaid Program and the reasons for lack of minority participation. Yet, despite these studies, the Commissioner implemented a policy that fosters and continues the egregious status of minority Medicaid patients within the program. In these circumstances, continued deference to the administrative agencies is inappropriate. To cure the effects of this policy, judicial intervention is necessary. To accomplish this, the Court ORDERS the Commissioner, in consultation with [the federal agency overseeing Medicaid], to submit a plan for court approval that will redress the disparate impact upon eligible minority Medicaid patients' access to qualified nursing home care due to the THDE certification policy and the State's past noncompliance with Title VI.

* * *

BRYAN V. KOCH
United States Court of Appeals, Second Circuit, 1980.
627 F.2d 612.

NEWMAN, CIRCUIT JUDGE:

This litigation challenges New York City's decision to close Sydenham Hospital, one of its 17 municipal hospitals, on the ground that the City's proposed action would constitute racial discrimination in the use of federal funds in violation of Title VI of the Civil Rights Act of 1964, *42 U.S.C. § 2000d* et seq. Like most American cities, New York has struggled mightily to provide adequate municipal services with limited financial resources; its difficulties have become particularly severe since its budget crisis that began in the mid-1970's. Closing Sydenham is one of the many painful steps that the City has undertaken or proposed in an effort to maintain financial stability. The discrimination claim in this case arises from the fact that Sydenham, located in central Harlem, serves a population that is 98% minority (Black and Hispanic). Three related cases have been brought to prevent the closing of the Hospital, or at least to ameliorate the effects that the closing would have on the minority population it serves. This consolidated appeal is from the denial of a preliminary injunction. . . .

* * *

In April, 1979, Mayor Koch appointed a Health Policy Task Force to examine ways of reducing costly excess hospital capacity while maintaining access to high quality health services. The Task Force report, issued June 20, 1979, recommended a series of steps that the City's Health and Hospital Corporation (HHC) estimated would save $ 30 million in fiscal year 1981. With respect to the 17 hospitals of the municipal hospital system, the report proposed that some hospitals be replaced, that some hospitals reduce the number of beds, and that two hospitals, Sydenham and Metropolitan, both located in Harlem, be closed. The HHC approved the report on June 28, 1979. On August 12, 1979, the first of the three cases in this litigation, Bryan v. Koch, was filed on behalf of a class of low income Black and Hispanic residents of New York City who use the municipal hospital system. Defendants are the City and State of New York, the HHC, the State Health Department, and various city and state officials, including Mayor Koch. The U. S. Department of Health and Human Services (HHS) (formerly Department of Health, Education and Welfare) was joined as a defendant, though not charged with any violation of law. . . . [The] suits, which have been consolidated, allege that the City's proposed plan for the municipal hospital system violates various federal civil rights statutes, primarily Title VI. . . .

* * *

. . . [T]he findings that the City acted without discriminatory intent are fully supported by the evidence.

* * *

. . . Even under the effects standard, we conclude that the judgment of the District Court should be affirmed. Applying that standard, we agree with the plaintiffs that they have sufficiently shown a disproportionate racial impact to require justification by the defendants. Disparity appears from comparing the 98% minority proportion of the Sydenham patients with the 66% minority proportion of the patients served by the City's municipal hospital system. Whether the impact of this disparity is sufficiently adverse to create a prima facie Title VI violation is a closer question. The District Court was satisfied with the City's estimates for the care of Sydenham patients in nearby municipal and voluntary hospitals. Nevertheless, the District Court acknowledged that at least a small number of patients, those admitted to the emergency room because of gunshot or knife wounds or drug overdoses, would suffer adverse consequences if the nearest emergency room treatment available were at even slightly more distant locations. Moreover, we share the plaintiffs' concern that the City's estimates for alternative care of Sydenham's patients rests on projections about the availability of bed space without sufficient assurance that the voluntary hospitals on which the City relies will admit all of Sydenham's patients in financial need. Whether these hospitals will admit Sydenham's Medicaid patients remains an issue in some dispute, and even if Medicaid patients will be admitted, it is also unclear what will happen with those eligible for Medicaid who are unable to establish their eligibility at the time hospital admission is needed.

* * *

We therefore consider it appropriate to complete an assessment of plaintiffs' Title VI claim by examining the justification advanced by the City for closing Sydenham. As the District Court found, that justification is both the reduction of expenditures and the increase in efficiency within the municipal hospital system. Though the plaintiffs dispute the amount of savings the City claims will be achieved, they acknowledge that there is sufficient evidence to support the District Court's finding that closing Sydenham will reduce expenditures to some extent. However, saving money, while obviously a legitimate objective of any governmental plan to close a public facility, cannot be a sufficient justification in a case like this where public officials have made a choice to close one of 17 municipal hospitals. In such circumstances it is the choice of this particular hospital that must be justified.

To provide a basis for making this choice the Task Force report initially assessed each of the municipal hospitals against four sets of criteria: (a) hospital size, scope of patient services, and extent of usage; (b) patient access to comparable alternative facilities; (c) quality of plant and operations; and (d) present and predicted fiscal performance. Among the several hospitals considered deficient in this assessment, recommendations for closure were made for those hospitals with disproportionately high operating deficits and obsolete plants, located within 30 minutes of other municipal hospitals with comparable or broader services. These criteria are reasonably related to the efficient operation of the City's municipal hospital system, and the evidence abundantly justifies the selection of Sydenham based on these criteria. . . .

If any of the municipal hospitals are to be closed, plaintiffs do not dispute that Sydenham is an appropriate choice for closing. Their claim is that the closing of a federally funded facility resulting in a disproportionate racial impact violates Title VI unless the defendants establish the unavailability of alternative measures that would save equivalent money with less disproportionate impact. Proceeding from this premise, plaintiffs further contend that instead of closing any municipal hospitals, the City could save as much money or more and avoid a disparate racial impact by such alternatives as hospital mergers, regionalization of services, increasing Sydenham's services to reduce its deficit, or increasing Medicaid reimbursement.

Neither Title VI nor the HHS regulations explicitly require a federal fund recipient to consider alternatives to a proposed placement or closing of a public facility. It is unlikely that challenges to such governmental actions were even contemplated when Title VI was enacted in 1964. The focus at that time was on federally aided facilities that denied access to minorities or admitted them only to segregated portions of a facility. [ ] The argument for consideration of alternatives in placing or closing facilities stems from two sources. HHS has expressed its view as the administrator of federal medical care funds that its regulations should be interpreted to require consideration of alternatives. On March 5, 1980, the Undersecretary of what was then HEW wrote to Mayor Koch, in connection with the proposed closing of municipal hospitals, that if closings were to create a disparate racial impact, the City would have an opportunity to establish that they are "necessary to achieve legitimate objectives unrelated to race or national origin and that these objectives cannot be achieved by other measures having a less disproportionate adverse effect." . . .

. . . [H]owever, the inquiry could frequently become too open-ended. If, for example, a court were to assess alternative ways of saving funds throughout the administration of a city or even throughout the administration of the health care function, it would seriously risk substituting its own judgment for that of the city's elected officials and appointed specialists. We are skeptical of the capacity and appropriateness of courts to conduct such broad inquiries concerning alternative ways to carry out municipal functions. Once a court is drawn into such a complex inquiry, it will inevitably be assessing the wisdom of competing political and economic alternatives. Moreover, such policy choices would be made without broad public participation and without sufficient assurance that the alternative selected will ultimately provide more of a benefit to the minority population.

. . . [W]e do not believe Title VI requires consideration of alternatives beyond an assessment of all the municipal hospitals in order to select one or more for closing. . . . The alternatives plaintiffs wish to have considered are more appropriate for examination by administrative, legislative, and other political processes than by the courts.

Without expressing any views as to the wisdom of closing Sydenham Hospital, we conclude that the plaintiffs have shown no likelihood of success in establishing that its closing would violate Title VI, and the District Court therefore did not err in denying a preliminary injunction.

* * *

NOTES AND QUESTIONS

1. *The Health Care Safety Net.* These cases introduce you to two essential parts of the U.S. health care safety net. In *Linton*, the plaintiffs depended on Medicaid, known primarily as the public insurance program for the poor. Medicaid, which is jointly funded by federal and state government, is the primary source of payment for institutional and long-term care. In *Bryan*, the plaintiffs depended on a local public hospital system, which could not refuse to treat anyone based on ability to pay. Although there is no universal right to health care in the U.S., states may require local governments to ensure a minimal level of service to indigent residents. See, e.g., Cal. Welf. & Inst. Code § 17000 ("[E]very city and county shall relieve and support all incompetent, poor, indigent persons, and those incapacitated by age, disease, or accident, lawfully resident therein, when such persons are not supported and relieved by their relatives or friends, by their own means, or by state hospitals or other state or private institutions.") But even in these states, local governments typically have great discretion in how they satisfy this duty, including the extent to which they rely on private actors subsidized with public funds.

2. *Proving Disparate Impact.* As noted in the cases above, Title VI regulations reach beyond intentional discrimination. They prohibit the use of criteria or methods of program administration which have discriminatory effects. (Recall the excerpt of 45 C.F.R. § 80.3(b)(2) reproduced in *Linton* above.) In both *Bryan* and *Linton*, there was no evidence of discriminatory intent, but plaintiffs were able to establish a prima facie case based on disparate racial impact. What evidence did plaintiffs offer to prove disparate impact?

Consider that race-based residential patterns originally established as a matter of *de jure* segregation in housing have produced a legacy of *de facto* racial segregation still engraved on the U.S. health care system today: Hospitals and nursing homes continue to avoid predominantly African-American neighborhoods; only one-quarter of pharmacies in these neighborhoods carry necessary prescription medications; and health insurers market different and more limited health plans, if they market at all, in these neighborhoods than in predominantly Caucasian neighborhoods. See Sidney Watson, Section 1557 of the Affordable Care Act: Civil Rights, Health Reform, Race, and Equity, 55 How. L. J. 855 (2012); David Barton Smith, Healthcare's

Hidden Civil Rights Legacy, supra; Ruqaiijah Yearby, African Americans Can't Win, Break Even, or Get Out of the System: The Persistence of "Unequal Treatment" in Nursing Home Care, 82 Temple L. Rev. 1177 (2010).

3. *"Sufficiently Adverse" Impact.* In *Bryan*, the court found that plaintiffs showed a disproportionate *racial* impact, but it noted that "[w]hether the impact of this disparity is *sufficiently adverse* to create a prima facie Title VI violation was a closer question." Because public hospitals are required by state or local law to treat indigent patients, they have been the primary source of treatment for patients who otherwise could not afford care or find providers willing to treat them. The City claimed that, after the closure, Sydenham's patients would still be able to get the care they needed from private hospitals in neighboring areas. The plaintiffs' challenged this assumption and the court shared plaintiffs' concern: Why? What additional information would be needed to determine the impact of Sydenham's closure on non-emergency care?

4. *Defenses to a Disparate Impact Claim.* Proving that a policy or practice has racially disparate effects is not enough to win a Title VI suit. In both *Bryan* and *Linton*, once the plaintiff made out a prima facie case for disparate impact, the burden shifted to the defendant to justify the challenged policy or practice. Compare the result in *Bryan* with that in *Linton*. Why did the court find the hospital closure in *Bryan* justifiable? What purpose does allowing a claim based on disparate effects or impact seem to serve? Is it designed merely to help root out pretext for intentional discrimination? Is the goal to eliminate health inequity more broadly? Or does the goal lie somewhere in-between: Is Title VI a vehicle for distinguishing *unjustifiable* policies that have discriminatory effects from *justifiable* policies that may have the unfortunate, but not illegal, effect of exacerbating health disparities? If so, then how do we determine what is justifiable?

5. *Private Disparate Impact Litigation?* In 2001, in Alexander v. Sandoval, 532 U.S. 275 (2001), the Supreme Court dealt a significant blow to private Title VI litigation. The Court held that only intentional discrimination is actionable through private suit under Title VI. The five-to-four majority strictly construed the text of Title VI, holding that there was no private cause of action to enforce regulations that went beyond the scope of the statutory prohibition. The Court held that enforcement of regulations prohibiting policies or criteria with racially disparate *effects* was exclusively within the purview of the agency charged with Title VI enforcement. In the case of health care, this would be the HHS Office for Civil Rights (OCR), discussed further in Notes 6 & 7 below. Although *Sandoval* was not a health care case, it has been widely understood as foreclosing private disparate impact claims in health care such as those in *Linton* and *Bryan*—that is, until enactment of Section 1557 of the ACA seemed to re-open this question.

An early case raised the possibility that the ACA's antidiscrimination provision would breathe new life into private disparate impact litigation. In Rumble v. Fairview Health Services, No. 14-CV-2037 SRN/FLN, 2015 WL 1197415 (D. Minn. Mar. 16, 2015), the court held that Congress intended to create a new, health-specific, anti-discrimination cause of action that is subject to a singular standard, regardless of a plaintiff's protected class status, because holding otherwise "would lead to an illogical result." The court explained that applying different standards would mean a plaintiff bringing a race discrimination claim under Section 1557 could allege only disparate treatment, but plaintiffs bringing a disability claim under the same statute could allege disparate treatment or disparate impact. *Rumble's* interpretation created the possibility that private disparate impact claims based on race may once again be available under Section 1557. It was also consistent with the 2016 Rule interpretation that "Section 1557 authoriz[ed] a private right of action *for claims of disparate impact discrimination on the basis of any of the criteria enumerated in the legislation.*" 81 Fed. Reg. at 31,439–40 (emphasis added). Subsequently, however, this interpretation was rejected in the 2020 Rule and it has not gained much traction with courts.

To date, most courts have found that although Congress created a private right of action for Section 1557 violations, it did not create a single standard for such claims. Rather, courts have held that Congress' express incorporation of the enforcement mechanisms from different federal civil rights statutes "manifests an intent to import the various different standards and burdens of proof into a Section 1557 claim." See Southeastern Penn. Transp. Auth. v. Gilead Sciences, 102 F. Supp. 3d 688 (E.D. Pa. 2015). *Southeastern Penn.* dealt specifically with a lawsuit challenging an insurer's denial of coverage for a Hepatitis C drug as a violation of Section 1557's prohibition of race discrimination, among other claims. The court dismissed the plaintiffs' Section 1557 claims for failure to state a viable cause of action. The court set out the required elements for a race discrimination claim under Section 1557 based on Title VI:

> To state a claim under Title VI of the Civil Rights Act, a plaintiff must show that he or she (1) was a member of a protected class, (2) qualified for the benefit or program at issue, (3) suffered an adverse action, and (4) the adverse action occurred under circumstances giving rise to an inference of discrimination. Private rights of action under Title VI itself are available only for allegations of intentional discrimination and not disparate impact. Plaintiffs suing under Title VI may show intentional discrimination with evidence demonstrating either discriminatory animus or deliberate indifference. Discriminatory animus requires plaintiff establish prejudice, spite, or ill will. To establish deliberate indifference, a plaintiff must show that a defendant (1) knew that a harm to a federally protected right was substantially likely and (2) failed to act.

Id. at 701 (citations omitted). Other courts have agreed that the respective standards from each of four incorporated statutes should be applied based on the category of discrimination. See, e.g., York v. Wellmark, 2017 WL 11261026 (S.D. Iowa 2017) (applying Title IX standards to a Section 1557 sex discrimination claim); Briscoe v. Health Care Service Corp. 281 F. Supp. 3d 725 (N.D. Ill. 2017) (same).

6. *Public Enforcement Efforts.* Although *Sandoval* eliminated private disparate impact claims under Title VI, this ruling does not preclude administrative enforcement on disparate impact grounds. OCR can pursue cases under both intentional discrimination and disparate effects theories, and, at times, has used its power to challenge a broad range of discriminatory actions. See Speech by Thomas Perez, Director, Office for Civil Rights, Discrimination and Health Disparities (Apr. 13, 1999) (describing actions targeting home health care chains engaging in medical redlining—a policy of refusing to serve people in predominantly minority neighborhoods; a hospital's policy of not giving epidurals to women who did not speak English; a hospital's segregated maternity wards; and a pharmacy's repeated failure to fill the prescription of an African-American Medicaid recipient).

OCR has also used its enforcement authority to intervene in hospital closures or relocations that could have a discriminatory impact. For example, in 2010 the OCR responded to complaints about a decision by the University of Pittsburgh Medical Center (UPMC) to close its hospital in Braddock, Pennsylvania, which was in a predominantly African-American community, and to relocate services to a new facility in a predominantly white area that was a great distance away. Although regulators did not stop the closure, they did reach an agreement with UPMC to ameliorate the effects on access by providing door-to-door transportation services from Braddock to its new outpatient facilities and expanding support for primary and urgent care services locally. See U.S. Department of Health and Human Services, Press Release, UPMC Agrees to Expand Access to Care After Closure of UPMC Braddock (Sep. 2, 2010).

Although OCR relies heavily on voluntary settlements, the above cases remain useful for understanding how a court might evaluate disparate impact claims if an OCR action proceeded to litigation.

7. *Critique of Public Enforcement Efforts.* Even with OCR enforcement, the lack of a private cause of action for disparate impact claims has been a huge blow to efforts to combat race discrimination. OCR's enforcement priorities depend on its leadership, and persistent resource, organizational, and political challenges have made administrative enforcement of Title VI uneven at best. Indeed, a 1999 report by the United States Commission on Civil Rights confirmed many of the longstanding complaints lodged by patients' advocates about OCR's "timid and ineffectual enforcement" and concluded that the structure and operation of HHS/OCR actually fostered and exacerbated health care discrimination, rather than combatting it. See Sara Rosenbaum & Joel Teitelbaum, Civil Rights Enforcement in the Modern Healthcare System: Reinvigorating the Role of the Federal Government in the Aftermath of Alexander v. Sandoval, 3 Yale J. Health Pol'y, L. & Ethics 215 (2003). For most of the period between Title VI enactment and *Sandoval*, it was private litigants—not the OCR—who primarily challenged public funding decisions having racially discriminatory effects, including hospital closures in and relocations from predominantly minority communities.

8. *Continuing Inequities*. Racial disparities in health care treatment and outcomes persist despite antidiscrimination efforts. As noted above, some disparities are linked to geographic segregation or socioeconomic status. During the COVID-19 pandemic, for example, some minority neighborhoods did not have equal access to COVID-19 testing, with Black and Latino communities more likely to experience longer wait times and understaffed testing centers. See Soo Rin Kim et al., Which Cities Have the Biggest Racial Gaps in COVID-19 Testing Access?, FiveThirtyEight.com (Jul. 22, 2020). And a recent study has found that the higher mortality of Black COVID-19 patients was attributable to the different hospitals to which Blacks and Whites were admitted, not only to patient characteristics. See David A. Asch et al., Patient and Hospital Factors Associated with Differences in Mortality Rates Among Black and White US Medicare Beneficiaries Hospitalized With COVID-19 Infection, 4 JAMA Network Open (Jun. 17, 2021).

These factors do not capture the full story, however. In 2003, a report by the Institute of Medicine (now the National Academy of Medicine) found serious racial and ethnic disparities, defined as "racial or ethnic differences in the quality of healthcare that are not due to access-related factors or clinical needs, preferences and appropriateness of intervention." Institute of Medicine, Unequal Treatment: Confronting Racial and Ethnic Disparities in Health Care (2003). Disparities in treatment decisions appear for a great variety of medical conditions. For example, studies show that African-American and Latino patients are categorized as needing less urgent care by emergency department personnel than whites with the same cardiac symptoms; are less likely to receive aspirin upon discharge after heart attack; and are less likely to receive appropriate pain medication in the emergency department for problems like bone fractures. See Rene Bowser, The Affordable Care Act and Beyond: Opportunities for Advancing Health Equity and Social Justice, 10 Hastings Race & Poverty L. J. 69 (2013). See also National Health Care Quality and Disparities Reports (2011), issued by the federal Agency on Healthcare Research and Quality, reporting that African Americans and Latinos receive lower quality care than do non-Hispanic whites on approximately 40 percent of quality measures.

There is also growing research that more directly investigates the role of racism in health care. See, e.g., David R. Williams, Jourdyn Lawrence, and Brigette Davis, Racism and Health: Evidence and Needed Research, 45 Annual Review Public Health 105 (2019) (describing the growing attention of scientific researchers to the operation of racism in health at the individual, structural, and cultural levels). See also Dayna Bowen Matthew, Just Medicine: A Cure for Racial Inequality in American Health Care (2015). Matthew provides a comprehensive review of the social science literature showing how implicit bias leads to unequal treatment and entrenches racial and ethnic health disparities in access and outcomes. She then proposes structural reforms to combat implicit bias based on social scientific evidence concerning the presence, function, and malleability of implicit bias that has largely been ignored. Her proposals include reforming antidiscrimination laws, such as Title VI and 1557, so that intentional discrimination is not required for private causes of action, as well as creating a new negligence-based cause of action that would allow institutions to be held legally accountable for failing to address evidence that implicit bias is impacting treatment decisions. Under this negligence theory, institutions would be able to satisfy their duty through relevant training programs and other evidence-based anti-bias interventions.

9.   *Recognizing or Re-Creating Race?* Although civil rights laws have been important for combatting the type of exclusionary or animus-based overt discrimination that animated de jure segregation, antidiscrimination law has not eliminated the use of race in medical decision-making. Indeed, some in the medical community continue to justify the explicit use of race as medically justifiable or relevant. See, e.g., Sally Satel, I am a Racially Profiling Doctor, N.Y. Times Magazine (May 5, 2002) (physician describing why she factors in race for diagnosis and treatment decisions); Pamela Sankar and Jonathan Kahn, BiDil: Race Medicine or Race Marketing, 24 Health Aff. (2005) (describing the controversy around the FDA's first approved drug with a race-specific indication, BiDil, to treat heart failure in Black patients); Darshali A. Vyas et al., Hidden in Plain Sight—Reconsidering the Use of Race Correction in Clinical Algorithms, 383 NEJM 874 (2020) (identifying the continued use of race as a "corrective factor" in medical determinations about the safety of vaginal birth after a C-section, how best to manage heart failure, and the measurement of kidney function; and noting that these practices merit greater scrutiny because of their potential to expose Black patients to riskier interventions or delayed care). This not only contributes to race-based health inequities. Vyas, supra. Such uses of race in medicine have also been criticized as the product of, and as a tool for reinforcing, the longstanding myth of race as a genetic or biological factor that has shaped our understanding of disease and treatment. Such beliefs have been used to support notions of racial inferiority and racialized disease susceptibility, which, in turn, have impeded the recognition and targeting of structural forces, like modern forms of racism, that continue to have an outsize impact on health. See Dorothy E. Roberts, Fatal Invention: How Science, Politics, and Big Business Re-create Race in the Twenty-First Century (2011).

NOTE: NATIONAL ORIGIN DISCRIMINATION

Discussions of race discrimination tend to treat race and ethnicity as interchangeable concepts because of the commonality in the forms of discrimination and disparities experienced by groups identifying as racial and ethnic minorities. Nonetheless, discrimination based on national origin presents distinct legal issues, such as language barriers to access. According to the 2015 American Community Survey from the U.S. Census Bureau, a record 64.7 million people spoke a language other than English at home, and more than 25.9 million people were "limited English proficient" (LEP). Data from this annual survey is available at the U.S. Census Bureau's official website. LEP individuals experience linguistic barriers to care that impact both the ability to access care and the quality of care received. In 1974, the Supreme Court in Lau v. Nichols, 414 U.S. 563, interpreted Title VI regulations as prohibiting conduct that has a disproportionate effect on individuals with limited English proficiency as a form of national origin discrimination. (As noted above, however, *Sandoval* subsequently limited enforcement of such violations to agencies.) In 2000, Executive Order 13166, titled "Improving Access to Services for Persons with Limited English Proficiency," was issued, explaining that federal funding recipients were "required to take reasonable steps to ensure meaningful access" for LEP persons. 65 Fed. Reg. 50121 (Aug. 11, 2000).

This "meaningful access requirement" was reaffirmed in both the 2016 Rule and the 2020 Rule. The 2020 Rule, however, rescinded the more specific regulatory requirements that had been created under the Obama administration to ensure access, for example, by eliminating the requirement that non-discrimination notices must include the availability of language assistance services and taglines in the top 15 languages spoken by LEP individuals in the state. The 2020 Rule, instead, revived an earlier balancing test that had been used for Title VI compliance, noting that OCR would assess whether an entity took reasonable steps to ensure meaningful access for LEP individuals based on how the entity balances four factors: "(i) The number or proportion of limited English proficient individuals eligible to be served or likely to be encountered in the eligible service population; (ii) The frequency with which LEP individuals come in contact with the entity's health program, activity, or service; (iii) The nature and importance of the entity's health program, activity, or service; and (iv) The resources available to the entity and costs." 42 C.F.R. 92.101(b).

Perhaps the most challenging aspect of national origin discrimination is conduct motivated by or linked with immigration status. Since enactment of the Personal Responsibility and Work Opportunity Reconciliation Act of 1996 (PRWORA), Pub. L. No. 104–193, 110 Stat. 2105, federal law has severely curtailed public health care benefits for immigrants who are here legally or illegally, or whose legal status is uncertain. The ACA excludes all U.S. residents not "lawfully present" in the U.S. from access to the health insurance exchanges and subsidies for insurance. The ACA also continues the five-year waiting period for Medicaid applied to lawfully present immigrants. See Health Coverage for Immigrants, The Henry J. Kaiser Family Foundation (Dec. 2017).

Funding restrictions on immigrants' access to health care are often part of a broader package of immigration-related initiatives limiting immigrants' access to public services as a means of discouraging illegal immigration. Despite evidence to the contrary, this perceived link between benefits and illegal immigration persists in popular culture, negatively impacting the willingness of some providers to treat immigrants. See Brietta R. Clark, The

Immigrant Health Care Narrative and What It Tells Us About the U.S. Health Care System, 17 Annals of Health Law 229 (2008). Such restrictions have also contributed to a climate of fear that makes some immigrants less willing to seek care, regardless of their status. See Wendy E. Parmet & Elisabeth Ryan, New Dangers for Immigrants and the Health Care System, Health Aff. Blog (Apr. 20, 2018).

Evidence of the chilling effect of benefits restrictions has long been available. But concerns about the health implications of this effect have increased in light of actions by the Trump administration to further restrict immigrants' use of public benefits under the "public charge" test. Some background: The Immigration and Nationality Act requires some categories of non-citizens seeking a visa or change in status to demonstrate they are not likely to become a public charge—that is, not likely to become dependent on government for cash assistance or long-term care. 8 U.S.C.A. § 1182(a)(4). Certain groups, such as asylees and refugees, are excluded from this public charge test. Under longstanding policy, health benefits, such as Medicaid and CHIP, have not been considered in the public charge test. This is because such support has been viewed as helping people stay healthy and productive.

In 2019, however, the Trump administration issued a rule that would allow such benefits to be considered in making a public charge determination. Many predicted that this change would likely discourage qualified immigrants from seeking Medicaid or CHIP benefits for themselves or their children out of fear that doing so would make them vulnerable to deportation, and these predictions seem to have been realized. Indeed, the chilling effect may have occurred even before the rule was finalized, as early as the beginning of President Trump's term when he signaled his intent to define public charge more broadly. According to U.S. Census data from 2016–2019, there was significant disenrollment by eligible immigrants from Medicaid, as well as from marketplace programs that were not even subject to the public charge rule. Randy Capps et al., Anticipated "Chilling Effects" of the Public-Charge Rule Are Real: Census Data Reflect Steep Decline in Benefits Use by Immigrant Families, Migration Policy Institute (Dec. 2020), at Migration Policy.org. See also Carol L. Galletly et al., Assessment of COVID-19-Related Immigration Concerns Among Latinx Immigrants in the US, JAMA Network Open (Jul. 19, 2021), describing the results of a survey of 336 adult Latinx immigrants in the U.S. The survey showed that 27 percent of the participants believed that hospital emergency departments provided the only source for COVID-19-related testing or treatment for uninsured immigrants, 32 percent agreed that using public COVID-19-related testing and treatment services could jeopardize an individual's immigration prospects, and 29 and 34 percent, respectively, would not identify an undocumented household member or coworker during contact tracing. The Trump rule was stayed temporarily by litigation, and in March 2021, the Biden administration rescinded the rule. 86 Fed. Reg. 14221 (Mar. 15, 2021).

Finally, lack of health insurance has also been linked to the practice of medical repatriation, which occurs when a hospital sends critically injured or ill immigrant patients back to their native country without their consent. In these cases, hospitals sometimes also fail to ensure that the patient is being properly discharged to a facility that can appropriately meet the patient's post-discharge medical needs. See Center for Social Justice at Seton Hall Law School and the Health Justice Program at New York Lawyers for the Public Interest Joint Project Report, Discharge, Deportation, and Dangerous Journeys: A Study on the Practice of Medical Repatriation (2012).

B. SEX OR GENDER

The last few decades have brought increased attention to gender-based disparities in health care, with a particular focus on the insurance and treatment barriers experienced by women and transgender individuals. Title VII and Section 1557 are the main federal statutes helping to combat sex discrimination in health care, and courts and regulators have been engaged in disputes about the scope of this protection. These disputes have centered on two themes.

The first theme involves the definition of sex discrimination under the federal laws applicable to health care. Important questions that have been confronting courts include: Does a prohibition of sex discrimination include discrimination on the basis of gender identity or sexual orientation? What kind of activity does antidiscrimination law prohibit (or require) in health care? When does the refusal to provide or pay for care that is sex-linked or has a disparate impact on a particular gender constitute sex discrimination?

The second theme involves the scope of available defenses or exceptions. Religious entities, for example, have expressed concerns that antidiscrimination law may be interpreted to require health care treatment or coverage that is inconsistent with their religious mission or beliefs. This issue is especially prominent in cases involving reproductive or gender-affirming health care—insurance exclusions or denials of care are frequently characterized as an exercise of religious liberty that should be legally protected. Such conflicts raise important questions about how religious objections should be balanced against the health and equity interests advanced by nondiscrimination protections.

Both of these themes are explored in the cases and notes below.

ERICKSON V. THE BARTELL DRUG COMPANY

United States District Court, Washington, 2001.
141 F. Supp. 2d 1266.

LASNIK, DISTRICT JUDGE.

The parties' cross-motions for summary judgment in this case raise an issue of first impression[:] whether the selective exclusion of prescription contraceptives from defendant's generally comprehensive prescription plan constitutes discrimination on the basis of sex. In particular, plaintiffs assert that Bartell's decision not to cover prescription contraceptives such as birth control pills, Norplant, Depo-Provera, intra-uterine devices, and diaphragms under its Prescription Benefit Plan for non-union employees violates Title VII, 42 U.S.C. § 2000e *et seq.,* as amended by the Pregnancy Discrimination Act, 42 U.S.C. § 2000e(k).

A. APPLICATION OF TITLE VII

Title VII makes it unlawful for an employer "to fail or refuse to hire or to discharge any individual, or otherwise to discriminate against any individual with respect to his compensation, terms, conditions, or privileges of employment, because of such individual's race, color, religion, sex, or national origin." 42 U.S.C. § 2000e–2(a)(1). Unfortunately, the legislative history of the Civil Rights Act of 1964, of which Title VII is a part, is not particularly helpful in determining what Congress had in mind when it added protection from discrimination based on sex. . . .

. . . What is clear from the law itself, its legislative history, and Congress' subsequent actions, is that the goal of Title VII was to end years of discrimination in employment and to place all men and women, regardless of race, color, religion, or national origin, on equal footing in how they were treated in the workforce.

In 1978, Congress had the opportunity to expound on its view of sex discrimination by amending Title VII to make clear that discrimination because of "pregnancy, childbirth, or related medical conditions" is discrimination on the basis of sex. 42 U.S.C. § 2000e(k). The amendment, known as the Pregnancy Discrimination Act ("PDA"), was not meant to alter the contours of Title VII: rather, Congress intended to correct what it felt was an erroneous interpretation of Title VII by the United States Supreme Court in *General Elec Co. v. Gilbert,* 429 U.S. 125, 50 L. Ed. 2d 343, 97 S. Ct. 401 (1976). In Gilbert, the Supreme Court held that an otherwise comprehensive short-term disability policy that excluded pregnancy-related disabilities from coverage did not discriminate on the basis of sex. The Gilbert majority based its decision on two findings: (a) pregnancy discrimination does not adversely impact all women and therefore is not the same thing as gender discrimination; and (b) disability insurance which covers the same illnesses and conditions for both men and women is equal coverage. To the Gilbert majority, the fact that pregnancy-related disabilities were an uncovered risk unique to women did not destroy the facial parity of the coverage. The dissenting justices, Justice Brennan, Justice Marshall, and Justice Stevens, took issue with these findings, arguing that: (a) women, as the only sex at risk for pregnancy, were being subjected to unlawful discrimination; and (b) in determining whether an employment policy treats the sexes equally, the court must look at the comprehensiveness of the coverage provided to each sex. . . .

The language of the PDA was chosen in response to the factual situation presented in Gilbert, namely a case of overt discrimination toward pregnant employees. Not surprisingly, the amendment makes no reference whatsoever to prescription contraceptives. Of critical importance to this case, however, is the fact that, in enacting the PDA, Congress embraced the dissent's broader interpretation of Title VII which not only recognized that there are sex-based differences between men and women employees, but also required employers to provide women-only benefits or otherwise incur additional expenses on behalf of women in order to treat the sexes the same. . . .

Although this litigation involves an exclusion for prescription contraceptives rather than an exclusion for pregnancy-related disability costs, the legal principles established by Gilbert and its legislative reversal govern the outcome of this case. An employer has chosen to offer an employment benefit which excludes from its scope of coverage services which are available only to women. [T]he intent of Congress in enacting the PDA, even if not the exact language used in the amendment, shows that mere facial parity of coverage does not excuse or justify an exclusion which carves out benefits that are uniquely designed for women.

The fact that equality under Title VII is measured by evaluating the relative comprehensiveness of coverage offered to the sexes has been accepted and amplified by the Supreme Court. . . .

The other tenet reaffirmed by the PDA (*i.e.,* that discrimination based on any sex-based characteristic is sex discrimination) has also been considered by the courts. The Supreme Court has found that classifying employees on the basis of their childbearing capacity, regardless of whether they are, in fact, pregnant, is sex-based discrimination. *International Union v. Johnson Controls, Inc.,* 499 U.S. 187 (1991). The court's analysis turned primarily on Title VII's prohibition on sex-based classifications, using the PDA merely to bolster a conclusion that had already been reached. To the extent that a woman's ability to get pregnant may not fall within the literal language of the PDA, the court was not overly concerned. Rather, the court focused on the fact that disparate treatment based on unique, sex-based characteristics, such as the capacity to bear children, is sex discrimination prohibited by Title VII.

[T]he Court finds that Bartell's exclusion of prescription contraception from its prescription plan is inconsistent with the requirements of federal law. The PDA is not a begrudging recognition of a limited grant of rights to a strictly defined group of women who happen to be pregnant. Read in the context of Title VII as a whole, it is a broad acknowledgment of the intent of Congress to outlaw any and all discrimination against any and all women in the terms and conditions of their employment, including the benefits an employer provides to its employees. Male and female employees have different, sex-based disability and healthcare needs, and the law is no longer blind to the fact that only women can get pregnant, bear children, or use prescription contraception. The special or increased healthcare needs associated with a woman's unique sex-based characteristics must be met to the same extent, and on the same terms, as other healthcare needs. Even if one were to assume that Bartell's prescription plan was not the result of intentional discrimination,[7] the exclusion of women-only benefits from a generally comprehensive prescription plan is sex discrimination under Title VII.

Title VII does not require employers to offer any particular type or category of benefit. However, when an employer decides to offer a prescription plan covering everything except a few specifically excluded drugs and devices, it has a legal obligation to make sure that the resulting plan does not discriminate based on sex-based characteristics and that it provides equally comprehensive coverage for both sexes. . . .

B. SPECIFIC ARGUMENTS RAISED BY DEFENDANT-EMPLOYER

Bartell argues that opting not to provide coverage for prescription contraceptive devices is not a violation of Title VII because: (1) treating contraceptives differently from other prescription drugs is reasonable in that contraceptives are voluntary, preventative, do not treat or prevent an illness or disease, and are not truly a "healthcare" issue; (2) control of one's fertility is not "pregnancy, childbirth, or related medical conditions" as those terms are used in the PDA; (3) employers must be permitted to control the costs of employment benefits by limiting the scope of coverage; (4) the exclusion of all "family planning" drugs and devices is facially neutral; (5) in the thirty-seven years Title VII has been on the books, no court has found that excluding contraceptives constitutes sex discrimination; and (6) this issue should be determined by the legislature, rather than the courts. Each of these arguments is considered in turn.

(1)   Contraceptives as a health care need

An underlying theme in Bartell's argument is that a woman's ability to control her fertility differs from the type of illness and disease normally treated with prescription drugs in such significant respects that it is permissible to treat prescription contraceptives differently than all other prescription medicines. The evidence submitted by plaintiffs shows,

---

[7]   There is no evidence or indication that Bartell's coverage decisions were intended to hinder women in their ability to participate in the workforce or to deprive them of equal treatment in employment or benefits. The most reasonable explanation for the current state of affairs is that the exclusion of women-only benefits is merely an unquestioned holdover from a time when employment-related benefits were doled out less equitably than they are today. The lack of evidence of bad faith or malice toward women does not affect the validity of plaintiffs' Title VII claim. Where a benefit plan is discriminatory on its face, no inquiry into subjective intent is necessary.

however, that the availability of affordable and effective contraceptives is of great importance to the health of women and children because it can help to prevent a litany of physical, emotional, economic, and social consequences. . . .

Unintended pregnancies, the condition which prescription contraceptives are designed to prevent, are shockingly common in the United States and carry enormous costs and health consequences for the mother, the child, and society as a whole. Over half of all pregnancies in this country are unintended. A woman with an unintended pregnancy is less likely to seek prenatal care, more likely to engage in unhealthy activities, more likely to have an abortion, and more likely to deliver a low birthweight, ill, or unwanted baby. Unintended pregnancies impose significant financial burdens on the parents in the best of circumstances. If the pregnancy results in a distressed newborn, the costs increase by tens of thousands of dollars. In addition, the adverse economic and social consequences of unintended pregnancies fall most harshly on women and interfere with their choice to participate fully and equally in the "marketplace and the world of ideas." *Stanton v. Stanton,* 421 U.S. 7 (1975). See also *Planned Parenthood v. Casey,* 505 U.S. 833, 856 (1992) ("The ability of women to participate equally in the economic and social life of the nation has been facilitated by their ability to control their reproductive lives.").

. . . Although there are many factors which help explain the unusually high rate of unintended pregnancies in the United States, an important cause is the failure to use effective forms of birth control. [ ] Insurance policies and employee benefit plans which exclude coverage for effective forms of contraception contribute to the failure of at-risk women to seek a physician's assistance in avoiding unwanted pregnancies. [ ]

The fact that prescription contraceptives are preventative appears to be an irrelevant distinction in this case: Bartell covers a number of preventative drugs under its plan. The fact that pregnancy is a "natural" state and is not considered a disease or illness is also a distinction without a difference. Being pregnant, though natural, is not a state that is desired by all women or at all points in a woman's life. Prescription contraceptives, like all other preventative drugs, help the recipient avoid unwanted physical changes. . . .

(2) Pregnancy Discrimination Act

[I]t is clear that in 1978 Congress had no specific intent regarding coverage for prescription contraceptives. The relevant issue, however, is whether the decision to exclude drugs made for women from a generally comprehensive prescription plan is sex discrimination under Title VII, with or without the clarification provided by the PDA. The Court finds that, regardless of whether the prevention of pregnancy falls within the phrase "pregnancy, childbirth, or related medical conditions," Congress' decisive overruling of [*Gilbert*] evidences an interpretation of Title VII which necessarily precludes the choices Bartell has made in this case.

(3) Business Decision to Control Costs

Bartell also suggests that it should be permitted to limit the scope of its employee benefit programs in order to control costs. Cost is not, however, a defense to allegations of discrimination under Title VII. [ ]While it is undoubtedly true that employers may cut benefits, raise deductibles, or otherwise alter coverage options to comply with budgetary constraints, the method by which the employer seeks to curb costs must not be discriminatory. Bartell offers its employees an admittedly generous package of healthcare benefits [ ]. It cannot, however, penalize female employees in an effort to keep its benefit costs low. The cost savings Bartell realizes by excluding prescription contraceptives from its healthcare plans are being directly borne by only one sex in violation of Title VII. Although Bartell is permitted, under the law, to use non-discriminatory cuts in benefits to control costs, it cannot balance its benefit books at the expense of its female employees.

(4) Neutrality of Exclusions

Prescription contraceptives are not the only drugs or devices excluded from coverage under Bartell's benefit plan. Bartell argues that it has chosen to exclude from coverage all drugs for "family planning," and that this exclusion is neutral and non-discriminatory. There is no "family planning" exclusion in the benefit plan, however, and the contours of such a theoretical exclusion are not clear. On the list of excluded drugs and devices, contraceptive devices and infertility drugs are the two categories which might be considered "family planning" measures. Contrary to defendant's explanation, there appear to be some drugs which fall under the "family planning" rubric which are covered by the plan. Prenatal vitamins, for example, are frequently prescribed in anticipation of a woman becoming pregnant and are expressly covered under the plan.

* * *

Although the Court's decision is a matter of first impression for the judiciary, it is not the first tribunal to consider the lawfulness of a contraception exclusion. On December 14, 2000, the EEOC made a finding of reasonable cause on the same issue which is entitled to some deference. Although the Commission's analysis focused primarily on the PDA, it [came] to the same conclusion reached by this Court.

(6) Legislative Issue

Although this litigation involves politically charged issues with far-reaching social consequences, the parties' dispute turns on the interpretation of an existing federal statute. . . . The normal rules of statutory construction, not the give and take of a legislative body, will guide this determination. Contrary to defendant's suggestion, it is the role of the judiciary, not the legislature, to interpret existing laws and determine whether they apply to a particular set of facts. While it is interesting to note that Congress and some state legislatures are considering proposals to require insurance plans to cover prescription contraceptives, that fact does not alter this Court's constitutional role in interpreting Congress' legislative enactments in order to resolve private disputes.

C. CONCLUSION

. . . For all of the foregoing reasons, the Court finds that Bartell's prescription drug plan discriminates against Bartell's female employees by providing less complete coverage than that offered to male employees.

NOTES AND QUESTIONS

1. *Why Title VII Is Still Relevant After Section 1557.* Title VII was the earliest federal antidiscrimination statute applicable to sex discrimination in health care, though it was limited to employment-based benefit plans. Despite the broader health care discrimination provision in Section 1557, Title VII remains important for a couple of reasons. First, Title VII may serve as a crucial gap filler for a subset of employees whose health plans are not subject to Section 1557. Although the 2016 Rule interpreted Section 1557's nondiscrimination provision as applying broadly to *all operations* of an entity principally engaged in the provision of health insurance, 81 Fed. Reg. at 31467 (formerly codified at 45 C.F.R. s 92.4), the 2020 Rule narrowed 1557's application to cover *only the parts of an insurer's operations that receive federal funding.* As HHS explained: "To the extent that employer-sponsored group health plans do not receive Federal financial assistance . . . they would not be covered entities." 85 Fed. Reg. at 37173. Second, although Section 1557 expressly incorporates sex by reference using Title IX, which governs educational settings, regulators and courts frequently acknowledge the relevance of Title VII jurisprudence for defining the contours of sex discrimination in the context of health care.

2. *Discrimination "Because of Sex." Erickson* explains the evolution of Title VII to include the Pregnancy Discrimination Act, as well as Congress's express endorsement of the *Gilbert* dissent's approach to defining sex discrimination. Essentially, the court notes that "an exclusion which carves out benefits that are uniquely designed for women" is clearly sex-linked and thus a form of disparate treatment on the basis of sex. The court emphasized the importance of looking beyond facial parity to determine relative comprehensiveness of coverage, but such an inquiry requires identifying an appropriate benchmark for purposes of comparison. In *Erikson*, the court was unconvinced by the employer's attempt to characterize the contraception exclusion as one that was consistent with a gender-neutral policy. But in a different case, the employer fared better. See, e.g., In re Union Pacific Railroad

Employment Practices Litigation, 479 F.3d 936 (8th Cir. 2007) (holding that an employer health plan that excluded contraception when used for purposes of contraception, but not for non-contraceptive medically necessary purposes, did not violate Title VII, because it excluded coverage of all contraception for men and women, both surgical and prescription).

3. *Discrimination "Because of Pregnancy, Childbirth, or Related Medical Conditions."* There has been some debate about what exactly is required by the PDA. Courts have consistently held that the PDA does not create a substantive health care mandate; rather, it clarifies the type of classifications that establish a prima facie case of sex discrimination for purposes of Title VII. See, e.g., Krauel v. Iowa Methodist Medical Center, 95 F.3d 674 (8th Cir. 1996) (noting that the PDA only requires employers to treat pregnancy, childbirth, or related medical conditions in a "neutral way"). What if a challenged exclusion from an otherwise comprehensive insurance policy is not "uniquely" sex-linked, but is nonetheless related to pregnancy and thus has a disparate impact on women, or is otherwise motivated by gender-based assumptions?

A few courts have confronted this question in the context of infertility treatment. Infertility causes and treatments tend to be characterized as female-specific or male-specific, or they may involve a combination of both. For this reason, many employers with blanket exclusions for infertility treatment have successfully characterized these exclusions as gender neutral when challenged under Title VII and the PDA. In *Krauel*, the Eighth Circuit rejected such a challenge even though plaintiffs offered evidence that the employer believed infertility benefits were used primarily, if not solely, by women, and that the employer had concerns about the costs attributable to a resulting pregnancy. The court's treatment of the exclusion as gender neutral meant the burden never shifted to the employer to justify the exclusion. Id.

The *Krauel* approach has been criticized in light of the individual and structural gender discrimination that has long shaped discriminatory benefit design by employers and insurers. With regard specifically to infertility treatment, there is evidence that such exclusions may disproportionately harm women by resulting in the implantation of more embryos at one time and thus increasing the health risk of pregnancy. See Brietta R. Clark, Erickson v. Bartell Drug Co.: A Roadmap for Gender Equality in Reproductive Health Care or an Empty Promise? 23 Law & Inequality 299 (2005).

Based on the reasoning in *Erikson*, do you think the court would have found the exclusion of prescription contraception to be sex-linked if the exclusion affected both men and women? Consider that reversible prescription contraception may soon be available for men, too. NIH to Evaluate Effectiveness of Male Contraceptive Skin Gel, National Institutes of Health, Nov. 28, 2018, https://www.nih.gov/news-events/news-releases/nih-evaluate-effectiveness-male-contraceptive-skin-gel. How, if at all, should this development impact our understanding of the gender equity implications of prescription contraception exclusions? As described more fully in Note 5 below, this development may not impact most coverage because currently the ACA's preventive health mandate includes prescription contraception.

4.   *Private Disparate Impact Claims: Title VII vs. Section 1557.* Although Title VII allows private causes of action based on disparate impact, Section 1557 may not. Some courts have already held that Section 1557 does not permit private disparate impact claims, based on federal court decisions extending *Sandoval's* limit on private disparate impact claims under Title VI to Title IX challenges. See Briscoe v. Health Care Svc. Corp., 281 F. Supp. 3d 725, 738–739 (2017); York v. Wellmark Inc., 2017 WL 11261026, slip op. at 17–18.

5.   *Gender Equity and ACA Insurance Mandates.* Through Title VII and the PDA, Congress made clear that employers could not treat medical costs associated with pregnancy or childbirth differently from other medical costs covered by their health insurance plans. Regulations have elaborated on the contours of Title VII as well. For example, employers must treat maternity-related medical conditions the same as other medical conditions with respect to terms of reimbursement (including payment maximums, deductibles, copayments, coinsurance, and out-of-pocket maximums); preexisting condition limitations; extension of benefits following termination of employment; and limitations on freedom of choice. See 29 C.F.R. App. to Pt. 1604, Questions 25–29. But Title VII only applies to employment-based policies. Before the ACA, insurance companies in the individual markets were able to treat women differently from men, for example, by charging them higher rates and excluding woman-specific benefits—that is, unless there were state laws prohibiting this discrimination.

The ACA has been a crucial tool for eradicating these barriers—not only because of its formal antidiscrimination prohibition in Section 1557, but also because of its many other substantive insurance regulations: prohibitions on risk-rating based on sex; prohibitions on insurers from denying coverage or benefits based on preexisting conditions or health factors, such as a prior caesarian section or a history of domestic violence; and its mandate that preventive health care be covered without cost-sharing.

In defining the scope of services to be included in this preventive mandate, the relevant regulatory agencies adopted guidelines from the Health Resources Services Administration, based on recommendations by the independent Institute of Medicine (IOM). The IOM recommended that preventive services include all forms of contraception approved by the FDA, which meant that most health insurance policies would cover hormonal methods of contraception (i.e., birth control pills and post-intercourse emergency contraception like Plan B and ella), as well as intrauterine devices (IUDs), sterilization, and patient education and counseling. See IOM, Clinical Preventive Services for Women: Closing the Gaps (National Academies Press 2011). In adopting these recommendations, federal regulators acknowledged the gender disparities that existed in the insurance market prior to the ACA and noted that "[t]he contraceptive coverage requirement is . . . designed to serve . . . compelling

public health and gender equity goals. . . ." Group Health Plans and Health
Insurance Issuers Relating to Coverage of Preventive Services Under the
Patient Protection and Affordable Care Act, 77 Fed. Reg. 8725, 8727–8729
(Feb. 15, 2012).

6. *Unequal Health Care Treatment.* There is evidence that, in addition
to insurance discrimination, women experience discrimination in medical
treatment. For example, studies in the U.S. have demonstrated that women
tend to receive less intensive treatment for acute myocardial infarction, have
an increased chance of misdiagnosis of stroke, and suffer higher rates of
mortality from sepsis than men. See Viola Vaccarino et al., Sex and Racial
Differences in the Management of Acute Myocardial Infarction, 1994–2002,
353 NEJM 671 (2005); David E. Newman-Toker et al., Missed Diagnosis of
Stroke in the Emergency Department: A Cross-Sectional Analysis of a Large
Population-Based Sample, 1 Diagnosis 155 (2014); Anthony P. Pietropaoli et
al., Gender Differences in Mortality in Patients with Severe Sepsis and Septic
Shock, 7 Gender Med. 422 (2010). Women are also more likely to be
undertreated for, or inappropriately diagnosed and treated for, pain as
compared to men. See Diane E. Hoffman and Anita J. Tarzian, The Girl Who
Cried Pain: A Bias Against Women in the Treatment of Pain, 29 J. L. Med. &
Ethics 13 (2001).

As with insurance, treatment discrimination has also been linked to
pregnancy and reproductive capacity. For example, pregnant patients have
been excluded from drug treatment programs. See Rebecca Stone, Pregnant
Women and Substance Use: Fear, Stigma and Barriers to Care, 3 Health and
Justice (2015). And the exclusion of pregnant women from early COVID-19
vaccine trials led to incomplete and conflicting guidance among providers
advising pregnant patients about whether to get the vaccine. See Rita Rubin,
Pregnant People's Paradox—Excluded from Vaccine Trials Despite Having a
Higher Risk of COVID-19 Complications, 325 JAMA 1027 (2021). Pregnancy—
more specifically, concern about potential fetal health risk—has also been used
to override pregnant patients' autonomy rights and force them to undergo
unwanted medical interventions, such as cesarean sections. See Chapter 3,
Section VI, discussing government and medical control of decision-making
during pregnancy, including pregnancy exclusions in advanced directives. For
more on the regulation of pregnancy and reproduction generally, see Chapter
3. While growing attention is being paid to the equity and health harms of
these disparate practices, antidiscrimination law has yet to be used effectively
to address them.

7. *Transgender Pregnancy.* Pregnancy-based discrimination is typically
understood as a form of discrimination against women, as reflected in the use
of terms like "maternal health care." In addition, cis-gender women—women
whose gender identity conforms to their female sex assignment at birth—seem
to be the primary focus of much research on pregnancy care. Nonetheless, there
is growing recognition of the fact that people giving birth may not necessarily

identify as female. See Margaret Besse et al., Experiences with Achieving Pregnancy and Giving Birth Among Transgender Men: A Narrative Literature Review, 93 Yale J. Biology & Med. 517–528 (2020) (describing research revealing that "transgender men face substantial obstacles to achieving pregnancy and significant challenges during pregnancy and birth, which are informed by institutionalized cisnormativity embedded within medical norms and practices"). The next case, *Flack v. Wisconsin Department of Health Services*, and the following notes more fully explore the implication of sex discrimination prohibitions for transgender individuals.

8.   *Abortion Exceptionalism.* Do denials of abortion coverage or care constitute discrimination on the basis of sex? Like pregnancy, abortion has been traditionally understood as uniquely affecting women. Moreover, the gender equity and health effects of unwanted pregnancy that result from denying access to contraception, as described in *Erickson*, would be the same in the case of barriers to abortion. Indeed, a number of religious organizations have expressed concern that Section 1557's new sex discrimination prohibition would be interpreted to require abortion coverage or provision, and some have brought preemptive legal challenges. See Religious Sisters of Mercy v. Azar, 513 F. Supp. 3d 1113 (D.N.D. 2021).

While this may seem like a logical application of sex discrimination jurisprudence, numerous federal laws create special exceptions or protections for those opposed to covering or providing abortion, in light of the profound ethical issues raised by the termination of life and the deep ideological divides over abortion access. For example, the PDA expressly provides that employers are not required to pay for abortion, except when necessary to save the woman's life or when medical complications arise from an abortion. Federal funding cannot be used to pay for abortions in most instances—whether through Medicaid, other family funding, or federal subsidies for private insurance under the ACA. And HHS, under both Democratic and Republican administrations, has acknowledged extensive federal conscience protections— some specific to abortion, some grounded in religious liberty—that allow insurers to exclude abortion coverage and allow providers to refuse to perform abortions.

To date, Section 1557 has not been used to require regulated entities to provide or cover abortions, and courts agree this is unlikely. Religious Sisters of Mercy v. Azar, supra. at 1136 ("As interpreted today, Section 1557 does not proscribe refusal to perform or insure abortions."). Indeed, abortion jurisprudence has been criticized for its failure to acknowledge the role that gender discrimination and sex stereotyping continue to play in certain kinds of abortion regulation. Reva B. Siegel, The New Politics of Abortion: An Equality Analysis of Woman-Protective Abortion Restriction, 2007 U. Ill. L. Rev. 991. For further discussion of the legal and ethical issues involved in abortion regulation, see Chapter 3.

FLACK V. WIS. DEPT. OF HEALTH SERVS.

United States District Court, W.D. Wisconsin, 2018.
328 F. Supp. 3d 931.

WILLIAM M. CONLEY, DISTRICT JUDGE.

As a group, transgender individuals have been subjected to harassment and discrimination in virtually every aspect of their lives, including in housing, employment, education, and health care. Their own families, acquaintances and larger communities can be sources of harassment. For some transgender individuals, though certainly not all, the dissonance between their gender identity and their natally assigned sex can manifest itself in the form of "gender dysphoria," a serious medical condition recognized by both sides' experts and the larger medical community as a whole. [Wisconsin Medicaid categorically denies coverage for medically prescribed gender-affirming care. Plaintiffs Cody Flack and Sara Ann Makenzie both have long-term gender dysphoria and have filed suit challenging this exclusion under the Affordable Care Act. They seek to preliminarily enjoin defendants from enforcing the exclusion against their requests for insurance coverage.]

UNDISPUTED FACTS

A. Gender Dysphoria

Every person has a "gender identity." For most people, their gender identity matches the natal sex assigned at birth. For transgender individuals, however, that is not true: their gender identity differs from the sex they were assigned at birth. Specifically, a transgender woman's birth-assigned sex is male, but she has a female gender identity; a transgender man's birth-assigned sex is female, but he has a male gender identity.

Gender dysphoria is a serious medical condition, which if left untreated or inadequately treated can cause adverse symptoms.

> *Gender dysphoria* refers to the distress that may accompany the incongruence between one's experienced or expressed gender and one's assigned gender. Although not all individuals will experience distress as a result of such incongruence, many are distressed if the desired physical interventions by means of hormones and/or surgery are not available. The current term is more descriptive than the previous DSM-IV term *gender identity disorder* and focuses on dysphoria as the clinical problem, not identity per se.

(DSM-5 [ ]).[3] It is worth emphasizing that not every transgender person has gender dysphoria. Adults with gender dysphoria "often" have "a desire to be rid of primary and/or secondary sex characteristics and/or a strong

---

[3]   DSM-5 is the fifth edition of the American Psychiatric Association's *Diagnostic and Statistical Manual of Mental Disorders*.

desire to acquire some primary and/or secondary sex characteristics of the other gender." [ ] Untreated, gender dysphoria can result in psychological distress: "preoccupation with cross-gender wishes often interferes with daily activities." [ ] Impairment—such as the development of substance abuse, anxiety and depression—is also a possible "consequence of gender dysphoria." [ ] Finally, gender dysphoria "is associated with high levels of stigmatization, discrimination, and victimization, leading to negative self-concept, increased rates of mental disorder comorbidity, school dropout, and economic marginalization, including unemployment, with attendant social and mental health risks. . . ." [ ]

Gender dysphoria can be alleviated through living consistently with one's gender identity, including being treated by others accordingly. Likewise, "appropriate individualized medical care as part of their gender transitions" can mitigate or prevent symptoms of gender dysphoria. [ ] In 2011, the World Professional Association of Transgender Health published the seventh version of *Standards of Care for the Health of Transsexual, Transgender, and Gender Nonconforming People* (the "WPATH Standards of Care"), which identifies psychotherapy, hormone therapy and various surgical procedures as treatment possibilities for gender dysphoria. . . .

B. Wisconsin Medicaid

. . . The Wisconsin Department of Health Services ("DHS") is the state agency charged with administering the Wisconsin Medicaid Program consistent with state and federal requirements. . . Wisconsin Medicaid provides coverage for "[p]hysician services," including "any medically necessary diagnostic, preventative, therapeutic, rehabilitative or palliative services . . . within the scope of the practice of medicine and surgery" that are "in conformity with generally accepted good medical practice" and provided by a physician or under one's direct supervision, unless otherwise excluded. [ ] Wisconsin Medicaid has a budget of approximately $9.7 billion to provide for its roughly 1.2 million enrollees. Approximately 5,000 of those enrollees are transgender, and some subset of this population suffers from gender dysphoria.

. . . Wisconsin Medicaid is [also] governed by [state law]. Included in the governing regulations is the "Challenged Exclusion," [Wis. Admin. Code] § 107.03(23)–(24), at issue in this case. [It] provides that "The following services are not covered . . . (23) Drugs, including hormone therapy, associated with transsexual surgery or medically unnecessary alteration of sexual anatomy or characteristics; [and] (24) Transsexual surgery." The Challenged Exclusion was adopted in 1996.[7] . . .

---

[7]   As its terms suggest, if inartfully, defendants contend that the Challenged Exclusion does not "prohibit[ ] all transition-related medical treatments, as hormone therapy is still provided to Flack and Makenzie for their gender dysphoria." [ ]

C. Plaintiffs' Medical Needs

1. Cody Flack

Cody Flack is an adult resident of Green Bay who suffers from gender dysphoria and identifies as male. He is unable to work because of cerebral palsy and other disabilities for which he receives Supplemental Security Income and Wisconsin Medicaid. At birth, Cody was assigned the sex of female and subsequently raised as a girl. He became aware of his male gender identity when he was about four or five years old. As a teenager, Cody began his gender transition by seeing a gender therapist, adopting the traditionally male name Cody, and presenting as a man. However, he was unable to complete his transition for several years because he lacked financial resources, was without emotional support, and feared isolation.

After relocating to Wisconsin in 2012, Cody found the wherewithal to resume his gender transition as he felt his gender identity was more supported, and he increasingly lived and presented as a man. . . . He also legally changed his name to Cody Jason Flack to align with his male gender identity and obtained a Wisconsin state identification card, identifying him as male. His Medicaid enrollment now matches his gender identity as well.

. . . . Since 2015, Cody has been seeing psychotherapist Daniel Bergman for his gender dysphoria and other mental health conditions. Since August 2016, Cody has also been receiving testosterone hormone therapy under the supervision of an endocrinologist [causing development of] facial and body hair, a more masculine appearance, and a deeper voice. In October 2016, Cody had his uterus, fallopian tubes, ovaries and cervix removed through a hysterectomy with bilateral salpingo-oophorectomy. This procedure was paid for by Wisconsin Medicaid to treat dysmenhorrhea (lower abdominal or pelvic pain during menstruation) and premenstrual dysphoric disorder (a severe form of premenstrual syndrome). . . .

Despite these changes, Cody still has female-appearing breasts. Plaintiffs and their experts, as well as Cody's treating physicians, contend this causes him severe gender dysphoria. . . . At minimum, it appears undisputed that Cody's breasts cause him significant, personal distress, as they are a marker of the female sex often contributing to his being perceived as female. Cody is particularly ashamed of his breasts when in public and routinely avoids social situations as a result. In an effort to conceal his breasts, Cody has engaged in "binding," which flattens or reduces their appearance, but has difficulty binding his breasts himself due to his disabilities and finds the technique extremely painful. Binding has caused him sores, skin irritation and respiratory distress.

Since early 2017, therefore, Cody has sought a double mastectomy and male chest reconstruction. He consulted with . . . a plastic surgeon [and provided him] with letters of support from his primary care physician, his therapist, his endocrinologist, and the surgeon who performed his hysterectomy. [The letters] detailed that Cody has gender dysphoria, and he met the criteria for surgery. . . .

On August 2, 2017, DHS denied [the] prior authorization request [for the surgery] without reviewing the medical necessity of his requested surgery as "a non-covered service" and a "not covered benefit" based on the Challenged Exclusion. [On] appeal, DHS noted that "gender dysphoria . . . is an accepted medical indication for the surgical treatment requested." [ ] Nevertheless, an administrative law judge dismissed the appeal [based on the statutory exclusion]. . . .

2.   Sara Ann Makenzie

Sara Ann Makenzie is an adult . . . and lifelong resident of Wisconsin. She also suffers from gender dysphoria after being assigned the male sex at birth and subsequently raised as a boy. [Sara Ann has] been found to be disabled and . . . has been enrolled in Wisconsin Medicaid for many years.

Despite being assigned the male sex at birth, Sara Ann first identified as female as a child, and she has been diagnosed with gender dysphoria for most of her life. She has legally changed her name to Sara Ann Makenzie, uses feminine pronouns, and wears women's clothing to conform with her female gender identity. Her birth certificate, passport, driver's license and Medicaid enrollment all reflect her name and female identity.

[Sara Ann] has been on hormone therapy since 2013, which has helped lessen her gender dysphoria. In 2014, she consulted with . . . her then primary care physician, about genital reconstruction surgery. . . .

Sara Ann Makenzie reports great distress upon seeing her male-appearing genitalia, which negatively affects her occupational functioning, sexuality and social life. She finds showering or seeing her body in a mirror painful; she lives in constant fear that someone will be able to see her male genitals through her clothing; and she is concerned that she may be attacked or mistreated by someone who recognizes her as transgender. Accordingly, she tries to minimize the appearance of her genitals by wearing multiple pairs of underwear at a time and engaging in "tucking," which is uncomfortable and very painful. She also does not have sex with her fiancée, which adds to her depression and anxiety. Sara Ann's treating physicians have recommended that she have surgery to create female-appearing external genitalia, specifically a bilateral orchiectomy and vaginoplasty. . . .

[In February 2018, Sara Ann's] primary care physician . . . referred Sara Ann to a plastic surgeon . . . for possible genital reconstruction surgery. Dr. Katherine Gast specializes in treating transgender individuals, and she informed Sara Ann of her eligibility for genital reconstruction after submitting two letters of support from mental health providers. However, she also advised that Wisconsin Medicaid would not pay for the procedure. [Sara Ann] was greatly distressed that Medicaid would not pay for the surgery because she could not afford it [and] has thoughts of removing her genitals on her own and of committing suicide. As a result, plaintiffs contend that her gender dysphoria has worsened.

Her psychotherapist, Jessica Bellard, notes that Sara Ann "continues to report symptoms of anxiety, depression, anger, and distress in response to the stressors of transitioning prior to completing gender reassignment surgery" and that she "has expressed a persistent desire for surgery since our original meeting." [ ] Sara Ann's independent evaluating therapist, Chelsea O'Neal Karcher, opined that "Sara's hope that the surgery will help lessen symptoms of anxiety and depression, increase happiness, help to increase her confidence, and align her body more fully with her identity" were "realistic expectations for the procedure" and that she "has met all the eligibility and readiness criteria outline[d] in the [WPATH Standards of Care]." [ ] Her former primary care physician opined that "genital reconstruction is a *medically necessary* treatment for Ms. Makenzie's gender dysphoria as it would treat the excessive mental distress that she experiences every day. . . ." [ ] . . .

OPINION

Plaintiffs seek a preliminary injunction enjoining defendants from enforcing the Challenged Exclusion against them. As "an extraordinary remedy," [t]he moving party must "mak[e] a threshold showing: (1) that he will suffer irreparable harm absent preliminary injunctive relief during the pendency of his action; (2) inadequate remedies at law exist; and (3) he has a reasonable likelihood of success on the merits." [ ] Once the moving party has done so, the court "determine[s] whether the balance of harm favors the moving party or whether the harm to other parties or the public sufficiently outweighs the movant's interests." [ ]

I.   Irreparable Harm & Inadequate Remedy at Law

While the moving party must show "more than a mere possibility of harm" to establish that it likely will suffer irreparable harm absent injunctive relief, this does not mean that the harm must occur or be certain to occur before the merits can be addressed. [ ] To be irreparable, the harm "cannot be prevented or fully rectified by the final judgment after trial." [ ]

As noted, plaintiffs are being denied coverage for medically necessary treatment that was prescribed by their doctors and meets the prevailing standards of care. They contend that allowing them to obtain their surgeries is necessary to protect their well-being and health because they "are at high risk of worsening mental health, exacerbated gender dysphoria, self-harm and stigma"—"none of which has an adequate remedy at law." [ ]

Importantly, [their] treating doctors agree. [Eds. Note: The Court then recounts much of the evidence presented in the Undisputed Facts and other testimony by treating physicians, respectively, demonstrating that the plaintiffs have a diagnosis of gender dysphoria, that plaintiffs meet the WPATH Standards of Care criteria for reconstructive surgery,* that such surgery is medically necessary, and that denial of surgery will increase the risk of harm.]

Accordingly, these plaintiffs have advanced more than enough evidence to establish that they face a possibility of irreparable harm at this point. [ ] Defendants' principal response to all this evidence is that "there is no proven medical benefit to the procedures for which Plaintiffs seek Medicaid coverage, and so Plaintiffs here will not face irreparable harm absent a preliminary injunction." [ ] [First, Defendants' expert Dr. Lawrence Mayer] opines that "[m]edical and surgical treatments have not been demonstrated to be safe and effective for gender dysphoria." [ ] More specifically, he opines based on a survey of published reports that there is only "minimal" evidence that these treatments are effective, safe and optimal, and further that "[t]here is even less evidence that [medical and surgical interventions] would be cost effective compared to social and psychological interventions." [ ]

Even Dr. Mayer acknowledges, however, that gender dysphoria "is a serious medical condition that deserves to be treated" and that such treatment "must be borne of medical necessity. [ ] He agrees that "reducing or eliminating" the very real distress associated with this condition is the "[o]ptimality consideration[ ]" for treating gender dysphoria.[ ] As outlined above, plaintiffs have the support of their treating physicians—to say nothing of their retained experts—who confirm that the surgeries they seek (1) are medically necessary and (2) will reduce their distress and gender dysphoria. Perhaps Dr. Mayer's opinions will prevail on the *general* efficacy of surgical interventions for gender dysphoria, although the apparent endorsement in DSM-5 and by the larger medical community would appear to make this a decidedly uphill battle.[17] Still, Dr. Mayer lays at most the

---

* The court notes the following WPATH Standards of Care: For male chest reconstruction, the standards of care include: "1. Persistent, well-documented gender dysphoria; 2. Capacity to make a fully informed decision and to consent for treatment; 3. Age of majority in a given country . . .; 4. If significant medical or health concerns are present, they must be reasonably well controlled." For genital reconstructive surgery, patients must meet all the preceding criteria plus: "12 continuous months of hormone therapy as appropriate to the patient's gender goals (unless the patient has a medical contraindication or is otherwise unable or unwilling to take hormones)," and for vaginoplasty, "12 continuous months of living in a gender role that is congruent with their gender identity."

17 (*See* Am. Psychiatric Assoc. 3 ("[A]ppropriately evaluated transgender and gender variant individuals can benefit greatly from medical and surgical gender transition treatments"); Am. Med. Assoc. 2 ("[M]edical and surgical treatments for gender dysphoria, as determined by shared decision making between the patient and physician, are medically necessary as outlined by generally-accepted standards of medical and surgical practice"); Am. Endocrine Soc'y 3 ("Medical intervention for transgender individuals (including both hormone therapy and medically indicated

foundation for defendants' general policy, while the only question at this point is whether Cody Flack and Sara Ann Makenzie have a medical need for these surgeries such that denial will be detrimental to *their* health. On the current record, the answer clearly is yes.

Second, Dr. Chester Schmidt opines that "there is an insufficient clinical basis to conclude that either Flack or Makenzie will suffer imminent, irreparable harm if they do not receive gender reassignment surgery prior to the conclusion of this case." [ ] Dr. Schmidt's opinion is based on what he describes as: (1) a lack of current mental status examinations for plaintiffs, creating "an insufficient basis for any clinician to conclude that either Flack or Makenzie faces an imminent risk of suicide or other self-harm"; (2) the insufficiency of plaintiffs' self-reports "to conclude that a serious risk of self-harm exists, let alone that receiving the surgical procedures . . . will reduce or eliminate that risk"; (3) Cody Flack's outpatient notes "do not indicate that he is so destabilized such that a substantial risk of imminent self-harm exists"; and (4) Flack's transition has been ongoing for several years, without any evidence of prior self-harm, indicating no substantial short-term risk of self-harm. [ ] While at least focused on the medical needs of the individual plaintiffs now before the court, Schmidt's opinion fundamentally misses the mark. . . . Plaintiffs were not required to prove "a substantial risk of imminent self-harm," but rather to show a likelihood of irreparable injury. Moreover, his opinion that a clinician could not conclude that these plaintiffs need these surgeries because of missing mental status reports misreads the record since both plaintiffs have opinions from their treatment providers supporting the proposed surgical interventions. Regardless, Dr. Schmidt's 10,000-foot review of the medical record and criticisms of the course of treatment proposed by plaintiffs' doctors, without offering a viable option to relieve their ongoing gender dysphoria, pales in comparison with the informed opinions of plaintiffs' treating physicians. To discount the opinions of plaintiffs' treating physicians as to the need for surgery to relieve plaintiffs' suffering at the preliminary injunction stage on the current, limited record would be tantamount to this court playing doctor.

Accordingly, plaintiffs have established that they are at risk of irreparable harm, and this factor weighs strongly in favor of injunctive relief. [T]he Seventh Circuit already held in *Whitaker* that serious, ongoing impact on plaintiffs' health "demonstrate[s] that any award would be 'seriously deficient as compared to the harm suffered.' " . . .

---

surgery) is effective, relatively safe (when appropriately monitored), and has been established as the standard of care.").)

II. Reasonable Likelihood of Success

. . . Plaintiffs claim that the Challenged Exclusion violates § 1557 of the ACA by unlawfully discriminating on the basis of sex—being transgender. . . .

There is no dispute that Wisconsin Medicaid is "a health program or activity" that "receiv[es] Federal financial assistance"; nor is there any dispute that Title IX prohibits discrimination "on the basis of sex." Instead, the parties dispute whether plaintiffs' transgender status falls under "sex." Facially, the answer would appear to be "yes," but because the Challenged Exclusion discriminates against coverage for "transsexual surgery," defendants argue that this no longer involves sex discrimination. Even though "sex" would seem to encompass "trans*sex*ual," defendants would distinguish between "gender identity"—what plaintiffs would define as an "internal sense of one's sex," which is innate—from one's "sex"—which is, according to defendants, "assigned at birth, refer[ring] to one's biological status as either male or female, and is associated primarily with physical attributes such as chromosomes, hormone prevalence, and external and internal anatomy," and is "immutable." [ ]

Even accepting defendants' definition of sex, however, the Challenged Exclusion certainly denies coverage for medically necessary surgical procedures based on a patient's *natal* sex, the same "immutable" sex the defendants claim the ACA intends to cover. Moreover, as plaintiffs' expert Dr. Loren Schechter explains, "[w]hen performing gender confirming surgery, surgeons use many of the same procedures that they use to treat other medical conditions." [ ] For example, if a natal female were born without a vagina, she could have surgery to create one, which would be covered by Wisconsin Medicaid if deemed medically necessary.[21] However, a natal male suffering from gender dysphoria would be denied the same medically necessary procedure because of her sex. Likewise, if a natal male were in a car accident and required a phalloplasty, that surgery would be covered if deemed medically necessary. However, a natal female seeking that same medically necessary procedure for gender dysphoria would be denied because of his sex. In this case, if plaintiffs' natally assigned sexes had *matched* their gender identities, their requested, medically necessary surgeries to reconstruct their genitalia or breasts would be covered by

---

[21] This is not just a hypothetical example. Approximately 1/100 people have bodies with anatomy differing from the typical male or female form, and one out of every 6,000 births involves a baby with vaginal agenesis (an undeveloped vagina). *See How Common Is Intersex?*, Intersex Society of North America, http://www.isna.org/faq/frequency (last visited July 24, 2018) ([sic]; *see also* Schechter Decl. (dkt. # 27) ¶ 38 ("[S]urgeons perform procedures to reconstruct male or female external genitalia for individuals who have certain medical conditions. . . . For the female genitalia, this would include procedures to correct conditions such as congenital absence of the vagina or reconstruction of the vagina/vulva following oncologic resection, traumatic injury, or infection.")).

Wisconsin Medicaid.[22] Here, plaintiffs have instead been denied coverage because of their natal sex, which would appear to be a straightforward case of sex discrimination.

Even if defendants' more tortured interpretation of the Challenged Exclusion prevailed, and "sex" needed to include transgender status, plaintiffs still have more than a reasonable likelihood of success on the merits under United States Supreme Court and Seventh Circuit precedent. This is because the scope of what qualifies as prohibited sex discrimination has changed over time. *See Hively v. Ivy Tech Cmty. Coll. of Ind.*, 853 F.3d 339, 345 (7th Cir. 2017) (en banc). As the *Hively* court explained, Title VII's prohibition on sex discrimination "has been understood to cover far more than the simple decision of an employer not to hire a woman for Job A, or a man for Job B"; it now "reaches sexual harassment in the work place, including same-sex workplace harassment; it reaches discrimination based on actuarial assumptions about a person's longevity; and it reaches discrimination based on a person's failure to conform to a certain set of gender stereotypes," which may "have surprised some who served in the 88th Congress."[ ] As the Seventh Circuit explained in *Whitaker*, "[b]y definition, a transgender individual does not conform to the sex-based stereotypes of the sex that he or she was assigned at birth." [ ] . . .

Indeed, the Seventh Circuit further concluded in *Whitaker* that a policy subjecting a transgender student—because he was transgender—"to different rules, sanctions, and treatment [compared to] non-transgender students" violated the prohibition against sex discrimination under Title IX. [ ] This is what the Challenged Exclusion does as well: it creates a different rule governing the medical treatment of transgender people. Specifically, Wisconsin Medicaid covers medically necessary treatment for other health conditions, yet the Challenged Exclusion expressly *singles out and bars* a medically necessary treatment solely for transgender people suffering from gender dysphoria. In fact, by excluding "transsexual surgery" from coverage, the Challenged Exclusion directly singles out a Medicaid claimant's transgender status as the basis for denying medical treatment. [ ] As defendants conceded at the preliminary injunction hearing, this means that if breast reduction surgery were deemed medically necessary due to back, neck or shoulder pain, a natal female's surgery would be covered by Wisconsin Medicaid. However, if breast reduction surgery were deemed medically necessary due to gender dysphoria, a natal female's surgery would *not* be covered. . . . This is text-book discrimination *based on sex*. . . .

---

[22] Despite the possible implication of the opinions of one of their experts, defendants do *not* argue that the proposed surgical procedures are excluded because they remain experimental, perhaps out of recognition that they are now commonly offered and performed across the country to ease the suffering of those with gender dysphoria. In fact, during oral argument, defendants acknowledged this type of surgery was not experimental in nature.

Aside from being unsympathetic to a medical condition which they acknowledge both plaintiffs suffer from, defendants fail to grasp that not every transgender person requires, or even desires, gender-confirming surgery. [A]s in other areas of health care covered by Wisconsin Medicaid, individuals should be allowed to decide in consultation with their treatment providers what treatment is best and then ultimately whether to pursue it. If anything, the Challenged Exclusion feeds into sex stereotypes by requiring all transgender individuals . . . to keep genitalia and other prominent sex characteristics consistent with their natal sex no matter how painful and disorienting it may prove for some. [ ]

Accordingly, plaintiffs have made a persuasive evidentiary showing, albeit a preliminary one, that the Challenged Exclusion prevents them from getting medically necessary treatments on the basis of both their natal sex *and* transgender status, which surely amounts to discrimination on the basis of sex in violation of the ACA. . . .

III. Balance of Harms & Public Interest

. . . Here, the likelihood of ongoing, irreparable harm facing these two, individual plaintiffs outweighs any marginal impacts on the defendants' stated concerns regarding public health or limiting costs. As addressed above, defendants' concern about protecting the public health by limiting Wisconsin Medicaid to "medically necessary purposes" is misplaced here, given the substantial likelihood that this interest would be served, rather than hindered, by covering plaintiffs' recommended surgical procedures. As to the latter concern, the court readily acknowledges the state's ongoing interest in reasonably reducing medical expenditures where appropriate, but as defendants also acknowledged the Challenged Exclusion is "expected to result in *nominal* savings for state government." . . . As such, the irreparable harm these plaintiffs face without injunctive relief substantially outweighs the harm the state will suffer by preliminarily enjoining the application of the Challenged Exclusion to their individual prior authorization requests. [ ] . . .

NOTES AND QUESTIONS

1. *Does Sex Discrimination Include Discrimination Based on Transgender Status or Gender Identity?* As noted above, perhaps the most anticipated development of Section 1557 was the protection it would provide transgender individuals. Indeed, in the 2016 Rule interpreting Section 1557 issued under the Obama-Biden administration, HHS defined sex discrimination to include "discrimination on the basis of . . . sex stereotyping[ ] and gender identity." It also spent significant time and attention discussing how Section 1557 would address the various forms of health care discrimination experienced by transgender people, including by prohibiting the types of categorical exclusions of gender-affirming care challenged in *Flack*. Under the Trump administration's 2020 Rule, however, regulators repealed

this definition and expressly rejected the notion that transgender people would be protected by Section 1557's sex discrimination prohibition. The 2020 Rule distinguished "biological sex" from "gender identity" and pointed to Title IX's legislative history focusing on equal opportunities for *women*. It interpreted the absence of an explicit reference to gender identity in either Title IX or Section 1557 as evidence that discrimination on the basis of transgender status was not intended to be prohibited as a form of sex discrimination. Ongoing litigation challenges to the 2020 Rule, and the Biden-Harris win in 2020, have created a kind of regulatory limbo as of the writing of this edition; that said, the new administration has, unsurprisingly, already signaled its embrace of the more protective antidiscrimination approach reflected in the 2016 Rule. HHS Office for Civil Rights, HHS Announces Prohibition on Sex Discrimination Includes Discrimination on the Basis of Sexual Orientation and Gender Identity, May 10, 2021.

Disputes about what constitutes sex discrimination have also been playing out in the courts, in ways that may constrain future regulatory interpretations. Although the Supreme Court has not yet faced the question of whether Section 1557 protects transgender individuals, the Court has done so in the context of Title VII. *Bostock v. Clayton County, Georgia,* involved three consolidated cases—one of them challenging the firing of a long-time employee for being transgender. 140 S. Ct. 1731 (2020). Justice Gorsuch delivered the opinion of the Court, in which Chief Justice Roberts and Justices Ginsburg, Breyer, Sotomayor, and Kagan joined. The majority conceded the distinction between sex and gender identity and assumed for the sake of argument that "sex" referred only to biological distinctions between male and female. The majority nonetheless concluded that sex discrimination included discrimination based on gender identity and sexual orientation. Justice Gorsuch explained that Title VII's language is clear in that "it prohibits employers from taking certain actions 'because of' sex," which "incorporates the 'simple' and 'traditional' standard of but-for causation." *Bostock,* 140 S. Ct. at 1739. In short, "an employer who intentionally treats a person worse because of sex—such as by firing the person for actions or attributes it would tolerate in an individual of another sex—discriminates against that person in violation of Title VII." Id. at 1740. Notably, the majority held that "it is impossible to discriminate against a person for being . . . transgender without discriminating against that individual based on sex." Id. at 1741–1742.

Given the majority's textualist approach to what it concluded was a rather straightforward meaning of "because of sex," its outcome did not depend on legislative history about which groups Congress intended to protect. Thus, it will almost certainly apply to similar sex discrimination prohibitions in other statutes, most notably Title IX and Section 1557. Indeed, lower courts have already begun acknowledging *Bostock's* relevance to Section 1557. See, e.g., Whitman-Walker Clinic, Inc. v. U.S. Dep't of Health & Hum. Servs., 485 F. Supp. 3d 1 (D.D.C. 2020) (enjoining the parts of the 2020 Rule that would have repealed the 2016 Rule definitions relating to discrimination based on sex stereotyping and gender identity, and requiring HHS to reconsider its repeal attempt in light of *Bostock*); Asapansa-Johnson Walker v. Azar, No. 20-CV-2834, 2020 WL 6363970 (E.D.N.Y. Oct. 29, 2020) (finding the 2020 Rule relating to sex discrimination contrary to *Bostock*, and staying repeal of the 2016 Rule definitions of "on the basis of sex," "gender identity," and "sex stereotyping" currently set forth in 45 C.F.R. § 92.4), *appeal filed* No. 20-3827 (2d Cir. Nov. 10, 2020). Even before *Bostock*, most courts agreed that Section 1557 prohibited discrimination against transgender individuals as a matter of statutory interpretation, which did not depend on the interpretation set forth in implementing rules. See, e.g., Flack v. Wisconsin Dept. of Health, 395 F. Supp. 3d 1001 (W.D. Wis. 2019); Tovar v. Essentia Health, 342 F. Supp. 3d 947 (D. Minn. 2018); Boyden v. Conlin, 341 F. Supp. 3d 979 (W.D. Wis. 2018); Prescott v. Rady, 265 F. Supp. 3d 1090 (S.D. Cal. 2017); Rumble v. Fairview Health Services, 2015 WL 1197415 (D. Minn. Mar. 16, 2015).

2. *Theory of Sex Discrimination in* Flack. In determining the plaintiffs' likelihood of success in proving sex discrimination under Section 1557, the *Flack* decision (which predates *Bostock*) offers a few different rationales for finding that the challenged Medicaid exclusion constitutes sex discrimination. Which of these rationales seem(s) most consistent with the reasoning in *Bostock*? (Note that the *Flack* court looked to Title IX and Title VII jurisprudence for guidance.) Recall from *Erickson* and Note 2, above, the importance of identifying the right comparator for determining when benefit exclusions are discriminatory. What comparator(s) were used in *Flack* to illustrate differential treatment based on transgender status or natal assigned sex?

Flack is consistent with the approach taken in the 2016 Rule issued by the Obama administration, which prohibited categorical limits on all gender-affirming care. The 2016 Rule also clarified that if a plan ordinarily covers a particular treatment, whether medically necessary or elective, it must apply the same standards to its coverage of comparable procedures related to gender transition. 81 Fed. Reg. at 31,377. Although the *Flack* excerpt above dealt with a preliminary injunction request, the next year the court permanently enjoined defendants from enforcing the challenged exclusion, adopting the same analysis with respect to plaintiffs' Section 1557 claim. Flack v. Wis. Dept. of Health Servs, 395 F. Supp. 3d 1001, 1015 (W.D. Wis. 2019).

3. *Categorical Exclusions vs. Individual Insurance Denials*. Even if categorical insurance exclusions of gender-affirming care constitute prohibited sex discrimination, this does not necessarily guarantee coverage for requested care. Coverage may still be denied if the particular treatment is not found to be medically necessary. For example, in a Title VII case, Baker v. Aetna Life Insurance Co., 228 F. Supp. 3d 764 (N.D. Tex. 2017), an employee alleging that she suffered from gender dysphoria brought an action against her employer for denying coverage for breast augmentation surgery. The court initially denied the employer's motion to dismiss because the plaintiff plausibly alleged that she was denied coverage for medically necessary surgery based on her gender in violation of Title VII. Ultimately, however, the court rejected the plaintiff's discrimination claim for insufficient evidence. After noting that the plan did

cover surgeries and other medically necessary care to treat gender dysphoria, the court highlighted the fact that the plan made an individualized determination that the plaintiff's request for breast augmentation surgery was not medically necessary, in light of the breast enhancement already achieved through hormone therapy. Baker v. Aetna Life Insurance Co. & L-3, No. 3:15-CV-3679-D, 2018 WL 572907 (Jan. 26, 2018).

In *Flack*, the parties' dispute about the plaintiffs' individual medical need and risk of harm for purposes of the preliminary injunction provides a glimpse of how such disputes may occur on an individualized basis. Given the pervasive discrimination against transgender individuals in health care and throughout society, how will courts be able to determine whether an individualized denial of specific care as not medically necessary is truly an application of the kind of neutral, nondiscriminatory criteria used for other types of conditions or care, or is instead the product of covert or implicit bias? How much weight should treating physicians' judgment be given?

4. *Discrimination in Health Care Treatment.* Transgender individuals experience many forms of unequal treatment and barriers to care. In a 2010 survey on discrimination against LGBT people, nearly 27 percent of transgender respondents reported being denied medically necessary care outright because of their transgender status. See Lamba Legal, When Health Care Isn't Caring: Survey on Discrimination Against LGBT People and People Living with HIV (2010). In the same survey, 70 percent of the respondents reported being treated poorly by health care providers, in ways that created serious dignitary harms and potentially jeopardized health. Survey participants described a broad range of such treatment by physicians: refusing to touch them or using excessive precautions; using harsh or abusive language; being physically rough or abusive; or blaming the patients for their health status. Examples of poor treatment by hospital staff were also reported, including laughing, pointing, taunting, using slurs and making other negative comments; violating confidentiality; using an improper name and/or pronoun for the patient; exceptionally long waits for care; inappropriate questions or exams, including needless viewing of genitals; and failure to follow standards of care. See Lamba Legal, supra.

One of the earliest cases to interpret Section 1557 involved alleged treatment by health care staff that mirrored many of the findings from the Lamba survey above. In *Rumble v. Fairview Health Services*, supra, the plaintiff, who self-identified as a "female-to-male transgender man," alleged hostile and abusive touching by a physician, misclassification of the plaintiff's gender, unnecessary disclosures revealing his status, and subjection to unnecessary and improper exams. There was also evidence the plaintiff did not receive adequate pain treatment or a timely diagnosis, and that the delay could have been life-threatening. Based on this evidence, the court refused to grant defendants' motion to dismiss and found that the nature and degree of defendants' "unprofessional" and "assaultive" actions were sufficient to raise an inference of discriminatory intent under Section 1557.

Rumble is consistent with the approach taken in the 2016 Final Rule. Relying on Title IX's definition of sex discrimination as prohibiting harassment that creates a hostile environment, the 2016 Rule noted that "the persistent and intentional refusal to use a transgender individual's preferred name and pronoun and insistence on using those corresponding to the individual's sex assigned at birth constitutes illegal sex discrimination if such conduct is sufficiently serious to create a hostile environment." Although the rule interpreted Section 1557 as requiring covered entities to treat individuals in accordance with their gender identity or expression, it contained an important clarification: this requirement could not be used to deny medically necessary care. (See former 45 C.F.R. § 92.206.) For example, the rule explained that a covered entity could not deny treatment for ovarian cancer based on an individual's identification as a transgender male, if that treatment is medically indicated.

Individuals in need of gender-affirming care, such as hormone therapy, surgery, or counseling, have trouble accessing such care due not only to insurance exclusions and routine denials, but also to provider refusals to provide care, lack of qualified providers, and inadequate training of medical professionals generally with respect to the medical needs of transgender patients. See National Center for Transgender Equality and National Gay and Lesbian Task Force, Injustice at Every Turn: A Report of the National Transgender Discrimination Survey 5–6 (2011). As with insurance, where a provider is qualified and typically provides a certain type of care for some medical needs, such as a hysterectomy to treat fibroids, refusal to provide care to treat gender dysphoria may constitute discrimination on the basis of sex. Anti-discrimination law has been less effective at addressing other structural barriers to care, such as an inadequate supply of qualified providers.

5. *Discrimination on the Basis of Sexual Orientation.* Although discrimination on the basis of sexual orientation has received comparatively less attention, lesbian, gay, and bisexual (LGB) patients experience many of the same forms of discrimination as transgender patients. In a 2009 survey, almost 56 percent of LGB respondents reported encountering at least one of the following types of discrimination in care: being refused needed care; health care professionals refusing to touch them or using excessive precautions; health care professionals using harsh or abusive language; being blamed for their health status; or health care professionals being physically rough or abusive. Lamba Legal, When Health Care Isn't Caring: Lambda Legal's Survey on Discrimination Against LGBT People and People Living with HIV (2010).

Section 1557 likely provides important antidiscrimination protections for LGB individuals as well. In *Bostock*, supra, the Supreme Court was also confronted with the question of whether Title VII's sex discrimination provision prohibits discrimination based on sexual orientation. The other two consolidated cases involved challenges to the firing of two long-time employees for being gay, and the majority applied the same "simple" "but-for" test, finding it "impossible to discriminate against a person for being [gay] without discriminating against that individual based on sex." *Bostock*, 140 S. Ct. at 1741.

6.   *Discrimination on the Basis of Atypical Sex Characteristics.* Does discrimination "on the basis of sex" protect individuals born with atypical sex characteristics or traits? The 2016 Rule answered this question affirmatively. It explained that sex discrimination includes discrimination on the basis of "the presence of atypical sex characteristics and intersex traits (i.e., people born with variations in sex characteristics, including in chromosomal, reproductive, or anatomical sex characteristics that do not fit the typical characteristics of binary females or males)." See supra. Such individuals, who may identify as "intersex" individuals or as individuals with "disorders of sex development," can experience discrimination similar to that experienced by transgender people. For example, the rule noted that such individuals have been denied care or coverage for medical care that is classified as "sex-specific" where the treatment classification did not match the person's registered sex; this denial would be prohibited under Section 1557. (See the *Flack* court's discussion of treatment for those with atypical sex characteristics in footnote 21.)

The 2016 Rule failed to address, however, concerns that have been raised about medical interventions performed on intersex infants and children to conform their sexual or reproductive anatomy to a binary sex norm. Although the most extreme examples of these interventions have been rejected by the mainstream medical community and have largely disappeared, medically unnecessary surgical alteration of infants still occurs. See Julie A. Greenberg, Intersexuality and the Law: Why Sex Matters (N.Y.U. Press 2012), arguing that this practice should be considered a form of sex discrimination, and discussing the medical, ethical, and legal issues arising out of this practice.

7.   *Cost.* In both *Erickson* and *Flack*, defendants claimed cost was at least one reason for excluding the challenged treatment. How did the courts address that proffered justification? In *Flack,* the defendants acknowledged that any cost savings from denying care would be nominal. Should this matter?

8.   *Gender Dysphoria as a Disability*. There may be additional protection afforded to a subset of transgender individuals under the ADA, discussed below. "Gender identity disorders" are excluded from ADA protections unless the disorder results from a physical impairment. 42 U.S.C. § 12211(b)(1). However, since enactment of the ADA, "gender dysphoria" was added to the DSM-5 as a formal diagnosis. Some courts have determined that gender dysphoria is a distinct diagnosis that could be considered an ADA-protected condition. See Doe v. Triangle Doughnuts, 472 F. Supp. 3d 115 (E.D. Pa 2020); Doe v. Mass. Dep't of Corr., No. 17-12255-RGS, 2018 WL 2994403 (D. Mass. June 14, 2018); Blatt v. Cabela's Retail, Inc., No. 5:14-CV-04822, 2017 WL 2178123 (E.D. Pa. May 18, 2017). But see Doe v. Northrop Grumman Sys. Corp., 418 F. Supp. 3d 921 (N.D. Ala. 2019).

NOTE: RELIGIOUS OBJECTIONS TO
ANTIDISCRIMINATION REQUIREMENTS

Religious liberty and other conscience protections have long played an important role in shaping health care access. Providers, insurers, and employers have raised religious objections to providing or covering certain kinds of care, and concerns about religious and conscience objections have motivated lawmakers and regulators to craft legislative or regulatory exemptions or accommodations to permit refusals under certain circumstances. As you will see elsewhere in the casebook, religious or conscience objections are perhaps most prominent in areas of health care that touch profound questions about the beginning and end of life. See Chapter 3, Reproduction and Birth, and Chapter 5, Life and Death Decisions. But they also impact health care relating to gender identity, sexual health, pregnancy prevention, and assisted reproduction. These areas implicate and challenge traditional norms or assumptions about sex and gender, heteronormativity, women's sexuality, and parenting.

As we have seen in this part of the chapter, when care or coverage is denied based on such norms, it can implicate sex antidiscrimination protections. But when such denials are linked to one's religious beliefs, they can also implicate religious liberty protections. Disputes arising out of religious objections to sex antidiscrimination protections are certainly not new—such conflicts have long arisen in states that had robust sex antidiscrimination protections before enactment of the ACA. These clashes have become more visible, however, with the new federal focus on gender equity in health seen in the ACA.

The Contraceptive Mandate & Gender Equity

To date, this conflict has been most visible in the context of insurance coverage for contraception. Because the contraceptive mandate was grounded in the preventive health mandate in the ACA, it was an early focus of religious objections to health reform. The Obama administration emphasized the importance of the mandate for ensuring gender equity in health care, but it attempted to resolve the conflict by crafting a regulatory accommodation. This accommodation would allow certain religious entities to refuse to pay for coverage but still ensured access to contraception by requiring such coverage to be provided by a third-party.

Rather than fending off litigation, however, this accommodation, along with subsequent regulatory action by the Trump administration dramatically expanding exemptions to the contraceptive mandate, has led to a flurry of litigation. Obama-era rules have been challenged as providing insufficient protection for religious liberty, and Trump-era rules have been challenged as going too far in the other direction and undermining the ACA's equity mandate. This litigation has resulted in an unusually large number of Supreme Court opinions addressing the conflict in a relatively short amount of time. Given the rather complicated and unique regulatory and jurisprudential path of this conflict in the contraceptive mandate context, a more complete account of these developments and unresolved questions is presented in Chapter 3, Section IV (Contraception).

Antidiscrimination Prohibitions

More recently, litigation has centered on religious objections to federal sex antidiscrimination laws, especially Section 1557 and Title VII, insofar as such laws would be interpreted to require religious entities to cover or provide gender-affirming care for transgender individuals. To date, several Catholic organizations and their members have brought pre-enforcement challenges to prevent this interpretation on religious liberty grounds. Specifically, the plaintiffs have argued such interpretation would violate the federal Religious Freedom Restoration Act of 1993 (RFRA), which prohibits federal action that imposes a substantial burden on religious exercise, unless that action is narrowly tailored to serve a compelling government interest. So far, two courts have agreed, enjoining the challenged implementation against the religious plaintiffs. See Religious Sisters of Mercy v. Azar, 513 F. Supp. 3d 1113 (D.N.D. 2021) (Section 1557 and Title VI); Franciscan Alliance v. Azar, 414 F. Supp. 3d 928 (N.D. Tex. 2019) (Section 1557). On May 10, 2021, the HHS Office for Civil Rights announced its intent to comply with these orders.

In both cases, the courts held that the challenged implementation would impose a substantial burden on the Catholic plaintiffs' exercise of religion, because noncompliance could result in a significant loss of federal funding and/or liability, and compliance would require plaintiffs to violate their religious beliefs as they sincerely understand them. In *Religious Sisters*, for example, the court pointed to the fact that the plaintiffs "explained that their religious beliefs regarding human sexuality and procreation prevent them from facilitating gender transitions through either medical services or insurance coverage." 513 F. Supp. 3d at 1147. Defendants have not disputed the sincerity of these beliefs, and courts have made clear that it is not their domain to question them.

The courts then applied strict scrutiny to determine whether the challenged implementation was the least restrictive means to serve a compelling interest. They rejected the government's interest in ensuring nondiscriminatory access to health care as too broad, noting that the government must articulate how granting specific exemptions for the Catholic plaintiffs will harm the asserted interest in preventing discrimination. They then found that even if the government could show a compelling interest, less restrictive alternatives to accomplishing this interest exist. For example, in *Religious Sisters*, the court suggested other ways the government could ensure access, such as providing subsidies, reimbursements, tax credits or tax deductions to employees, or paying for services at community health centers, public clinics, and hospitals with income-based support. It also pointed to the possibility that the government could more directly "assist transgender individuals in finding and paying for transition procedures available from the growing number of health care providers who offer and specialize in those services." Id. at 22–23. Because the government had not shown these alternatives were unfeasible, the court held, the challenged implementation fails strict scrutiny.

Not all courts have applied strict scrutiny analysis in this way. In a state court case, *Minton v. Dignity Health Care*, 39 Cal.App.5th 1155 (2019), the court rejected a Catholic hospital's arguments that the religious freedom protections in the state constitution should protect its refusal to provide a hysterectomy to a transgender male patient, despite subjecting the state's antidiscrimination law to a strict scrutiny analysis. The patient sued Dignity Health Care for a discriminatory refusal to treat by one of its Catholic hospitals—Mercy Hospital. The complaint alleged discrimination on the basis of his gender identity in violation of California's Unruh Civil Rights Act, Cal. Civ. Code § 51, because his scheduled hysterectomy was canceled the day before when hospital staff learned the plaintiff was transgender. According to the complaint, Mercy routinely provided hysterectomies for other medical diagnoses, such as for chronic pelvic pain and uterine fibroids, but Mercy's president said the procedure would "never" be allowed to treat gender dysphoria.

One of the arguments raised by Dignity Health Care in the suit was that the discrimination claim should be barred by the guarantees of religious freedom in the California Constitution. (RFRA was not raised as a defense because it does not apply to state laws). Dignity claimed that Mercy, as a Catholic hospital, was bound to follow the Ethical and Religious Directives for Catholic Health Care Services (the Directives) issued by the United States Conference of Catholic Bishops. Dignity noted that under the Directives "[a]ll persons served by Catholic health care have the right and duty to protect and preserve their bodily and functional integrity." Id. at 1162. It also pointed to a provision prohibiting the "direct sterilization of either men or women, whether permanent or temporary," but they do permit "procedures that induce sterility . . . when their direct effect is the cure or alleviation of a present and serious pathology and a simpler treatment is not available." Id. Although Mercy Hospital was bound by these directives, the defendant, Dignity Health Care, was ultimately able to arrange for the procedure to be rescheduled at one of its non-Catholic hospitals. Plaintiffs alleged that this only occurred after the plaintiff and the plaintiffs' surgeon used the media and political channels to bring public attention to the refusal. The delay caused the plaintiff anxiety and grief, in part because the timing of the surgery was sensitive—it had to be performed three months before another surgery already scheduled.

The trial court dismissed Minton's claim, but the appellate court reversed. It found that Minton had stated a claim for discrimination, despite the fact that the defendant did ultimately reschedule the procedure at another hospital in response to publicity. Specifically, the court found that "it cannot constitute full equality under [the state antidiscrimination law] to cancel his procedure for a discriminatory purpose, wait to see if his doctor complains, and only then attempt to reschedule the procedure at a different hospital. 'Full and equal' access requires avoiding discrimination, not merely remedying it after it has occurred." Id. at 1165.

The court also rejected the religious freedom defense. It noted that "Minton's claim does not compel Dignity Health to violate its religious principles if it can provide all persons with full and equal medical care at comparable facilities not subject to the same religious restrictions." Id. To the extent there was compulsion, however, the court applied a strict scrutiny analysis to find that any burden on the exercise of religion was justified by California's compelling interest in ensuring full and equal access to medical treatment for all its residents, and that no less restrictive means existed to accomplish that goal. In applying strict scrutiny, it cited to an earlier California Supreme Court case coming to a similar conclusion in a sexual orientation discrimination claim brought by a lesbian patient who, based on religious grounds, was denied access to an artificial insemination procedure by a physician. North Coast Women's Care Medical Group, Inc. v. Superior Court, 44 Cal.4th 1145 (2008). The court in *North Coast* specifically noted that the defendant could comply with the antidiscrimination law and still avoid violating its religious beliefs by ensuring that patients receive "full and equal" access to the medical procedure through a North Coast physician who did not have defendants' religious objections.

Future Questions

There is no question that the antidiscrimination protections in both Section 1557 and Title VII are subject to the strict scrutiny test established in RFRA, but important questions remain about how future courts will resolve religiously based refusals to treat or pay for care. How should the government's interest in preventing the equity and health harms that result from discriminatory refusals be balanced against religious objections? In the *Minton* and *North Coast* cases, the courts emphasized the importance of having a policy to ensure full and equal access to care despite religious objections, noting that access could be achieved through a referral or an arrangement with a qualified and willing provider. Does this approach achieve the right balance of competing interests? What if the provider, employer, or insurer has religious objections to even making such referrals or arrangements, on the grounds that this would be a facilitation of sin? (For example, religious objections based on facilitation have been used to challenge the special accommodation process for the contraceptive mandate created under the Obama administration. See Chapter 3.IV.C.)

In terms of the least restrictive alternative test, is there any limit to the list of alternatives that a court can consider as evidence that a challenged law is not narrowly tailored? The list of alternative possibilities suggested by the courts in *Religious Sisters* and *Franciscan Alliance* was quite long. Should the fact that the government can theoretically create a different or new program to fill in gaps created by religious refusal be enough to prevent the government from prohibiting discriminatory actions by federal funding recipients as part of an existing comprehensive health care program? Or does the fact that the government was able and willing to create a special accommodation process for religious objections to insurance coverage of contraception support the result in *Religious Sisters* and *Franciscan Alliance*—if the government could do it for the prescription contraceptive mandate, why not for gender-affirming coverage? But consider whether it is as easy for the federal government to create an accommodation process that would prevent disruption in health care *treatment* due to provider denials of care, as opposed to *insurance denials*, especially in light of the local character of health care delivery organization and regulation.

Recent Supreme Court decisions on RFRA in the contraceptive mandate context provide important clues as to how such conflicts may be resolved. See Chapter 3 Section IV.C (The Clash Between Government Mandates and Religious or Moral Beliefs). In addition, the appointments of Justice Kavanaugh to replace Justice Kennedy, and Justice Barrett to replace Justice Ginsburg, have been widely perceived as cementing a shift in the Court that is increasingly protective of religious liberty claims, even in the face of compelling health and public health interests. Of course, the law merely establishes a legal floor in terms of one's obligations. Many professional organizations have established ethical guidelines for how their members should handle requests for care that violate their religious beliefs, encouraging them to take actions to minimize any disruption in care. See Chapter 3 Section IV.C (Note: The Ethics of Conscientious Refusals).

C. DISABILITY

People with disabilities have experienced a history of unequal treatment in medicine and health care, including denial of needed services; inadequate or inappropriate treatment; unnecessary institutionalization, often under unacceptable conditions; involuntary sterilization; and other harms. The legislative history of the ADA documents widespread discrimination against individuals with disabilities across society, including in medicine and health care. The treatment of individuals with disabilities is also discussed in later chapters addressing sterilization and reproductive care (Chapter 3), organ transplantation and the determination of death (Chapter 4), life and death decisions (Chapter 5), medically assisted dying (Chapter 6), and research (Chapter 7).

Despite passage of the ADA in 1990, inequities and barriers to care for individuals with disabilities persist. Fifteen years after ADA passage, in 2005, the Surgeon General's Call to Action to Improve the Health and Wellness of Persons with Disabilities highlighted inequities such as poorer reported health status and specific risk of secondary conditions, as well as barriers to health and health care including lack of transportation, provider attitudes and misconceptions, and inaccessible facilities and services. A few years later, the National Council on Disability published its evaluation of decades of research on health inequities and barriers experienced by people with disabilities. Its 2009 report, The Current State of Health Care for People with Disabilities, found that people with disabilities report poorer health status than people without disabilities, use fewer preventative services despite using health care services at a significantly higher rate overall, and face more problems accessing health care than other groups. The report identified lack of training on disability competence issues and on legal requirements for health care practitioners as among the most significant barriers. Reports from governmental agencies, advocacy organizations, academic researchers, and others continue to document inequities in health and health care. See, e.g., Gloria Krahn, Persons with Disabilities as an Unrecognized Health Disparity Population, 105 Am. J. Pub. Health S198 (2015); Jana J. Peterson-Besse et al., Clinical Preventative Service Use Disparities Among Subgroups of People with Disabilities: A Scoping Review, 7 Disability & Health J. 373 (2014).

While there are many reasons for these inequities, discrimination and unequal treatment play a role. For decades, research has shown that physicians hold negative views of people with disabilities and fail to fully appreciate the value and quality of life with a disability. More recently, a series of reports from the National Council on Disability have explored how persistent devaluation of the lives of people with disabilities by the medical community, legislators, researchers, and others have perpetuated inequities in health and access to health care, including life-saving care.

The ADA was enacted to end a history of discrimination based on disability and to ensure equal opportunities, integration in the community, and full participation across all areas of American life. Title I of the ADA (42 U.S.C. § 12111) applies to employment, including employer-sponsored health and wellness plans; Title II (42 U.S.C. § 12131) applies to state and local government services, programs, and activities, including state and local public hospitals and health programs such as Medicaid; Title III (42 U.S.C. § 12181) applies to public accommodations, including private health care facilities and offices open to the public; and Title IV (42 U.S.C. § 12201) includes miscellaneous provisions, including important provisions regarding insurance.

The ADA expanded the protections of an earlier law, Section 504 of the Rehabilitation Act of 1973 (29 U.S.C. § 749), which prohibits discrimination on the basis of disability only in federal employment and in programs and activities that are funded by federal agencies. The ADA and Section 504 are quite similar in most respects, and courts have used cases under the Rehabilitation Act, which applies only to programs and services receiving federal funding, to assist in interpreting the more widely applicable ADA. There are some significant differences, however, that are highlighted in the notes below. As also discussed below, Section 1557 of the ACA broadens the reach of Section 504 significantly.

The HHS Office for Civil Rights (OCR) is responsible for enforcing Title II of the ADA, Section 504 of the Rehabilitation Act, and Section 1557 of the ACA. The U.S. Department of Justice (DOJ) is also charged with enforcing Section 504, Title II and III of the ADA, and Section 1557 of the ACA. The DOJ and OCR have broad authority to investigate, mediate, litigate, and settle individual and class-based claims under these laws. Private individuals and groups can also bring actions under these laws. Public and private ADA enforcement actions and settlement agreements have addressed a wide range of barriers to health care services for individuals with disabilities. However, many experts claim that the ADA and Section 504 are underenforced, and research also shows lack of knowledge of, and noncompliance with, disability nondiscrimination laws in health care settings. See, e.g., Nicole D. Agaronnik et al., Knowledge of Practicing Physicians about Their Legal Obligations When Caring for Patients with Disability, 38(4) Health Aff. 545 (April 1, 2019).

The following materials explore the impact of the ADA, Section 504, and Section 1557 on health care for individuals with disabilities and the promise and challenge of using these laws to address different forms of discrimination based on disability in health care services, programs and activities, and in health insurance.

BRAGDON V. ABBOTT
Supreme Court of the United States, 1998.
524 U.S. 624.

KENNEDY, J., delivered the opinion of the Court, in which STEVENS, SOUTER, GINSBERG, and BREYER, JJ., joined. STEVENS, J., filed a concurring opinion. REHNQUIST, C.J., filed an opinion concurring in the judgment in part and dissenting in part, in which SCALIA and THOMAS, JJ., joined, and in Part II of which O'CONNOR, J., joined. O'CONNOR, J., filed an opinion concurring in the judgment in part and dissenting in part.

. . . We granted certiorari to review . . . whether the Court of Appeals, in affirming a grant of summary judgment, cited sufficient material in the record to determine, as a matter of law, that respondent's infection with HIV posed no direct threat to the health and safety of her treating dentist.

I

Respondent Sidney Abbott has been infected with HIV since 1986. When the incidents we recite occurred, her infection had not manifested its most serious symptoms. On September 16, 1994, she went to the office of petitioner Randon Bragdon in Bangor, Maine, for a dental appointment. She disclosed her HIV infection on the patient registration form. Petitioner completed a dental examination, discovered a cavity, and informed respondent of his policy against filling cavities of HIV-infected patients. He offered to perform the work at a hospital with no added fee for his services, though respondent would be responsible for the cost of using the hospital's facilities. Respondent declined.

* * *

. . . Notwithstanding the protection given respondent by the ADA's definition of disability, petitioner could have refused to treat her if her infectious condition "posed a direct threat to the health or safety of others."[ ] The ADA defines a direct threat to be "a significant risk to the health or safety of others that cannot be eliminated by a modification of policies, practices, procedures, or by the provision of auxiliary aids or services."[ ] . . .

The ADA's direct threat provision stems from the recognition in School Bd. of Nassau Cty. v. Arline[ ] of the importance of prohibiting discrimination against individuals with disabilities while protecting others from significant health and safety risks, resulting, for instance, from a contagious disease. In *Arline,* the Court reconciled these objectives by construing the Rehabilitation Act not to require the hiring of a person who posed "a significant risk of communicating an infectious disease to others."[ ] . . . [The ADA's] direct threat provision codifies *Arline.* Because few, if any, activities in life are risk free, *Arline* and the ADA do not ask whether a risk exists, but whether it is significant.[ ]

The existence, or nonexistence, of a significant risk must be determined from the standpoint of the person who refuses the treatment or accommodation, and the risk assessment must be based on medical or other objective evidence.[ ] As a health care professional, petitioner had the duty to assess the risk of infection based on the objective, scientific information available to him and others in his profession. His belief that a significant risk existed, even if maintained in good faith, would not relieve him from liability. To use the words of the question presented, petitioner receives no special deference simply because he is a health care professional. It is true that *Arline* reserved "the question whether courts should also defer to the reasonable medical judgments of private physicians on which an employer has relied."[ ] At most, this statement reserved the possibility that employers could consult with individual physicians as objective third-party experts. It did not suggest that an individual physician's state of mind could excuse discrimination without regard to the objective reasonableness of his actions.

. . . In assessing the reasonableness of petitioner's actions, the views of public health authorities, such as the U.S. Public Health Service, CDC, and the National Institutes of Health, are of special weight and authority.[ ] The views of these organizations are not conclusive, however. A health care professional who disagrees with the prevailing medical consensus may refute it by citing a credible scientific basis for deviating from the accepted norm.[ ]

* * *

[An] illustration of a correct application of the objective standard is the Court of Appeals' refusal to give weight to the petitioner's offer to treat respondent in a hospital.[ ] Petitioner testified that he believed hospitals had safety measures, such as air filtration, ultraviolet lights, and respirators, which would reduce the risk of HIV transmission.[ ] Petitioner made no showing, however, that any area hospital had these safeguards or even that he had hospital privileges.[ ] His expert also admitted the lack of any scientific basis for the conclusion that these measures would lower the risk of transmission.[ ] Petitioner failed to present any objective, medical evidence showing that treating respondent in a hospital would be safer or more efficient in preventing HIV transmission than treatment in a well-equipped dental office.

We are concerned, however, that the Court of Appeals might have placed mistaken reliance upon two other sources. In ruling no triable issue of fact existed on this point, the Court of Appeals relied on the CDC Dentistry Guidelines and the 1991 American Dental Association Policy on HIV.[ ] This evidence is not definitive. ... [T]he CDC Guidelines recommended certain universal precautions which, in CDC's view, "should reduce the risk of disease transmission in the dental environment."[ ] The Court of Appeals determined that, "[w]hile the guidelines do not state explicitly that no further risk-reduction measures are desirable or that routine dental care for HIV-positive individuals is safe, those two conclusions seem to be implicit in the guidelines' detailed delineation of procedures for office treatment of HIV-positive patients."[ ] In our view, the Guidelines do not necessarily contain implicit assumptions conclusive of the point to be decided. The Guidelines set out CDC's recommendation that the universal precautions are the best way to combat the risk of HIV transmission. They do not assess the level of risk.

Nor can we be certain, on this record, whether the 1991 American Dental Association Policy on HIV carries the weight the Court of Appeals attributed to it. The Policy does provide some evidence of the medical community's objective assessment of the risks posed by treating people infected with HIV in dental offices. It indicates:

"Current scientific and epidemiologic evidence indicates that there is little risk of transmission of infectious diseases through dental treatment if recommended infection control procedures are routinely followed. Patients with HIV infection may be safely treated in private dental offices when appropriate infection control procedures are employed. Such infection control procedures provide protection both for patients and dental personnel."[ ]

We note, however, that the Association is a professional organization, which, although a respected source of information on the dental profession, is not a public health authority. It is not clear the extent to which the Policy was based on the Association's assessment of dentists' ethical and professional duties in addition to its scientific assessment of the risk to which the ADA refers. Efforts to clarify dentists' ethical obligations and to encourage dentists to treat patients with HIV infection with compassion may be commendable, but the question under the statute is one of statistical likelihood, not professional responsibility. Without more information on the manner in which the American Dental Association formulated this Policy, we are unable to determine the Policy's value in evaluating whether petitioner's assessment of the risks was reasonable as a matter of law.

* * *

We acknowledge the presence of other evidence in the record before the Court of Appeals which, subject to further arguments and examination, might support affirmance of the trial court's ruling. For instance, the record contains substantial testimony from numerous health experts indicating that it is safe to treat patients infected with HIV in dental offices.[ ] We are unable to determine the import of this evidence, however. The record does not disclose whether the expert testimony submitted by respondent turned on evidence available in September 1994.[ ]

There are reasons to doubt whether petitioner advanced evidence sufficient to raise a triable issue of fact on the significance of the risk. Petitioner relied on two principal points: First, he asserted that the use of high-speed drills and surface cooling with water created a risk of airborne HIV transmission. The study on which petitioner relied was inconclusive, however, determining only that "further work is required to determine whether such a risk exists."[ ] Petitioner's expert witness conceded, moreover, that no evidence suggested the spray could transmit HIV. His opinion on airborne risk was based on the absence of contrary evidence, not on positive data. Scientific evidence and expert testimony must have a traceable, analytical basis in objective fact before it may be considered on summary judgment.[ ]

[P]etitioner argues that, as of September 1994, CDC had identified seven dental workers with possible occupational transmission of HIV.[ ] These dental workers were exposed to HIV in the course of their employment, but CDC could not determine whether HIV infection had resulted.[ ] It is now known that CDC could not ascertain whether the seven dental workers contracted the disease because they did not present themselves for HIV testing at an appropriate time after their initial exposure.[ ] It is not clear on this record, however, whether this information was available to petitioner in September 1994. If not, the seven cases might have provided some, albeit not necessarily sufficient, support for petitioner's position. Standing alone, we doubt it would meet the objective, scientific basis for finding a significant risk to the petitioner.

* * *

We conclude the proper course is to give the Court of Appeals the opportunity to determine whether our analysis of some of the studies cited by the parties would change its conclusion that petitioner presented neither objective evidence nor a triable issue of fact on the question of risk.

JUSTICE STEVENS, with whom JUSTICE BREYER joins, concurring.

. . . I do not believe petitioner has sustained his burden of adducing evidence sufficient to raise a triable issue of fact on the significance of the risk posed by treating respondent in his office. . . . I join the opinion even though I would prefer an outright affirmance.[ ]

CHIEF JUSTICE REHNQUIST, with whom JUSTICE SCALIA and JUSTICE THOMAS join, and with whom JUSTICE O'CONNOR joins as to Part II, concurring in the judgment in part and dissenting in part.

* * *

II

I agree with the Court that "the existence, or nonexistence, of a significant risk must be determined from the standpoint of the person who refuses the treatment or accommodation," as of the time that the decision refusing treatment is made.[ ] I disagree with the Court, however, that "in assessing the reasonableness of petitioner's actions, the views of public health authorities . . . are of special weight and authority."[ ] Those views are, of course, entitled to a presumption of validity when the actions of those authorities themselves are challenged in court, and even in disputes between private parties where Congress has committed that dispute to adjudication by a public health authority. But in litigation between private parties originating in the federal courts, I am aware of no provision of law or judicial practice that would require or permit courts to give some scientific views more credence than others simply because they have been endorsed by a politically appointed public health authority (such as the Surgeon General). In litigation of this latter sort, which is what we face here, the credentials of the scientists employed by the public health authority, and the soundness of their studies, must stand on their own. The Court cites no authority for its limitation upon the courts' truth-finding function, except the statement in School Bd. of Nassau Cty. v. Arline,[ ] that in making findings regarding the risk of contagion under the Rehabilitation Act, "courts normally should defer to the reasonable medical judgments of public health officials." But there is appended to that dictum the following footnote, which makes it very clear that the Court was urging respect for *medical* judgment, and not necessarily respect for "official" medical judgment over "private" medical judgment: "This case does not present, and we do not address, the question whether courts should also defer to the reasonable medical judgments of private physicians on which an employer has relied."[ ]

Applying these principles here, it is clear to me that petitioner has presented more than enough evidence to avoid summary judgment on the "direct threat" question . . . . Given the "severity of the risk" involved here, i.e., near certain death, and the fact that no public health authority had outlined a protocol for *eliminating* this risk in the context of routine dental treatment, it seems likely that petitioner can establish that it was objectively reasonable for him to conclude that treating respondent in his office posed a "direct threat" to his safety.

* * *

NOTES AND QUESTIONS

1. *Refusals to Treat.* The ADA and Section 504 prohibit refusals to treat based on disability. However, empirical studies demonstrate that a significant number of health care professionals refuse to provide care for persons with HIV/AIDS. See Brad Sears et al., HIV Discrimination in Dental Care: Results of a Testing Study in Los Angeles County, 45 Loy. L.A. L. Rev. 909 (2012), including results of studies from 2003–2008 demonstrating that 46 percent of skilled nursing facilities, 55 percent of OB/GYNs, and 26 percent of plastic surgeons in Los Angeles County refused to treat persons with HIV. See also Ronda Goldfein, From the Streets of Philadelphia: The AIDS Law Project of Pennsylvania's How-To Primer on Mitigating Health Disparities, 82 Temp. L. Rev. 1205 (2010). The DOJ and OCR have negotiated agreements with providers, some quite recently, concerning refusal to treat patients with HIV. The agreements are available on their respective websites. The agreements show that refusals to treat based on disability are not limited to persons with HIV/AIDS. See also *Lyons* in Section II; Chapter 4, discussing refusals to treat based on disability in the context of organ transplantation.

2. *Denials of Care During the COVID-19 Pandemic.* During the COVID-19 pandemic, concerns arose regarding scarce medical resource allocation policies that explicitly or implicitly disadvantaged people with disabilities and others. The death of Michael Hickson, a forty-six-year-old Black man with a disability, exemplified these concerns for many. According to Mr. Hickson's family, his death resulted from a hospital's refusal to provide him with life-saving care for COVID-19 and its withholding of nutrition and hydration. According to Melissa Hickson, Mr. Hickson's wife, she recorded a conversation with a doctor who told her that the hospital would not provide her husband with medical treatment because of his low quality of life due to preexisting quadriplegia and head injury. See Ariana Eunjung Cha, Quadriplegic Man's Death from COVID-19 Spotlights Questions of Disability, Race and Family, The Wash. Post, July 5, 2020. See discussion of medical futility in Chapter 5.

In March 2020, the OCR issued a bulletin affirming that federal disability nondiscrimination laws, like other civil rights laws, remained in effect during the pandemic. It provided that people with disabilities should not be denied medical care on the basis of "stereotypes, assessments of quality of life," or "judgments about a person's relative 'worth' based on the presence or absence of disabilities or age." Instead, decisions concerning whether an individual is a candidate for treatment should be "based on an individualized assessment of the patient based on the best available objective medical evidence." Dep't Health & Hum. Serv.'s, Bulletin: Civil Rights, HIPAA, and the Coronavirus Disease 2019 (COVID-19) (Mar. 28, 2020). The bulletin also emphasized the requirements of Section 1557 and the Rehabilitation Act, discussed later in this section, including the obligation to ensure effective communication with individuals who are deaf, hard of hearing, or blind, or who have low vision or speech disabilities, and to make reasonable modifications to address the needs of individuals with disabilities. See Samuel R. Bagenstos, Who Gets the Ventilator? Disability Discrimination in COVID-19 Medical-Rationing Protocols, 130 Yale L. J. F. 1 (2020).

3. *Assessment of "Direct Threat." Bragdon* established that the analysis of "direct threat" as a defense to a claim under the ADA must rely on an individualized assessment of scientific evidence rather than stereotyping or misconceptions. On remand, the First Circuit upheld the District Court's grant of summary judgment in favor of the plaintiff:

> The [American Dental] Association formulates scientific and ethical policies by separate procedures, drawing on different member groups and different staff complements. The Association's Council on Scientific Affairs, comprised of 17 dentists (most of whom hold advanced dentistry degrees), together with a staff of over 20 professional experts and consultants, drafted the Policy at issue here. By contrast, ethical policies are drafted by the Council on Ethics, a wholly separate body. Although the Association's House of Delegates must approve policies drafted by either council, we think that the origins of the Policy satisfy any doubts regarding its scientific foundation.

> For these reasons, we are confident that we appropriately relied on the Guidelines and the Policy.... Thus, we again conclude, after due reevaluation, that Ms. Abbott served a properly documented motion for summary judgment.

> We next reconsider whether Dr. Bragdon offered sufficient proof of direct threat to create a genuine issue of material fact and thus avoid the entry of summary judgment .... The Supreme Court suggested that one such piece of evidence—the seven cases that the CDC considered "possible" HIV patient-to-dental worker transmissions—should be reexamined. Since an objective standard pertains here, the existence of the list of seven "possible" cases does not create a genuine issue of material fact as to direct threat.... Each piece of evidence to which [defendant directs] us is still "too speculative or too tangential (or, in some instances, both) to create a genuine issue of material fact."

Abbott v. Bragdon, 163 F.3d 87 (1st Cir. 1998), cert. denied, 526 U.S. 1131 (1999).

4. *Risk of Transmission.* According to the CDC there has been only one confirmed case of occupational transmission of HIV to a health care worker reported since 1999. CDC, Occupational HIV Transmission and Prevention Among Health Care Workers (2015). The CDC recommends that "universal precautions" (using barriers such as gloves; handwashing; and design and use of sharps to reduce accidental needle sticks) against transmission of infectious diseases (including hepatitis, which is much more prevalent than HIV). For HIV in particular, the CDC recommends a specific protocol of post-exposure treatment following any exposure to prevent seroconversion in the exposed health care worker.

5. *Definition of "Disability."* The ADA and Section 504 protect individuals who meet the statutory definition of "disability." An individual with a disability is one who has "a physical or mental impairment that substantially limits one or more major life activities of such individual; a record of such impairment; or [is] regarded as having such an impairment." 42 U.S.C. §§ 12101–12213. Disabilities are diverse, and can be physical, sensory, cognitive, intellectual, or developmental. Mental health conditions, substance use disorder, and chronic illness can also be disabilities. In *Bragdon*, a deeply divided Court concluded that asymptomatic HIV could be considered a disability based on the specific facts of the case. In a series of cases after *Bragdon*, however, the Court established a very narrow interpretation of the ADA statutory definition of disability. Congress amended the ADA in 2008 to clarify that the statutory definition of disability should be construed in favor of broad coverage of individuals.

HOWE V. HULL
U.S. District Court, Northern District, Ohio, 1994.
874 F. Supp.779.

JOHN W. POTTER, SENIOR DISTRICT JUDGE.

* * *

Plaintiff brought suit in the current action alleging that on April 17, 1992, defendants refused to provide Charon medical treatment because he was infected with HIV. Plaintiff claims that defendants' actions violate the Americans with Disabilities Act (ADA) [and] the Federal Rehabilitation Act of 1973 (FRA).... The defendants vehemently dispute these claims and allegations and have moved for summary judgment on all of plaintiff's claims.

* * *

On April 17, 1992, Charon and plaintiff Howe were travelling through Ohio, on their way to vacation in Wisconsin. Charon was HIV positive. That morning Charon took a floxin tablet for the first time. Floxin is a prescription antibiotic drug. Within two hours of taking the drug, Charon began experiencing fever, headache, nausea, joint pain, and redness of the skin.

Due to Charon's condition, Charon and plaintiff checked into a motel and, after consulting with Charon's treating physician in Maine, sought medical care at the emergency room of Fremont Memorial Hospital. Charon was examined by the emergency room physician on duty, Dr. Mark Reardon. There is some dispute over what Dr. Reardon's initial diagnosis of Charon's condition was.

Dr. Reardon testified that Charon suffered from a severe drug reaction, and that it was his diagnosis that this reaction was probably Toxic Epidermal Necrolysis (TEN).[2] This diagnosis was also recorded in Charon's medical records. Dr. Reardon also testified regarding Charon's condition that "possibly it was an early stage of toxic epidermal necrolysis, although I had never seen one." Dr. Reardon had no prior experience with TEN, other than what he had read in medical school.

Plaintiff's medical expert Calabrese, however, testified that, after reviewing the medical records and Reardon's deposition, while Charon did appear to be suffering from a severe allergic drug reaction, Calabrese "did not believe that [TEN] was the likely or even probable diagnosis. . . . "

Prior to Charon's eventual transfer to the Medical College of Ohio, Dr. Reardon called Dr. Lynn at MCO and asked Lynn if he would accept the transfer of Charon. Dr. Lynn testified that at no time did Dr. Reardon mention that plaintiff had been diagnosed with the extremely rare and deadly TEN. Dr. Reardon also did not inform the ambulance emergency medical technicians that plaintiff was suffering from TEN.

Dr. Reardon determined that Charon "definitely needed to be admitted" to Memorial Hospital. Since Charon was from out of town, procedure required that Charon be admitted to the on-call physician, Dr. Hull. Dr. Reardon spoke with Dr. Hull on the telephone and informed Dr. Hull that he wanted to admit Charon, who was HIV-positive and suffering from a non-AIDS related severe drug reaction.

While Dr. Reardon and Dr. Hull discussed Charon's situation, the primary area of their discussion appears to have been whether Charon's condition had advanced from HIV to full-blown AIDS. Dr. Hull inquired neither into Charon's physical condition nor vital signs, nor did he ask Dr. Reardon about the possibility of TEN. During this conversation, it is undisputed that Dr. Hull told Dr. Reardon that "if you get an AIDS patient in the hospital, you will never get him out," and directed that plaintiff be sent to the "AIDS program" at MCO. When Dr. Hull arrived at the hospital

---

[2]   TEN is a very serious, very rare, and often lethal skin condition that causes an individual's skin to slough off the body.

after Dr. Reardon's shift but prior to Charon's transfer, he did not attempt to examine or meet with Charon.

* * *

Charon was transferred to the Medical College of Ohio some time after 8:45 P.M. on April 17. After his conversation with Dr. Hull and prior to the transfer, Dr. Reardon told Charon and plaintiff that "I'm sure you've dealt with this before. . . . " Howe asked, "What's that, discrimination?" Dr. Reardon replied, "You have to understand, this is a small community, and the admitting doctor does not feel comfortable admitting [Charon]."

* * *

Charon was admitted and treated at the Medical College of Ohio (MCO). Despite the TEN diagnosis, Charon was not diagnosed by MCO personnel as having TEN and, in fact, was never examined by a dermatologist. After several days, Charon recovered from the allergic drug reaction and was released from MCO.

* * *

Before examining the merits of defendants' contentions, the Court must look at and compare the applicable parameters of the ADA and FRA. There are three basic criteria plaintiff must meet in order to establish a prima facie case of discrimination under the ADA:

a)   the plaintiff has a disability;

b)   the defendants discriminated against the plaintiff; and

c)   the discrimination was based on the disability.

42 U.S.C. § 12182(a); 42 U.S.C. § 12182(b). The discrimination can take the form of the denial of the opportunity to receive medical treatment, segregation unnecessary for the provision of effective medical treatment, unnecessary screening or eligibility requirements for treatment, or provision of unequal medical benefits based upon the disability. [ ] A defendant can avoid liability by establishing that it was unable to provide the medical care that a patient required. [ ]

Similarly, to establish a prima facie case under the FRA the plaintiff must show

a)   the plaintiff has a disability;

b)   plaintiff was otherwise qualified to participate in the program;

c)   defendants discriminated against plaintiff solely on the basis of the disability; and

d)   the program received federal funding.

29 U.S.C. § 794(a).

[A] reasonable jury could conclude that the TEN diagnosis was a pretext and that Charon was denied treatment solely because of his disability. Further, there is no evidence to support the conclusion that Memorial Hospital was unable to treat a severe allergic drug reaction. In fact, the evidence indicates that Dr. Reardon initially planned to admit Charon for treatment. Therefore, Charon was "otherwise qualified" for treatment within the meaning of the FRA. . . .

The FRA states that "no otherwise qualified individual with a disability . . . shall, solely by reason of his or her disability . . . be subjected to discrimination. . . . ". 29 U.S.C. § 794(a). The equivalent portion of the ADA reads "No individual shall be discriminated against on the basis of disability. . . . " 42 U.S.C. § 12182(a). It is abundantly clear that the exclusion of the "solely by reason of . . . disability" language was a purposeful act by Congress and not a drafting error or oversight . . . .

The inquiry under the ADA, then, is whether the defendant, despite the articulated reasons for the transfer, improperly considered Charon's HIV status. More explicitly, was Charon transferred for the treatment of a non-AIDS related drug reaction because defendant unjustifiably did not wish to care for an HIV-positive patient? Viewing the evidence in the light most favorable to the plaintiff, the Court finds plaintiff has presented sufficient evidence to preclude a grant of summary judgment on these claims. Defendant Memorial Hospital's motion for summary judgment on the plaintiff's ADA and FRA claims will be denied.

NOTES AND QUESTIONS

1.   *Disability and Health Care Decision-Making. Howe* correctly notes that the ADA and Section 504 differ significantly on what the plaintiff must prove as the reason for the defendant's action. It may be difficult in a particular case to prove the reason for the refusal of treatment or other medical decision, much less meet the requirement that the disability be the "sole" reason for denial of adequate treatment. Recall the facts of *Lyons* in Section II above. Early cases brought under Section 504 suggested that disability nondiscrimination laws could not easily be applied to medical decision-making, or could only be applied where the underlying disability is *unrelated to* the medical decision at issue. See, e.g., U.S. v. Univ. Hosp., State Univ. at Stony Brook, 729 F.2d 144 (2d Cir. 1984). However, the ADA, and Supreme Court cases such as *Bragdon,* make clear that these laws do apply to medical services, programs, and activities. See also *Choate* and *Olmstead*, below.

How much should courts defer to a doctor's judgment as to the best course of treatment for a disabled patient in the context of a discriminatory denial of treatment claim? How would you go about proving or defending against a claim that a medical judgment defense is a subterfuge? What does the Court's analysis in *Bragdon* suggest? Does the requirement that individuals with disabilities be assessed individually based on the best available medical or scientific evidence, rather than based on stereotypes or perceived quality or value of life, help with this inquiry? See E. Haavi Morreim, Futilitarianism, Exoticare, and Coerced Altruism: The ADA Meets Its Limits, 25 Seton Hall L. Rev. 883 (1995); Mary A. Crossley, Of Diagnoses and Discrimination: Discriminatory Nontreatment of Infants with HIV Infection, 93 Colum. L. Rev. 1581 (1993).

2. *ACA Section 1557 Adopts Existing Protections.* The ADA and Section 504 address a range of barriers in health care settings. These laws require: physical access to services and facilities, including the provision of accessible spaces and the removal of barriers; effective communication, including auxiliary aids and services such as the provision of sign language interpreters or materials in alternative formats; and reasonable modification of policies, practices, and procedures when necessary to accommodate individual needs. 28 C.F.R. 35.130(b)(7); 42 U.S.C. 12182(b)(2)(A), 12183(a). Section 1557 also prohibits discrimination on the basis of disability, and the final rule incorporates many of the ADA's regulatory requirements, including those for reasonable modifications, effective communication, and readily accessible buildings and information technology. 45 C.F.R. § 92.202–.205.

3. *Effective Communication.* Provision of sign language interpreters for deaf patients has been an active issue in ADA and Section 504 litigation. See e.g., Loeffler v. Staten Island Univ. Hosp., 582 F.3d 268 (2d Cir. 2009), holding that an action for damages under Section 504 could proceed where facts alleged by plaintiff husband (the patient) and wife, both deaf, could support a conclusion that the hospital acted with deliberate indifference in refusing repeated requests for a sign language interpreter for the patient, forcing his minor children to interpret for him. The DOJ has brought a number of suits against providers for failing to provide sign language interpreters, and settlement agreements are available on the DOJ website.

The 2016 Rule implementing Section 1557 adopted the requirements applicable to Title II of the ADA, which require that covered entities ensure effective communication with people with disabilities and give "primary consideration" to that person's choice of auxiliary aid or service (such as a qualified interpreter, an assistive technology device, or materials in alternative formats). 45 C.F.R. § 92.202. The 2020 Rule requested public comment on communication requirements, but it did not adopt significant changes in that area.

4. *Accessible Facilities and Equipment.* Access to medical care for persons with disabilities has been greatly compromised by providers' failure to choose medical diagnostic equipment, such as examination tables, dental and eye examination chairs, weight scales, and mammography equipment that is accessible for persons with mobility limitations. For a thorough discussion, see Elizabeth Pendo, Reducing Disparities Through Health Reform: Disability and Accessible Medical Equipment, 2010 Utah L. Rev. 1057 (2010), which includes a review of litigation approaches and provisions of the ACA requiring promulgation of standards for accessible medical equipment. Class action

litigation has targeted health care facilities for violation of the ADA regarding physical access. See, e.g., Settlement Agreement in United Spinal Association et al. v. Beth Israel Medical Center et al., No. 13-cv-5131 (S.D.N.Y., filed Nov. 17, 2017).

Regulations implementing Section 1557 incorporate existing requirements regarding readily accessible buildings, and, for most health care facilities, they adopt the 2010 ADA Standards for Accessible Design. 45 C.F.R. § 92.203. In accordance with other provisions of the ACA, the U.S. Access Board, an independent federal agency responsible for developing accessibility guidelines and standards under disability nondiscrimination laws, issued a final rule that delineates minimum technical criteria for the accessibility of examination tables, examination chairs, weight scales, mammography equipment, and other diagnostic imaging equipment. Architectural and Transportation Barriers Compliance Board, Standards for Accessible Medical Diagnostic Equipment. 82 Fed. Reg. 2810 (Jan. 9, 2017) (codified at 36 C.F.R. Part 1195). The standards become mandatory when adopted by a federal agency.

5. *Health Care Decision-Making for Newborns with Disabilities.* Questions about medical care for newborns with disabilities revolve around whether denying treatment could constitute discrimination based on disability, the role of parental decision-making in this analysis, and whether treatment is medically appropriate. For a discussion of medical decision-making for children, see Chapter 5.

ALEXANDER V. CHOATE

Supreme Court of the United States, 1985.
469 U.S. 287.

JUSTICE MARSHALL delivered the opinion of the Court.

In 1980, Tennessee proposed reducing the number of annual days of inpatient hospital care covered by its state Medicaid program. The question presented is whether the effect upon the handicapped that this reduction will have is cognizable under § 504 of the Rehabilitation Act of 1973 or its implementing regulations. We hold that it is not.

I

Faced in 1980–1981 with projected state Medicaid costs of $42 million more than the State's Medicaid budget of $388 million, the directors of the Tennessee Medicaid program decided to institute a variety of cost-saving measures. Among these changes was a reduction from 20 to 14 in the number of inpatient hospital days per fiscal year that Tennessee Medicaid would pay hospitals on behalf of a Medicaid recipient. Before the new measures took effect, respondents, Tennessee Medicaid recipients, brought a class action for declaratory and injunctive relief in which they alleged, *inter alia,* that the proposed 14-day limitation on inpatient coverage would have a discriminatory effect on the handicapped. Statistical evidence, which petitioners do not dispute, indicated that in the 1979–1980 fiscal year, 27.4% of all handicapped users of hospital services who received Medicaid required more than 14 days of care, while only 7.8% of nonhandicapped users required more than 14 days of inpatient care.

Based on this evidence, respondents asserted that the reduction would violate § 504 of the Rehabilitation Act of 1973, 87 Stat. 394, as amended, 29 U.S.C. § 794, and its implementing regulations. Section 504 provides:

> "No otherwise qualified handicapped individual . . . shall, solely by reason of his handicap, be excluded from the participation in, be denied the benefits of, or be subjected to discrimination under any program or activity receiving Federal financial assistance . . . ." 29 U.S.C. § 794.

Respondents' position was twofold. First, they argued that the change from 20 to 14 days of coverage would have a disproportionate effect on the handicapped and hence was discriminatory.[3] The second, and major, thrust of respondents' attack was directed at the use of *any* annual limitation on the number of inpatient days covered, for respondents acknowledged that, given the special needs of the handicapped for medical care, any such limitation was likely to disadvantage the handicapped disproportionately. Respondents noted, however, that federal law does not require States to impose any annual durational limitation on inpatient coverage, and that the Medicaid programs of only 10 States impose such restrictions.[4] Respondents therefore suggested that Tennessee follow these other States and do away with any limitation on the number of annual inpatient days covered. Instead, argued respondents, the State could limit the number of days of hospital coverage on a per-stay basis, with the number of covered days to vary depending on the recipient's illness (for example, fixing the number of days covered for an appendectomy); the period to be covered for each illness could then be set at a level that would keep Tennessee's Medicaid program as a whole within its budget. The State's refusal to adopt this plan was said to result in the imposition of gratuitous costs on the handicapped and thus to constitute discrimination under § 504.

A divided panel of the Court of Appeals for the Sixth Circuit held that respondents had indeed established a prima facie case of a § 504 violation. *Jennings v. Alexander,* 715 F.2d 1036 (1983). The majority apparently concluded that any action by a federal grantee that disparately affects the handicapped states a cause of action under § 504 and its implementing regulations. Because both the 14-day rule and any annual limitation on

---

[3] The evidence indicated that, if 19 days of coverage were provided, 16.9% of the handicapped, as compared to 4.2% of the nonhandicapped, would not have their needs for inpatient care met.

[4] As of 1980 the average ceiling in those States was 37.6 days. Six States also limit the number of reimbursable days per admission, per spell of illness, or per benefit period. See App. B to Brief for United States as *Amicus Curiae.*

inpatient coverage disparately affected the handicapped, the panel found that a prima facie case had been made out, and the case was remanded to give Tennessee an opportunity for rebuttal. According to the panel majority, the State on remand could either demonstrate the unavailability of alternative plans that would achieve the State's legitimate cost-saving goals with a less disproportionate impact on the handicapped, or the State could offer "a substantial justification for the adoption of the plan with the greater discriminatory impact." *Id.*, at 1045. We granted certiorari to consider whether the type of impact at issue in this case is cognizable under § 504 or its implementing regulations, 465 U.S. 1021, 104 S.Ct. 1271, 79 L.Ed.2d 677 (1984), and we now reverse.

II

The first question the parties urge on the Court is whether proof of discriminatory animus is always required to establish a violation of § 504 and its implementing regulations, or whether federal law also reaches action by a recipient of federal funding that discriminates against the handicapped by effect rather than by design. The State of Tennessee argues that § 504 reaches only purposeful discrimination against the handicapped. . . .

* * *

Discrimination against the handicapped was perceived by Congress to be most often the product, not of invidious animus, but rather of thoughtlessness and indifference—of benign neglect.[12] Thus, Representative Vanik, introducing the predecessor to § 504 in the House, described the treatment of the handicapped as one of the country's "shameful oversights," which caused the handicapped to live among society "shunted aside, hidden, and ignored." [ ]. Similarly, Senator Humphrey, who introduced a companion measure in the Senate, asserted that "we can no longer tolerate the invisibility of the handicapped in America . . . ." [ ] And Senator Cranston, the Acting Chairman of the Subcommittee that drafted § 504, described the Act as a response to "previous societal neglect."[ ] . . . . Federal agencies and commentators on the plight of the handicapped similarly have found that discrimination against the handicapped is primarily the result of apathetic attitudes rather than affirmative animus.

In addition, much of the conduct that Congress sought to alter in passing the Rehabilitation Act would be difficult if not impossible to reach were the Act construed to proscribe only conduct fueled by a discriminatory

---

[12] To be sure, well-cataloged instances of invidious discrimination against the handicapped do exist. See, *e.g.,* United States Commission on Civil Rights, Accommodating the Spectrum of Individual Abilities, Ch. 2 (1983); Wegner, The Antidiscrimination Model Reconsidered: Ensuring Equal Opportunity Without Respect to Handicap Under Section 504 of the Rehabilitation Act of 1973, 69 Cornell L.Rev. 401, 403, n. 2 (1984).

intent. For example, elimination of architectural barriers was one of the central aims of the Act, [ ], yet such barriers were clearly not erected with the aim or intent of excluding the handicapped. Similarly, Senator Williams, the chairman of the Labor and Public Welfare Committee that reported out § 504, asserted that the handicapped were the victims of "[d]iscrimination in access to public transportation" and "[d]iscrimination because they do not have the simplest forms of special educational and rehabilitation services they need . . . ." [ ]. And Senator Humphrey, again in introducing the proposal that later became § 504, listed, among the instances of discrimination that the section would prohibit, the use of "transportation and architectural barriers," the "discriminatory effect of job qualification . . . procedures," and the denial of "special educational assistance" for handicapped children. [ ] These statements would ring hollow if the resulting legislation could not rectify the harms resulting from action that discriminated by effect as well as by design.

At the same time, the position urged by respondents—that we interpret § 504 to reach all action disparately affecting the handicapped—is also troubling. Because the handicapped typically are not similarly situated to the nonhandicapped, respondents' position would in essence require each recipient of federal funds first to evaluate the effect on the handicapped of every proposed action that might touch the interests of the handicapped, and then to consider alternatives for achieving the same objectives with less severe disadvantage to the handicapped. The formalization and policing of this process could lead to a wholly unwieldy administrative and adjudicative burden. [ ] Had Congress intended § 504 to be a National Environmental Policy Act for the handicapped, requiring the preparation of "Handicapped Impact Statements" before any action was taken by a grantee that affected the handicapped, we would expect some indication of that purpose in the statute or its legislative history. Yet there is nothing to suggest that such was Congress' purpose. Thus, just as there is reason to question whether Congress intended § 504 to reach only intentional discrimination, there is similarly reason to question whether Congress intended § 504 to embrace all claims of disparate-impact discrimination.

Any interpretation of § 504 must therefore be responsive to two powerful but countervailing considerations—the need to give effect to the statutory objectives and the desire to keep § 504 within manageable bounds. Given the legitimacy of both of these goals and the tension between them, we decline the parties' invitation to decide today that one of these goals so overshadows the other as to eclipse it. While we reject the boundless notion that all disparate-impact showings constitute prima facie cases under § 504, we assume without deciding that § 504 reaches at least some conduct that has an unjustifiable disparate impact upon the handicapped. On that assumption, we must then determine whether the disparate effect of which respondents complain is the sort of disparate impact that federal law might recognize.

III

To determine which disparate impacts § 504 might make actionable, the proper starting point is *Southeastern Community College v. Davis,* 442 U.S. 397, 99 S.Ct. 2361, 60 L.Ed.2d 980 (1979), our major previous attempt to define the scope of § 504. *Davis* involved a plaintiff with a major hearing disability who sought admission to a college to be trained as a registered nurse, but who would not be capable of safely performing as a registered nurse even with full-time personal supervision. We stated that, under some circumstances, a "refusal to modify an existing program might become unreasonable and discriminatory. Identification of those instances where a refusal to accommodate the needs of a disabled person amounts to discrimination against the handicapped [is] an important responsibility of HEW." *Id.,* at 413, 99 S.Ct., at 2370. We held that the college was not required to admit Davis because it appeared unlikely that she could benefit from any modifications that the relevant HEW regulations required, *id.,* at 409, 99 S.Ct., at 2368, and because the further modifications Davis sought—full-time, personal supervision whenever she attended patients and elimination of all clinical courses—would have compromised the essential nature of the college's nursing program, *id.,* at 413–414, 99 S.Ct., at 2370–2371. Such a "fundamental alteration in the nature of a program" was far more than the reasonable modifications the statute or regulations required. *Id.,* at 410, 99 S.Ct., at 2369. *Davis* thus struck a balance between the statutory rights of the handicapped to be integrated into society and the legitimate interests of federal grantees in preserving the integrity of their programs: while a grantee need not be required to make "fundamental" or "substantial" modifications to accommodate the handicapped, it may be required to make "reasonable" ones.

The balance struck in *Davis* requires that an otherwise qualified handicapped individual must be provided with meaningful access to the benefit that the grantee offers. The benefit itself, of course, cannot be defined in a way that effectively denies otherwise qualified handicapped individuals the meaningful access to which they are entitled; to assure meaningful access, reasonable accommodations in the grantee's program or benefit may have to be made.[21] In this case, respondents argue that the

---

[21] As the Government states: "Antidiscrimination legislation can obviously be emptied of meaning if every discriminatory policy is 'collapsed' into one's definition of what is the relevant benefit." Brief for United States as *Amicus Curiae* 29, n. 36. At oral argument, the Government also acknowledged that "special measures for the handicapped, as the *Lau* case shows, may sometimes be necessary . . . ." Tr. of Oral Arg. 14–15 (referring to *Lau v. Nichols,* 414 U.S. 563, 94 S.Ct. 786, 39 L.Ed.2d 1 (1974)).

The regulations implementing § 504 are consistent with the view that reasonable adjustments in the nature of the benefit offered must at times be made to assure meaningful access. See, *e.g.,* 45 CFR § 84.12(a) (1984) (requiring an employer to make "reasonable accommodation to the known physical or mental limitations" of a handicapped individual); 45 CFR § 84.22 and § 84.23 (1984)

14-day rule, or any annual durational limitation, denies meaningful access to Medicaid services in Tennessee. We examine each of these arguments in turn.

A

The 14-day limitation will not deny respondents meaningful access to Tennessee Medicaid services or exclude them from those services. The new limitation does not invoke criteria that have a particular exclusionary effect on the handicapped; the reduction, neutral on its face, does not distinguish between those whose coverage will be reduced and those whose coverage will not on the basis of any test, judgment, or trait that the handicapped as a class are less capable of meeting or less likely of having. Moreover, it cannot be argued that "meaningful access" to state Medicaid services will be denied by the 14-day limitation on inpatient coverage; nothing in the record suggests that the handicapped in Tennessee will be unable to benefit meaningfully from the coverage they will receive under the 14-day rule.[22] The reduction in inpatient coverage will leave both handicapped and nonhandicapped Medicaid users with identical and effective hospital services fully available for their use, with both classes of users subject to the same durational limitation. The 14-day limitation, therefore, does not exclude the handicapped from or deny them the benefits of the 14 days of care the State has chosen to provide.

To the extent respondents further suggest that their greater need for prolonged inpatient care means that, to provide meaningful access to Medicaid services, Tennessee must single out the handicapped for *more* than 14 days of coverage, the suggestion is simply unsound. At base, such a suggestion must rest on the notion that the benefit provided through state Medicaid programs is the amorphous objective of "adequate health care." But Medicaid programs do not guarantee that each recipient will receive that level of health care precisely tailored to his or her particular needs. Instead, the benefit provided through Medicaid is a particular package of health care services, such as 14 days of inpatient coverage. That package of services has the general aim of assuring that individuals will receive necessary medical care, but the benefit provided remains the individual services offered—not "adequate health care."

(requiring that new buildings be readily accessible, building alterations be accessible "to the maximum extent feasible," and existing facilities eventually be operated so that a program or activity inside is, "when viewed in its entirety," readily accessible); 45 CFR § 84.44(a) (1984) (requiring certain modifications to the regular academic programs of secondary education institutions, such as changes in the length of time permitted for the completion of degree requirements, substitution of specific courses required for the completion of degree requirements, and adaptation of the manner in which specific courses are conducted).

[22] The record does not contain any suggestion that the illnesses uniquely associated with the handicapped or occurring with greater frequency among them cannot be effectively treated, at least in part, with fewer than 14 days' coverage. In addition, the durational limitation does not apply to only particular handicapped conditions and takes effect regardless of the particular cause of hospitalization.

The federal Medicaid Act makes this point clear. The Act gives the States substantial discretion to choose the proper mix of amount, scope, and duration limitations on coverage, as long as care and services are provided in "the best interests of the recipients." 42 U.S.C. § 1396a(a)(19). The District Court found that the 14-day limitation would fully serve 95% of even handicapped individuals eligible for Tennessee Medicaid, and both lower courts concluded that Tennessee's proposed Medicaid plan would meet the "best interests" standard. That unchallenged conclusion indicates that Tennessee is free, as a matter of the Medicaid Act, to choose to define the benefit it will be providing as 14 days of inpatient coverage.

Section 504 does not require the State to alter this definition of the benefit being offered simply to meet the reality that the handicapped have greater medical needs. To conclude otherwise would be to find that the Rehabilitation Act requires States to view certain illnesses, *i.e.,* those particularly affecting the handicapped, as more important than others and more worthy of cure through government subsidization. Nothing in the legislative history of the Act supports such a conclusion. Cf. Doe v. Colautti, 592 F.2d 704 (CA3 1979) (State may limit covered-private-inpatient-psychiatric care to 60 days even though State sets no limit on duration of coverage for physical illnesses). Section 504 seeks to assure evenhanded treatment and the opportunity for handicapped individuals to participate in and benefit from programs receiving federal assistance. Southeastern Community College v. Davis, 442 U.S. 397, 99 S.Ct. 2361, 60 L.Ed.2d 980 (1979). The Act does not, however, guarantee the handicapped equal results from the provision of state Medicaid, even assuming some measure of equality of health could be constructed. *Ibid.*

Regulations promulgated by the Department of Health and Human Services (HHS) pursuant to the Act further support this conclusion. These regulations state that recipients of federal funds who provide health services cannot "provide a qualified handicapped person with benefits or services that are not as effective (as defined in § 84.4(b)) as the benefits or services provided to others." 45 CFR § 84.52(a)(3) (1984). The regulations also prohibit a recipient of federal funding from adopting "criteria or methods of administration that have the purpose or effect of defeating or substantially impairing accomplishment of the objectives of the recipient's program with respect to the handicapped." 45 CFR § 84.4(b)(4)(ii) (1984).

While these regulations, read in isolation, could be taken to suggest that a state Medicaid program must make the handicapped as healthy as the nonhandicapped, other regulations reveal that HHS does not contemplate imposing such a requirement. Title 45 CFR § 84.4(b)(2) (1984), referred to in the regulations quoted above, makes clear that

"[f]or purposes of this part, aids, benefits, and services, to be equally effective, are not required to produce the identical result or level of achievement for handicapped and nonhandicapped persons, but must afford handicapped persons equal opportunity to obtain the same result, to gain the same benefit, or to reach the same level of achievement . . . ."

This regulation, while indicating that adjustments to existing programs are contemplated, also makes clear that Tennessee is not required to assure that its handicapped Medicaid users will be as healthy as its nonhandicapped users. Thus, to the extent respondents are seeking a distinct durational limitation for the handicapped, Tennessee is entitled to respond by asserting that the relevant benefit is 14 days of coverage. Because the handicapped have meaningful and equal access to that benefit, Tennessee is not obligated to reinstate its 20-day rule or to provide the handicapped with more than 14 days of inpatient coverage.

B

We turn next to respondents' alternative contention, a contention directed not at the 14-day rule itself but rather at Tennessee's Medicaid *plan* as a whole. Respondents argue that the inclusion of any annual durational limitation on inpatient coverage in a state Medicaid plan violates § 504. The thrust of this challenge is that all annual durational limitations discriminate against the handicapped because (1) the effect of such limitations falls most heavily on the handicapped and because (2) this harm could be avoided by the choice of other Medicaid plans that would meet the State's budgetary constraints without disproportionately disadvantaging the handicapped. Viewed in this light, Tennessee's current plan is said to inflict a gratuitous harm on the handicapped that denies them meaningful access to Medicaid services.

Whatever the merits of this conception of meaningful access, it is clear that § 504 does not require the changes respondents seek. In enacting the Rehabilitation Act and in subsequent amendments, Congress did focus on several substantive areas—employment, education, and the elimination of physical barriers to access—in which it considered the societal and personal costs of refusals to provide meaningful access to the handicapped to be particularly high. But nothing in the pre- or post-1973 legislative discussion of § 504 suggests that Congress desired to make major inroads on the States' longstanding discretion to choose the proper mix of amount, scope, and duration limitations on services covered by state Medicaid, see *Beal v. Doe,* 432 U.S. 438, 444, 97 S.Ct. 2366, 2370, 53 L.Ed.2d 464 (1977). And, more generally, we have already stated, *supra,* at 719–720, that § 504 does not impose a general NEPA-like requirement on federal grantees.

The costs of such a requirement would be far from minimal, and thus Tennessee's refusal to pursue this course does not, as respondents suggest, inflict a "gratuitous" harm on the handicapped. On the contrary, to require that the sort of broad-based distributive decision at issue in this case always be made in the way most favorable, or least disadvantageous, to the handicapped, even when the same benefit is meaningfully and equally offered to them, would be to impose a virtually unworkable requirement on state Medicaid administrators. Before taking any across-the-board action affecting Medicaid recipients, an analysis of the effect of the proposed change on the handicapped would have to be prepared. Presumably, that analysis would have to be further broken down by class of handicap—the change at issue here, for example, might be significantly less harmful to the blind, who use inpatient services only minimally, than to other subclasses of handicapped Medicaid recipients; the State would then have to balance the harms and benefits to various groups to determine, on balance, the extent to which the action disparately impacts the handicapped. In addition, respondents offer no reason that similar treatment would not have to be accorded other groups protected by statute or regulation from disparate-impact discrimination.

It should be obvious that administrative costs of implementing such a regime would be well beyond the accommodations that are required under *Davis*. As a result, Tennessee need not redefine its Medicaid program to eliminate durational limitations on inpatient coverage, even if in doing so the State could achieve its immediate fiscal objectives in a way less harmful to the handicapped.

IV

The 14-day rule challenged in this case is neutral on its face, is not alleged to rest on a discriminatory motive, and does not deny the handicapped access to or exclude them from the particular package of Medicaid services Tennessee has chosen to provide. The State has made the same benefit—14 days of coverage—equally accessible to both handicapped and nonhandicapped persons, and the State is not required to assure the handicapped "adequate health care" by providing them with more coverage than the nonhandicapped. In addition, the State is not obligated to modify its Medicaid program by abandoning reliance on annual durational limitations on inpatient coverage. Assuming, then, that § 504 or its implementing regulations reach some claims of disparate-impact discrimination, the effect of Tennessee's reduction in annual inpatient coverage is not among them. For that reason, the Court of Appeals erred in holding that respondents had established a prima facie violation of § 504. The judgment below is accordingly reversed.

It is so ordered.

NOTES AND QUESTIONS

1.   *"Meaningful Access."* The Supreme Court in *Choate* rejected the idea that Section 504 prohibits only intentional discrimination and reaffirmed the availability of disparate impact claims. However, it also held that the mandate

of Section 504 "to assure evenhanded treatment and the opportunity for handicapped individuals to participate in and benefit from programs receiving federal assistance" is met when people with disabilities are provided "meaningful access" to such programs. "Meaningful access" is a key concept, and courts have interpreted this standard in different ways. For a discussion of "meaningful access" in the context of health care, see Leslie Pickering Francis & Anita Silvers, Debilitating Alexander v. Choate: "Meaningful Access" to Health Care for People with Disabilities, 35 Fordham Urb. L. J. 447 (2008). See also Mark C. Weber, Meaningful Access and Disability Discrimination: The Role of Social Science and Other Empirical Evidence, 39 Cardozo L. Rev. 649 (2017). There also is uncertainty as to whether a claim of disparate impact requires a showing of intent (e.g., "deliberate indifference"). Some courts have borrowed the analysis of intent from Title VI, while others have not. For a discussion, see Mark C. Weber, Accidentally on Purpose: Intent in Disability Discrimination Law, 56 B.C. L. Rev. 1417 (2015).

2. *Distinguishing Choate.* How might you distinguish facially neutral treatment that falls more heavily on people with disabilities from treatment that "targets" people with disabilities? In the early 2000s, a class action lawsuit was filed under Title II of the ADA challenging a decision by Los Angeles County to close the Rancho Los Amigos National Rehabilitation Center, a facility dedicated primarily to providing inpatient and outpatient care to people with disabilities. Rodde v. Bonta, 357 F.3d 988 (9th Cir. 2004). In 2002, the county consolidated its services for people with certain severe disabilities at Rancho, one of six public facilities at the time. One year later, the county proposed closing Rancho as a means to save money. The plaintiffs were current and future participants in California's Medicaid program, Medi-Cal, with disabilities and conditions that required medical services offered by Rancho. In granting a preliminary injunction, the district court concluded that Rancho patients with disabilities could not easily be served elsewhere and that the county had made no transition plans to accommodate them.

The Ninth Circuit affirmed, distinguishing *Choate* on several grounds. First, unlike in *Choate*, the decision to close Rancho was not a facially neutral, "across-the-board" cut, but was instead the elimination of the one facility out of six that, due to the county's prior consolidation of services, provided services disproportionately required by people with disabilities. Second, there was evidence that closing Rancho would disproportionately harm people with disabilities because the services provided could not be provided elsewhere in the county system at that time. The parties reached a settlement in which the county agreed that Rancho would remain open for at least another three years, maintaining at least the same level of service.

3. *Private Disparate Impact Claims.* The Court in *Choate* assumed that Section 504 permits private actions challenging facially neutral policies and programs that have a disparate impact on people with disabilities. In support of such a right, the Court stated that "much of the conduct that Congress sought to alter in passing the Rehabilitation Act would be difficult if not impossible to reach were the Act construed to proscribe only conduct fueled by

a discriminatory intent." It reasoned that, in enacting Section 504, Congress intended to reach discrimination that was not only the result of "invidious animus," but also of "thoughtlessness," "indifference," and "benign neglect."

Relying on *Choate*, at least four circuits have recognized a disparate impact theory based on lack of meaningful access under Section 504, and by extension, Section 1557. The Sixth Circuit, however, has held that a disparate impact theory is inconsistent with the text of the Rehabilitation Act, although it did not address the meaningful access standard in the context of the ACA. Doe v. BlueCross BlueShield of Tenn., Inc., 926 F.3d 235 (6th Cir. 2019); see also discussion of *Sandoval*, above.

In 2021, the U.S. Supreme Court granted a petition for certiorari on this question. In *Doe v. CVS Pharmacy*, the Ninth Circuit joined the Second, Seventh, and Tenth Circuits in recognizing a disparate impact claim under Section 504. Doe v. CVS Pharmacy, 982 F.3d. 1204 (9th Cir. 2020). In that case, HIV-positive class members challenged a provision of their health plan under which they were eligible for in-network prices for specialty medications, including HIV/AIDS medications, only if they accepted the medications by mail or through a specialty pick-up service at a CVS pharmacy, which did not include interaction with a pharmacist. Any prescriptions filled in-person were subject to out-of-network prices. The enrollees challenged this plan benefit under Section 1557, alleging that it denied them meaningful access to their prescription drug benefit within the meaning of the Section 504, as incorporated in Section 1557 of the ACA, because they did not receive effective treatment for HIV/AIDS, including medically appropriate dispensing of their medications and access to consultation with a pharmacist.

In holding that the enrollees had stated a claim, the Ninth Circuit held:

> Following Choate, we recognized that the unique impact of a facially-neutral policy on people with disabilities may give rise to a disparate impact claim where state "services, programs, and activities remain open and easily accessible to others." . . . Here, Does have alleged that even though the Program applies to specialty medications that may not be used to treat conditions associated with disabilities, the Program burdens HIV/AIDS patients differently because of their unique pharmaceutical needs. Specifically, they claim that changes in medication to treat the continual mutation of the virus requires pharmacists to review all of an HIV/AIDS patient's medications for side effects and adverse drug interactions, a benefit they no longer receive under the Program.

Id. at 1211. The Supreme Court granted CVS's petition for certiorari on the question of whether Section 504, and by extension the ACA, provides a disparate impact cause of action for plaintiffs alleging disability discrimination. CVS Pharmacy v. Doe, 141 S. Ct. 2882 (2021). CVS agreed to withdraw its petition in November 2021, effectively preserving existing access to private disparate impact actions. However, this issue, along with legal challenges and anticipated regulatory activity surrounding Section 1557, should be watched closely.

4. *Integration Mandate.* In the landmark 1999 case *Olmstead v. L.C,* the Supreme Court held that unnecessary segregation of persons with disability constitutes discrimination in violation of Title II of the ADA. Olmstead v. L.C. ex rel. Zimring, 527 U.S. 581, 592 (1992). The Court held that public entities, including state Medicaid programs, must provide community-based services to persons with disabilities when such services are appropriate, desired by the recipient, and can be reasonably accommodated by the public entity. The Court relied on the "integration mandate" in the Title II regulations that requires public entities to "administer services, programs, and activities in the most integrated setting appropriate to the needs of qualified individuals with disabilities." The regulations define the "most integrated setting" as one that "enables individuals with disabilities to interact with nondisabled persons to the fullest extent possible. . . ."

The *Olmstead* holding means individuals with a disability, including persons with cognitive and intellectual disabilities and persons with mental illness, cannot be required to live in institutions or group settings to obtain the health care and other services they need.

NOTE: DISABILITY DISCRIMINATION IN HEALTH INSURANCE

People with disabilities face challenges obtaining adequate and affordable health insurance. The 2009 report, The Current State of Health Care for People with Disabilities, found that the complex, fragmented, and often overly restrictive U.S. health insurance system leaves some people with disabilities with no health care coverage and others with exclusions, limits, and cost-sharing obligations that prevent them from obtaining needed medications, medical equipment, specialty care, dental and vision care, long-term care, and care coordination. Nat'l Council on Disability, The Current State of Health Care for People with Disabilities (2009). See also Valarie Blake, An Opening for Civil Rights in Health Insurance After the Affordable Care Act, 36 Boston College J. of L. & Soc. Justice 235 (2016), discussing disability discrimination in public and private health insurance.

Challenges in Applying the ADA

On its face, several provisions of the ADA appear to prohibit insurers and employers administering benefit plans from imposing coverage terms and conditions that discriminate against persons with particular disabilities. Title I of the ADA prohibits discrimination "against a qualified individual with a disability because of the disability of such individual in regard to . . . [the] terms, conditions, and privileges of employment." 42 U.S.C. § 12112. The statute prohibits discrimination in nearly all aspects of work, including the receipt of "fringe benefits." 42 U.S.C. § 12112(b)(4); 29 C.F.R. § 1630.4(f). Title II similarly prohibits public entities from discriminating. 42 U.S.C. § 12132. Title III proscribes discrimination "on the basis of disability in the full and equal enjoyment of the goods, services, facilities, privileges, advantages, or accommodations of any place of public accommodation . . . ." 42 U.S.C. § 12182. "Public accommodation" is specifically defined to include an "insurance office."

42 U.S.C. § 12181(7)(F). Finally, Title V of the ADA contains a specific "safe harbor" provision that states that the ADA does not restrict insurers, HMOs, employers, plans, or administrators from "underwriting risks, classifying risks, or administering such risks that are based on or not inconsistent with State law," as long as the entity does not use this provision "as a subterfuge to evade the purposes" of the ADA. 42 U.S.C. § 12201(c).

The ADA has been applied to health insurance, but with significant limitations. Cases have been brought under the ADA challenging policies that provided less coverage for treatment of mental illnesses than for treatment of physical conditions, Rogers v. Dep't of Health & Envtl. Control, 174 F.3d 431 (4th Cir. 1999); Fletcher v. Tufts Univ., 367 F. Supp. 2d 99 (D. Mass. 2005); that capped coverage for AIDS but not for other conditions, Doe v. Mutual of Omaha Ins. Co., 179 F.3d 557 (7th Cir. 1999); or that excluded coverage for particular services, such as heart transplants, Lenox v. Healthwise of Kentucky, Ltd., 149 F.3d 453 (6th Cir. 1998), and infertility, Krauel v. Iowa Methodist Med. Ctr., 95 F.3d 674 (8th Cir. 1996). Although some of these cases have succeeded, they have encountered increasingly serious obstacles. First, most courts have held that the ADA does not require employers or insurers to offer any particular form of coverage; it merely prohibits them from offering different terms and conditions to disabled persons. See, e.g., EEOC v. Staten Island Sav. Bank, 207 F.3d 144 (2d Cir. 2000). The courts have held that the ADA does not demand all disabilities be treated similarly, but only that disabled persons not be disfavored in comparison to nondisabled persons. Providing different coverage for different conditions, moreover, is not even necessarily prohibited unless the condition itself is a disability or unless discrimination in the coverage of a particular condition disproportionately affects disabled persons. Even then, such distinctions may be permitted under the "safe harbor" provisions discussed below.

Second, whether and when the ADA applies to insurance policies has sparked much debate. Though Title III clearly covers insurance offices, several courts have held that it only applies to physical places and not to the terms and conditions of the products offered independent of these places. See, e.g., Weyer v. Twentieth Century Fox Film Corp., 198 F.3d 1104 (9th Cir. 2000). A number of other courts and the EEOC Guidelines, on the other hand, have held that Title III might extend to the contents of insurance policies. See, e.g., Doe v. Mut. Of Omaha Ins. Co., 179 F.3d 557, 558–59 (7th Cir. 1999); Fletcher v. Tufts Univ., 367 F. Supp. 2d 99, 114–115 (D. Mass. 2005). See Jeffrey S. Manning, Are Insurance Companies Liable Under the Americans With Disabilities Act?, 88 Calif. L. Rev. 607 (2000). Of course, if an employer offers insurance, Title I prohibits discrimination against an employee, even if Title III does not cover the insurer. However, several courts have limited Title I actions to current employees, contending that former employees, such as retirees, have no rights under the statute. See, e.g., EEOC v. CNA Ins. Cos., 96 F.3d 1039, 1045 (7th Cir. 1996). But see Castellano v. City of N.Y., 142 F.3d 58 (2d Cir. 1998). Title I of the ADA also limits the questions that employers can ask their employees about health issues, which affects workplace wellness programs.

Third, several courts have read Title V's insurance "safe harbor" broadly to protect insurer practices that are not designed to discriminate, following the Supreme Court's interpretation of the term "subterfuge" in the Age Discrimination in Employment Act of 1967, 29 U.S.C. §§ 621–630, ("ADEA") in the case Pub. Emps. Ret. Sys. Of Ohio v. Betts, 492 U.S. 158 (1989). See Ford v. Schering-Plough Corp., 145 F.3d 601 (3d Cir. 1998) and Krauel v. Iowa Methodist Med. Ctr., 95 F.3d 674, 678–9 (8th Cir. 1996). Other courts, however, have required actuarial support for treating different conditions differently, particularly when the insurance practice is also suspect under state law. Morgenthal v. Am. Tel. & Tel. Co., 1999 WL 187055 (S.D.N.Y. 1999); Chabner v. United of Omaha Life Ins. Co., 994 F. Supp. 1185 (N.D. Cal. 1998).

Section 1557

As a result of these limitations, prior to the ACA, many people with disabilities were denied health insurance, or charged higher prices, often with coverage limitations and exclusions. The ACA's insurance reforms address some of these problems. For example, prior to the ACA, many insurers were permitted to exclude or restrict coverage for individuals with a preexisting condition such as cancer, asthma, or other chronic conditions or disabilities under the ADA's "safe harbor" exception.

In addition to broad insurance reforms, Section 1557 was initially viewed as an important new way to challenge discriminatory benefit exclusions or denials based on disability and other protected categories. The 2016 Rule implementing Section 1557 defined health programs and activities to include health insurance and the provision of health-related services. It provides that covered entities may not deny, cancel, limit, or refuse to issue or renew a health insurance policy, deny or limit coverage of a health insurance claim, impose additional cost sharing or other limitations or restrictions on coverage, or use discriminatory marketing practices or insurance benefit designs on the basis of any protected category. The 2016 Rule also clarifies that if one part of an entity that is principally engaged in providing or administering health services or health insurance coverage receives federal financial assistance, the entire entity is forbidden to discriminate.

What constitutes discrimination in insurance benefit design? HHS declined to identify specific practices in the 2016 Rule, in favor of a fact-specific inquiry. The preamble did identify factors that OCR will consider when assessing whether a plan benefit design is discriminatory: whether a covered entity utilized, in a nondiscriminatory manner, a neutral rule or principle when deciding to adopt the design feature or take the challenged action; whether the reason for its coverage decision is a pretext for discrimination; and whether coverage for the same or a similar service or treatment is available to individuals outside of that protected class or those with different health conditions. It will also evaluate the reasons for any differences in coverage. See 81 Fed. Reg. at 31433.

Would placing all drugs used to treat a specific disability such HIV/AIDS in a plan's highest cost-sharing tier be considered discriminatory? See Jane Perkins and Wayne Turner, NHeLP and The AIDS Institute Complaint to HHS Re HIV/AIDS Discrimination by Florida Insurers (May 29, 2014), available at the NHeLP website, healthlaw.org. In separate guidance, CMS identified this practice as an example of potential discrimination, along with applying age limits to services that have been found clinically effective at all ages, and requiring prior authorization and/or step therapy for all or most medications in drug classes such as anti-HIV protease inhibitors and/or immune suppressants, regardless of medical evidence. 81 Fed. Reg. at 31434, n. 258.

The future of Section 1557 as a vehicle to challenge discriminatory benefit design is uncertain. As discussed above, the 2020 Rule significantly limits the scope of the 2016 Rule and eliminates the provisions prohibiting discrimination in plan benefit design. In light of an anticipated new notice of proposed rulemaking to revise the Section 1557 regulations, and other steps to reverse Trump Administration policy and regulations that significantly narrowed the reach of Section 1557, developments in the interpretation and enforcement of Section 1557 should be watched closely.

D. OTHER NONDISCRIMINATION REQUIREMENTS

This section identifies additional federal nondiscrimination laws that may apply to health care insurers and others. State nondiscrimination laws may also apply, as Section 1557 does not limit their application. 42 U.S.C. § 18116(b). State laws often parallel the protections of federal law but can differ in terms of entities covered, specific protections, and enforcement processes. State law can also address gaps or provide more protections than federal law.

1. Genetic Information

The ability to read (and, perhaps, change) a person's genetic characteristics has created tremendous hopes and terrible fears. There is a fear that the terrifying history of eugenics will repeat itself. In Buck v. Bell, 274 U.S. 200 (1927), for example, the Supreme Court tragically misunderstood and misused genetics and decided that the forced sterilization of a woman perceived to be intellectually disabled, whose mother and grandmother were also believed to be so, was not a violation of her constitutional rights. Justice Holmes, in a now infamous declaration, stated that it is desirable "to prevent those who are manifestly unfit from continuing their kind . . . . Three generations of imbeciles are enough." For a full history of that case and the family, see Paul Lombardo, Three Generations, No Imbeciles: New Light on Buck v. Bell, 60 N.Y.U. L. Rev. 30 (1985). About 8,000 Virginians who were low-income, uneducated, and believed to have intellectual disabilities were sterilized as part of a eugenics program in that state between 1924 and 1979. Thirty states engaged in such programs, and 65,000 individuals nationwide were involuntarily sterilized.

Similarly, government efforts targeted people with sickle cell based on mistaken genetic assumptions:

> In the 1970s, large scale screening [for sickle cell] was undertaken with the goal of changing African American mating behavior. Unfortunately, the initiative promoted confusion regarding the difference between carriers and those with the disease. This confusion resulted in widespread discrimination against African Americans. Some states passed legislation requiring all African American children entering school to be screened for the sickle-cell trait, even though there was no treatment or cure for the sickle-cell disease. Some states required prisoners to be tested, even though there would be no opportunity for them to pass on the trait. Job and insurance discrimination were both real and attempted. The military considered banning all African Americans from the armed services. African American airline stewardesses were fired. Insurance rates went up for carriers. Some companies refused to insure carriers. During that period, many African Americans came to believe that the sickle-cell screening initiative was merely a disguised genocide attempt, since often the only advice given to African Americans with the trait was, "Don't have kids."

Vernellia R. Randall, Trusting the Health Care System Ain't Always Easy! An African American Perspective on Bioethics, 15 St. Louis U. Public L. Rev. 191 (1996). See also Norman-Bloodsaw v. Lawrence Berkeley Lab., 135 F.3d 1260 (9th Cir. 1998), involving employment-based testing for sickle cell.

The potential for discrimination against people based on genetic information was recognized by the passage of the Genetic Information Nondiscrimination Act of 2008 (GINA) which regulates the collection and use of genetic information by group health plans, by those marketing individual policies for health insurance, and by employers. GINA amends ERISA, HIPAA, and the Internal Revenue Code, and adds a provision to the federal employment discrimination statutes. GINA prohibits health insurers, in both group health plans and in the individual market, from making insurance decisions based on "genetic information." Genetic information is defined broadly as genetic tests, the genetic tests of family members, and the manifestation of a disease or disorder in family members, with exceptions. See 42 U.S.C. § 2000ff. Insurers may consider genetic diseases that have manifested themselves in ways other than genetic tests. However, the ADA may apply for manifested diseases that meet the statutory definition of disability. The EEOC has promulgated regulations to enforce the statute in the employment context, while the

OCR and other agencies have issued regulations to enforce the statute in other contexts.

Unlike the other federal nondiscrimination laws discussed in this chapter, GINA does not regulate the delivery of health care services. However, it addresses related concerns. The Preamble of the Act describes the rationale for its enactment:

> Deciphering the sequence of the human genome and other advances in genetics open major new opportunities for medical progress. New knowledge about the genetic basis of illness will allow for earlier detection of illnesses, often before symptoms have begun. Genetic testing can allow individuals to take steps to reduce the likelihood that they will contract a particular disorder. New knowledge about genetics may allow for the development of better therapies that are more effective against disease or have fewer side effects than current treatments. These advances give rise to the potential misuse of genetic information to discriminate in health insurance and employment.

> The early science of genetics became the basis of State laws that provided for the sterilization of persons having presumed genetic "defects" such as mental retardation, mental disease, epilepsy, blindness, and hearing loss, among other conditions. . . . [T]he current explosion in the science of genetics, and the history of sterilization laws by the States based on early genetic science, compels Congressional action in this area.

> Although genes are facially neutral markers, many genetic conditions and disorders are associated with particular racial and ethnic groups and gender. Because some genetic traits are most prevalent in particular groups, members of a particular group may be stigmatized or discriminated against as a result of that genetic information. . . . To alleviate some of this stigma, Congress in 1972 passed the National Sickle Cell Anemia Control Act, which withholds Federal funding from States unless sickle cell testing is voluntary.

NOTES AND QUESTIONS

1. *Limitations of GINA and Genetic Privacy Laws.* GINA does not reach discrimination issues other than genetic discrimination in employment and health insurance. It generally provides no protection against discrimination in the life, long-term care, or disability insurance markets, for example. See Ellen Wright Clayton et al., The Law of Genetic Privacy: Applications, Implications, and Limitations, 6 J. of L. and the Biosciences 1 (2019) and Jarrod O. Anderson, et al., The Problem with Patchwork: State Approaches to Regulating Insurer Use of Genetic Information, 22 DePaul J. of Health L. (2021) discussing limitations of existing genetic information privacy protections.

2.   *Future of GINA.* While some of the insurance protections mandated under GINA are also included in the ACA, GINA will continue to be an additional source of support for those who face discrimination in health insurance eligibility, price, or conditions because of their genetic status. GINA also applies to workplace wellness plans, which have increasingly been used as a way to collect employees' health and genetic information. GINA's protections may become more important as genetic information becomes more informative, more widely used in medicine and research, and more easily re-identified. Barbara J. Evans, The Genetic Information Nondiscrimination Act at 10: GINA's Controversial Assertion that Data Transparency Protects Privacy and Civil Rights, 60 William & Mary L. Rev. 2017 (2019).

3.   *Devaluing "Imperfect" Lives?* Given our often-moralistic attitudes toward sickness and the continuing human history of exclusion and discrimination against minorities and people with disabilities, will "imperfect humans" be devalued and penalized? This is a familiar dynamic in history, but the new power of genetics may increase the potential for stigma and discrimination. See Nat'l Council on Disability, Genetic Testing and the Rush to Perfection (2019). Increased opportunities for control of genetic traits in reproductive decision making, for example, raise questions of choice and consequences. See Eric Rakowski, Who Should Pay for Bad Genes, 90 Cal. L. Rev. 1345 (2002), arguing that parents who choose to bear such a "genetically disadvantaged" child should incur a greater liability for the costs of the child's care. On the other hand, others are concerned about the drive toward "genetic enhancement" and the creation of a genetically perfect person (or society). See Maxwell Mehlman, Law of Above Averages: Leveling the New Genetic Enhancement Playing Field, 85 Iowa L. Rev. 517 (2000); Michael Malinowski, Choosing the Genetic Makeup of Our Children: Our Eugenics Past—Present, and Future, 36 Conn. L. Rev. 125 (2003). Reproductive decision-making is discussed in detail in Chapter 3.

4.   *Implications for Groups.* Genetic information is likely to have implications for groups. The idea that groups may have concerns, interests, and perhaps rights is one that is hotly debated in genetic research. See, e.g., Joan L. McGregor, Population Genomics and Research Ethics, 35 J.L. Med. & Ethics 356 (2007); Laura Underkuffler, Human Genetics Studies: The Case for Group Rights, 35 J.L. Med. & Ethics 383 (2007). See also the discussion of research with the Havasupai tribe in Chapter 7.

Sometimes the distribution of genetic traits may appear to parallel what have been called "folk notions" of race or ethnicity. Pilar Ossorio, Race, Genetic Variation, and the Haplotype Mapping Project, 66 La. L. Rev. 131 (2005). For example, the sickle cell trait is more common in persons with ancestors from sub-Saharan Africa, Latin America, Saudi Arabia, India, and Mediterranean countries. Tay-Sachs is more common in Jews of Eastern European extraction (although 1 in 250 persons in the general population also carries the gene for the disease and it is found in higher-than-general rates in French Canadians and Cajuns). Each of these common claims of association of genetic traits with particular geographically based populations is extraordinarily imprecise and overgeneralized, however. Keith Wailoo & Stephen Pemberton, The Troubled Dream of Genetic Medicine: Ethnicity and Innovation in Tay-Sachs, Cystic Fibrosis, and Sickle Cell Disease (2006). See also Mark Rothstein, Legal Conceptions of Equality in a Genomic Age, 25 Law & Ineq. 429 (2007).

5. *International Views.* Genetic discrimination is of substantial concern beyond the U.S., and international human rights law has taken notice of the possibility of inappropriate genetic discrimination. In 1997, UNESCO issued a Universal Declaration on the Human Genome and Human Rights that requires that "no-one shall be subjected to discrimination based on genetic characteristics that is intended to infringe or has the effect of infringing human rights, fundamental freedoms and human dignity." United Nations Educational, Scientific, and Cultural Organization, Universal Declaration on the Human Genome and Human Rights art. 6 (Nov. 11, 1997).

UNESCO's International Declaration on Human Data notes that "every effort should be made to ensure that human genetic data are not used for purposes that are discriminatory or in any way that would lead to the stigmatization of an individual, a family, or a group." United Nations Educational, Scientific, and Cultural Organization, International Declaration on Human Data art. 7 (Oct. 16, 2003). Many other countries have laws regulating such discrimination, as do international regional organizations. For an interesting account of Australian practices with regard to employment, see M. Otlowski et al., Practices and Attitudes of Australian Employers with Regard to the Use of Genetic Information, 31 Comp. Lab. L. & Pol'y J. 637 (2010).

2. Mental Health Parity and Addiction Equity

Historically, stigmatization of, and misunderstandings about, mental illness have led to a devaluation of mental health care in society generally, and within the health care system more specifically. This longstanding neglect of mental health resulted in less financing and other structural resources essential for ensuring access to mental health care. For example, while some health insurers failed to provide any coverage for mental health treatment, others imposed greater restrictions on mental health services, such as higher co-payments or limits on the number of visits covered, than for medical/surgical services.

Over time, however, mental health has become a more important and visible priority in health law and policy. This is due, in part, to growing research demonstrating the significant individual and societal impact of untreated mental and behavioral health conditions, as well as important advancements in mental and behavioral science. Stigma has also decreased, especially as high-profile figures have increasingly shared their experiences, and patient advocacy groups have become more vocal in demanding access to care.

In 1996, the federal government took a limited step toward addressing discrimination against mental health care by enacting the Mental Health Parity Act. It requires group health plans to use the same aggregate lifetime and annual dollar limits for mental health benefits that the plans impose on medical/surgical benefits.

In 2008, Congress passed the Mental Health Parity and Addiction Equity Act (the "Parity Act"), which extended parity requirements to treatment limitations. It requires insurers and employers to treat benefits for mental health conditions and substance use disorders in the same manner as benefits for physical conditions. For example, limits on the frequency of treatment, number of visits, days of coverage, or other limits on the scope or duration of treatment cannot differ as between mental health/substance use and medical/surgical benefits. Additionally, financial requirements such as copays and coinsurance cannot be greater for mental health/substance use than for medical/surgical benefits. Finally, if a health plan allows patients to go out-of-network for medical/surgical benefits, it must also do so for mental health/substance use benefits. The Parity Act also includes a cost exemption that allows group health plans to receive a waiver exempting them from some of the law's requirements if they demonstrate that costs increased at least one percent as a result of compliance, but the exemption only lasts one year. Mental Health Parity and Addiction Equity Act, Pub. L. 110–343, 122 Stat. 3765, amending 29 U.S.C. 1185a, § 712 (ERISA); 42 U.S.C. 300gg–5, § 2705 (Public Health Service Act); and I.R.C. § 9812 (Internal Revenue Code) (2008).

In 2010, the ACA applied the Parity Act to insurers in the individual market and qualified health plans offered through the marketplaces. The ACA also helped close an important gap in mental health coverage, because the Parity Act did not require insurers to cover mental health and substance use conditions; it only required parity in the event the plan provided mental health or substance use coverage. The ACA went further, however, to require the inclusion of mental health and substance use conditions as part of the essential health benefits ("EHBs") insurers must cover. Patient Protection and Affordable Care Act § 1302(b)(1), 42 U.S.C. § 18022(b) (2010). As a result, all health insurance plans in the individual and employer market must include coverage for treatment of mental health and substance use conditions. In order to satisfy the EHB requirement, insurers must comply with the Parity Act. 78 Fed. Reg. 68,239 (2013); Centers for Medicare & Medicaid Services, The Mental Health Parity and Addiction Equity Act, www.cms.gov; Health Aff., Mental Health Parity (2014).

While the Parity Act and the ACA mandate mental health parity, they have failed to offer guidance on how to determine whether a plan achieves parity with respect to practices that do not lend themselves to quantitative measurement. It is simple to evaluate parity for quantitative limits like number of covered visits. But how does one determine parity in the case of nonquantitative limits, such as limits on certain types of treatments, restrictions on geographic location and provider specialty, and methods of determining provider reimbursement? What if a plan excludes a particular type of mental health treatment that may not have a direct medical/surgical analog for comparison? Compare Joseph F. v. Sinclair Servs. Co., 158 F. Supp. 3d 1239, 1262 (D. Utah 2016) (holding that the health plan's exclusion for residential treatment violated the Parity Act, because it imposed a treatment limit that applied only to mental health conditions and not to medical/surgical services provided at analogous skilled nursing facilities), with Roy C. v. Aetna Life Ins. Co., No. 2:17-CV-1216, 2018 WL 4511972, at *3 (D. Utah Sept. 20, 2018) (plaintiffs failed to allege facts sufficient to support a Parity Act violation based on the plan's exclusion of wilderness therapy, also known as outdoor behavioral health care, because plaintiffs failed to sufficiently identify a comparison or analogue to wilderness therapy in the medical and surgical fields of treatment). But see Michael D. v. Anthem Health Plans of Kentucky, Inc., 369 F. Supp. 3d 1159 (D. Utah 2019) (noting in dicta that Parity Act challenges to the exclusion of wilderness camps are complicated because there is no clear analog to wilderness camps in the medical or surgical field, but also arguing that plans should not be able to exclude mental health treatments only because a clear analog does not exist).

Ambiguity concerning non-quantitative coverage limits has created ample opportunity for consumers to test the contours of the Parity Act through litigation. For an in-depth study of the types of issues raising questions about what constitutes mental health parity, see Kelsey N. Berry et al., Litigation Provides Clues to Ongoing Challenges in Implementing Insurance Parity, 42 J. Health Polit., Pol'y, & L. 1065 (2017) (highlighting common areas of dispute, including limits or exclusions on certain habilitative treatments such as applied behavioral analysis for autism, credentialing standards for providers, and medical necessity determinations). The Consolidated Appropriations Act, 2021, Pub. L. No. 116–260, amends the Parity Act to require group health plans and issuers that provide both mental health and substance use benefits and medical/surgical benefits to prepare a comparative analysis of any nonquantitative treatment limits that apply to each category of benefits.

3.   Age

Age discrimination in health care has received comparatively less attention than the other categories discussed in this chapter. But the COVID-19 pandemic has brought renewed attention to this problem, especially as early proposals for rationing treatment during the COVID-19 pandemic used age, either explicitly or indirectly, as a basis for determining access to ventilators and ICU beds. See Timothy W. Farrell et al., Rationing Limited Healthcare Resources in the COVID-19 Era and Beyond: Ethical Considerations Regarding Older Adults, 68 J. Am. Geriatr. Soc. 1143–1149 (2020). In March 2020, OCR issued a bulletin on medical resource allocation policies (discussed earlier in this Section in the notes after *Bragdon*), providing that people should not be denied medical care on the basis of "stereotypes, assessments of quality of life," or "judgments about a person's relative "worth" based on the presence or absence of disabilities *or age*." (emphasis added). Dep't Health & Hum. Serv.'s, Bulletin, Civil Rights, HIPAA, and the Coronavirus Disease 2019 (COVID-19) (Mar. 28, 2020). For information on other OCR actions involving allegations of age discrimination in crisis standards of care, see Civil Rights and COVID-19, at HHS.gov. In addition, the advocacy group Justice in Aging has actively monitored states' crisis standards of care addressing the rationing of COVID-19 treatment, with the goal of identifying and fighting to remove age-based criteria.

The problem of age discrimination goes beyond COVID-19-specific issues, however. Ageism has long been used to deny other kinds of care and has been linked to poorer outcomes and increased mortality. See also Sharon K. Inouye, Creating an Anti-Ageist Healthcare System to Improve Care for Our Current and Future Selves, 1 Nature Aging 150 (2021).

A few laws address age discrimination related to health care. The Age Discrimination Act of 1975, 29 U.S.C. §§ 6101 et seq. ("Age Act") prohibits discrimination against any individual on the basis of age in programs and activities that receive federal financial assistance. The Age Act does permit the use of age distinctions in limited circumstances, however. For example, it permits the use of age distinctions in legislation designed to provide benefits or assistance to persons based on age, such as age requirements for Medicare eligibility. As described in the introductory note, Section 1557 of the ACA extends antidiscrimination protection for age by prohibiting discrimination on the grounds set forth in the Age Act.

Age discrimination in employment is also prohibited by the Age Discrimination in Employment Act of 1967, 29 U.S.C. §§ 621–630, ("ADEA"). The ADEA limits the ability of covered employers to discriminate among employees 40 years of age and older with respect to the provision of health insurance benefits. However, in April 2004, the EEOC issued a rule stating that employers could reduce or eliminate health benefits for Medicare-eligible beneficiaries without violating the ADEA. 68 Fed. Reg. 41,542 (2003). See also Am. Ass'n of Retired Persons v. EEOC, 489 F.3d 558 (3d Cir. 2007) (upholding the EEOC rule).

CHAPTER 3

REPRODUCTION AND BIRTH

■ ■ ■

I. INTRODUCTION

As the other chapters in this book have demonstrated, health policy can be motivated by diverse economic, political, cultural, and social justice goals. These goals, in turn, are often shaped by deeply held beliefs about the ethical principles and moral values that should inform government's role in society. Perhaps no area of health law stimulates as much passionate debate and reveals as sharply diverging views on the appropriate role of government, as the regulation of reproduction and birth.

One reason for this is that regulation of reproduction implicates questions of personhood. The rights, obligations, privileges, and relationships previously described in this book are generally rights, obligations, privileges, and relationships of people. But who ought to be recognized as a person, subject to the legal principles that regulate our rights and obligations? When does a person, entitled to formal legal respect as such, come into existence? And how much work does this do to resolve difficult disputes involving competing fundamental interests? Can such answers be resolved through an analysis of medicine, philosophy, ethics, law, social history, anthropology, theology and other disciplines which seek to answer the basic questions of human existence?

Physicians and lawyers have necessarily been involved in these questions, as demonstrated by conflicts over government regulation of abortion and contraception, compelled medical intervention during pregnancy, and legal questions about the treatment of embryos arising out of the growing use of assisted reproductive technology. It is therefore appropriate to consider these issues in this text. The deeply contested and complex legal and philosophical questions around personhood are surveyed in the first portion of this chapter, and this issue is an important theme in the legal and ethical questions considered throughout the rest of the chapter.

The issues surveyed in this chapter also raise important gender, racial, disability, and socioeconomic justice concerns, which have motivated complaints of both too much and too little government regulation in the area of reproduction. Gender concerns are most obviously implicated, because although all people have reproductive health care needs, the ability to get pregnant has been linked to stereotypes about women's proper role in society, including presumptions of motherhood, limits on sexual liberty, exclusion from employment, and many other forms of discrimination. Socioeconomic concerns are implicated because the ability to control the timing and manner of pregnancy impacts individual and family well-being and socioeconomic opportunities. Many of the reforms combatting pregnancy-based discrimination have been driven by concerns about gender equity, as well as the particular health and social consequences of denying care that enables, prevents, or terminates pregnancy.

While reproductive *rights* advocates have traditionally focused on women's ability to prevent pregnancy and birth, a more recent movement around reproductive *justice* has shone a light on government actions that regulate and control reproduction more broadly. This includes government actions that create barriers to procreation, attempt to police behavior during pregnancy, and undermine the right to safe and dignified parenting for historically marginalized groups. For example, assumptions about who *ought* to be a parent, or who can be trusted to parent effectively, have motivated government sterilization programs targeting people with developmental disabilities, the poor, racial and ethnic minorities, and women in prison. In addition, state and federal laws have traditionally made it more difficult for some groups to realize their procreative dreams through assisted reproductive technology, especially those who do not conform to traditional gender and sexual norms.

The remaining parts of this chapter are not intended to be a comprehensive analysis of all of the legal and ethical questions raised by reproduction. The goal of this chapter is to provide structure for an analysis of issues surrounding procreation and reproduction that are likely to be of special concern to patients, the health care professionals and institutions entrusted with their care, and the attorneys helping to navigate disputes when they arise.

II. WHEN DOES HUMAN LIFE BECOME A "PERSON"?

This society has had difficulty defining who is a "person." In part, this arises out of the different and inconsistent purposes for which we seek a definition. The "person" from whom we wish to harvest a kidney for transplantation may be defined differently from the "person" who is protected by the Fourteenth Amendment, federal civil rights laws, and various other federal and state laws. Even when the purpose of the definition is settled—such as who is a person able to bring an action under state tort law—there is no consensus on when the status of "personhood" first attaches. The most obvious definition of personhood is a recursive one:

a human being (and, thus, a "person") is the reproductive product of other human beings. Even if we accept this "human stock" definition of person, however, the inquiry remains open. Does that human stock become a person, for tort law or other purposes, upon conception? Upon quickening? Upon viability? Upon birth? A year after birth? Upon physical maturity?

The definition of "person" is not limited to various stages in the development of human stock. "Personhood" could commence upon ensoulment, upon the development of self-concept, upon the development of a sense of personal history, or upon the ability to communicate through language although none of these points is easy to identify in individual cases. The resolution of the question appears to require a resort to first principles.

In the vast majority of cases, it is not difficult to distinguish a person from something else. You are easily distinguishable from your arms, your dog, your insurance company and your smart phone, as close as you may feel to each of them. The most difficult questions tend to arise at the very beginning and at the very end of human life. Just as you may be able to identify the fact that you were in love, but not be able to identify exactly when it began, or the moment when it ended, the beginnings and the endings of "personhood" are the fuzzy portions.

There are limits to what may reasonably be considered a "person," even when we limit our consideration to human stock. Few suggest that anything independent of the unified sperm and ovum, or its consequences, ought to be considered a person, although advances in cloning and cellular manipulation may challenge this assumption. A great many religious groups consider "personhood" to attach at conception. Aristotle viewed the development of the person as a three-stage process, going from vegetable (at conception) to animal (in utero), to rational (sometime after birth). For many centuries, Christian theology fixed the point of "immediate animation" when the fetus was "ensouled" as forty days after conception for males and eighty days after conception for females. St. Thomas Aquinas determined that the ensoulment took place at the time of quickening, usually fourteen to eighteen weeks after conception, and his determination had a very substantial effect on the development of the common law in England and in this country. Recently some philosophers have suggested that "personhood," at least to the extent that it includes a right to life, depends on attributes that are not likely to be developed until sometime after birth.

One thoughtful and oft-cited set of attributes of personhood has been developed by Joseph Fletcher, a bioethicist. Consider his fifteen criteria, described below, and determine whether some or all of them can be used to properly define who is your colleague in personhood and who is not. Consider whether the fact that many of these criteria disqualify fetuses, newborns, and those with serious cognitive and intellectual disabilities, including those with severe brain injuries and disorders of consciousness (discussed in Chapter 4), should or does affect their acceptability as standards. Further, does the fact that some animal or machine might eventually fulfill all of these criteria cause you to doubt their validity? What are the consequences of our failure to define as a person a clone, a highly intelligent and communicative ape, a robot, or a manufactured cell that can be brought to term in an artificial womb, with regard to our conceptions of "democracy" and "slavery," for example?

JOSEPH FLETCHER, "HUMANNESS," IN HUMANHOOD: ESSAYS IN BIOMEDICAL ETHICS
12–16 (1979).

Synthetic concepts such as human and man and person require operational terms, spelling out the which and what and when. Only in that way can we get down to cases—to normative decisions. There are always some people who prefer to be visceral and affective in their moral choices, with no desire to have any rationale for what they do. But ethics is precisely the business of rational, critical reflection (encephalic and not merely visceral) about the problems of the moral agent—in biology and medicine as much as in law, government, education, or anything else.

To that end, then, for the purposes of biomedical ethics, I now turn to a *profile of man* in concrete and discrete terms. . . . There is time only to itemize the inventory, not to enlarge upon it, but I have fifteen positive propositions. Let me set them out, in no rank order at all, and as hardly more than a list of criteria or indicators, by simple title.

1.   Minimum Intelligence

Mere biological life, before minimal intelligence is achieved or after it is lost irretrievably, is without personal status.

2.   Self-awareness

* * *

3.   Self-control

If an individual is not only not controllable by others (unless by force) but not controllable by the individual himself or herself, a low level of life is reached about on a par with that of a paramecium. . . .

4.   A Sense of Time

* * *

5.   A Sense of Futurity

How "truly human" is any man who cannot realize there is a time yet to come as well as the present? Subhuman animals do not look forward in time; they live only on what we might call visceral strivings, appetites. Philosophical anthropologies (one recalls that of William Temple, the Archbishop of Canterbury, for instance) commonly emphasize purposiveness as a key to humanness. Chesterton once remarked that we would never ask a puppy what manner of dog it wanted to be when it grows up. . . .

6.   A Sense of the Past

* * *

7.   The Capability to Relate to Others

Interpersonal relationships, of the sexual-romantic and friendship kind, are of the greatest importance for the fullness of what we idealize as being truly personal. . . .

8.   Concern for Others

Some people may be skeptical about our capacity to care about others (what in Christian ethics is often distinguished from romance and friendship as "neighbor love" or "neighbor concern"). . . . But whether concern for others is disinterested or inspired by enlightened self-interest, it seems plain that a conscious extra-ego orientation is a trait of the species . . . .

9.   Communication

Utter alienation or disconnection from others, if it is irreparable, is de-humanization. . . .

10.   Control of Existence

It is of the nature of man that he is not helplessly subject to the blind workings of physical or physiological nature. He has only finite knowledge, freedom, and initiative, but what he has of it is real and effective. . . .

11.   Curiosity

To be without affect, sunk in *anomie,* is to be not a person. Indifference is inhuman. Man is a learner and a knower as well as a tool maker and user. . . .

12.   Change and Changeability

To the extent that an individual is unchangeable or opposed to change, he denies the creativity of personal beings. It means not only the fact of biological and physiological change, which goes on as a condition of life, but the capacity and disposition for changing one's mind and conduct as well. Biologically, human beings are developmental: birth, life, health, and death are processes, not events, and are to be understood progressively, not episodically. All human existence is on a continuum, a matter of becoming . . . .

13. Balance of Rationality and Feeling

. . . . As human beings we are not coldly rational or cerebral, nor are we merely creatures of feeling and intuition. It is a matter of being both, in different combinations from one individual to another. . . .

14. Idiosyncrasy

The human being is idiomorphous, a distinctive individual. . . . To be a person is to have an identity, to be recognizable and callable by name.

15. Neocortical Function

In a way, this is the cardinal indicator, the one all the others are hinged upon. Before cerebration is in play, or with its end, in the absence of the synthesizing function of the cerebral cortex, the person is nonexistent. Such individuals are objects but not subjects. This is so no matter how many other spontaneous or artificially supported functions persist in the heart, lungs, neurologic and vascular systems. Such noncerebral processes are not personal. . . .

NOTES AND QUESTIONS

1. *Capacity for Consciousness.* Some commentators have concluded that the real consensus requirement of personhood is the capacity for conscious experience. Do you think that this is a necessary attribute? A sufficient attribute? Does it encompass many—or all—of the attributes described by Fletcher? Capacity for conscious experience is also relevant to questions about the determination of death. See Chapter 4.

2. *A Bundle of Attributes?* Fletcher commenced a serious debate over whether the persons protected by law ought to be defined in terms of attributes we wish to protect or in terms of the human stock from which the person is created. Both forms of definition may be valuable for different purposes. We provide some rights to people because they possess many or all of the attributes that distinguish human beings. The right to make medical decisions, based on the autonomy of individuals, is not accorded to those without some "minimum intelligence." See discussion of competence in Chapter 5. On the other hand, we provide minimally adequate housing, food, medical care, and other necessities for those of human stock, even when they do not meet some of Fletcher's criteria, and even when we do not provide those same benefits to others, (e.g., animals) who fail the same criteria. In the end, the Fletcher propositions may be useful in determining some of the rights of persons and the "human stock" definition may be helpful in determining others. Just as property is often described as a bundle of rights, it may turn out that "personhood" is a bundle of attributes that need to be separated out and individually analyzed.

PROBLEM: DOES THINKING MAKE IT HUMAN?

Assume that a laboratory has learned enough about neuroscience and the operation of the brain, artificial intelligence, robotics and psychology to create a machine that acts in a way that is indistinguishable from a human being. Assume that without actually seeing the mechanical interior of this machine it would be effectively impossible for a person to distinguish the machine from any another human being. It speaks well, it "sees" accurately, and it has reactions that we would expect from one with all of the attributes, including the emotional attributes, of a human being. Further, it looks like any other human being—at least from the outside—and it has an ability to learn and change the way human beings do. It does not have the ability to reproduce the way other human beings do, but it does have the ability to reconstruct itself in the workshop, and thus can continue its line. Those of you with a background in philosophy or artificial intelligence will recognize this as a slightly updated version of the renowned "Turing test" first proposed by the brilliant computer scientist Alan Turing in one of the most influential philosophical works of the last century. A. Turing, Computing Machinery and Intelligence, 49 Mind 433 (1950). See also Stuart Russell and Peter Norving, Artificial Intelligence: A Modern Approach (2d ed. 2003).

For legal and other purposes, should this machine be treated as a human being? Should it be able to vote or otherwise be protected by the Fourteenth Amendment? If you don't think it should be treated like a human being, why is that? What attribute is it missing? What if it had that attribute, too? Is there some attribute of the human mind (or the human being more generally) that is not a part of the physical human brain? What are the consequences of treating this machine as a person? What are the consequences of failing to do so?

What if a natural human being whose brain had failed as a result of an utterly disastrous stroke had her brain replaced by an electronic version. Would that still be a person? Would it be the same person? What is the relationship between the mind, the brain, our individual histories and values, and our personhood? Is there a place for the consideration of the "soul" in this discussion? Are the questions raised here the same as the basic questions raised by the "mind vs. brain" debate over the past millennia? There has been a great deal of discussion of the issue of machine-as-human recently, with several provocative books, articles, and movies exploring both the future possibilities and ethical implications of continued evolution in AI. There has also been discussion of intellectual property issues that may grow out of nontraditional human beings. See, e.g., Eric Zylstra, Presumed Sapient: A Proposed Test for the Constitutional Personhood and Patentability of Human-Animal Chimeras and Hybrids, 46 U. S. F. L. Rev. 1075 (2012).

Is the only way out of this dilemma to define a human being as one created through human stock—that is, one who is born (in more or less the traditional way) of other human beings? Of course, as we saw earlier, even the adoption of a "human stock" definition does not answer the question of when that human stock becomes a person. What is the attribute of the human stock that makes it a person—genetic uniqueness? Responsiveness? The potential to be born? The appearance of a human being? Consider the following list of the alternative medical points of personhood.

C.R. AUSTIN, "WHEN DOES A PERSON'S LIFE REALLY BEGIN?" IN HUMAN EMBRYOS: THE DEBATE ON ASSISTED REPRODUCTION
22–31 (1989).

. . . . Probably most people who were asked this question would answer "at fertilization" (or "conception"). Certainly, several interesting and unusual things happen then—it is really the most *obvious* event to pick—but for biologists the preceding and succeeding cellular processes are *equally* important. Nevertheless, "fertilization" continues to be the cry of many religious bodies and indeed also of the august World Medical Association, who, in 1949, adopted the Geneva Convention Code of Medical Ethics, which contains the clause: "I will maintain the utmost respect for human life from the time of conception." So we do need to look more closely at this choice, for a generally acceptable "beginning" for human life would be a great help in reaching ethical and legal consensus.

In the first place human *life,* as such, obviously begins before fertilization, since the egg or oocyte is alive before sperm entry, as were innumerable antecedent cells, back through the origin of species into the mists of time. A more practical starting point would be that of the life of the human *individual,* so it is individuality that we should be looking for, at least as one of the essential criteria. Now the earliest antecedents of the eggs, as of sperm, are the primordial germ cells, which can be seen as a group of distinctive little entities migrating through the tissues of the early embryo. When they first become recognizable, they number only about a dozen or two, but they multiply fast and soon achieve large numbers, reaching a peak of 7–10 million about 6 months after conception. Then, despite continued active cell division, there is a dramatic decline in the cell population, which has tempted people to suggest that some sort of "selection of the fittest" occurs, but there is no good evidence in support of this idea; nor is there any good reason to look for individuality in that mercurial population. In due course, the primordial germ cells, while still undergoing cell divisions, settle down in the tissues of the future ovary, change subtly in their characteristics, and thus become oogonia; and then, soon after birth, *cell division ceases,* the cells develop large nuclei and are now recognizable as primary oocytes. From now on, there are steady cell losses but no further cell divisions . . . ; it is the same entity that was a primary oocyte, becomes a fertilized egg, and then develops as an embryo. The primary oocytes are very unusual cells, for they have the capacity to live for much longer than most other body cells; the *same* oocytes can be seen in the ovaries of women approaching the menopause—cells that have lived for about 40 years or longer. And it is with the emergence of the primary oocytes that we can hail the start of *individuality*. Then, in those oocytes that are about to be ovulated, the first meiotic division takes place—another important step, for the "shuffling" of genes that occurs at that point bestows *genetic uniqueness* on the oocyte. So both individuality and genetic uniqueness are established before sperm penetration and fertilization; these processes have distinctly different actions—providing the stimulus that initiates cleavage and contributing to biparental inheritance. Thus, the preferred choice for the start of the human individual should surely be the formation of the primary oocyte, but there is certainly no unanimity on this score.

Passing over now the popularity of fertilization, for many people it is instead the emergence of the embryonic disc and primitive streak that most appeals as the stage in which to identify the start of "personhood" (one or more persons, in view of the imminent possibility of twinning), and there is much to support this opinion. Here, for the first time, are structures that are designed to have a different destiny than *all the rest of the embryo*— they represent the primordium of the fetus, and the developmental patterns of embryo and fetus progressively diverge from this stage onwards. An additional point is that this new emergence is not inevitable, for in around one in two-thousand pregnancies the embryo grows, often to quite a large size, but there is no fetus . . . .

At the time of appearance of the embryonic disc, and shortly beforehand, the process of implantation is occurring, and this is considered by many to have special significance in relation to embryonic potential—so far as we know, implantation cannot occur once the development of the embryo has passed the stage when interaction with . . . the uterus normally takes place.

But despite all that has been said, there are still many folk who remain unconvinced—is the being at this stage sufficiently "human" to qualify as the start of a person? After all, the disc is just a collection of similar cells, virtually undifferentiated, poorly delineated from its surroundings, about a fifth of a millimetre long, non-sentient, and without the power of movement. It is in no way a "body" and it does not bear the faintest resemblance to a human being—*and* the soul cannot enter yet, for the disc may yet divide in the process of twinning, and the soul being unique is indivisible. Also, it is argued that we should be looking for some spark of personality, and a moral philosopher has proposed that some sort of responsiveness is an essential feature.

One of the earliest succeeding changes in the direction of humanness could be the development of the heart primordium, and soon after that the beginnings of a circulatory system; the first contractions of the heart muscle occur possibly as early as day 21, with a simple tubular heart at that stage, and in the fourth week a functional circulation begins. With the heart beats we have the first movements initiated within the embryo (?fetus) and thus in a way the first real "sign of life." The conceptus is now about 6 mm long. During the fifth and sixth weeks, nerve fibres grow out from the spinal cord and make contact with muscles, so that at this time or soon afterwards, a mechanical or electrical stimulus might elicit a muscle twitch; this is important for it would be the first indication of sentient existence—of "responsiveness." At this stage, too, the embryo could possibly feel pain. But, still, some would find cause to demur: only an expert could tell that this embryo/fetus, now 12–13 mm in length, with branchial arches (corresponding to the "gill-slits" in non-mammalian embryos), stubby limbs, and a prominent tail, is human. A marginally more acceptable applicant is the fetus at 7½ weeks, when the hands and feet can be seen to have fingers and toes, and thereafter physical resemblance steadily improves; also at this time, a special gene on the Y-chromosome (the "testis-determining factor" or TDF) is switched on, and the fetuses that have this chromosome, the males, proceed thenceforward to develop *as* males, distinguishable from females.

At about 12 weeks, electrical activity can be detected in the brain of the fetus, which could signal the dawn of consciousness. Here, we would seem to have a very logical stage marking the *start* of a person, for the cessation of electrical activity in the brain ("brain death") is accepted in both medical and legal circles as marking the *termination* of a person—as an indication that life no longer exists in victims of accidents or in patients with terminal illnesses. Around the fourth or fifth month of pregnancy, the mother first experiences movements of the fetus ("quickening"), which were regarded by St. Thomas Aquinas as the first indication of life, for he believed that life was distinguished by two features, knowledge and movement; moreover, it would seem logical that the fetus would move when the *animus* (life or soul) took up residence. . . .

At about 24 weeks, the fetus reaches a state in which it can commonly survive outside the maternal body, with assistance. . . . Just which stage marks the start of a person's life is a matter of personal opinion. Much of the foregoing argumentation may seem to some people difficult to comprehend, especially if they have not had formal training in biology, and to others may even seem irrelevant, in view of the firm line taken by many church authorities. But it really is important that we should try to reach a consensus on just when a person's life should be held to begin, for the decision does have important practical consequences—it directly affects the rights of other embryos, of fetuses, and of people.

III. ABORTION

Abortion is the area where the most high profile and controversial debates have occurred over questions about when life begins, as well as how the answer to this question should inform government regulation of reproduction. We begin this section with the seminal case, *Roe v. Wade*, defining abortion as a privacy right. This right has been subject to several decades of legal contests. In most of these, a state has enacted legislation that regulates or restricts access to abortions, and patients or providers challenge the laws on constitutional grounds. Abortion restrictions or regulations have tested, and often narrowed the boundaries of the constitutional right to abortion established in *Roe*. As this edition goes to print near the end of 2021, *Roe* still stands, even if in a somewhat altered form. But as described further in the Notes and Questions following *Whole Woman's Health v. Hellerstedt*, recent developments have called the fate of *Roe* into question once again.

A. ABORTION AS A PRIVACY RIGHT

ROE V. WADE
Supreme Court of the United States, 1973.
410 U.S. 113.

MR. JUSTICE BLACKMUN delivered the opinion of the Court.

* * *

We forthwith acknowledge our awareness of the sensitive and emotional nature of the abortion controversy, of the vigorous opposing views, even among physicians, and of the deep and seemingly absolute convictions that the subject inspires. One's philosophy, one's experiences, one's exposure to the raw edges of human existence, one's religious training, one's attitudes toward life and family and their values, and the moral standards one establishes and seeks to observe, are all likely to influence and to color one's thinking and conclusions about abortion.

In addition, population growth, pollution, poverty, and racial overtones tend to complicate and not to simplify the problem.

Our task, of course, is to resolve the issue by constitutional measurement, free of emotion and of predilection. We seek earnestly to do this, and, because we do, we have inquired into, and in this opinion place some emphasis upon, medical and medical-legal history and what that history reveals about man's attitudes toward the abortion procedure over the centuries. We bear in mind, too, Mr. Justice Holmes' admonition in his now-vindicated dissent in Lochner v. New York[ ]:

> [The Constitution] is made for people of fundamentally differing views, and the accident of our finding certain opinions natural and familiar

or novel and even shocking ought not to conclude our judgment upon the question whether statutes embodying them conflict with the Constitution of the United States.

* * *

The principal thrust of appellant's attack on the Texas statutes is that they improperly invade a right, said to be possessed by the pregnant woman, to choose to terminate her pregnancy. Appellant would discover this right in the concept of personal "liberty" embodied in the Fourteenth Amendment's Due Process Clause; or in personal, marital, familial, and sexual privacy said to be protected by the Bill of Rights or its penumbras,[ ]; or among those rights reserved to the people by the Ninth Amendment[ ]. Before addressing this claim, we feel it desirable briefly to survey, in several aspects, the history of abortion, for such insight as that history may afford us, and then to examine the state purposes and interests behind the criminal abortion laws.

VI

It perhaps is not generally appreciated that the restrictive criminal abortion laws in effect in a majority of States today are of relatively recent vintage. Those laws, generally proscribing abortion or its attempt at any time during pregnancy except when necessary to preserve the pregnant woman's life, are not of ancient or even of common-law origin. Instead, they derive from statutory changes effected, for the most part, in the latter half of the 19th century.

[The Court then reviewed, in great detail, ancient attitudes, the Hippocratic Oath, the common law, English statutory law, American Law, the position of the American Medical Association, the position of the American Public Health Association, and the position of the American Bar Association.]

VII

Three reasons have been advanced to explain historically the enactment of criminal abortion laws in the 19th century and to justify their continued existence.

It has been argued occasionally that these laws were the product of a Victorian social concern to discourage illicit sexual conduct. Texas, however, does not advance this justification in the present case, and it appears that no court or commentator has taken the argument seriously. . . .

A second reason is concerned with abortion as a medical procedure. When most criminal abortion laws were first enacted, the procedure was a hazardous one for the woman. . . . Thus, it has been argued that a State's real concern in enacting a criminal abortion law was to protect the pregnant woman, that is, to restrain her from submitting to a procedure that placed her life in serious jeopardy.

Modern medical techniques have altered this situation. Appellants and various amici refer to medical data indicating that abortion in early pregnancy, this is, prior to the end of the first trimester, although not without its risk, is now relatively safe. Mortality rates for women undergoing early abortions, where the procedure is legal, appear to be as low as or lower than the rates for normal childbirth. Consequently, any interest of the State in protecting the woman from an inherently hazardous procedure, except when it would be equally dangerous for her to forgo it, has largely disappeared. Of course, important state interests in the area of health and medical standards do remain. The State has a legitimate interest in seeing to it that abortion, like any other medical procedure, is performed under circumstances that assure maximum safety for the patient. . . .

The third reason is the State's interest—some phrase it in terms of duty—in protecting prenatal life. Some of the argument for this justification rests on the theory that a new human life is present from the moment of conception. . . . Logically, of course, a legitimate state interest in this area need not stand or fall on acceptance of the belief that life begins at conception or at some other point prior to live birth. [R]ecognition may be given to the less rigid claim that as long as at least potential life is involved, the State may assert interests beyond the protection of the pregnant woman alone.

* * *

It is with these interests, and the weight to be attached to them, that this case is concerned.

VIII

The Constitution does not explicitly mention any right of privacy. In a line of decisions, however, going back perhaps as far as[ ] 1891 the Court has recognized that a right of personal privacy, or a guarantee of certain areas or zones of privacy, does exist under the Constitution. In varying contexts, the Court or individual Justices have, indeed, found at least the roots of that right in the First Amendment,[ ] in the Fourth and Fifth Amendments,[ ] in the penumbras of the Bill of Rights, Griswold v. Connecticut,[ ] the Ninth Amendment,[ ] or in the concept of liberty guaranteed by the first section of the Fourteenth Amendment.[ ] These decisions make it clear that only personal rights that can be deemed "fundamental" or "implicit in the concept of ordered liberty,"[ ] are included in this guarantee of personal privacy. They also make it clear that the right has some extension to activities relating to marriage,[ ] family relationships,[ ] and child rearing and education[ ].

This right of privacy, whether it be founded in the Fourteenth Amendment's concept of personal liberty and restrictions upon state action, as we feel it is, or, as the District Court determined, in the Ninth Amendment's reservation of rights to the people, is broad enough to encompass a woman's decision whether or not to terminate her pregnancy. The detriment that the State would impose upon the pregnant woman by denying this choice altogether is apparent. Specific and direct harm medically diagnosable even in early pregnancy may be involved. Maternity, or additional offspring, may force upon the woman a distressful life and future. Psychological harm may be imminent. Mental and physical health may be taxed by child care. There is also the distress, for all concerned, associated with the unwanted child, and there is the problem of bringing a child into a family already unable, psychologically and otherwise, to care for it. In other cases, as in this one, the additional difficulties and continuing stigma of unwed motherhood may be involved. All these are factors the woman and her responsible physician necessarily will consider in consultation.

On the basis of elements such as these, appellant and some amici argue that the woman's right is absolute and that she is entitled to terminate her pregnancy at whatever time, in whatever way, and for whatever reason she alone chooses. With this we do not agree. . . . [A] State may properly assert important interests in safeguarding health, in maintaining medical standards, and in protecting potential life. At some point in pregnancy, these respective interests become sufficiently compelling to sustain regulation of the factors that govern the abortion decision.

* * *

IX

* * *

Appellee argues that the State's determination to recognize and protect prenatal life from and after conception constitutes a compelling state interest. [W]e do not agree fully with [this] formulation.

[ ] The appellee and certain amici argue that the fetus is a "person" within the language and meaning of the Fourteenth Amendment. In support of this, they outline at length and in detail the well-known facts of fetal development. If this suggestion of personhood is established, the appellant's case, of course, collapses, for the fetus' right to life is then guaranteed specifically by the Amendment. The appellant conceded as much on reargument. On the other hand, the appellee conceded on reargument that no case could be cited that holds that a fetus is a person within the meaning of the Fourteenth Amendment.

The Constitution does not define "person" in so many words. Section 1 of the Fourteenth Amendment contains three references to "person." The first, in defining "citizens," speaks of "persons born or naturalized in the United States." The word also appears both in the Due Process Clause and in the Equal Protection Clause. "Person" is used in other places in the Constitution: in the listing of qualifications for Representatives and Senators [ ]; in the Apportionment Clause [ ][53] in the Migration and Importation provision [ ]; in the Emolument Clause [ ]; in the Electors provisions [ ]; in the provision outlining qualifications for the office of President [ ]; in the Extradition provisions, [ ], and the superseded Fugitive Slave Clause; and in the Fifth, Twelfth, and Twenty-second Amendments, as well as in §§ 2 and 3 of the Fourteenth Amendment. But in nearly all these instances, the use of the word is such that it has application only postnatally. None indicates, with any assurance, that it has any possible prenatal application.[54]

* * *

All this, together with our observation . . . that throughout the major portion of the 19th century prevailing legal abortion practices were far freer than they are today, persuades us that the word "person," as used in the Fourteenth Amendment, does not include the unborn. . . .

* * *

Texas urges that, apart from the Fourteenth Amendment, life begins at conception and is present throughout pregnancy, and that, therefore, the State has a compelling interest in protecting that life from and after conception. We need not resolve the difficult question of when life begins. When those trained in the respective disciplines of medicine, philosophy, and theology are unable to arrive at any consensus, the judiciary, at this point in the development of man's knowledge, is not in a position to speculate as to the answer.

* * *

---

[53] We are not aware that in the taking of any census under this clause, a fetus has ever been counted.

[54] When Texas urges that a fetus is entitled to Fourteenth Amendment protection as a person, it faces a dilemma. Neither in Texas nor in any other State are all abortions prohibited. Despite broad proscription, an exception always exists. The exception contained in Art. 1196, for an abortion procured or attempted by medical advice for the purpose of saving the life of the mother, is typical. But if the fetus is a person who is not to be deprived of life without due process of law, and if the mother's condition is the sole determinant, does not the Texas exception appear to be out of line with the Amendment's command?

There are other inconsistencies between Fourteenth Amendment status and the typical abortion statute. It has already been pointed out [ ] that in Texas the woman is not a principal or an accomplice with respect to an abortion upon her. If the fetus is a person, why is the woman not a principal or an accomplice? Further, the penalty for criminal abortion specified by Art. 1195 is significantly less than the maximum penalty for murder prescribed by Art. 1257 of the Texas Penal Code. If the fetus is a person, may the penalties be different?

X

In view of all this, we do not agree that, by adopting one theory of life, Texas may override the rights of the pregnant woman that are at stake.

* * *

With respect to the State's important and legitimate interest in the health of the mother, the "compelling" point, in the light of present medical knowledge, is at approximately the end of the first trimester. This is so because of the now-established medical fact . . . that until the end of the first trimester mortality in abortion may be less than mortality in normal childbirth. It follows that, from and after this point, a State may regulate the abortion procedure to the extent that the regulation reasonably relates to the preservation and protection of maternal health. . . .

This means, on the other hand, that, for the period of pregnancy prior to this "compelling" point, the attending physician, in consultation with his patient, is free to determine, without regulation by the State, that, in his medical judgment, the patient's pregnancy should be terminated. If that decision is reached, the judgment may be effectuated by an abortion free of interference by the State.

With respect to the State's important and legitimate interest in potential life, the "compelling" point is at viability. This is so because the fetus then presumably has the capability of meaningful life outside the mother's womb. State regulation protective of fetal life after viability thus has both logical and biological justifications. If the State is interested in protecting fetal life after viability, it may go so far as to proscribe abortion during that period, except when it is necessary to preserve the life or health of the mother.

* * *

XI

To summarize and to repeat:

1.   A state criminal abortion statute of the current Texas type, that excepts from criminality only a *lifesaving* procedure on behalf of the mother, without regard to pregnancy stage and without recognition of the other interests involved, is violative of the Due Process Clause of the Fourteenth Amendment.

(a)   For the stage prior to approximately the end of the first trimester, the abortion decision and its effectuation must be left to the medical judgment of the pregnant woman's attending physician.

(b)   For the stage subsequent to approximately the end of the first trimester, the State, in promoting its interest in the health of the mother, may, if it chooses, regulate the abortion procedure in ways that are reasonably related to maternal health.

(c) For the stage subsequent to viability, the State in promoting its interest in the potentiality of human life may, if it chooses, regulate, and even proscribe, abortion except where it is necessary, in appropriate medical judgment, for the preservation of the life or health of the mother.

* * *

NOTES AND QUESTIONS

1. *State Regulation of Abortion.* The Court rejected Texas' ability to apply its "theory of life" to "override the rights of the pregnant woman at stake," but the Court also acknowledged the state's interest in protecting prenatal life. How did the Court understand the various interests at stake and what balance did it strike?

2. *The State's "Theory of Life."* In *Roe*, the Court was confronted with the question of when personhood begins for purposes of determining rights under the Fourteenth Amendment of the Constitution, but Justice Blackmun refused to answer "the difficult question of when life begins." Indeed, the Supreme Court has never formally determined when a fetus becomes a "person" for Constitutional purposes. This has left an open question about whether the substantive provisions of the Constitution (as interpreted by the Court) put any limit on the way that states may define "person" for *other* purposes? In 1989, the Supreme Court faced this question in Webster v. Reproductive Health Services, 492 U.S. 490, 109 S.Ct. 3040, 106 L.Ed.2d 410 (1989). The Supreme Court reviewed a Missouri statute that restricted the availability of abortions in several ways. The statute also included a preamble that defined personhood:

1. The general assembly of this state finds that:

   (1) The life of each human being begins at conception;

   (2) Unborn children have protectable interests in life, health, and well-being;

   * * *

2. [T]he laws of this state shall be interpreted and construed to acknowledge on behalf of the unborn child at every stage of development, all the rights, privileges, and immunities available to other persons, citizens, and residents of this state, subject only to the Constitution of the United States, and decisional interpretations thereof . . . .

3. As used in this section, the term "unborn children" or "unborn child" shall include all unborn child or children or the offspring of human beings from the moment of conception until birth at every stage of biological development.

* * *

Vernon's Ann.Mo.Stat. § 1.205. This preamble was attacked on the grounds that it was beyond the constitutional authority of the state legislature to define personhood, at least to the extent that the definition extended personhood to pre-viable fetuses. The Supreme Court sidestepped that question by concluding that the preamble was nothing more than a state value judgment favoring childbirth over abortion, and that such a value judgment was clearly within the authority of the legislature. It noted that the preamble itself did not regulate abortions and could be interpreted to do no more than offer protections to unborn children in tort and probate law.

Questions about personhood have influenced states' decisions to expand statutory and common law rights for fetuses outside the abortion context. The debate over the common law recognition of the personhood of the fetus has been waged primarily over whether a fetus may recover under state wrongful death and survival statutes. Today, many states allow such claims as a matter of common law or through statute. See Ex Parte Ankrom, 152 So.3d 397, 425–428 (Ala. 2013) (Parker, J. concurring) (noting that forty states and D.C. allow recovery for the wrongful death of an unborn child when post-viability injuries to that child cause its death before birth, and eleven of these states allow recovery regardless of stage of pregnancy). One reason given is the perception that a live birth requirement is an arbitrary line that serves no purpose of the wrongful death and survival statutes. Others, more influenced by the arguments over abortion, have urged that such actions should be recognized because the fetus is a full human being entitled to all protections accorded to all other human beings. Indeed, debates about extending protections to fetuses in these contexts are often shaped by the fear or hope that doing so will influence abortion jurisprudence, allowing for greater restrictions on a woman's right to have an abortion.

Over the past several years, some states have made the penalties for feticide commensurate with the penalties for homicide. Typically such statutes were used to allow recovery or impose punishment on third parties who harmed a fetus. Id. at 423–424 (noting that a majority of states treat killing an unborn child as criminal homicide unless it occurs as the result of a medical abortion). Increasingly, however feticide laws are being used to punish pregnant women who want to carry a pregnancy term, but who suffer a miscarriage or stillbirth believed to be the result of harmful or risky behavior by the woman. For a discussion of how feticide and other "fetal protection" laws are used to regulate and punish women's behavior during pregnancy, see Section VI of this chapter.

3.   *The Role of Sexual Morality in Reproductive Regulation.* In exploring the reasons for the emergence of abortion bans, the Court mentions, but quickly dismisses, the argument that such laws were the product of a "Victorian social concern to discourage illicit sexual conduct." But does the Court give this short shrift? When it says no court or commentator has taken the argument seriously, is it that the Court does not believe that this is an underlying motivation of such laws? Or is it dismissing its legitimacy as a state interest for purposes of this analysis? If there is evidence that a state's desire

to regulate sexual morality motivated abortion regulation, what impact, if any, should that have on a court's constitutional analysis? For a history of the relationship between sex, religion, and reproductive regulation, see Geoffrey R. Stone, Sex and the Constitution (2017).

4.   *Bans on Use of Government Resources for Abortion. Roe* invalidated a criminal ban on abortion, but access to health care also depends on having the resources to access a particular service. A 2014 survey found that three-fourths of abortion patients were low income—49% living at less than the federal poverty level, and 26% living at 100–199% of the poverty level. See Jenna Jerman et al., Characteristics of U.S. Abortion Patients in 2014 and Changes Since 2008, Guttmacher Institute (May 2016). Traditionally, the poor have been less likely to be employed in jobs that offered health insurance, and they have been more likely to depend on public resources, such as public hospitals or public insurance. Pregnant women are eligible for Medicaid coverage only if their income is low enough to qualify. Immediately after *Roe*, the battle over abortion quickly moved into the realm of funding. To the extent that federal and state governments were funding or providing health care services, did the privacy right in *Roe* have anything to say about government's ability to exclude abortion from these programs?

In a series of cases, the Supreme Court has made clear that the Constitution does not prevent the government from excluding abortion from public insurance coverage or from banning abortions in government run facilities. Beal v. Doe, 432 U.S. 438 (1977); Maher v. Roe, 432 U.S. 464 (1977); Poelker v. Doe, 432 U.S. 519 (1977). In 1980, in Harris v. McRae, 448 U.S. 297 (1980), the Court addressed the constitutionality of the Hyde Amendment, which limits federal Medicaid funding for abortions. The Hyde Amendment only funds abortions in cases of life endangerment, rape or incest. It excludes funding for abortions that may not be viewed as life threatening but are otherwise medically necessary because carrying a pregnancy to term would pose a serious danger to a woman's health or may even shorten her life. Examples of cases where abortions may be deemed medically necessary, though not immediately life threatening, are where pregnancy would exacerbate medical risky conditions, such as heart conditions, diabetes, or cancer, and could lead to permanent damage to the woman's health. The Supreme Court upheld the Hyde amendment against due process and equal protection challenges, with Justice Stewart writing for the court:

> . . . . [R]egardless of whether the freedom of a woman to choose to terminate her pregnancy for health reasons lies at the core or the periphery of the due process liberty recognized in Roe v. Wade, 410 U.S. 113, 93 S.Ct. 705, 35 L.Ed.2d 147 (1973), it simply does not follow that a woman's freedom of choice carries with it a constitutional entitlement to the financial resources to avail herself of the full range of protected choices. [A]lthough government may not place obstacles in the path of a woman's exercise of her freedom of choice, it need not remove those not of its own creation. Indigency falls in the latter category. The financial constraints that restrict an indigent woman's ability to enjoy the full

range of constitutionally protected freedom of choice are the product not of governmental restrictions on access to abortions, but rather of her indigency. Although Congress has opted to subsidize medically necessary services generally, but not certain medically necessary abortions, the fact remains that the Hyde Amendment leaves an indigent woman with at least the same range of choice [ ] as she would have had if Congress had chosen to subsidize no health care costs at all. . . . Although the liberty protected by the Due Process Clause affords protection against unwarranted government interference with freedom of choice in the context of certain personal decisions, it does not confer an entitlement to such funds as may be necessary to realize all the advantages of that freedom. . . .

Quoting another case that upheld a similar abortion restriction by a state, the Court made clear that " 'the constitutional freedom recognized in Roe v. Wade and its progeny, did not prevent [government] from making a value judgment favoring childbirth over abortion, and . . . implement[ing] that judgment by the allocation of public funds.' " Id. (quoting Maher v. Roe, 432 U.S. 464 (1977).

5. *Impact of the ACA.* The ACA reinforced Hyde Amendment restrictions, which have been attached to Medicaid and other relevant appropriations by the U.S. Department of Health and Human Services (HHS) since the 1970s. The ACA extended federal funding limits on abortion in significant ways. It excludes abortion from the required essential health benefits, requires each health insurance exchange to offer at least one multi-state plan that does not include abortion coverage, and allows states to forbid exchange plans from covering abortion. Most state laws banning abortion coverage provide narrow exceptions for life endangerment, rape, or incest, but Louisiana and Tennessee do not provide any exceptions.

The ACA permits plans to cover abortion but establishes strict rules for how this can be done. 42 U.S.C.A. § 18023. Plans that offer abortion coverage must essentially do so by selling a separate abortion rider; that is, they must charge a separate, actuarially segregated premium for that service, which is paid separately by each insured who desires that coverage (and each insured's employer, if the employer contributes to the employee's health coverage). The actuarial value of this coverage may not count savings arising from the performance of an abortion (such as prenatal care or delivery). These restrictions have created administrative barriers to abortion coverage, as well as confusion among enrollees as to whether certain plans even cover abortion. For example, a 2019 review of marketplace plans found that eight states that did not have laws restricting abortion coverage (and thus allowed plans to cover it) nonetheless had no marketplace plans that included coverage. See Alina Salganicoff et al., Coverage for Abortion Services in Medicaid, Marketplace Plans and Private Plans, The Henry J. Kaiser Family Foundation (Jun. 2019). Before the ACA, most private health insurance plans covered abortion. Now, "as a combined result of the state laws and insurance company choices, women in 34 states . . . do not have access to insurance coverage for abortions through a Marketplace plan—the only place where consumers can qualify for tax subsidies to help pay for the cost of health insurance premiums if they are income eligible." Id.

Finally, the ACA contains a provision providing that nothing in the ACA shall be construed as preempting state law regarding the prohibition of or requirement of coverage, funding, or procedural requirements on abortion. 42 U.S.C.A. § 18023(c)(1). This has been understood to preserve state power to regulate abortion coverage, outside of the specific limits mentioned above. But see the Note on Federal Conscience Protections at the end of Section III of this chapter.

6.   *State Approaches to Funding Abortion.* In the face of federal funding restrictions on abortion (pre- and post-ACA), states have responded very differently. Most states have followed the federal government's lead in restricting abortion coverage. In Medicaid, thirty-five states (including D.C.) prohibit coverage of abortions excluded by the Hyde Amendment, and one of those states is even more restrictive, only covering abortion in cases of life-endangerment. For private insurance, twenty-six states prohibit marketplace plans from offering abortion coverage, and eleven of those prohibit all insurers from providing abortion coverage. Again, most laws allow narrow exceptions similar to those in the Hyde Amendment. See Salganicoff, supra.

On the other hand, some states have responded by filling in the gaps. Sixteen states use only state funds for abortions excluded by the Hyde Amendment. Id. Some state legislatures have done this voluntarily, while others have been required to do so by the courts. See, e.g., Right to Choose v. Byrne, 450 A.2d 925 (N.J. 1982); Moe v. Secretary of Administration and Finance, 417 N.E.2d 387 (Mass. 1981); Committee to Defend Reproductive Rights v. Myers, 625 P.2d 779 (Cal. 1981). Since *Harris v. McRae*, thirteen state supreme courts have recognized a state constitutional right to have state Medicaid programs pay for abortions as it does other medical procedures. In the case of private insurance, however, only three states require nearly all insurance plans to provide abortion coverage, including marketplace plans.

7.   *Beyond Abortion Funding: Attempts to Defund Planned Parenthood & the Fungibility Theory.* For decades, anti-abortion advocates have gone further than simply seeking bans on funding for abortion services; they have proposed absolute government funding bans on clinics that perform abortions or have an affiliation with such clinics. Planned Parenthood, widely recognized as a provider of abortion and advocate for access to legal abortion services, has been the primary target of such defunding attempts. It is important to distinguish this from bans on funding used for abortion: as noted above, there have long been prohibitions on the use of federal funding for abortion except in extreme cases, and many states have taken this approach with state funding as well. Funding bans targeting clinics, like Planned Parenthood, would deprive such clinics of government funding for *any* service—they would not be able to receive funding to provide the non-abortion services they typically provide, such as health checkups, cancer screenings, disease prevention, and contraception. In addition, some of the bans proposed to date would apply to

clinics that do not even perform abortions. The theory behind this more sweeping funding ban is the "fungibility" of money—the presumed interchangeability of government and private funds. The argument is that any amount of government money that funds such clinics frees up money to be used to provide abortion, thus funding it indirectly.

At the federal level, bans targeting Planned Parenthood by name have been introduced, but have not passed both houses of Congress. See, e.g., Defund Planned Parenthood Act of 2017, H.R. 354, 115th Cong. (2017) (providing that for one-year, no federal funding "may be made available for *any purpose* to Planned Parenthood Federation of America, Inc. [(PPFA)] or *any affiliate* or *clinic* of [PPFA], unless such entities certify that [PPFA and its] affiliates and clinics will not perform, and will not provide any funds to any other entity that performs, an abortion during such period.") (emphasis added). A similar provision was part of the ACA repeal and replace bill approved by the House in May 2017. See American Health Care Act, H.R. 1628, 115th Cong. (2017).

State bans have been easier to enact but have been challenged on several grounds. The most successful challenges have come in the context of exclusions from the Medicaid program. Most circuits hearing these challenges have struck down states' attempts to exclude Planned Parenthood and similar clinics from receiving Medicaid payments, as violating a provision in the Medicaid Act that guarantees patients the freedom to choose their provider; but recently two circuits have allowed the exclusion. Challenges to state exclusions from other federally funded programs have yielded mixed results. See, e.g., Planned Parenthood Ass'n of Utah (PPAU) v. Herbert, 828 F.3d 1245 (10th Cir. 2016) (finding PPAU established a likelihood of success on its claim that the Governor's directive to a state health department official to stop distributing federal grant funds to PPAU violated the First and Fourteenth Amendments under the unconstitutional conditions doctrine, where the denial of funding was likely retaliation for PPAU's role in providing abortions, advocacy for abortion rights, and its affiliation with Planned Parenthood Federation of America, Inc.). But see Planned Parenthood of Greater Ohio v. Hodges, 201 F. Supp. 3d 898 (S.D. Ohio 2016), *aff'd*, 888 F.3d 224 (6th. Cir. 2018), *rehearing en banc*, 917 F.3d 908 (6th Cir. 2019) (upholding a state law prohibiting federal funds to entities that performed or promoted nontherapeutic abortions, or contracted with or became an affiliate of such an entity); Planned Parenthood Ass'n of Hidalgo County Texas, Inc. v. Suehs, 692 F.3d 343 (5th Cir. 2012) (upholding a narrower state ban on Women's Health Program (WHP) funds to entities that perform or promote elective abortions, in light of WHP's stated goal of subsidizing non-abortion family planning).

8. *After Roe v. Wade*. The Court's decision invigorated political forces opposed to abortion, encouraging state legislatures to seek creative ways to discourage abortions without directly running afoul of the right articulated in *Roe*. Outside of the funding context, *Roe* initially proved to be a formidable check on state abortion restrictions. The "compelling interest" standard was used to strike down a number of state regulations of abortion. For example, a

requirement that all second-trimester abortions be performed in a hospital was struck down as an unreasonable health regulation in light of evidence that such abortions could be safely performed in an outpatient clinic. City of Akron v. Akron Center for Reproductive Health, Inc., 462 U.S. 416 (1983), *overruled by* Planned Parenthood v. Casey, 505 U.S. 833 (1992). In another case, the Court invalidated a state's requirement that physicians must provide certain information to women before an abortion, such as descriptions of fetal characteristics or locations of agencies providing alternatives to abortion. The Court viewed the information as medically irrelevant, emphasizing that "[s]tates are not free, under the guise of protecting maternal health or potential life, to intimidate women into continuing pregnancies." Thornburgh v. American College of Obstetricians and Gynecologists, 476 U.S. 747 (1986), *overruled by* Planned Parenthood v. Casey, 505 U.S. 833 (1992). Other state abortion restrictions repeatedly landed in court and were used to test the boundaries of this privacy right.

Gradually, however, the number of justices supporting this approach declined over time. *Roe* was reaffirmed more than a dozen times in its first decade, but by 1986 the 7–2 majority was down to 5–4, Thornburgh v. American College of Obstetricians and Gynecologists, 476 U.S. 747 (1986), and by 1989 the Court appeared to be evenly divided, with Justice O'Connor unwilling to confront the issue. Webster v. Reproductive Health Services, 492 U.S. 490 (1989). Eventually some of the same types of laws that had been invalidated by the Court under the reasoning of *Roe v. Wade*, made it back to the Supreme Court in 1992, in *Planned Parenthood v. Casey* (below). Although the Court explicitly affirmed *Roe's* "essential holding," *Casey's* reasoning and outcome signaled a significant re-shaping of abortion jurisprudence that would open the door to increased state regulation.

B. *ROE* REVISITED: THE "UNDUE BURDEN" TEST

PLANNED PARENTHOOD OF SOUTHEASTERN PENNSYLVANIA V. CASEY

Supreme Court of the United States, 1992.
505 U.S. 833.

JUSTICE O'CONNOR, JUSTICE KENNEDY, and JUSTICE SOUTER announced the judgment of the Court and delivered the opinion of the Court with respect to Parts I, II, III, V-A, V-C, and VI, an opinion with respect to Part V-E, in which JUSTICE STEVENS joins, and an opinion with respect to Parts IV, V-B, and V-D.

I.

Liberty finds no refuge in a jurisprudence of doubt. Yet 19 years after our holding that the Constitution protects a woman's right to terminate her pregnancy in its early stages,[ ] that definition of liberty is still questioned. . . .

At issue in these cases are five provisions of the Pennsylvania Abortion Control Act of 1982. . . . The Act requires that a woman seeking an abortion give her informed consent prior to the abortion procedure, and specifies that she be provided with certain information at least 24 hours before the abortion is performed.[ ] For a minor to obtain an abortion, the Act requires the informed consent of one of her parents, but provides for a judicial bypass option if the minor does not wish to or cannot obtain a parent's consent.[ ] Another provision of the Act requires that, unless certain exceptions apply, a married woman seeking an abortion must sign a statement indicating that she has notified her husband of her intended abortion.[ ] The Act exempts compliance with these three requirements in the event of a "medical emergency," which is defined in the Act.[ ] In addition to the above provisions regulating the performance of abortions, the Act imposes certain reporting requirements on facilities that provide abortion services.[ ]

* * *

After considering the fundamental constitutional questions resolved by *Roe*, principles of institutional integrity, and the rule of stare decisis, we are led to conclude this: the essential holding of Roe v. Wade should be retained and once again reaffirmed.

It must be stated at the outset and with clarity that *Roe's* essential holding, the holding we reaffirm, has three parts. First is a recognition of the right of the woman to choose to have an abortion before viability and to obtain it without undue interference from the State. Before viability, the State's interests are not strong enough to support a prohibition of abortion or the imposition of a substantial obstacle to the woman's effective right to elect the procedure. Second is a confirmation of the State's power to restrict abortions after fetal viability, if the law contains exceptions for pregnancies which endanger a woman's life or health. And third is the principle that the State has legitimate interests from the outset of the pregnancy in protecting the health of the woman and the life of the fetus that may become a child. These principles do not contradict one another; and we adhere to each.

II.

* * *

Men and women of good conscience can disagree, and we suppose some always shall disagree, about the profound moral and spiritual implications of terminating a pregnancy, even in its earliest stage. Some of us as individuals find abortion offensive to our most basic principles of morality, but that cannot control our decision. Our obligation is to define the liberty of all, not to mandate our own moral code. The underlying constitutional issue is whether the State can resolve these philosophic questions in such a definitive way that a woman lacks all choice in the matter, except perhaps in those rare circumstances in which the pregnancy is itself a danger to her own life or health, or is the result of rape or incest. . . . Abortion is a unique act. It is an act fraught with consequences for others: for the woman who must live with the implications of her decision; for the persons who perform and assist in the procedure; for the spouse, family, and society which must confront the knowledge that these procedures exist, procedures some deem nothing short of an act of violence against innocent human life; and, depending on one's beliefs, for the life or potential life that is aborted. Though abortion is conduct, it does not follow that the State is entitled to proscribe it in all instances. That is because the liberty of the woman is at stake in a sense unique to the human condition and so unique to the law. The mother who carries a child to full term is subject to anxieties, to physical constraints, to pain that only she must bear. That these sacrifices have from the beginning of the human race been endured by woman with a pride that ennobles her in the eyes of others and gives to the infant a bond of love cannot alone be grounds for the State to insist she make the sacrifice. Her suffering is too intimate and personal for the State to insist, without more, upon its own vision of the woman's role, however dominant that vision has been in the course of our history and our culture. The destiny of the woman must be shaped to a large extent on her own conception of her spiritual imperatives and her place in society.

* * *

While we appreciate the weight of the arguments made on behalf of the State in the case before us, arguments which in their ultimate formulation conclude that *Roe* should be overruled, the reservations any of us may have in reaffirming the central holding of *Roe* are outweighed by the explication of individual liberty we have given combined with the force of stare decisis. We turn now to that doctrine.

III.

A.

[In this section, the court discussed the conditions under which it is appropriate for the Court to reverse its own precedent.]

So in this case we may inquire whether *Roe's* central rule has been found unworkable; whether the rule's limitation on state power could be removed without serious inequity to those who have relied upon it or significant damage to the stability of the society governed by the rule in question; whether the law's growth in the intervening years has left *Roe's* central rule a doctrinal anachronism discounted by society; and whether *Roe's* premises of fact have so far changed in the ensuing two decades as to render its central holding somehow irrelevant or unjustifiable in dealing with the issue it addressed.

* * *

The sum of the precedential inquiry to this point shows *Roe's* underpinnings unweakened in any way affecting its central holding. While it has engendered disapproval, it has not been unworkable. An entire generation has come of age free to assume *Roe's* concept of liberty in defining the capacity of women to act in society, and to make reproductive decisions; no erosion of principle going to liberty or personal autonomy has left *Roe's* central holding a doctrinal remnant; *Roe* portends no developments at odds with other precedent for the analysis of personal liberty; and no changes of fact have rendered viability more or less appropriate as the point at which the balance of interests tips. Within the bounds of normal stare decisis analysis, then, and subject to the considerations on which it customarily turns, the stronger argument is for affirming *Roe's* central holding, with whatever degree of personal reluctance any of us may have, not for overruling it.

B.

[The Court next distinguished the rule in the abortion cases from the rules in *Lochner* and the "separate but equal" cases, two areas in which the Supreme Court did reverse its well settled precedents this century. The Court also explained that it should not expend its political capital and put the public respect for the Court and its processes at risk by reversing *Roe*.]

IV.

From what we have said so far it follows that it is a constitutional liberty of the woman to have some freedom to terminate her pregnancy. We conclude that the basic decision in *Roe* was based on a constitutional analysis which we cannot now repudiate. The woman's liberty is not so unlimited, however, that from the outset the State cannot show its concern for the life of the unborn, and at a later point in fetal development the State's interest in life has sufficient force so that the right of the woman to terminate the pregnancy can be restricted.

* * *

We conclude the line should be drawn at viability, so that before that time the woman has a right to choose to terminate her pregnancy. We adhere to this principle for two reasons. First, as we have said, is the doctrine of stare decisis. . . .

The second reason is that the concept of viability, as we noted in *Roe,* is the time at which there is a realistic possibility of maintaining and nourishing a life outside the womb, so that the independent existence of the second life can in reason and all fairness be the object of state protection that now overrides the rights of the woman. . . .

The woman's right to terminate her pregnancy before viability is the most central principle of Roe v. Wade. It is a rule of law and a component of liberty we cannot renounce.

* * *

Yet it must be remembered that Roe v. Wade speaks with clarity in establishing not only the woman's liberty but also the State's "important and legitimate interest in potential life." [ ] That portion of the decision in *Roe* has been given too little acknowledgment and implementation by the Court in its subsequent cases. Those cases decided that any regulation touching upon the abortion decision must survive strict scrutiny, to be sustained only if drawn in narrow terms to further a compelling state interest. [ ] Not all of the cases decided under that formulation can be reconciled with the holding in *Roe* itself that the State has legitimate interests in the health of the woman and in protecting the potential life within her. In resolving this tension, we choose to rely upon *Roe,* as against the later cases.

* * *

We reject the trimester framework, which we do not consider to be part of the essential holding of *Roe.* [ ] Measures aimed at ensuring that a woman's choice contemplates the consequences for the fetus do not necessarily interfere with the right recognized in *Roe*, although those measures have been found to be inconsistent with the rigid trimester framework announced in that case. A logical reading of the central holding in *Roe* itself, and a necessary reconciliation of the liberty of the woman and the interest of the State in promoting prenatal life, require, in our view, that we abandon the trimester framework as a rigid prohibition on all previability regulation aimed at the protection of fetal life.

* * *

The fact that a law which serves a valid purpose, one not designed to strike at the right itself, has the incidental effect of making it more difficult or more expensive to procure an abortion cannot be enough to invalidate it. Only where state regulation imposes an undue burden on a woman's ability to make this decision does the power of the State reach into the heart of the liberty protected by the Due Process Clause.

* * *

Not all burdens on the right to decide whether to terminate a pregnancy will be undue. In our view, the undue burden standard is the appropriate means of reconciling the State's interest with the woman's constitutionally protected liberty.

* * *

A finding of an undue burden is a shorthand for the conclusion that a state regulation has the purpose or effect of placing a substantial obstacle in the path of a woman seeking an abortion of a nonviable fetus. A statute with this purpose is invalid because the means chosen by the State to further the interest in potential life must be calculated to inform the woman's free choice, not hinder it. And a statute which, while furthering the interest in potential life or some other valid state interest, has the effect of placing a substantial obstacle in the path of a woman's choice cannot be considered a permissible means of serving its legitimate ends. . . .

Some guiding principles should emerge. What is at stake is the woman's right to make the ultimate decision, not a right to be insulated from all others in doing so. Regulations which do no more than create a structural mechanism by which the State, or the parent or guardian of a minor, may express profound respect for the life of the unborn are permitted, if they are not a substantial obstacle to the woman's exercise of the right to choose. [ ]

[The Justices then summarized their new undue burden test:]

(a) To protect the central right recognized by Roe v. Wade while at the same time accommodating the State's profound interest in potential life, we will employ the undue burden analysis as explained in this opinion. An undue burden exists, and therefore a provision of law is invalid, if its purpose or effect is to place a substantial obstacle in the path of a woman seeking an abortion before the fetus attains viability.

(b) We reject the rigid trimester framework of Roe v. Wade. To promote the State's profound interest in potential life, throughout pregnancy the State may take measures to ensure that the woman's choice is informed, and measures designed to advance this interest will not be invalidated as long as their purpose is to persuade the woman to choose childbirth over abortion. These measures must not be an undue burden on the right.

(c) As with any medical procedure, the State may enact regulations to further the health or safety of a woman seeking an abortion. Unnecessary health regulations that have the purpose or effect of presenting a substantial obstacle to a woman seeking an abortion impose an undue burden on the right.

(d) Our adoption of the undue burden analysis does not disturb the central holding of Roe v. Wade, and we reaffirm that holding. Regardless of whether exceptions are made for particular circumstances, a State may not prohibit any woman from making the ultimate decision to terminate her pregnancy before viability.

(e) We also reaffirm Roe's holding that "subsequent to viability, the State in promoting its interest in the potentiality of human life may, if it chooses, regulate, and even proscribe, abortion except where it is necessary, in appropriate medical judgment, for the preservation of the life or health of the mother." [ ]

* * *

V.

* * *

A.

Because it is central to the operation of various other requirements, we begin with the statute's definition of medical emergency. Under the statute, a medical emergency is "that condition which, on the basis of the physician's good faith clinical judgment, so complicates the medical condition of a pregnant woman as to necessitate the immediate abortion of her pregnancy to avert her death or for which a delay will create serious risk of substantial and irreversible impairment of a major bodily function." [ ]

Petitioners argue that the definition is too narrow, contending that it forecloses the possibility of an immediate abortion despite some significant health risks.

[The Justices accepted the Court of Appeals interpretation of the statute, which assured that the "abortion regulation would not in any way pose a significant threat to the life or health of a woman," and determined that the definition imposed no undue burden on a woman's right to an abortion.]

B.

We next consider the informed consent requirement. [ ] Except in a medical emergency, the statute requires that at least 24 hours before performing an abortion a physician inform the woman of the nature of the procedure, the health risks of the abortion and of childbirth, and the "probable gestational age of the unborn child." The physician or a qualified nonphysician must inform the woman of the availability of printed materials published by the State describing the fetus and providing information about medical assistance for childbirth, information about child support from the father, and a list of agencies which provide adoption and other services as alternatives to abortion. An abortion may not be performed unless the woman certifies in writing that she has been informed of the availability of these printed materials and has been provided them if she chooses to view them.

* * *

. . . If the information the State requires to be made available to the woman is truthful and not misleading, the requirement may be permissible. [The Court then rejects the argument that the physician's first amendment speech rights trump the state-mandated obligation to provide patients with identified truthful information.]

* * *

The Pennsylvania statute also requires us to reconsider the holding [ ] that the State may not require that a physician, as opposed to a qualified assistant, provide information relevant to a woman's informed consent. [ ] . . . Our cases reflect the fact that the Constitution gives the States broad latitude to decide that particular functions may be performed only by licensed professionals, even if an objective assessment might suggest that those same tasks could be performed by others. [ ] Thus, we uphold the provision as a reasonable means to insure that the woman's consent is informed.

Our analysis of Pennsylvania's 24-hour waiting period between the provision of the information deemed necessary to informed consent and the performance of an abortion under the undue burden standard requires us to reconsider the premise behind the decision in *Akron I* invalidating a parallel requirement. In *Akron I* we said: "Nor are we convinced that the State's legitimate concern that the woman's decision be informed is reasonably served by requiring a 24-hour delay as a matter of course." [ ] We consider that conclusion to be wrong. The idea that important decisions will be more informed and deliberate if they follow some period of reflection does not strike us as unreasonable, particularly where the statute directs that important information become part of the background of the decision.

* * *

C.

. . . Pennsylvania's abortion law provides, except in cases of medical emergency, that no physician shall perform an abortion on a married woman without receiving a signed statement from the woman that she has notified her spouse that she is about to undergo an abortion.

* * *

This information and the District Court's findings reinforce what common sense would suggest. In well-functioning marriages, spouses discuss important intimate decisions such as whether to bear a child. But there are millions of women in this country who are the victims of regular physical and psychological abuse at the hands of their husbands. Should these women become pregnant, they may have very good reasons for not wishing to inform their husbands of their decision to obtain an abortion. Many may have justifiable fears of physical abuse, but may be no less fearful of the consequences of reporting prior abuse to the Commonwealth of Pennsylvania. Many may have a reasonable fear that notifying their husbands will provoke further instances of child abuse; these women are not exempt from [the] notification requirement. Many may fear devastating forms of psychological abuse from their husbands, including verbal harassment, threats of future violence, the destruction of possessions, physical confinement to the home, the withdrawal of financial support, or the disclosure of the abortion to family and friends. . . .

The spousal notification requirement is thus likely to prevent a significant number of women from obtaining an abortion. It does not merely make abortions a little more difficult or expensive to obtain; for many women, it will impose a substantial obstacle. We must not blind ourselves to the fact that the significant number of women who fear for their safety and the safety of their children are likely to be deterred from procuring an abortion as surely as if the Commonwealth had outlawed abortion in all cases.

Respondents attempt to avoid the conclusion that [the spousal notification provision] is invalid by pointing out that it imposes almost no burden at all for the vast majority of women seeking abortions. . . . Respondents argue that since some of [the 20% of women who seek abortions who are married] will be able to notify their husbands without adverse consequences or will qualify for one of the exceptions, the statute affects fewer than one percent of women seeking abortions. For this reason, it is asserted, the statute cannot be invalid on its face. [ ] We disagree with respondents' basic method of analysis.

The analysis does not end with the one percent of women upon whom the statute operates; it begins there. Legislation is measured for consistency with the Constitution by its impact on those whose conduct it affects. . . . [A]s we have said, [the Act's] real target is narrower even than the class of women seeking abortions . . . : it is married women seeking abortions who do not wish to notify their husbands of their intentions and who do not qualify for one of the statutory exceptions to the notice requirement. The unfortunate yet persisting conditions . . . will mean that in a large fraction of the cases . . . , [the statute] will operate as a substantial obstacle to a woman's choice to undergo an abortion. It is an undue burden, and therefore invalid.

* * *

[The spousal notification provision] embodies a view of marriage consonant with the common-law status of married women but repugnant to our present understanding of marriage and of the nature of the rights secured by the Constitution. Women do not lose their constitutionally protected liberty when they marry. . . .

D.

* * *

Our cases establish, and we reaffirm today, that a State may require a minor seeking an abortion to obtain the consent of a parent or guardian, provided that there is an adequate judicial bypass procedure. [ ] Under these precedents, in our view, the [Pennsylvania] one-parent consent requirement and judicial bypass procedure are constitutional.

* * *

E.

[The Justices upheld all of the record keeping and reporting requirements of the statute, except for that provision requiring the reporting of a married woman's reason for failure to give notice to her husband.]

VI.

Our Constitution is a covenant running from the first generation of Americans to us and then to future generations. It is a coherent succession. Each generation must learn anew that the Constitution's written terms embody ideas and aspirations that must survive more ages than one. We accept our responsibility not to retreat from interpreting the full meaning of the covenant in light of all of our precedents. We invoke it once again to define the freedom guaranteed by the Constitution's own promise, the promise of liberty.

* * *

[In addition to those parts of the statute found unconstitutional in the three-justice opinion, Justice Stevens would find unconstitutional the requirement that the doctor deliver state-produced materials to a woman seeking an abortion, the counseling requirements, and the 24-hour-waiting requirement. His concurring and dissenting opinion is omitted. Justice Blackman's opinion, concurring in the judgment in part and dissenting in part, is also omitted.]

CHIEF JUSTICE REHNQUIST, with whom JUSTICE WHITE, JUSTICE SCALIA, and JUSTICE THOMAS join, concurring in the judgment in part and dissenting in part.

The joint opinion, following its newly-minted variation on stare decisis, retains the outer shell of Roe v. Wade, [ ] but beats a wholesale retreat from the substance of that case. We believe that *Roe* was wrongly decided, and that it can and should be overruled consistently with our traditional approach to stare decisis in constitutional cases.

* * *

The end result of the joint opinion's paeans of praise for legitimacy is the enunciation of a brand new standard for evaluating state regulation of a woman's right to abortion—the "undue burden" standard. As indicated above, Roe v. Wade adopted a "fundamental right" standard under which state regulations could survive only if they met the requirement of "strict scrutiny." While we disagree with that standard, it at least had a recognized basis in constitutional law at the time *Roe* was decided. The same cannot be said for the "undue burden" standard, which is created largely out of whole cloth by the authors of the joint opinion. It is a standard which even today does not command the support of a majority of this Court. And it will not, we believe, result in the sort of "simple limitation," easily applied, which the joint opinion anticipates. [ ] In sum, it is a standard which is not built to last.

In evaluating abortion regulations under that standard, judges will have to decide whether they place a "substantial obstacle" in the path of a woman seeking an abortion. [ ] In that this standard is based even more on a judge's subjective determinations than was the trimester framework, the standard will do nothing to prevent "judges from roaming at large in the constitutional field" guided only by their personal views. [ ]

* * *

The sum of the joint opinion's labors in the name of stare decisis and "legitimacy" is this: Roe v. Wade stands as a sort of judicial Potemkin Village, which may be pointed out to passers by as a monument to the importance of adhering to precedent. But behind the facade, an entirely new method of analysis, without any roots in constitutional law, is imported to decide the constitutionality of state laws regulating abortion. Neither stare decisis nor "legitimacy" are truly served by such an effort.

* * *

JUSTICE SCALIA, with whom the CHIEF JUSTICE, JUSTICE WHITE, and JUSTICE THOMAS join, concurring in the judgment in part and dissenting in part.

* * *

The States may, if they wish, permit abortion-on-demand, but the Constitution does not require them to do so. The permissibility of abortion, and the limitations upon it, are to be resolved like most important questions in our democracy: by citizens trying to persuade one another and then voting. As the Court acknowledges, "where reasonable people disagree the government can adopt one position or the other." [ ] The Court is correct in adding the qualification that this "assumes a state of affairs in which the choice does not intrude upon a protected liberty," [ ]—but the crucial part of that qualification is the penultimate word. A State's choice between two positions on which reasonable people can disagree is constitutional even when (as is often the case) it intrudes upon a "liberty" in the absolute sense. Laws against bigamy, for example—which entire societies of reasonable people disagree with—intrude upon men and women's liberty to marry and live with one another. But bigamy happens not to be a liberty specially "protected" by the Constitution.

That is, quite simply, the issue in this case: not whether the power of a woman to abort her unborn child is a "liberty" in the absolute sense; or even whether it is a liberty of great importance to many women. Of course it is both. The issue is whether it is a liberty protected by the Constitution of the United States. I am sure it is not. I reach that conclusion not because of anything so exalted as my views concerning the "concept of existence, of meaning, of the universe, and of the mystery of human life." [ ] I reach it for the same reason I reach the conclusion that bigamy is not constitutionally protected—because of two simple facts: (1) the Constitution says absolutely nothing about it, and (2) the longstanding traditions of American society have permitted it to be legally proscribed.

* * *

I am certainly not in a good position to dispute that the Court has saved the "central holding" of *Roe*, since to do that effectively I would have to know what the Court has saved, which in turn would require me to understand (as I do not) what the "undue burden" test means. . . . I thought I might note, however, that the following portions of *Roe* have not been saved:

•   Under *Roe*, requiring that a woman seeking an abortion be provided truthful information about abortion before giving informed written consent is unconstitutional, if the information is designed to influence her choice [ ]. Under the joint opinion's "undue burden" regime (as applied today, at least) such a requirement is constitutional [ ].

•   Under *Roe*, requiring that information be provided by a doctor, rather than by nonphysician counselors, is unconstitutional [ ]. Under the "undue burden" regime (as applied today, at least) it is not [ ].

•   Under *Roe*, requiring a 24-hour waiting period between the time the woman gives her informed consent and the time of the abortion is unconstitutional. Under the "undue burden" regime (as applied today, at least) it is not [ ].

•   Under *Roe*, requiring detailed reports that include demographic data about each woman who seeks an abortion and various information about each abortion is unconstitutional [ ]. Under the "undue burden" regime (as applied today, at least) it generally is not [ ].

* * *

NOTES AND QUESTIONS

1.   Casey *vs.* Roe. At the beginning of 2018, *Casey's* holding remains the governing law with regard to abortion. Does *Casey* really affirm *Roe*? As Justice Scalia points out, several laws that had been struck down under *Roe* are being upheld under *Casey*. Which aspects of *Roe's* reasoning survived and which did not?

2.   *Parental Consent and Notification Requirements.* After *Casey*, states may require minors to obtain the consent of a parent to an abortion, as long as the state permits a court to dispense with that requirement under some circumstances. As of mid-2021, thirty-eight states require some kind of parental involvement (with three states requiring both parents' consent, and one requiring notification of both parents). See Parental Involvement in Minors' Abortions, Guttmacher Institute (Jul. 2021). The Supreme Court has never determined if a two-parent consent requirement (with a proper judicial bypass provision) would pass Constitutional muster, and courts have reached different conclusions. Compare Barnes v. Mississippi, 992 F.2d 1335 (5th Cir. 1993), cert. denied, 510 U.S. 976 (1993) (upholding two-parent consent requirement because involvement of both parents will increase "reflection and deliberation" on the process, and if one parent denies consent the other will be able to go to court in support of the child) and S.H. v. D.H., 796 N.E.2d 1243 (Ind. App. 2003) (family court judge cannot require consent of both parents before an abortion is performed, even if the divorced parents have joint custody). The Supreme Court has suggested that a state requirement of mere parental notification (rather than consent) also requires a judicial bypass, and lower courts have generally assumed that it does. Ohio v. Akron Ctr. for Reproductive Health, 497 U.S. 502 (1990). But states may have more robust state constitutional protections that limit their ability to require parental notification or consent. For example, one state supreme court has found that a parental notification requirement violates the state constitution's equal protection clause when it is imposed only on those who seek abortions, not on pregnant children making other, equally medically risky, reproductive decisions. Planned Parenthood of Central New Jersey v. Farmer, 762 A.2d 620 (N.J. 2000).

3.   *Judicial Bypass.* It is not clear what constitutional standard a court must apply in a "bypass" case. Under what circumstance is the court required to waive consent or notification: if it finds that the minor is sufficiently mature that she should be able to make the decision herself; if it finds that if she seeks consent from (or notifies) her parents she will be subject to abuse; if it finds that it is in her best interest to have the abortion; if it finds that it is in her best interest not to be required to notify, or get consent from, her parent; or some combination of these? See Lambert v. Wicklund, 520 U.S. 292 (1997) (per curiam).

4.   *Two-Trip Requirement and Waiting Period.* Many states require patients to receive counseling before getting an abortion and impose a waiting period between the time a pregnant woman is provided information and the

time of the medical procedure. This typically requires the pregnant woman make two trips to the medical facility. In *Casey* the Court announced that a 24-hour waiting period did not constitute an unconstitutional undue burden, because it considered "[t]he idea that important decisions will be more informed and deliberate if they follow some period of reflection" to be reasonable, the statute was understood as not requiring the waiting period in a medical emergency, and "the record evidence show[ed] that in the vast majority of cases, a 24-hour delay does not create any appreciable health risk." One state supreme court had found that a three-day waiting period violated the *state* constitution's protected right of privacy, citing evidence of the burdens such a requirement could impose:

> [P]atient mortality rates for abortions increase as the length of pregnancy increases. Studies also suggest that a large majority of women who have endured waiting periods prior to obtaining an abortion have suffered increased stress, nausea and physical discomfort, but very few have reported any benefit from having to wait. [Evidence also] indicates that the waiting period increases a woman's financial and psychological burdens, since many women must travel long distances and be absent from work to obtain an abortion . . . . [which] is especially problematic for women who suffer from poverty or abusive relationships.

Planned Parenthood of Middle Tennessee v. Sundquist, 38 S.W.3d 1 (Tenn. 2000). In response, the Tennessee legislature amended the State's constitution to clarify that it does not protect the right to an abortion, and the 48-hour waiting period was restored. Bristol Regional Women's Center, P.C. v. Slatery, 7 F.4th 478 (2021).

As of early 2021, about half of the states imposed waiting periods of some sort. See Counseling and Waiting Periods for Abortion, Guttmacher Institute (Jul. 2021). What if the patient must travel a very far distance to the facility and has difficulty accessing transportation? Should this impact an undue burden analysis of a waiting period requirement? A 2014 study of the distance that patients would have to travel for an abortion revealed spatial disparities in abortion access. The median distance to the nearest clinic providing abortion services was less than 15 miles in twenty-three states, 15–29 miles in sixteen states, 30–89 miles in eight states. In three states, at least half of all women had to travel more than 90 miles (168.49 miles in Wyoming; 151.58 in North Dakota; and 92.06 miles in South Dakota). The study also found that in states with a relatively low median travel distance (like Alaska), a sizeable percentage of women still had to travel over 150 miles to reach the nearest abortion clinic. See Jonathan M. Bearak et al., Disparities and Change Over Time in Distance Women Would Need to Travel to Have an Abortion in the USA: A Spatial Analysis, 2 Lancet Public Health 493–500 (Nov. 2017).

5.   *State "Informed Consent" Requirements for Abortion.* In *Casey* the Court upheld a statute requiring that a state-approved packet of material be given to the pregnant woman as a part of the informed consent process, and that the consent process be obtained by the physician, not by anyone else. The

only requirement the majority in *Casey* appears to put on the distributed material is that it is "truthful and not misleading." How should this be understood in the context of the summary of the "undue burden" test (offered in Part IV of the opinion by Justices O'Connor, Justice Kennedy, and Justice Souter), which allows the State to enact regulations that "further the health or safety of a woman seeking an abortion" but prohibits "[u]necessary health regulations that have the purpose or effect of presenting a substantial obstacle to a woman seeking an abortion." This has led to a number of conflicts as state legislatures have enacted more expansive information requirements at the behest of groups that want to discourage women from having an abortion. Consider the following examples:

- *Information About Medical Risk.* In early 1996 a couple of epidemiological studies arguably suggested a very weak relationship between having an abortion and the subsequent risk of developing breast cancer and the severity of that cancer. Subsequent reports cast doubt on this relationship. See P.A. Newcomb, et al., Pregnancy Termination in Relation to Risk of Breast Cancer, 275 JAMA 283 (1996). A handful of states require that a woman seeking an abortion be told of the purported breast cancer risk. A handful of states also require that women be told that an abortion can jeopardize a woman's future fertility and can have serious mental health consequences, despite weak or contrary scientific support.

- *Information About Fetal Pain.* Nearly a dozen states require that a woman seeking an abortion be told of the possibility that the fetus can feel pain. Should courts scrutinize the scientific basis of this assertion? (For discussion of how fetal pain theories are being used to justify previability criminal bans on abortion, see Note 5, after *Gonzales v. Carhart* below).

- *Moral Information Laws.* How far can states go in promoting their preference for pregnancy over abortion? Should a state be able to require physicians to provide materials highlighting moral objections to abortion? Five states require that a woman be told that personhood begins at conception.

- *Mandatory Ultrasound & Viewing Laws.* Several states have added requirements that before a woman can have an abortion, she must undergo an ultrasound and be required or given the opportunity to see the fetus, as well as to hear the fetal heartbeat and a description of the fetus (such as dimensions of the embryo or fetus, presence of cardiac activity, and the presence of external members and internal organs). Abortion is the only medical procedure in which the state requires a patient to undergo a particular diagnostic test even if the doctor does not believe the tests will have any diagnostic or therapeutic value and the patient does not want it. The Fourth Circuit struck down North Carolina's mandatory ultrasound law requiring the physician to display the sonogram, and describe the fetus to women seeking abortions even if the woman actively "avert[s] her eyes" and "refus[es] to hear." Stuart v. Camnitz, 774 F.3d 238 (4th Cir. 2014), *cert. denied* 576 U.S. 1028 (2015). The court found this "compelled speech" to be a violation of the First Amendment because it "interfere[d] with the physician's right to free speech beyond the extent permitted for reasonable regulation of the medical profession, while simultaneously threatening harm to the patient's psychological health, interfering with the physician's professional judgment, and compromising the doctor-patient relationship." But other courts have upheld similar statutes. See EMW Women's Surgical Center v. Behear, 920 F.3d 421 (6th Cir. 2019) (finding Kentucky's mandatory "speech-and-display" ultrasound law to be a constitutionally permissible regulation of medicine consistent with Supreme Court jurisprudence allowing mandates of "truthful, non-misleading information); see also Texas Medical Providers Performing Abortion Services v. Lakey, 667 F.3d 570 (5th Cir. 2012) (upholding a Texas law on similar reasoning).

6.   *State Disclosure Requirements on Pregnancy Centers.* The battle over what information should be provided to women seeking abortions has taken a different turn recently. There is growing attention to entities often referred to as "Pregnancy Centers," "Crisis Pregnancy Centers," or "Limited-Service Pregnancy Centers." These Centers only provide support and assistance for those who want to carry a pregnancy to term. Some help women who want to keep their children but feel like they cannot afford to do so by connecting the women with resources and support, such as counseling, drug abuse treatment, and vocational training. See Michelle Oberman, The Women the Abortion War Leaves Out, N.Y. Times (Jan. 11, 2018). There is evidence, however, that some centers do not provide this kind of support and instead are focused on persuading women not to have abortions through false and misleading information about the health risks of abortion. See U.S. House of Representatives, Committee on Government Reform—Minority Staff, False and Misleading Information Provided by Federally Funded Pregnancy Resource Centers (July 2006). Findings from this report and some state investigations assert that some of these Centers intentionally "pose as full-service women's health clinics" despite their aim to discourage women from having an abortion, and then use "counseling practices that often confuse, misinform, and even intimidate women from making fully-informed, time-sensitive decisions about critical health care." See National Institute of Family & Life Advocates (NIFLA) v. Harris, 839 F.3d 823 (9th Cir. 2016).

Some states have responded to these concerns by requiring Centers to make disclosures about the limits of the services they provide as well as information about where more comprehensive family planning, including abortion, can be accessed. But the Supreme Court recently invalidated such disclosures required under California law on free speech grounds in National Institute of Family and Life Advocates (NIFLA) v. Becerra, 138 S.Ct. 2361 (2018). This surprised many legal scholars because of the deference typically given to state provider disclosure requirements for the purposes of ensuring informed patient decision-making, as well as *Casey*'s decision allowing states to mandate a broad range of information that can impact health care decision-making, including resource information about how to access alternative options. In striking down the required disclosures, Justice Thomas tried to distinguish *Casey*, reaffirming states' ability to mandate disclosures designed to discourage women from having an abortion. Although the *NIFLA* decision did not explicitly turn on religious liberty arguments, the majority was clearly sensitive to the Centers' religious objections to California's required disclosure and those objections informed their free speech challenge. For more on the First Amendment issues raised by informed consent laws in the context of abortion, see Aziza Ahmed, Informed Decision Making and Abortion: Crisis Pregnancy Centers, Informed Consent, and the First Amendment, 43 J. L. Med. & Ethics 51 (2015). For more on the tension between government regulation and religious objections, see the Note on Federal Conscience Protections at the end of Section III.

To the extent *NIFLA* limits states' ability to proactively address potential deception through disclosure requirements, what legal avenues remain available? Every state has a consumer protection law that prohibits businesses from engaging in fraud, deception, and unfair business practices. Health care consumers who encounter licensed or unlicensed health providers engaging in deceptive practices can file complaints with the state, which is then investigated by a state consumer protection agency or the State Attorney General. It is not clear how effective this will be, in part, because women most at risk for harm from delay often do not know about their rights, do not have the time or resources to file a complaint, and/or fear the stigma they believe will come from revealing their need for an abortion. Tort law also allows patients to sue providers who fail to ensure informed consent in treatment, but this may not be very effective either. First, there is the question of the appropriate standard to apply, especially where pregnancy centers are unlicensed, or licensed and expressly provide limited care. Second, harm due to delay may be difficult to prove, especially if the delay is likely one of multiple factors contributing to the woman's increased risk of harm.

GONZALES V. CARHART
Supreme Court of the United States, 2007.
550 U.S. 124.

JUSTICE KENNEDY delivered the opinion of the Court.

These cases require us to consider the validity of the Partial-Birth Abortion Ban Act of 2003 (Act), [ ] a federal statute regulating abortion procedures. . . . We conclude the Act should be sustained against the objections lodged by the broad, facial attack brought against it.

* * *

I

A

[The Medical Procedure]

The Act proscribes a particular manner of ending fetal life, so it is necessary here [ ] to discuss abortion procedures in some detail. . . .

Abortion methods vary depending to some extent on the preferences of the physician and, of course, on the term of the pregnancy and the resulting stage of the unborn child's development. Between 85 and 90 percent of the approximately 1.3 million abortions performed each year in the United States take place in the first three months of pregnancy, which is to say in the first trimester. [ ] The most common first-trimester abortion method is vacuum aspiration (otherwise known as suction curettage) in which the physician vacuums out the embryonic tissue. Early in this trimester an alternative is to use medication, such as mifepristone (commonly known as RU-486), to terminate the pregnancy. [ ] The Act does not regulate these procedures.

Of the remaining abortions that take place each year, most occur in the second trimester. The surgical procedure referred to as "dilation and evacuation" or "D & E" is the usual abortion method in this trimester. [ ] Although individual techniques for performing D & E differ, the general steps are the same.

A doctor must first dilate the cervix at least to the extent needed to insert surgical instruments into the uterus and to maneuver them to evacuate the fetus. [ ] The steps taken to cause dilation differ by physician and gestational age of the fetus. . . .After sufficient dilation the surgical operation can commence. The woman is placed under general anesthesia or conscious sedation. The doctor, often guided by ultrasound, inserts grasping forceps through the woman's cervix and into the uterus to grab the fetus. The doctor grips a fetal part with the forceps and pulls it back through the cervix and vagina, continuing to pull even after meeting resistance from the cervix. The friction causes the fetus to tear apart. For example, a leg might be ripped off the fetus as it is pulled through the cervix and out of the woman. The process of evacuating the fetus piece by piece continues until it has been completely removed. A doctor may make 10 to 15 passes with the forceps to evacuate the fetus in its entirety, though sometimes removal is completed with fewer passes. Once the fetus has been evacuated, the placenta and any remaining fetal material are suctioned or scraped out of the uterus. The doctor examines the different parts to ensure the entire fetal body has been removed. [ ]

Some doctors, especially later in the second trimester, may kill the fetus a day or two before performing the surgical evacuation. They inject digoxin or potassium chloride into the fetus, the umbilical cord, or the amniotic fluid. Fetal demise may cause contractions and make greater dilation possible. Once dead, moreover, the fetus' body will soften, and its removal will be easier. . . .

The abortion procedure that was the impetus for the numerous bans on "partial-birth abortion," including the Act, is a variation of this standard D & E. [ ] The medical community has not reached unanimity on the appropriate name for this D & E variation. It has been referred to as "intact D & E," "dilation and extraction" (D & X), and "intact D & X." [ ] For discussion purposes this D & E variation will be referred to as intact D & E. The main difference between the two procedures is that in intact D & E a doctor extracts the fetus intact or largely intact with only a few passes. There are no comprehensive statistics indicating what percentage of all D & Es are performed in this manner.

Intact D & E, like regular D & E, begins with dilation of the cervix. Sufficient dilation is essential for the procedure. . . . In an intact D & E procedure the doctor extracts the fetus in a way conducive to pulling out its entire body, instead of ripping it apart. . . .

Intact D & E gained public notoriety when, in 1992, Dr. Martin Haskell gave a presentation describing his method of performing the operation. [ ] In the usual intact D & E the fetus' head lodges in the cervix, and dilation is insufficient to allow it to pass. [ ] Haskell explained the next step as follows:

> " 'At this point, the right-handed surgeon slides the fingers of the left [hand] along the back of the fetus and "hooks" the shoulders of the fetus with the index and ring fingers (palm down).

> " 'While maintaining this tension, lifting the cervix and applying traction to the shoulders with the fingers of the left hand, the surgeon takes a pair of blunt curved Metzenbaum scissors in the right hand. He carefully advances the tip, curved down, along the spine and under his middle finger until he feels it contact the base of the skull under the tip of his middle finger.

> " 'The surgeon then forces the scissors into the base of the skull or into the foramen magnum. Having safely entered the skull, he spreads the scissors to enlarge the opening.

> " 'The surgeon removes the scissors and introduces a suction catheter into this hole and evacuates the skull contents. With the catheter still in place, he applies traction to the fetus, removing it completely from the patient.' "[ ]

This is an abortion doctor's clinical description. Here is another description from a nurse who witnessed the same method performed on a 26-week fetus and who testified before the Senate Judiciary Committee:

" 'Dr. Haskell went in with forceps and grabbed the baby's legs and pulled them down into the birth canal. Then he delivered the baby's body and the arms—everything but the head. The doctor kept the head right inside the uterus. . . .

" 'The baby's little fingers were clasping and unclasping, and his little feet were kicking. Then the doctor stuck the scissors in the back of his head, and the baby's arms jerked out, like a startle reaction, like a flinch, like a baby does when he thinks he is going to fall.

" 'The doctor opened up the scissors, stuck a high-powered suction tube into the opening, and sucked the baby's brains out. Now the baby went completely limp. . . .

" 'He cut the umbilical cord and delivered the placenta. He threw the baby in a pan, along with the placenta and the instruments he had just used.' " [ ]

[JUSTICE KENNEDY next describes other varieties of this procedure in equally vivid terms.]

* * *

D & E and intact D & E are not the only second-trimester abortion methods. . . .

B

[The Statute]

After Dr. Haskell's procedure received public attention, with ensuing and increasing public concern, bans on " 'partial birth abortion' " proliferated. By the time of the *Stenberg* decision [finding Nebraska's statutory partial birth abortion ban to be unconstitutional in 2003], about 30 States had enacted bans designed to prohibit the procedure. [ ] In 1996, Congress also acted to ban partial-birth abortion. President Clinton vetoed the congressional legislation, and the Senate failed to override the veto. Congress approved another bill banning the procedure in 1997, but President Clinton again vetoed it. In 2003, after this Court's decision in *Stenberg*, Congress passed the Act at issue here. [ ] President Bush signed the Act into law. It was to take effect the following day. [ ]

The Act responded to *Stenberg* in two ways. First, Congress made factual findings. Congress determined that this Court in *Stenberg* "was required to accept the very questionable findings issued by the district court judge," [ ], but that Congress was "not bound to accept the same factual findings," [ ]. Congress found, among other things, that "[a] moral, medical, and ethical consensus exists that the practice of performing a partial-birth abortion . . . is a gruesome and inhumane procedure that is never medically necessary and should be prohibited." [ ]

Second, and more relevant here, the Act's language differs from that of the Nebraska statute struck down in *Stenberg*. [ ] The operative provisions of the Act provide in relevant part:

"(a) Any physician who, in or affecting interstate or foreign commerce, knowingly performs a partial-birth abortion and thereby kills a human fetus shall be fined under this title or imprisoned not more than 2 years, or both. This subsection does not apply to a partial-birth abortion that is necessary to save the life of a mother whose life is endangered by a physical disorder, physical illness, or physical injury, including a life-endangering physical condition caused by or arising from the pregnancy itself. * * *

"(b) As used in this section—

"(1) the term 'partial-birth abortion' means an abortion in which the person performing the abortion—

"(A) deliberately and intentionally vaginally delivers a living fetus until, in the case of a head-first presentation, the entire fetal head is outside the body of the mother, or, in the case of breech presentation, any part of the fetal trunk past the navel is outside the body of the mother, for the purpose of performing an overt act that the person knows will kill the partially delivered living fetus; and

"(B) performs the overt act, other than completion of delivery, that kills the partially delivered living fetus;

* * *

"(e) A woman upon whom a partial-birth abortion is performed may not be prosecuted under this section, for a conspiracy to violate this section * * *.

The Act also includes a provision authorizing civil actions [for damages following a "Partial Birth Abortion"] that is not of relevance here. [ ]

* * *

II

[The Holding in *Casey*]

The principles set forth in the joint opinion in *Casey* [ ] did not find support from all those who join the instant opinion.[ ] Whatever one's views concerning the *Casey* joint opinion, it is evident a premise central to its conclusion—that the government has a legitimate and substantial interest in preserving and promoting fetal life—would be repudiated were the Court now to affirm the judgments of the Courts of Appeals. *Casey* involved a challenge to Roe v. Wade [ ]. [One of the three essential principles of *Roe* that was upheld in Casey was] the principle that the State has legitimate interests from the outset of the pregnancy in protecting the health of the woman and the life of the fetus that may become a child. . . . [W]e must determine whether the Act furthers the legitimate interest of the Government in protecting the life of the fetus that may become a child.

To implement its holding, *Casey* rejected both *Roe*'s rigid trimester framework and the interpretation of *Roe* that considered all previability regulations of abortion unwarranted. . . .

We assume the following principles for the purposes of this opinion. Before viability, a State "may not prohibit any woman from making the ultimate decision to terminate her pregnancy." [ ] It also may not impose upon this right an undue burden, which exists if a regulation's "purpose or effect is to place a substantial obstacle in the path of a woman seeking an abortion before the fetus attains viability." [ ] On the other hand, "regulations which do no more than create a structural mechanism by which the State, or the parent or guardian of a minor, may express profound respect for the life of the unborn are permitted, if they are not a substantial obstacle to the woman's exercise of the right to choose." [ ] *Casey*, in short, struck a balance. The balance was central to its holding. We now apply its standard to the cases at bar.

III

We begin with a determination of the Act's operation and effect. A straightforward reading of the Act's text demonstrates its purpose and the scope of its provisions: It regulates and proscribes, with exceptions or qualifications to be discussed, performing the intact D & E procedure.

Respondents agree the Act encompasses intact D & E, but they contend its additional reach is both unclear and excessive. Respondents assert that, at the least, the Act is void for vagueness because its scope is indefinite. In the alternative, respondents argue the Act's text proscribes all D & Es. Because D & E is the most common second-trimester abortion method, respondents suggest the Act imposes an undue burden. In this litigation the Attorney General does not dispute that the Act would impose an undue burden if it covered standard D & E.

We conclude that the Act is not void for vagueness, does not impose an undue burden from any overbreadth, and is not invalid on its face.

A

[Interpreting the Statute]

The Act punishes "knowingly performing" a "partial-birth abortion." [ ] It defines the unlawful abortion in explicit terms. [ ]

First, the person performing the abortion must "vaginally deliver a living fetus." [ ] The Act does not restrict an abortion procedure involving the delivery of an expired fetus. The Act, furthermore, is inapplicable to abortions that do not involve vaginal delivery (for instance, hysterotomy or hysterectomy). The Act does apply both previability and postviability because, by common understanding and scientific terminology, a fetus is a living organism while within the womb, whether or not it is viable outside the womb. . . .

Second, the Act's definition of partial-birth abortion requires the fetus to be delivered "until, in the case of a head-first presentation, the entire fetal head is outside the body of the mother, or, in the case of breech presentation, any part of the fetal trunk past the navel is outside the body of the mother." . . . .

Third, to fall within the Act, a doctor must perform an "overt act, other than completion of delivery, that kills the partially delivered living fetus." [ ] For purposes of criminal liability, the overt act causing the fetus' death must be separate from delivery. And the overt act must occur after the delivery to an anatomical landmark. This is because the Act proscribes killing "the partially delivered" fetus, which, when read in context, refers to a fetus that has been delivered to an anatomical landmark. [ ]

Fourth, the Act contains scienter requirements concerning all the actions involved in the prohibited abortion. To begin with, the physician must have "deliberately and intentionally" delivered the fetus to one of the Act's anatomical landmarks. . . . In addition, the fetus must have been delivered "for the purpose of performing an overt act that the [doctor] knows will kill [it]." . . . .

B

[Vagueness]

Respondents contend the language described above is indeterminate, and they thus argue the Act is unconstitutionally vague on its face. "As generally stated, the void-for-vagueness doctrine requires that a penal statute define the criminal offense with sufficient definiteness that ordinary people can understand what conduct is prohibited and in a manner that does not encourage arbitrary and discriminatory enforcement." [ ] The Act satisfies both requirements.

. . . Unlike the statutory language in *Stenberg* that prohibited the delivery of a " 'substantial portion' " of the fetus—where a doctor might question how much of the fetus is a substantial portion—the Act defines the line between potentially criminal conduct on the one hand and lawful abortion on the other. [ ]

* * *

C

[Undue Burden]

We next determine whether the Act imposes an undue burden, as a facial matter, because its restrictions on second-trimester abortions are too broad. A review of the statutory text discloses the limits of its reach. The Act prohibits intact D & E; and, notwithstanding respondents' arguments, it does not prohibit the D & E procedure in which the fetus is removed in parts.

* * *

A comparison of the Act with the Nebraska statute struck down in *Stenberg* confirms this point. The statute in *Stenberg* prohibited " 'deliberately and intentionally delivering into the vagina a living unborn child, or a substantial portion thereof, for the purpose of performing a procedure that the person performing such procedure knows will kill the unborn child and does kill the unborn child.' " [ ]. The Court concluded that this statute encompassed D & E because "D & E will often involve a physician pulling a 'substantial portion' of a still living fetus, say, an arm or leg, into the vagina prior to the death of the fetus." . . .

Congress, it is apparent, responded to these concerns because the Act departs in material ways from the statute in *Stenberg*. It adopts the phrase "delivers a living fetus," [ ] instead of " 'delivering . . . a living unborn child, or a substantial portion thereof,' " [ ]. The Act's language, unlike the statute in *Stenberg*, expresses the usual meaning of "deliver" when used in connection with "fetus," namely, extraction of an entire fetus rather than removal of fetal pieces. . . .

The identification of specific anatomical landmarks to which the fetus must be partially delivered also differentiates the Act from the statute at issue in *Stenberg*. . . .

By adding an overt-act requirement Congress sought further to meet the Court's objections to the state statute considered in *Stenberg*. . . . The fatal overt act must occur after delivery to an anatomical landmark, and it must be something "other than [the] completion of delivery." [ ] This distinction matters because, unlike intact D & E, standard D & E does not involve a delivery followed by a fatal act.

The canon of constitutional avoidance, finally, extinguishes any lingering doubt as to whether the Act covers the prototypical D & E procedure. " 'The elementary rule is that every reasonable construction must be resorted to, in order to save a statute from unconstitutionality.' " [ ] It is true this longstanding maxim of statutory interpretation has, in the past, fallen by the wayside when the Court confronted a statute regulating abortion. . . . *Casey* put this novel statutory approach to rest. . . .

* * *

IV

[Substantial Obstacle]

Under the principles accepted as controlling here, the Act, as we have interpreted it, would be unconstitutional "if its purpose or effect is to place a substantial obstacle in the path of a woman seeking an abortion before the fetus attains viability." *Casey* [ ] The abortions affected by the Act's regulations take place both previability and postviability; so the quoted language and the undue burden analysis it relies upon are applicable. The question is whether the Act, measured by its text in this facial attack, imposes a substantial obstacle to late-term, but previability, abortions. The Act does not on its face impose a substantial obstacle, and we reject this further facial challenge to its validity.

A

[Congressional Purpose]

The Act's purposes are set forth in recitals preceding its operative provisions. A description of the prohibited abortion procedure demonstrates the rationale for the congressional enactment. The Act proscribes a method of abortion in which a fetus is killed just inches before completion of the birth process. Congress stated as follows: "Implicitly approving such a brutal and inhumane procedure by choosing not to prohibit it will further coarsen society to the humanity of not only newborns, but all vulnerable and innocent human life, making it increasingly difficult to protect such life." [ ] The Act expresses respect for the dignity of human life. Congress was concerned, furthermore, with the effects on the medical community and on its reputation caused by the practice of partial-birth abortion. . . .

There can be no doubt the government "has an interest in protecting the integrity and ethics of the medical profession." Washington v. Glucksberg, [ ] [reprinted in Chapter 6]. Under our precedents, it is clear the State has a significant role to play in regulating the medical profession.

Casey reaffirmed these governmental objectives. The government may use its voice and its regulatory authority to show its profound respect for the life within the woman. . . . The three premises of *Casey* must coexist.[ ] The third premise, that the State, from the inception of the pregnancy, maintains its own regulatory interest in protecting the life of the fetus that may become a child, cannot be set at naught by interpreting *Casey's* requirement of a health exception so it becomes tantamount to allowing a doctor to choose the abortion method he or she might prefer. Where it has a rational basis to act, and it does not impose an undue burden, the State may use its regulatory power to bar certain procedures and substitute others, all in furtherance of its legitimate interests in regulating the medical profession in order to promote respect for life, including life of the unborn.

. . . The Court has in the past confirmed the validity of drawing boundaries to prevent certain practices that extinguish life and are close to actions that are condemned. *Glucksberg* found reasonable the State's "fear that permitting assisted suicide will start it down the path to voluntary and perhaps even involuntary euthanasia." [ ]

Respect for human life finds an ultimate expression in the bond of love the mother has for her child. The Act recognizes this reality as well. Whether to have an abortion requires a difficult and painful moral decision. [ ] While we find no reliable data to measure the phenomenon, it seems unexceptionable to conclude some women come to regret their choice to abort the infant life they once created and sustained. [ ] Severe depression and loss of esteem can follow. [ ]

In a decision so fraught with emotional consequence some doctors may prefer not to disclose precise details of the means that will be used, confining themselves to the required statement of risks the procedure entails. From one standpoint this ought not to be surprising. Any number of patients facing imminent surgical procedures would prefer not to hear all details, lest the usual anxiety preceding invasive medical procedures become the more intense. This is likely the case with the abortion procedures here in issue. [ ]

It is, however, precisely this lack of information concerning the way in which the fetus will be killed that is of legitimate concern to the State. [ ] The State has an interest in ensuring so grave a choice is well informed. It is self-evident that a mother who comes to regret her choice to abort must struggle with grief more anguished and sorrow more profound when she learns, only after the event, what she once did not know: that she allowed a doctor to pierce the skull and vacuum the fast-developing brain of her unborn child, a child assuming the human form.

* * *

It is objected that the standard D & E is in some respects as brutal, if not more, than the intact D & E, so that the legislation accomplishes little. . . . It was reasonable for Congress to think that partial-birth abortion, more than standard D & E, "undermines the public's perception of the appropriate role of a physician during the delivery process, and perverts a process during which life is brought into the world." . . . In sum, we reject the contention that the congressional purpose of the Act was "to place a substantial obstacle in the path of a woman seeking an abortion." [ ]

B

[Protecting the Health of the Mother]

The Act's furtherance of legitimate government interests bears upon, but does not resolve, the next question: whether the Act has the effect of imposing an unconstitutional burden on the abortion right because it does not allow use of the barred procedure where " 'necessary, in appropriate medical judgment, for [the] preservation of the . . . health of the mother.' " [ ] The prohibition in the Act would be unconstitutional, under precedents we here assume to be controlling, if it "subjected [women] to significant health risks." . . . Here, by contrast, whether the Act creates significant health risks for women has been a contested factual question. The evidence presented in the trial courts and before Congress demonstrates both sides have medical support for their position.

* * *

The question becomes whether the Act can stand when this medical uncertainty persists. The Court's precedents instruct that the Act can survive this facial attack. The Court has given state and federal legislatures wide discretion to pass legislation in areas where there is medical and scientific uncertainty.

* * *

In reaching the conclusion the Act does not require a health exception we reject certain arguments made by the parties on both sides of these cases. On the one hand, the Attorney General urges us to uphold the Act on the basis of the congressional findings alone. [ ] Although we review congressional factfinding under a deferential standard, we do not in the circumstances here place dispositive weight on Congress' findings. The Court retains an independent constitutional duty to review factual findings where constitutional rights are at stake. [ ]

As respondents have noted, and the District Courts recognized, some recitations in the Act are factually incorrect. . . . Uncritical deference to Congress' factual findings in these cases is inappropriate.

On the other hand, relying on the Court's opinion in *Stenberg*, respondents contend that an abortion regulation must contain a health exception "if 'substantial medical authority supports the proposition that banning a particular procedure could endanger women's health.' " . . . *Stenberg* has been interpreted to leave no margin of error for legislatures to act in the face of medical uncertainty. [ ]

A zero tolerance policy would strike down legitimate abortion regulations, like the present one, if some part of the medical community were disinclined to follow the proscription. This is too exacting a standard to impose on the legislative power, exercised in this instance under the

Commerce Clause, to regulate the medical profession. Considerations of marginal safety, including the balance of risks, are within the legislative competence when the regulation is rational and in pursuit of legitimate ends. . . . The Act is not invalid on its face where there is uncertainty over whether the barred procedure is ever necessary to preserve a woman's health, given the availability of other abortion procedures that are considered to be safe alternatives.

V

[Propriety of a Facial Attack on the Statute]

The considerations we have discussed support our further determination that these facial attacks should not have been entertained in the first instance. In these circumstances the proper means to consider exceptions is by as-applied challenge. . . . This is the proper manner to protect the health of the woman if it can be shown that in discrete and well-defined instances a particular condition has or is likely to occur in which the procedure prohibited by the Act must be used. In an as-applied challenge the nature of the medical risk can be better quantified and balanced than in a facial attack.

* * *

As the previous sections of this opinion explain, respondents have not demonstrated that the Act would be unconstitutional in a large fraction of relevant cases. . . .

The Act is open to a proper as-applied challenge in a discrete case. . . . No as-applied challenge need be brought if the prohibition in the Act threatens a woman's life because the Act already contains a life exception. [ ]

* * *

JUSTICE THOMAS, with whom JUSTICE SCALIA joins, concurring.

I join the Court's opinion because it accurately applies current jurisprudence, including *Casey* [ ]. I write separately to reiterate my view that the Court's abortion jurisprudence, including *Casey* and *Roe* [ ] has no basis in the Constitution. [ ] I also note that whether the Act constitutes a permissible exercise of Congress' power under the Commerce Clause is not before the Court. The parties did not raise or brief that issue; it is outside the question presented; and the lower courts did not address it. [ ]

JUSTICE GINSBURG, with whom JUSTICE STEVENS, JUSTICE SOUTER, and JUSTICE BREYER join, dissenting.

* * *

In reaffirming *Roe*, the *Casey* Court described the centrality of "the decision whether to bear . . . a child," [ ] to a woman's "dignity and autonomy," her "personhood" and "destiny," her "conception of . . . her place in society." [ ] Of signal importance here, the *Casey* Court stated with unmistakable clarity that state regulation of access to abortion procedures, even after viability, must protect "the health of the woman." [ ]

Seven years ago, in *Stenberg*, [ ] the Court invalidated a Nebraska statute criminalizing the performance of a medical procedure that, in the political arena, has been dubbed "partial-birth abortion." With fidelity to the *Roe-Casey* line of precedent, the Court held the Nebraska statute unconstitutional in part because it lacked the requisite protection for the preservation of a woman's health. [ ]

Today's decision is alarming. It refuses to take *Casey* and *Stenberg* seriously. It tolerates, indeed applauds, federal intervention to ban nationwide a procedure found necessary and proper in certain cases by the American College of Obstetricians and Gynecologists (ACOG). It blurs the line, firmly drawn in *Casey*, between previability and postviability abortions. And, for the first time since *Roe*, the Court blesses a prohibition with no exception safeguarding a woman's health.

I dissent from the Court's disposition. Retreating from prior rulings that abortion restrictions cannot be imposed absent an exception safeguarding a woman's health, the Court upholds an Act that surely would not survive under the close scrutiny that previously attended state-decreed limitations on a woman's reproductive choices.

I

A

As *Casey* comprehended, at stake in cases challenging abortion restrictions is a woman's "control over her [own] destiny." . . .Women, it is now acknowledged, have the talent, capacity, and right "to participate equally in the economic and social life of the Nation." [ ] Their ability to realize their full potential, the Court recognized, is intimately connected to "their ability to control their reproductive lives." [ ] Thus, legal challenges to undue restrictions on abortion procedures do not seek to vindicate some generalized notion of privacy; rather, they center on a woman's autonomy to determine her life's course, and thus to enjoy equal citizenship stature. [ ]

In keeping with this comprehension of the right to reproductive choice, the Court has consistently required that laws regulating abortion, at any stage of pregnancy and in all cases, safeguard a woman's health. [ ]

We have thus ruled that a State must avoid subjecting women to health risks not only where the pregnancy itself creates danger, but also where state regulation forces women to resort to less safe methods of abortion. . . .

In *Stenberg*, we expressly held that a statute banning intact D & E was unconstitutional in part because it lacked a health exception. [ ] We noted that there existed a "division of medical opinion" about the relative safety of intact D & E, [ ], but we made clear that as long as "substantial medical authority supports the proposition that banning a particular abortion procedure could endanger women's health," a health exception is required. [ ] We explained:

> "The word 'necessary' in *Casey*'s phrase 'necessary, in appropriate medical judgment, for the preservation of the life or health of the [pregnant woman],' cannot refer to an absolute necessity or to absolute proof. Medical treatments and procedures are often considered appropriate (or inappropriate) in light of estimated comparative health risks (and health benefits) in particular cases. Neither can that phrase require unanimity of medical opinion. Doctors often differ in their estimation of comparative health risks and appropriate treatment. And *Casey*'s words 'appropriate medical judgment' must embody the judicial need to tolerate responsible differences of medical opinion. . . . " [ ]

Thus, we reasoned, division in medical opinion "at most means uncertainty, a factor that signals the presence of risk, not its absence." [ ] "[A] statute that altogether forbids [intact D & E] . . . consequently must contain a health exception." [ ]

B

In 2003, a few years after our ruling in *Stenberg*, Congress passed the Partial-Birth Abortion Ban Act—without an exception for women's health. [ ] The congressional findings on which the Partial-Birth Abortion Ban Act rests do not withstand inspection, as the lower courts have determined and this Court is obliged to concede.

* * *

C

* * *

During the District Court trials, "numerous" "extraordinarily accomplished" and "very experienced" medical experts explained that, in certain circumstances and for certain women, intact D & E is safer than alternative procedures and necessary to protect women's health. [ ]

According to the expert testimony plaintiffs introduced, the safety advantages of intact D & E are marked for women with certain medical conditions . . . . Further, plaintiffs' experts testified that intact D & E is significantly safer for women with certain pregnancy-related conditions, such as placenta previa and accreta, and for women carrying fetuses with certain abnormalities, such as severe hydrocephalus. [ ]

Intact D & E, plaintiffs' experts explained, provides safety benefits over D & E by dismemberment for several reasons [which Justice Ginsburg lists and describes]. . . .

* * *

Nevertheless, despite the District Courts' appraisal of the weight of the evidence, and in undisguised conflict with *Stenberg*, the Court asserts that the Partial-Birth Abortion Ban Act can survive "when . . . medical uncertainty persists." [ ] This assertion is bewildering.

* * *

II

A

The Court offers flimsy and transparent justifications for upholding a nationwide ban on intact D & E *sans* any exception to safeguard a women's health. Today's ruling, the Court declares, advances "a premise central to [*Casey*'s] conclusion"—i.e., the Government's "legitimate and substantial interest in preserving and promoting fetal life." [ ] But the Act scarcely furthers that interest: The law saves not a single fetus from destruction, for it targets only a *method* of performing abortion. [ ] And surely the statute was not designed to protect the lives or health of pregnant women. [ ]

* * *

Ultimately, the Court admits that "moral concerns" are at work, concerns that could yield prohibitions on any abortion. [ ] Notably, the concerns expressed are untethered to any ground genuinely serving the Government's interest in preserving life. By allowing such concerns to carry the day and case, overriding fundamental rights, the Court dishonors our precedent. [ ]

Revealing in this regard, the Court invokes an antiabortion shibboleth for which it concededly has no reliable evidence: Women who have abortions come to regret their choices, and consequently suffer from "severe depression and loss of esteem." [ ] Because of women's fragile emotional state and because of the "bond of love the mother has for her child," the Court worries, doctors may withhold information about the nature of the intact D & E procedure. [ ] The solution the Court approves, then, is *not* to require doctors to inform women, accurately and adequately, of the different procedures and their attendant risks. [ ] Instead, the Court deprives women of the right to make an autonomous choice, even at the expense of their safety.

This way of thinking reflects ancient notions about women's place in the family and under the Constitution—ideas that have long since been discredited. . . .

Though today's majority may regard women's feelings on the matter as "self-evident," [ ] this Court has repeatedly confirmed that "the destiny of the woman must be shaped . . . on her own conception of her spiritual imperatives and her place in society." *Casey* [ ].

B

* * *

The Court's hostility to the right *Roe* and *Casey* secured is not concealed. Throughout, the opinion refers to obstetrician-gynecologists and surgeons who perform abortions not by the titles of their medical specialties, but by the pejorative label "abortion doctor." [ ] A fetus is described as an "unborn child," and as a "baby," [ ]; second-trimester, previability abortions are referred to as "late-term," [ ]; and the reasoned medical judgments of highly trained doctors are dismissed as "preferences" motivated by "mere convenience," [ ]. Instead of the heightened scrutiny we have previously applied, the Court determines that a "rational" ground is enough to uphold the Act [ ]. And, most troubling, *Casey*'s principles, confirming the continuing vitality of "the essential holding of *Roe*," are merely "assumed" for the moment [ ], rather than "retained" or "reaffirmed," *Casey* [ ].

III

A

The Court further confuses our jurisprudence when it declares that "facial attacks" are not permissible in "these circumstances," i.e., where medical uncertainty exists. . . .

Without attempting to distinguish *Stenberg* and earlier decisions, the majority asserts that the Act survives review because respondents have not shown that the ban on intact D & E would be unconstitutional "in a large fraction of relevant cases." [ ] But *Casey* makes clear that, in determining whether any restriction poses an undue burden on a "large fraction" of women, the relevant class is *not* "all women," nor "all pregnant women," nor even all women "seeking abortions." [ ] Rather, a provision restricting access to abortion, "must be judged by reference to those [women] for whom it is an actual rather than an irrelevant restriction."[ ] Thus the absence of a health exception burdens *all* women for whom it is relevant—women who, in the judgment of their doctors, require an intact D & E because other procedures would place their health at risk. [ ] It makes no sense to conclude that this facial challenge fails because respondents have not shown that a health exception is necessary for a large fraction of second-trimester abortions, including those for which a health exception is unnecessary: The very purpose of a health *exception* is to protect women in *exceptional* cases.

B

If there is anything at all redemptive to be said of today's opinion, it is that the Court is not willing to foreclose entirely a constitutional challenge to the Act. "The Act is open," the Court states, "to a proper as-applied challenge in a discrete case." [ ] But the Court offers no clue on what a "proper" lawsuit might look like. [ ] Nor does the Court explain why the injunctions ordered by the District Courts should not remain in place, trimmed only to exclude instances in which another procedure would safeguard a woman's health at least equally well. Surely the Court cannot mean that no suit may be brought until a woman's health is immediately jeopardized by the ban on intact D & E. A woman "suffering from medical complications" [ ] needs access to the medical procedure at once and cannot wait for the judicial process to unfold. [ ]

* * *

IV

As the Court wrote in *Casey*, "overruling *Roe*'s central holding would not only reach an unjustifiable result under principles of *stare decisis*, but would seriously weaken the Court's capacity to exercise the judicial power and to function as the Supreme Court of a Nation dedicated to the rule of law." . . .

Though today's opinion does not go so far as to discard *Roe* or *Casey*, the Court, differently composed than it was when we last considered a restrictive abortion regulation, is hardly faithful to our earlier invocations of "the rule of law" and the "principles of *stare decisis*." Congress imposed a ban despite our clear prior holdings that the State cannot proscribe an abortion procedure when its use is necessary to protect a woman's health. [ ] Although Congress' findings could not withstand the crucible of trial, the Court defers to the legislative override of our Constitution-based rulings. [ ] A decision so at odds with our jurisprudence should not have staying power.

In sum, the notion that the Partial-Birth Abortion Ban Act furthers any legitimate governmental interest is, quite simply, irrational. The Court's defense of the statute provides no saving explanation. In candor, the Act, and the Court's defense of it, cannot be understood as anything other than an effort to chip away at a right declared again and again by this Court—and with increasing comprehension of its centrality to women's lives. [ ] When "a statute burdens constitutional rights and all that can be said on its behalf is that it is the vehicle that legislators have chosen for expressing their hostility to those rights, the burden is undue." [ ]

NOTES AND QUESTIONS

1. *The Undue Burden Test.* The Court "assume[s]" for purposes of the opinion that a regulation would constitute an "undue burden" if the regulation's purpose or effect is to place a substantial obstacle in the path of a woman seeking an abortion of a nonviable fetus. How is a court expected to divine the purpose of challenged legislation? How does a court determine whether a challenged statute in "effect" creates a substantial obstacle? Is an obstacle substantial because it imposes a serious limitation on any identifiable woman, because it affects a large number of women, or because it affects a large percentage of the cases in which women would seek abortions?

2. *Health Justifications.* One justification offered by Justice Kennedy was based on the government's interest in protecting women from a lack of information about the specific nature of the abortion procedure because of the emotional harm that could result from finding out the details after the fact. Yet, Justice Kennedy admits "there is no reliable data to measure this phenomenon." If claims about potential health risks of a procedure are being asserted to ban that procedure, should those risks have to be substantiated with scientific or medical evidence in court? Or should it be enough that such harm is "self-evident" to the Court?

3. *The Health Exception.* Consider that, unlike the abortion regulations challenged in *Casey*, the Partial Birth Abortion Ban Act did not contain an exception for the preservation of the mother's health. (It only provided an exception for life-threatening emergencies). *Casey's* decision depended, in part, on its finding that "the abortion regulation would not in any way pose a significant threat to the life *or health* of a woman," and this was part of the reason why the partial-birth abortion ban in *Stenberg* had been struck down. Why was the absence of a "health" exception not fatal in *Carhart*? The Court notes that "there is uncertainty over whether the barred procedure is ever necessary to preserve a woman's health." As Justice Ginsburg points out in dissent, this was contradicted by numerous medical experts at trial; but even if it were true, does this type of "medical uncertainty" weigh in favor of upholding the ban without a health exception, or striking it down?

4. Stenberg *vs.* Carhart. The Court distinguished its earlier decision striking down Nebraska's partial birth abortion ban in *Stenberg*. What were the relevant differences between the partial birth abortion ban that was rejected in *Stenberg* and the one accepted in *Carhart*? Was it the legislative findings that saved the federal statute? The more precise definition of the banned procedure? The scienter requirement? The fact that D & E abortions are still permitted? Some combination of these? Or was the change in the Court's approach due perhaps to the replacement of Justice O'Connor by Justice Alito?

5. *Other Criminal Bans on Abortion.* Consider the ethical and legal questions raised by other state abortion bans based on *type of procedure*, *timing of procedure*, or *reason for the procedure*—laws which are currently being or have been challenged in court. For example, some states some have gone

further than the federal partial birth abortion ban, by not allowing an exception for life endangerment, and some have banned D & E (the most commonly used form of abortion) in the second trimester, with limited exceptions. Seventeen states have enacted previability bans on abortions at 20 weeks (with limited exceptions, such as life endangerment). This is often justified on the grounds the fetus may be able to feel pain at this point, despite contrary scientific evidence that the neural anatomy of a fetus is not sufficiently developed to be able to perceive pain until at least after 24 weeks of gestation. See I. Glenn Cohen & Sadath Sayeed, Fetal Pain, Abortion, Viability, and the Constitution, 39 J.L. Med. & Ethics 235 (2011). Eight states have enacted abortion previability bans for reasons of sex selection, and a few states have outlawed abortion where the fetus may have a genetic anomaly, including a recently enacted Ohio law banning abortion after prenatal tests reveal Down syndrome in a fetus.

6. *Selective Abortion & Disability Justice.* Laws attempting to prevent abortions on the basis of genetic anomaly have been justified as anti-eugenics or anti-discrimination measures, with states citing anecdotal reports of parents feeling coerced by physicians into terminating pregnancies where prenatal diagnosis suggests Down syndrome. These justifications played a prominent role in the Sixth Circuit's decision to uphold what it characterized as a limited ban in Cleveland v. Himes, 994 F.3d 512 (6th Cir. 2021). The court denied a preliminary injunction against an Ohio law based on its interpretation that the law does not actually ban women from getting an abortion based on a Down syndrome diagnosis. Rather, the court reasoned that the law prohibits a physician from performing an abortion *only if the physician knows* the patient is seeking the abortion because the fetus has Down syndrome. But in Planned Parenthood of Ind. & Ky., Inc. v. Comm'r Ind. State Dep't of Health, the Seventh Circuit affirmed a district court opinion finding unconstitutional a law that banned previability abortions where the reason for the abortion was based on some characteristic of the fetus, such as sex, race, or Down syndrome diagnosis. 888 F.3d 300 (7th Cir. 2018), *cert. granted in part and judgment reversed in part* 139 S.Ct. 1780 (2019) (per curiam). On appeal, the Supreme Court denied certiorari on this issue, but Justice Thomas authored a concurrence highlighting the disturbing history of abortion and sterilization as eugenics tools against those deemed "unfit" due to disability. Id. at 1782 (Thomas, J. concurring).

Disability rights activists have argued that selective abortion raises moral and justice concerns, because it reinforces a view that a life with disability is not as valuable or worth living, and because such decisions seem to be driven by misinformation and pressure by health professionals. But the history of reproductive control as a tool for injustice against people with disabilities, including denying them the right to make their own reproductive choices, has led many disability rights activists to oppose abortion bans. See, e.g., Samuel Bagenstos, Disability and Reproductive Justice, 14 Harv. L. & Pol'y Rev. 273 (2020) (arguing instead for an approach informed by disability and reproductive justice frameworks that would ensure patients had the information and structural supports needed to empower the choice to have a child that may have a genetic anomaly; this would include informed consent requirements designed to counteract negative stereotypes about disability). For more on reproductive control as a eugenics tool, see Section V, below.

7. *Regulating the Abortion Procedure. Carhart* invigorated attempts to target abortion providers for stricter regulations, effectively making access more difficult. Yet, important questions remained about exactly what kind of regulation or degree of impact would constitute an undue burden under *Casey*. After *Carhart*, restrictions targeting abortion providers were easily upheld in the lower courts; but when two such restrictions reached the Supreme Court in *Whole Woman's Health* (below), it seemed that, once again, the trajectory of abortion jurisprudence may have shifted.

C. AN UNDUE BURDEN TEST WITH TEETH?

WHOLE WOMAN'S HEALTH V. HELLERSTEDT
Supreme Court of the United States, 2016.
136 S.Ct. 2292.

JUSTICE BREYER delivered the opinion of the Court.

In *Planned Parenthood of Southeastern Pa. v. Casey* [ ] a plurality of the Court concluded that there "exists" an "undue burden" on a woman's right to decide to have an abortion, and consequently a provision of law is constitutionally invalid, if the *"purpose or effect"* of the provision *"is to place a substantial obstacle* in the path of a woman seeking an abortion before the fetus attains viability." [ ] The plurality added that "[u]nnecessary health regulations that have the purpose or effect of presenting a substantial obstacle to a woman seeking an abortion impose an undue burden on the right." [ ]

We must here decide whether two provisions of Texas' House Bill 2 violate the Federal Constitution as interpreted in *Casey*. The first provision, which we shall call the *"admitting-privileges requirement,"* says that

> "[a] physician performing or inducing an abortion . . . must, on the date the abortion is performed or induced, have active admitting privileges at a hospital that . . . is located not further than 30 miles from the location at which the abortion is performed or induced." [ ]

This provision amended Texas law that had previously required an abortion facility to maintain a written protocol "for managing medical emergencies and the transfer of patients requiring further emergency care to a hospital." [ ]

The second provision, which we shall call the *"surgical-center requirement,"* says that "the minimum standards for an abortion facility must be equivalent to the minimum standards adopted under [the Texas Health and Safety Code section] for ambulatory surgical centers." [ ]

We conclude that neither of these provisions confers medical benefits sufficient to justify the burdens upon access that each imposes. Each places a substantial obstacle in the path of women seeking a previability abortion, each constitutes an undue burden on abortion access, [ ], and each violates the Federal Constitution. [ ]

I

A

In July 2013, the Texas Legislature enacted House Bill 2 [which is the subject of this litigation. A part of the Act, the admitting-privileges provision, was challenged on its face in *Planned Parenthood of Greater Tex. Surgical Health Servs. v. Abbott*. Ultimately, the Fifth Circuit upheld that provision, and no review was sought in the Supreme Court.]

* * *

B

[One] week after the Fifth Circuit's decision, petitioners, a group of abortion providers (many of whom were plaintiffs in the previous lawsuit), filed the present lawsuit in Federal District Court. They sought an injunction preventing enforcement of the admitting-privileges provision as applied to physicians at two abortion facilities . . . . They also sought an injunction prohibiting enforcement of the surgical-center provision anywhere in Texas. They claimed that the admitting-privileges provision and the surgical-center provision violated the Constitution's Fourteenth Amendment, as interpreted in *Casey*.

The District Court subsequently received stipulations from the parties and depositions from the parties' experts. The court conducted a 4-day bench trial. It heard, among other testimony, the opinions from expert witnesses for both sides. On the basis of the stipulations, depositions, and testimony, that court reached the following conclusions:

1. Of Texas' population of more than 25 million people, "approximately 5.4 million" are "women" of "reproductive age," living within a geographical area of "nearly 280,000 square miles." [ ]

2. "In recent years, the number of abortions reported in Texas has stayed fairly consistent at approximately 15–16% of the reported pregnancy rate, for a total number of approximately 60,000–72,000 legal abortions performed annually." [ ]

3. Prior to the enactment of H.B. 2, there were more than 40 licensed abortion facilities in Texas, which "number dropped by almost half leading

up to and in the wake of enforcement of the admitting-privileges requirement that went into effect in late-October 2013." [ ]

4. If the surgical-center provision were allowed to take effect, the number of abortion facilities, after September 1, 2014, would be reduced further, so that "only seven facilities and a potential eighth will exist in Texas." [ ]

* * *

7. The suggestion "that these seven or eight providers could meet the demand of the entire state stretches credulity." [ ]

8. . . . After September 2014, should the surgical-center requirement go into effect, the number of women of reproductive age living significant distances from an abortion provider will increase [substantially]. [ ]

9. The "two requirements erect a particularly high barrier for poor, rural, or disadvantaged women." [ ]

10. "The great weight of evidence demonstrates that, before the act's passage, abortion in Texas was extremely safe with particularly low rates of serious complications and virtually no deaths occurring on account of the procedure." [ ]

11. "Abortion, as regulated by the State before the enactment of House Bill 2, has been shown to be much safer, in terms of minor and serious complications, than many common medical procedures not subject to such intense regulation and scrutiny." [ ].

12. "Additionally, risks are not appreciably lowered for patients who undergo abortions at ambulatory surgical centers as compared to nonsurgical-center facilities." [ ].

13. "[W]omen will not obtain better care or experience more frequent positive outcomes at an ambulatory surgical center as compared to a previously licensed facility." [ ]

* * *

On the basis of these and other related findings, the District Court determined that the surgical-center requirement "imposes an undue burden on the right of women throughout Texas to seek a previability abortion," and that the "admitting-privileges requirement . . . in conjunction with the ambulatory-surgical-center requirement, imposes an undue burden on the right of women . . . to seek a previability abortion." [ ] The District Court concluded that the "two provisions" would cause "the closing of almost all abortion clinics in Texas that were operating legally in the fall of 2013," and thereby create a constitutionally "impermissible obstacle as applied to all women seeking a previability abortion" by "restricting access to previously available legal facilities." [ ] [T]he court enjoined the enforcement of the two provisions.

C

* * *

[T]he Court of Appeals reversed the District Court on the merits. With minor exceptions, it found both provisions constitutional and allowed them to take effect. [ ]

* * *

III

Undue Burden—Legal Standard

We begin with the standard, as described in *Casey*. We recognize that the "State has a legitimate interest in seeing to it that abortion, like any other medical procedure, is performed under circumstances that insure maximum safety for the patient." *Roe v. Wade* [ ]. But, we added, "a statute which, while furthering [a] valid state interest, has the effect of placing a substantial obstacle in the path of a woman's choice cannot be considered a permissible means of serving its legitimate ends." *Casey* [ ]. Moreover, "[u]nnecessary health regulations that have the purpose or effect of presenting a substantial obstacle to a woman seeking an abortion impose an undue burden on the right." [ ]

The Court of Appeals wrote that a state law is "constitutional if: (1) it does not have the purpose or effect of placing a substantial obstacle in the path of a woman seeking an abortion of a nonviable fetus; and (2) it is reasonably related to (or designed to further) a legitimate state interest." [ ] The Court of Appeals went on to hold that "the district court erred by substituting its own judgment for that of the legislature" when it conducted its "undue burden inquiry," in part because "medical uncertainty underlying a statute is for resolution by legislatures, not the courts." [ ]

The Court of Appeals' articulation of the relevant standard is incorrect. The first part of the Court of Appeals' test may be read to imply that a district court should not consider the existence or nonexistence of medical benefits when considering whether a regulation of abortion constitutes an undue burden. The rule announced in *Casey,* however, requires that courts consider the burdens a law imposes on abortion access together with the benefits those laws confer. [ ] And the second part of the test is wrong to equate the judicial review applicable to the regulation of a constitutionally protected personal liberty with the less strict review applicable where, for example, economic legislation is at issue. [ ] The Court of Appeals' approach simply does not match the standard that this Court laid out in *Casey,* which asks courts to consider whether any burden imposed on abortion access is "undue."

The statement that legislatures, and not courts, must resolve questions of medical uncertainty is also inconsistent with this Court's case law. Instead, the Court, when determining the constitutionality of laws regulating abortion procedures, has placed considerable weight upon evidence and argument presented in judicial proceedings. In *Casey,* for example, we relied heavily on the District Court's factual findings and the research-based submissions of *amici* in declaring a portion of the law at issue unconstitutional. [ ] (discussing evidence related to the prevalence of spousal abuse in determining that a spousal notification provision erected an undue burden to abortion access). And, in *Gonzales* the Court, while pointing out that we must review legislative "factfinding under a deferential standard," added that we must not "place dispositive weight" on those "findings." [ ] *Gonzales* went on to point out that the *"Court retains an independent constitutional duty to review factual findings where constitutional rights are at stake."* [ ] (emphasis added). Although there we upheld a statute regulating abortion, we did not do so solely on the basis of legislative findings explicitly set forth in the statute, noting that "evidence presented in the District Courts contradicts" some of the legislative findings. [ ] In these circumstances, we said, "[u]ncritical deference to Congress' factual findings . . . is inappropriate." [ ]

Unlike in *Gonzales,* the relevant statute here does not set forth any legislative findings. Rather, one is left to infer that the legislature sought to further a constitutionally acceptable objective (namely, protecting women's health). [ ] For a district court to give significant weight to evidence in the judicial record in these circumstances is consistent with this Court's case law. As we shall describe, the District Court did so here. It did not simply substitute its own judgment for that of the legislature. It considered the evidence in the record—including expert evidence, presented in stipulations, depositions, and testimony. It then weighed the asserted benefits against the burdens. We hold that, in so doing, the District Court applied the correct legal standard.

IV

Undue Burden—Admitting-Privileges Requirement

Turning to the lower courts' evaluation of the evidence, we first consider the admitting-privileges requirement. Before the enactment of H.B. 2, doctors who provided abortions were required to "have admitting privileges *or* have a working arrangement with a physician(s) who has admitting privileges at a local hospital in order to ensure the necessary back up for medical complications." [ ] The new law changed this requirement by requiring that a "physician performing or inducing an abortion . . . must, on the date the abortion is performed or induced, have active admitting privileges at a hospital that . . . is located not further than 30 miles from the location at which the abortion is performed or induced." [ ] The District Court held that the legislative change imposed an "undue burden" on a woman's right to have an abortion. We conclude that there is adequate legal and factual support for the District Court's conclusion.

The purpose of the admitting-privileges requirement is to help ensure that women have easy access to a hospital should complications arise during an abortion procedure. [ ] But the District Court found that it brought about no such health-related benefit. The court found that "[t]he great weight of evidence demonstrates that, before the act's passage, abortion in Texas was extremely safe with particularly low rates of serious complications and virtually no deaths occurring on account of the procedure." [ ]. Thus, there was no significant health-related problem that the new law helped to cure.

[The Court then summarized the evidence that drew the trial court to that conclusion.]

* * *

At the same time, the record evidence indicates that the admitting-privileges requirement places a "substantial obstacle in the path of a woman's choice." [ ] [The Court pointed out that the evidence showed that many clinics closed when the admitting-privileges requirement went into effect.]

* * *

[There are many] common prerequisites to obtaining admitting privileges that have nothing to do with ability to perform medical procedures. See Brief for Medical Staff Professionals as *Amici Curiae* [ ] (listing, for example, requirements that an applicant has treated a high number of patients in the hospital setting in the past year, clinical data requirements, residency requirements, and other discretionary factors); see also Brief for American College of Obstetricians and Gynecologists [ ] ("[S]ome academic hospitals will only allow medical staff membership for clinicians who also . . . accept faculty appointments"). Again, returning to the District Court record, we note that Dr. Lynn of the McAllen clinic, a veteran obstetrics and gynecology doctor who estimates that he has delivered over 15,000 babies in his 38 years in practice was unable to get admitting privileges at any of the seven hospitals within 30 miles of his clinic. [ ] He was refused admitting privileges at a nearby hospital for reasons, as the hospital wrote, "not based on clinical competence considerations." [ ] The admitting-privileges requirement does not serve any relevant credentialing function.

In our view, the record contains sufficient evidence that the admitting-privileges requirement led to the closure of half of Texas' clinics, or thereabouts. Those closures meant fewer doctors, longer waiting times, and increased crowding. Record evidence also supports the finding that after the admitting-privileges provision went into effect, the "number of women of reproductive age living in a county . . . more than 150 miles from a provider increased from approximately 86,000 to 400,000 . . . and the number of women living in a county more than 200 miles from a provider from approximately 10,000 to 290,000." [ ]. We recognize that increased driving distances do not always constitute an "undue burden." [ ] But here, those increases are but one additional burden, which, when taken together with others that the closings brought about, and when viewed in light of the virtual absence of any health benefit, lead us to conclude that the record adequately supports the District Court's "undue burden" conclusion. [ ]

* * *

[T]he dissent suggests that one benefit of H.B. 2's requirements would be that they might "force unsafe facilities to shut down." [ ] To support that assertion, the dissent points to the Kermit Gosnell scandal. Gosnell, a physician in Pennsylvania, was convicted of first-degree murder and manslaughter [as a result of the operation of an abortion clinic]. . . . Gosnell's behavior was terribly wrong. But there is no reason to believe that an extra layer of regulation would have affected that behavior. Determined wrongdoers, already ignoring existing statutes and safety measures, are unlikely to be convinced to adopt safe practices by a new overlay of regulations. . . . The record contains nothing to suggest that H.B. 2 would be more effective than pre-existing Texas law at deterring wrongdoers like Gosnell from criminal behavior.

V

Undue Burden—Surgical-Center Requirement

The second challenged provision of Texas' new law sets forth the surgical-center requirement. Prior to enactment of the new requirement, Texas law required abortion facilities to meet a host of health and safety requirements [which are listed in the opinion by the court]. . . .

H.B. 2 added the requirement that an "abortion facility" meet the "minimum standards . . . for ambulatory surgical centers" under Texas law. [ ] The surgical-center regulations include, among other things, detailed specifications relating to the size of the nursing staff, building dimensions, and other building requirements. [The Court went on to list the many requirements imposed on surgical-centers.]

There is considerable evidence in the record supporting the District Court's findings indicating that the statutory provision requiring all abortion facilities to meet all surgical-center standards does not benefit patients and is not necessary. The District Court found that "risks are not appreciably lowered for patients who undergo abortions at ambulatory surgical centers as compared to nonsurgical-center facilities." [ ] The court added that women "will not obtain better care or experience more frequent positive outcomes at an ambulatory surgical center as compared to a previously licensed facility." [ ] And these findings are well supported.

The record makes clear that the surgical-center requirement provides no benefit when complications arise in the context of an abortion produced through medication. That is because, in such a case, complications would almost always arise only after the patient has left the facility. [ ] The record also contains evidence indicating that abortions taking place in an abortion facility are safe—indeed, safer than numerous procedures that take place outside hospitals and to which Texas does not apply its surgical-center requirements. . . . Nationwide, childbirth is 14 times more likely than abortion to result in death, [ ] but Texas law allows a midwife to oversee childbirth in the patient's own home. Colonoscopy, a procedure that typically takes place outside a hospital (or surgical center) setting, has a mortality rate 10 times higher than an abortion. [T]he mortality rate for liposuction, another outpatient procedure, is 28 times higher than the mortality rate for abortion. Medical treatment after an incomplete miscarriage often involves a procedure identical to that involved in a nonmedical abortion, but it often takes place outside a hospital or surgical center. And Texas partly or wholly grandfathers (or waives in whole or in part the surgical-center requirement for) about two-thirds of the facilities to which the surgical-center standards apply. But it neither grandfathers nor provides waivers for any of the facilities that perform abortions. [ ] These facts indicate that the surgical-center provision imposes "a requirement that simply is not based on differences" between abortion and other surgical procedures "that are reasonably related to" preserving women's health, the asserted "purpos[e] of the Act in which it is found." [ ]

* * *

At the same time, the record provides adequate evidentiary support for the District Court's conclusion that the surgical-center requirement places a substantial obstacle in the path of women seeking an abortion. The parties stipulated that the requirement would further reduce the number of abortion facilities available to seven or eight facilities . . . . In the District Court's view, the proposition that these "seven or eight providers could meet the demand of the entire State stretches credulity." [ ] We take this statement as a finding that these few facilities could not "meet" that "demand."

The Court of Appeals held that this finding was "clearly erroneous." . . . [W]e hold that the record provides adequate support for the District Court's finding.

* * *

More fundamentally, in the face of no threat to women's health, Texas seeks to force women to travel long distances to get abortions in crammedto-capacity superfacilities. Patients seeking these services are less likely to get the kind of individualized attention, serious conversation, and emotional support that doctors at less taxed facilities may have offered. Healthcare facilities and medical professionals are not fungible commodities. Surgical centers attempting to accommodate sudden, vastly increased demand, [ ] may find that quality of care declines. Another commonsense inference that the District Court made is that these effects would be harmful to, not supportive of, women's health. [ ].

* * *

We agree with the District Court that the surgical-center requirement, like the admitting-privileges requirement, provides few, if any, health benefits for women, poses a substantial obstacle to women seeking abortions, and constitutes an "undue burden" on their constitutional right to do so.

VI

* * *

JUSTICE GINSBURG, concurring.

The Texas law called H.B. 2 inevitably will reduce the number of clinics and doctors allowed to provide abortion services. Texas argues that H.B. 2's restrictions are constitutional because they protect the health of women who experience complications from abortions. In truth, "complications from an abortion are both rare and rarely dangerous." [ ] Many medical procedures, including childbirth, are far more dangerous to patients, yet are not subject to ambulatory-surgical-center or hospital admitting-privileges requirements. [ ] Given those realities, it is beyond rational belief that H.B. 2 could genuinely protect the health of women, and certain that the law "would simply make it more difficult for them to obtain abortions." [ ] When a State severely limits access to safe and legal procedures, women in desperate circumstances may resort to unlicensed rogue practitioners, *faute de mieux,* at great risk to their health and safety. [ ] So long as this Court adheres to *Roe v. Wade* [ ], and *Planned Parenthood of Southeastern Pa. v. Casey,* [ ] Targeted Regulation of Abortion Providers laws like H.B. 2 that "do little or nothing for health, but rather strew impediments to abortion," [ ] cannot survive judicial inspection.

JUSTICE THOMAS, dissenting.

* * *

II

Today's opinion . . . reimagines the undue-burden standard used to assess the constitutionality of abortion restrictions. Nearly 25 years ago, in *Planned Parenthood of Southeastern Pa. v. Casey* [ ] a plurality of this Court invented the "undue burden" standard as a special test for gauging the permissibility of abortion restrictions. *Casey* held that a law is unconstitutional if it imposes an "undue burden" on a woman's ability to choose to have an abortion, meaning that it "has the purpose or effect of placing a substantial obstacle in the path of a woman seeking an abortion of a nonviable fetus." [ ] *Casey* thus instructed courts to look to whether a law substantially impedes women's access to abortion, and whether it is reasonably related to legitimate state interests. As the Court explained, "[w]here it has a rational basis to act, and it does not impose an undue burden, the State may use its regulatory power" to regulate aspects of abortion procedures, "all in furtherance of its legitimate interests in regulating the medical profession in order to promote respect for life, including life of the unborn." [ ]

I remain fundamentally opposed to the Court's abortion jurisprudence. [ ] Even taking *Casey* as the baseline, however, the majority radically rewrites the undue-burden test in three ways. First, today's decision requires courts to "consider the burdens a law imposes on abortion access together with the benefits those laws confer." [ ] Second, today's opinion tells the courts that, when the law's justifications are medically uncertain, they need not defer to the legislature, and must instead assess medical justifications for abortion restrictions by scrutinizing the record themselves. [ ] Finally, even if a law imposes no "substantial obstacle" to women's access to abortions, the law now must have more than a "reasonabl[e] relat[ion] to . . . a legitimate state interest." [ ] (internal quotation marks omitted). These precepts are nowhere to be found in *Casey* or its successors, and transform the undue-burden test to something much more akin to strict scrutiny.

First, the majority's free-form balancing test is contrary to *Casey*. . . . Contrary to the majority's statements [ ] *Casey* did not balance the benefits and burdens of Pennsylvania's spousal and parental notification provisions, either. Pennsylvania's spousal notification requirement, the plurality said, imposed an undue burden because findings established that the requirement would "likely . . . prevent a significant number of women from obtaining an abortion"—not because these burdens outweighed its benefits. [ ]. And *Casey* summarily upheld parental notification provisions because even pre-*Casey* decisions had done so. [ ]

* * *

Second, by rejecting the notion that "legislatures, and not courts, must resolve questions of medical uncertainty," [ ] the majority discards another core element of the *Casey* framework. Before today, this Court had "given state and federal legislatures wide discretion to pass legislation in areas where there is medical and scientific uncertainty." . . .

Today, however, the majority refuses to leave disputed medical science to the legislature because past cases "placed considerable weight upon the evidence and argument presented in judicial proceedings." . . .

Finally, the majority overrules another central aspect of *Casey* by requiring laws to have more than a rational basis even if they do not substantially impede access to abortion. [ ] "Where [the State] *has a rational basis to act* and it does not impose an undue burden," this Court previously held, "the State may use its regulatory power" to impose regulations "in furtherance of its legitimate interests in regulating the medical profession in order to promote respect for life, including life of the unborn." [ ] No longer. Though the majority declines to say how substantial a State's interest must be, [ ] one thing is clear: The State's burden has been ratcheted to a level that has not applied for a quarter century. . . .

The majority's undue-burden test looks far less like our post-*Casey* precedents and far more like the strict-scrutiny standard that *Casey* rejected, under which only the most compelling rationales justified restrictions on abortion. [ ] One searches the majority opinion in vain for any acknowledgment of the "premise central" to *Casey*'s rejection of strict scrutiny: "that the government has a legitimate and substantial interest in preserving and promoting fetal life" from conception, not just in regulating medical procedures. . . . Moreover, by second-guessing medical evidence and making its own assessments of "quality of care" issues, [ ] the majority reappoints this Court as "the country's *ex officio* medical board with powers to disapprove medical and operative practices and standards throughout the United States." [ ] And the majority seriously burdens States, which must guess at how much more compelling their interests must be to pass muster and what "commonsense inferences" of an undue burden this Court will identify next.

III

. . . The undue-burden standard is just one variant of the Court's tiers-of-scrutiny approach to constitutional adjudication. And the label the Court affixes to its level of scrutiny in assessing whether the government can restrict a given right—be it "rational basis," intermediate, strict, or something else—is increasingly a meaningless formalism. As the Court applies whatever standard it likes to any given case, nothing but empty words separates our constitutional decisions from judicial fiat.

* * *

The Court should abandon the pretense that anything other than policy preferences underlies its balancing of constitutional rights and interests in any given case.

* * *

[Justice Alito's dissenting opinion, with whom the Chief Justice and Justice Thomas join, is omitted.]

* * *

NOTES AND QUESTIONS

1. *The Role of Science.* Was the Texas legislature's fatal flaw its failure to make formal findings regarding the health benefits of H.B. 2 and include them in the text of the statute? If the legislature had done so, would the result have been different? Recall the question after *Carhart*: How closely should courts analyze the medical or scientific basis for state maternal or fetal health justifications? Does *Whole Women's Health* reflect the Court's willingness to scrutinize such justifications more closely than in *Carhart*?

2. *Licensing Requirements & Scope of Practice.* Courts have had little trouble with the requirement (now in almost forty states) that only licensed physicians, not physicians' assistants or others, perform abortions, because the authority granted with professional licenses is a matter of state law. Such licensing requirements are seen as an uncontroversial way of protecting the pregnant woman, not imposing a burden on her decision to have an abortion. More recently, some states have expanded their scope of practice laws to allow non-physician certified or licensed health care professionals with the relevant specialized training to perform abortions in limited circumstances, and a growing number of professional organizations are calling for an expansion of the pool of clinicians able to provide medication abortions. Yet, significant regulatory and political barriers to doing this remain. See American Public Health Association, Policy Statement, Provision of Abortion Care by Advanced Practice Nurses and Physician Assistants (Nov. 1, 2011) (noting that the scope of primary and specialty practice of nurse practitioners, certified nurse midwives, and physician assistants includes management of conditions and procedures significantly more complex than medication or aspiration abortion; also noting evidence that with appropriate education and training, they can competently provide all components of medication abortion care). (Medication abortion is discussed further at Section IV.A., Note: The Distinction Between Contraception and Abortion).

3. Whole Woman's Health *Revisited.* After *Whole Woman's Health*, two new appointments were made to the Supreme Court. Justice Gorsuch was appointed to fill the vacancy left by Justice Scalia's passing, and Justice Kennedy—the conservative justice whose vote was key to cementing the *Whole Women's Health* majority opinion—retired and was replaced by Justice Kavanaugh. This fueled speculation about the possibility that abortion jurisprudence would take a decisive turn toward allowing even greater state restrictions on abortion access. This was tested just four years later in *June Medical Services L.L.C. v. Russo*, 140 S.Ct. 2103 (2020), when the Court heard a constitutional challenge to a Louisiana admitting privileges law that was virtually identical to the Texas law struck down in *Whole Woman's Health*.

In a surprising development, Chief Justice Roberts, who originally sided with the dissent in *Whole Woman's Health*, voted with Justices Breyer, Ginsburg, Sotomayor, and Kagan to strike down the Louisiana law based on *Whole Woman's Health*. Justices Thomas, Alito, Gorsuch, and Kavanaugh dissented. This was not a simple application of precedent, however. There was no majority opinion affirming the underlying reasoning of *Whole Woman's Health*, but there were several opinions—a concurring opinion from Chief Justice Roberts, and several other dissenting opinions—that cast doubt on *Whole Woman's Health* legacy.

Justice Breyer wrote the plurality opinion in *June Medical*, joined by Justices Ginsburg, Sotomayor, and Kagan. It mirrored the majority's approach in *Whole Woman's Health*, reiterating the standard that "requires courts independently to review the legislative findings upon which an abortion-related statute rests and to weigh the law's 'asserted benefits against the burdens' it imposes on abortion access." Id. at 2112. Chief Justice Roberts wrote a separate opinion concurring in the judgment only, noting that his vote did not reflect a change of heart about *Whole Woman's Health*, but rather was required by *stare decisis*. Id. at 2133 (Roberts, C.J., concurring). Roberts rejected the plurality's interpretation of *Casey* as requiring courts to weigh the law's asserted benefits against the burdens it imposes on access:

> Nothing about *Casey* suggested that a weighing of costs and benefits of an abortion regulation was a job for the courts. On the contrary, we have explained that the "traditional rule" that "state and federal legislatures [have] wide discretion to pass legislation in areas where there is medical and scientific uncertainty" is "consistent with *Casey*." *Gonzales v. Carhart*, 550 U.S. 124 (2007). *Casey* instead focuses on the existence of a substantial obstacle, the sort of inquiry familiar to judges across a variety of contexts. . . .

> To be sure, the Court at times discussed the benefits of the regulations . . . . But in the context of *Casey*'s governing standard, these benefits were not placed on a scale opposite the law's burdens. Rather, *Casey* discussed benefits in considering the threshold requirement that the State have a "legitimate purpose" and that the law be "reasonably related to that goal."

> So long as that showing is made, the only question for a court is whether a law has the "effect of placing a substantial obstacle in the path of a woman seeking an abortion of a nonviable fetus." [ ]

The lack of a majority opinion as to the proper theory of the undue burden test that should be applied to regulations justified on health grounds has created uncertainty for lower courts. For example, in EMW Women's Surgical Center, P.S.C. v. Friedlander, 978 F.3d 418 (6th Cir. 2020), *rehearing en banc denied* Dec. 31, 2020, the Sixth Circuit reversed a district court ruling that found a Kentucky abortion law unconstitutional based on the *Whole Woman's Health* balancing test. In reversing and remanding this decision, a divided panel of the Sixth Circuit held that the balancing test from *Whole Woman's Health* is no longer controlling law after *June Medical*. See also Hopkins v.

Jegley, 968 F.3d 912, 915 (8th Cir. 2020) (per curiam) (combining the concurrence and dissenting opinions in *June Medical* to determine controlling law). The Seventh Circuit, however, has rejected this approach. See Planned Parenthood of Indiana and Kentucky, Inc. v. Box, 991 F.3d 740, 741 (7th Cir. 2021), *appeal filed* Mar. 29, 2021 (holding that because a majority of justices has not held otherwise, the balancing test set forth in Whole Woman's Health remains binding precedent). How, if at all, does *June Medical* change your answer to the question in Note 1?

4. *The Future of Abortion Jurisprudence.* The cases in this section have revealed the more significant twists and turns that abortion jurisprudence has taken since abortion was established as a constitutional right in *Roe v. Wade*. Despite the qualifications and limits that have been fleshed out since then, the Court has repeatedly affirmed the "essential" holding of *Roe* that anchors judicial scrutiny of abortion laws to a fundamental right of procreation. But how secure is this right? People on different sides of the abortion debate have long hoped or feared that a continued shift in the Supreme Court would eventually bring about *Roe's* undoing, if not through an explicit overruling of *Roe's* recognition of a fundamental right to reproductive choice, then through an equally effective evisceration of the current constitutional framework used to scrutinize laws burdening this right. Indeed, Justice Thomas has consistently expressed his disagreement with the Court's identification of abortion as a protected constitutional right and called for overruling *Roe v. Wade*.

But a number of recent developments have revived and intensified speculation about *Roe's* fate. The first is the apparent significant shift in the Supreme Court, with Justices Gorsuch, Kavanaugh, and Barrett being appointed by President Trump, who promised as a candidate to nominate Supreme Court justices who would overturn *Roe v. Wade*. The second development is the Supreme Court's recent grant of certiorari in Dobbs v. Jackson Women's Health, 945 F.3d 265 (5th Cir. 2019), *cert. granted* 141 S.Ct. 2619 (2021). *Dobbs* is a challenge to Mississippi's Gestational Age Act, which prohibits abortions after 15 weeks' gestation—well before the viability of the fetus—except in medical emergency or in case of severe fetal abnormality. Both the District Court and Fifth Circuit found this to be a clear violation of longstanding abortion jurisprudence that prohibits bans on previability abortions, and there was no circuit court split that the Supreme Court needed to resolve. Rather, numerous scholars have observed that the only apparent reason for the Supreme Court granting certiorari on the question of whether all previability bans on abortion are unconstitutional would be to allow the state of Mississippi to present evidence challenging the viability framework used to protect abortion rights since *Roe* and *Casey*. Several states have already enacted so-called "trigger" bans—that is, laws creating previability abortion bans that are clearly prohibited under current abortion jurisprudence but will immediately take effect to the extent permitted by federal law if the Supreme Court overrules or substantially limits the constitutional right to abortion recognized in *Roe*. See State Legislation Tracker: Major

Developments in Sexual and Reproductive Health, Guttmacher Institute. Oral arguments have been scheduled for the October 2021 term, so a decision will not be issued until after the publication of this casebook. Developments in this area should be watched closely.

One final development has emerged rather suddenly as this casebook goes to print. In 2021, Texas enacted a law that would ban all abortions for pregnant patients if embryonic cardiac activity has been detected—activity that may be detectable as early as six weeks, before an embryo has a fully developed heart. Effectively, this means that abortion would be banned at six weeks, before most women know they are pregnant. Moreover, the law does not contain an exception for pregnancies that result from rape, sexual abuse, incest, or fetal defect incompatible with life after birth. Ordinarily, such a ban on previability abortions would be easily held unconstitutional under the framework established under *Roe* and *Casey,* and thus prevented from going into effect. The Texas law takes an unprecedented approach to enforcement, however, by deliberately failing to create a mechanism for state enforcement. Instead, the law effectively outsources enforcement by authorizing civil suits by private individuals against anyone who performs, aids and abets or intends to participate in a prohibited abortion. This absence of "state action" was designed to preclude those affected by the law from seeking an injunction against enforcement, and the strategy has been working so far.

The Fifth Circuit allowed the law to take effect in September 2021, and the U.S. Supreme Court initially refused to intervene at the request of providers. Whole Woman's Health v. Jackson, 141 S.Ct. 2494 (2021). Without a full argument or briefing, the Court issued a brief, late night order announcing its refusal to intervene, but Chief Justice Roberts and Justices Breyer, Sotomayor, and Kagan authored four dissenting opinions. Justice Sotomayor wrote a particularly scathing dissent that decried Texas' approach of effectively "deputizing [its] citizens as bounty hunters, offering them cash prizes for civilly prosecuting their neighbors' medical procedures" and that criticized the majority for "burying their heads in the sand" in the face of a "flagrantly unconstitutional law engineered to prohibit women from exercising their constitutional rights and evade judicial scrutiny." Id. at 2498. Not long thereafter, providers, along with the U.S. government, returned to the Supreme Court to request intervention again. This time the Court granted an expedited review with oral arguments set for November 2021–not to address the constitutionality of the Texas abortion ban, but to consider whether the design of the Texas law should be able to preclude providers, or the federal government, from enjoining enforcement of the law pending litigation. This time, the Court ruled that the providers may pursue a pre-enforcement challenge to the Texas ban against some government officials; but, once again, the Court refused to enjoin the Texas ban while the legal challenges in federal and state courts run their course. Whole Woman's Health v. Jackson, No. 21–463, slip op. (Dec. 10, 2021). Legal challenges to the Texas law are evolving quickly, but as of this writing, the Texas ban is still in effect.

NOTE: FEDERAL CONSCIENCE PROTECTIONS

Soon after *Roe v. Wade* the federal government enacted legislation known as the "Church Amendments" creating conscience protections for those objecting to providing abortion or sterilization. 42 U.S.C. § 300a–7. These Amendments apply to federal funds and grants provided pursuant to Population Research and Voluntary Family Planning Programs, and they make clear that such funds cannot be used to require individuals to perform or assist in abortions (or sterilizations), if contrary to the individual's religious or moral beliefs. Such funds also cannot be used to require entities to make their facilities available for the performance of abortions (or sterilizations), or to make personnel available for such procedures, if contrary to the moral or religious beliefs of the entity or personnel, respectively.

In 2004, Congress passed the Weldon Amendment, originally adopted as section 508(d) of the Labor-HHS Division of the 2005 Consolidated Appropriations Act, Public Law 108–447, 118 Stat. 2809, 3163 (Dec. 8, 2004), and readopted in each subsequent HHS appropriations act. The Weldon Amendment provides that "[n]one of the funds made available in this Act may be made available to a Federal agency or program, or to a state or local government, *if such agency, program, or government subjects any institutional or individual health care entity to discrimination on the basis that the health care entity does not provide, pay for, provide coverage of, or refer for abortions.*" The term "health care entity" is defined broadly to include health care professionals, hospitals and other health facilities, and health care organizations, such as health insurance plans. These statutory conscience protections remain in effect after the ACA, and the ACA extended conscience protection to providers participating in health plans offered through the exchange. The ACA language largely tracks the Weldon Amendment, prohibiting exchange plans from *discriminating against any individual health care provider or facility because of its unwillingness to provide, pay for, provide coverage of, or refer for abortions.*

When the Weldon Amendment was passed, legal challenges were brought based on concerns that it would conflict with other federal or state obligations to patients, putting funding recipients in a Catch 22: they feared that by obeying one law, they risked violating the other, and either way they risked losing funding. For example, in National Family Planning and Reproductive Health Assn. v. Gonzales, 468 F.3d 826 (D.C. Cir. 2006), the plaintiffs alleged a potential conflict between the Weldon Amendment, on the one hand, and plaintiffs' ethical and legal duties as family planning grant recipients under HHS regulations, on the other. HHS regulations required recipients to offer pregnant women the opportunity to be provided information and counseling regarding pregnancy termination. If a caregiver objected to providing this information, one way the plaintiff might respond would be to reassign that caregiver to another job; this would accommodate the worker's religious objections while ensuring patients get information that is legally and ethically required. In the lawsuit, the plaintiff expressed the concern that if the caregiver wanted to retain a pregnancy counseling job, such a reassignment may be viewed as "discrimination" under the Weldon Amendment. Without ruling on the merits, the court noted the absence of any indication that the federal government would view a reassignment or other attempts to navigate such a conflict as discrimination.

A California district court came to a similar conclusion in California v. U.S., 2008 WL 744840 (N.D. Cal. 2008). California officials claimed that the Weldon Amendment appeared to be in conflict with EMTALA, the federal law that obligates hospitals with emergency rooms to provide emergency care, as well as a state law that tried to balance provider conscience protections against duties to patients in emergency cases. California's state conscience law exempted medical personnel and facilities from participating in abortions if they objected on moral, ethical, or religious grounds, but it contained an exception for medical emergency situations and spontaneous abortions. In an unreported opinion, the court did not resolve the potential conflict. Rather, it held that California could not show an injury in fact (and thus did not have standing), because there was no clear indication from either the text of the Amendment or the federal regulator that California's emergency exception would be viewed as "discrimination" under the Weldon Amendment.

Although actual conflicts did not materialize in the above cases, during the Trump administration, HHS issued a rule that renewed concerns about how federal regulators would interpret the Weldon Amendment and other statutory conscience protections going forward. See Protecting Statutory Conscience Rights in Health Care; Delegations of Authority, 84 Fed. Reg. 23170 (May 21, 2019). The rule signaled the Trump administration's intention to interpret conscience statutes more broadly than in the past and to use its enforcement power against state laws intended to protect abortion access. But lawsuits delayed the rule's implementation, and since then the Biden Administration has signaled its intent to reverse the Trump-era approach.

For a discussion of the ethical considerations that inform conscience protections generally, see Section IV.C. below.

IV. CONTRACEPTION

Contraception, generally, has not been as controversial as abortion, and as we will in this section, the jurisprudence developed in response to contraceptive access questions has taken a somewhat different journey. This section begins with the constitutional origins of the right to procreative choice. Next, it explores how, over time, access to contraception has tended to receive broader public and private support, and even legal protection, based on health, economic, and gender equity concerns. Conflicts do arise, however, typically based on objections to certain forms of contraception as interfering with the reproductive process after "life" has begun, and often linked to religious freedom claims. The final part of this section reviews this conflict.

A. CONTRACEPTIVE USE AS A PRIVACY RIGHT

Government regulation of reproduction, in particular abortion and contraception, has ebbed and flowed throughout history. In Geoffrey Stone's book, Sex and the Constitution, he explains that during the 18th and early 19th centuries, there were no laws prohibiting either contraception or abortion before quickening. Indeed, during this time, he found that contraception and abortion were increasingly recognized as legitimate and morally acceptable practices to maintain and improve a family's economic welfare, and that advertisements for birth control products were commonplace. Social leaders concerned about overpopulation leading to poverty and social decline viewed birth control as a tool for promoting individual and societal well-being.

The increasing medicalization of the reproductive process, and a growing focus on the link between illicit sexual activity and reproductive control within powerful religious movements, brought greater public attention to reproduction, and this eventually led to greater regulation. In 1873, Congress enacted the Comstock Act, which prohibited anyone from mailing "any drug or medicine, or any article whatever, for the prevention of conception" or from mailing any information about how to obtain these articles. Anthony Comstock, the author of this act, considered contraceptive material "obscene" because it was believed to encourage immoral thoughts and behavior. According to Stone, Comstock wanted to outlaw birth control "because the availability of contraceptives reduces the risk that individuals who engage in premarital sex, extramarital sex, or prostitution will suffer the consequences of venereal disease or unwanted pregnancy."

In the next couple of decades, most states enacted similar criminal bans and some, like Connecticut went even further, criminalizing the *use* of contraception, as well as the act of helping someone to violate the law. This ban was ultimately invalidated by the Supreme Court in the seminal case Griswold v. Connecticut, 381 U.S. 479 (1965). In *Griswold,* an official of the Planned Parenthood League of Connecticut and a Yale physician were charged with aiding and abetting "the use of a drug, medicinal article, or instrument for the purpose of preventing conception," a crime under Connecticut law, by providing contraceptives to a married couple. The Supreme Court reversed their conviction. Justice Douglas, writing for the Court, concluded:

> The present case . . . concerns a relationship lying within the zone of privacy created by several fundamental constitutional guarantees. And it concerns a law which, in forbidding the *use* of contraceptives rather than regulating their manufacture or sale, seeks to achieve its goals by means having a maximum destructive impact upon that relationship. Such a law cannot stand in light of the familiar principle

so often applied by this Court, that a "governmental purpose to control or prevent activities constitutionally subject to state regulation may not be achieved by means which sweep unnecessarily broadly and thereby invade the area of protected freedoms." Would we allow the police to search the sacred precincts of marital bedrooms for telltale signs of the use of contraceptives? The very idea is repulsive to the notions of privacy surrounding the marriage relationship.

We deal with a right of privacy older than the Bill of Rights—older than our political parties, older than our school system. Marriage is a coming together for better or worse, hopefully enduring, and intimate to the degree of being sacred. It is an association that promotes a way of life, not causes; a harmony in living, not political faith; a bilateral loyalty, not commercial or social projects. Yet it is an association for as noble a purpose as any involved in our prior decisions.

381 U.S. at 484.

Griswold's articulation of the privacy right focused so heavily on preventing intrusions into the marital relationship that questions remained about the application of this right to non-married people. This was understandable in light of the moral aims of such laws to prevent pre-marital or extramarital sex. For example, three justices who concurred in *Griswold*, explained that they found the law invalid because they rejected the state's assertion of a rational relationship between the ban and the state's interest in discouraging extra-marital relations. But the justices went on to note that the Court's holding "in no way interferes with a State's proper regulation of sexual promiscuity or misconduct."

The Court was confronted with this question in 1972, in Eisenstadt v. Baird, 405 U.S. 438 (1972). The Court determined that a law that allowed married people, but not unmarried people, to have access to contraceptives violated the equal protection clause of the Fourteenth Amendment because there could be no rational basis for distinguishing between married and unmarried people in permitting access to contraceptives. The Court suggested that "if the right of privacy means anything, it is the right of the individual, married or single, to be free from unwarranted government intrusion into matters so fundamentally affecting a person as a decision whether to bear a child." In Carey v. Population Services International, 431 U.S. 678 (1977), the Court confirmed this right not only prohibits bans on the use of contraception, it prevents any *unjustified intrusion by* the State, such as a ban on the *distribution* or sale of contraceptives.

But what constitutes an unjustified intrusion? Relying on *Roe v. Wade*, the Court in *Carey v. Population Services International* made clear that recognizing a privacy right for procreative choice does not "automatically invalidate every state regulation in this area." Id. at 685-685. It noted that some regulations may not infringe on individual choice at all, and "even a burdensome regulation may be validated by a sufficiently compelling state interest." Id. at 686. For example, the FDA has authority to approve and monitor the safety of drugs and devices, including prescription contraception. States also have broad power to regulate health and safety. But in *Carey*, the Court struck down a state law limiting the distribution of *nonprescription* contraceptives to licensed pharmacists, noting that the limit on these "nonhazardous" forms of contraception bore no relation to the State's interest in protecting health. A plurality opinion also invalidated a provision banning distribution of nonprescription contraceptives to those under 16.

NOTE: THE DISTINCTION BETWEEN CONTRACEPTION AND ABORTION

Early objections to the use and distribution of contraceptives tended to focus on their use for pregnancy prevention and they often applied to any type of birth control. More recently, however, objections or restrictions tend to target certain forms of contraceptives because of how and when they work. (A detailed explanation of the process of human reproduction is provided at the beginning of Section VII.A (Assisted Reproductive Technologies) of this chapter.)

Contraception is designed to work by preventing fertilization. Some forms are taken or used prior to intercourse, while others may be taken after intercourse. Two examples of medication that work after intercourse but prior to fertilization are Plan B (levonorgestrel) and ella (ulipristal acetate), also commonly referred to as "morning after pills." They are designed to avoid the development of pregnancy by inhibiting ovulation or follicle rupture. Questions have been raised, however, about whether such pills may alter the uterine lining and, possibly, make it less likely that a fertilized ovum will enter the uterus or implant in the uterine wall, though recent studies do not seem to support this theory. For those who believe that life begins at fertilization, concern that morning after pills may prevent implantation of a fertilized ovum has fueled opposition to the pills.

Plan B became available in 2006 after a long and highly political history. In 1999 an FDA advisory panel had recommended that the drug be made available over the counter for all purchasers, but objections (arguably political and arguably medical) caused the FDA to delay approval until 2003 and then require a prescription for this medication. After a threat by some members of the Senate to withhold confirmation of a newly appointed FDA Commissioner in 2006, Plan B was finally made available over the counter for persons over 18, and that age was subsequently lowered to 17. See Gardiner Harris, Morning-After Pill: Politics and the F.D.A., N.Y. Times (Aug. 28, 2015). Plan B, a two-dose regimen, was eventually replaced by Plan B One Step, which as the name suggests, is a one-dose regimen. Plan B One Step is effective at reducing the risk of pregnancy if taken within 72 hours of intercourse, and its efficacy increases if taken within 24 hours. Approved in 2010 after its own long medical and political battle, ella works in largely the same way, but can be effective within five days of having sex.

Devices and medication that prevent fertilization or implantation are considered by the medical community to be *contraceptives*, that is a form of pregnancy prevention. Use of Mifepristone (RU-486), on the other hand, is classified as a *medication abortion* because it works after pregnancy. It is designed to disrupt implantation of the fertilized embryo and to change the uterus to become less hospitable to the fertilized ovum, eventually expelling it. Mifepristone is available only through prescription and dispensed only by a certified health care provider who meets certain qualifications.

The generally accepted medical definitions used to distinguish contraception from medication abortion may not be accepted by those who believe that life begins at fertilization. People who hold this belief may have no objections to contraception that prevents fertilization; but they view medications or devices that work after fertilization as terminating life and thus would characterize them as abortifacients. This definitional disagreement can be significant in a number of contexts. As described in this Note, it can fuel political opposition to government action that would facilitate access to certain drugs. But it may also be relevant to the question of how far government can go in requiring individuals or entities to provide contraception as part of health care, a question we revisit in Section C below.

B. ACCESS TO FUNDING FOR CONTRACEPTION

As noted in Section III, the Supreme Court has made clear that a constitutional right to have one's procreative choices protected from unwarranted government intrusion does not create an entitlement to government funding or subsidization of those services. Nonetheless, other legal and policy avenues have facilitated access to funding for contraception. Antidiscrimination law has been one such avenue. For example, in Erickson v. Bartell Drug Company, 141 F. Supp. 2d 1266 (2001), plaintiff employees successfully challenged the exclusion of prescription contraception from an otherwise comprehensive employment-based insurance plan, on the grounds that such an exclusion constituted prohibited discrimination on the basis of sex. Plaintiffs brought the claim under Title VII of the 1964 Civil Rights Acts, as amended by the 1978 Pregnancy Discrimination Act.

An excerpt of *Erickson* is reproduced at Chapter 2, Section IV.B. (Gender Discrimination) of this casebook. After reading *Erickson*, consider the following Notes and Questions.

NOTES AND QUESTIONS

1. *Gender Equity and the ACA's Contraceptive Mandate.* Although the *Erickson* plaintiffs won, Title VII could provide only limited protection because it did not apply to non-employment-based health plans or to health care

providers. The ACA filled this gap, by expanding gender equity protections in several ways. One is the inclusion of a preventive health mandate, interpreted as requiring coverage for prescription contraception. Notably, federal regulators have asserted that this mandate serves compelling health and gender equity goals. For more on the limits of Title VII and the importance of the contraceptive mandate in the ACA, see Notes 2 and 5, following the *Erickson* case.

2. *Title VII's Relationship to the ACA's Sex Nondiscrimination Provision.* The ACA also created a new health-specific prohibition on sex discrimination in its Section 1557 antidiscrimination provision. For more on the relevance of Title VII to defining the scope of Section 1557's antidiscrimination protections, see Chapter 2, Section IV.B.

3. *Federal & State Family Planning Programs.* Various government health programs also provide coverage for prescription contraception for low-income individuals. For example, Medicaid plays a critical role in providing family planning services, a federally mandated benefit, to low-income populations. Although states have some discretion to determine the specific services they will cover through Medicaid, contraception is one of the primary services covered and most states offer broad coverage. Contraception is also an important service in other family planning and disease prevention programs, such as the Title X Family Planning Program.

4. *Other State Actions to Facilitate Access.* By the time of the ACA, twenty-eight states already required some form of contraceptive coverage in at least some health insurance plans sold in their states. But many states have gone further, to also require some providers to provide access to certain kinds of contraceptive care. For example, eighteen states and the District of Columbia require emergency rooms to provide information about emergency contraception to sexual assault victims, and thirteen states and DC require emergency rooms to dispense the drug to sexual assault victims on request. Four states require either pharmacies or pharmacists to fill all valid prescriptions, and eight states allow pharmacists to dispense emergency contraception without a physician's prescription under certain conditions. States have also expanded minors' authority to consent to health care generally, but in particular relating to sexual and reproductive health care. See Guttmacher Institute, Fact Sheets on Emergency Contraception and Minors' Access to Contraceptive Services (Jul. 1, 2021).

C. THE CLASH BETWEEN GOVERNMENT MANDATES AND RELIGIOUS OR MORAL BELIEFS

In response to government action to expand contraceptive access, political and legal challenges have emerged. In particular, those who object to certain medication or devices on religious or moral grounds have sought exemptions from government mandates to provide or pay for contraception. As described below, this has led to a rather dramatic back-and-forth in the regulatory space and in the courts. Indeed, several challenges involving religious objections to the mandate have reached the Supreme Court in a very short span of time, and yet the conflict continues.

1. Religious Objections to the ACA Contraceptive Mandate

Early in the process of issuing regulatory rules for coverage of preventive services under the ACA, including the prescription contraception mandate, the Obama administration included a very narrow exemption for *religious employers* that (1) have the purpose of religious inculcation of religious values, (2) employ primarily people who share its religious values, (3) primarily serve people who share its religious tenets, and (4) are non-profit organizations established to be a church, association of churches or a religious order. See Coverage of Certain Preventive Services Under the Affordable Care Act, 78 Fed. Reg. 39870 (Jul. 2, 2013). Essentially, this exemption meant *churches* and other *houses of worship* did not have to provide contraception coverage, and it applied to some religious primary and secondary schools organized to inculcate religious views and designed for attendance by coreligionists only.

The exemption did not extend to most religious colleges and universities that had students of all faiths, nor to social service organizations that provided service to people of all faiths. These religious organizations would have to provide insurance that included contraception. Because the rule was later extended to cover student health plans, the fact that religious colleges (and, in particular, Catholic universities open to students of all faiths) were not exempt proved to be very significant. Although the scope of this exemption was consistent with the kinds of religious exemptions provided in state mandates at the time, it met with strong objections from religiously affiliated organizations that the exemption was too narrow. Opposition, led by the U.S. Conference of Catholic Bishops, evangelical Christians, and conservative politicians, claimed the rule was an unconscionable intrusion on religious liberty.

Obama's Compromise: Accommodation for Religious Nonprofits

In the face of deeply held and conflicting views on the issue, the Obama administration tried to fashion a compromise that would ensure women received coverage for contraception while allowing religious employers and universities to opt out of participation in the provision of any services they found to be morally objectionable. New regulations were issued, extending protection to the *religious nonprofits* (schools and social service organizations) previously left out, if they had religious objections to providing coverage for contraceptive services required by law. See Coverage of Certain Preventive Services Under the Affordable Care Act, supra. See also Robert Pear, Birth Control Rule Altered to Allay Religious Objections, N.Y. Times (Feb. 2, 2013).

Importantly, the regulations did not *exempt* these religious nonprofits from the mandate; instead, they established an *accommodation* that would allow these religious nonprofits (classified as *"eligible organizations"* in the regulation) to "opt-out" of providing contraceptive coverage for their employees by formally announcing that they had religious objections to offering this coverage. This would not impede access, though, because it would trigger a process by which the health insurer or other third party would be required to provide the contraceptive coverage, with no cost sharing by the employee and without the employer having to pay for it or help arrange for its provision.

Legal Challenges to the Obama Era Compromise

There were two types of challenges to the Obama era rules worthy of note. The first was by for-profit companies, whose owners operated their companies in accord with religious principles, but who were not entitled to an accommodation under the Obama compromise. Numerous companies filed legal challenges to the contraceptive mandate on religious freedom grounds, and the first suit to reach the Supreme Court was Burwell v. Hobby Lobby, Inc., 573 U.S. 682 (2014). In *Hobby Lobby*, the Supreme Court determined that closely held private corporations could assert the religious liberty interests of their owners, and that the ACA's mandate to provide contraception was a violation of the federal Religious Freedom Restoration Act (RFRA), 42 U.S.C.A. § 2000bb–1(a), (b).

Justice Alito, who wrote for the 5–4 majority finding that the contraceptive mandate violated RFRA, summarized his opinion this way:

> We must decide in these cases whether the Religious Freedom Restoration Act . . . (RFRA) [ ] permits the United States Department of Health and Human Services (HHS) to demand that three closely held corporations provide health-insurance coverage for methods of contraception that violate the sincerely held religious beliefs of the companies' owners. We hold that the regulations that impose this obligation violate RFRA, which prohibits the Federal Government from taking any action that substantially burdens the exercise of religion unless that action constitutes the least restrictive means of serving a compelling government interest.

> In holding that the HHS mandate is unlawful, we reject HHS's argument that the owners of the companies forfeited all RFRA protection when they decided to organize their businesses as corporations rather than sole proprietorships or general partnerships. The plain terms of RFRA make it perfectly clear that Congress did not discriminate in this way against men and women who wish to run their businesses as for-profit corporations in the manner required by their religious beliefs.

Since RFRA applies in these cases, we must decide whether the challenged HHS regulations substantially burden the exercise of religion, and we hold that they do. The owners of the businesses have religious objections to abortion, and according to their religious beliefs the four contraceptive methods at issue are abortifacients. If the owners comply with the HHS mandate, they believe they will be facilitating abortions, and if they do not comply, they will pay a very heavy price—as much as $1.3 million per day, or about $475 million per year, in the case of one of the companies. If these consequences do not amount to a substantial burden, it is hard to see what would.

Under RFRA, a Government action that imposes a substantial burden on religious exercise must serve a compelling government interest, and we assume that the HHS regulations satisfy this requirement. But in order for the HHS mandate to be sustained, it must also constitute the least restrictive means of serving that interest, and the mandate plainly fails that test. There are other ways in which Congress or HHS could equally ensure that every woman has cost-free access to the particular contraceptives at issue here and, indeed, to all FDA-approved contraceptives.

In fact, HHS has already devised and implemented a system that seeks to respect the religious liberty of religious nonprofit corporations while ensuring that the employees of these entities have precisely the same access to all FDA-approved contraceptives as employees of companies whose owners have no religious objections to providing such coverage. [Eds. Note: This refers to the opt-out scheme, described above.] The employees of these religious nonprofit corporations still have access to insurance coverage without cost sharing for all FDA-approved contraceptives; and according to HHS, this system imposes no net economic burden on the insurance companies that are required to provide or secure the coverage.

Although HHS has made this system available to religious nonprofits that have religious objections to the contraceptive mandate, HHS has provided no reason why the same system cannot be made available when the owners of for-profit corporations have similar religious objections. We therefore conclude that this system constitutes an alternative that achieves all of the Government's aims while providing greater respect for religious liberty. And under RFRA, that conclusion means that enforcement of the HHS contraceptive mandate against the objecting parties in these cases is unlawful.

As this description of our reasoning shows, our holding is very specific. We do not hold, as the principal dissent alleges, that for-profit corporations and other commercial enterprises can "opt out of any law (saving only tax laws) they judge incompatible with their sincerely held religious beliefs." [ ] (opinion of Ginsburg, J.). Nor do we hold, as the dissent implies, that such corporations have free rein to take steps that impose "disadvantages . . . on others" or that require "the general public [to] pick up the tab." [ ] And we certainly do not hold or suggest that "RFRA demands accommodation of a for-profit corporation's religious beliefs no matter the impact that accommodation may have on . . . thousands of women employed by Hobby Lobby." [ ] The effect of the HHS-created accommodation on the women employed by Hobby Lobby and the other companies involved in these cases would be precisely zero. Under that accommodation, these women would still be entitled to all FDA-approved contraceptives without cost sharing.

Burwell v. Hobby Lobby, Inc., 573 U.S. 682, 688–693 (2014).

The second type of challenge brought against the Obama era rule was based on objections to the opt-out process itself. Numerous religious organizations and employers filed suit alleging that the requirement they submit a formal written document asking to be relieved of the mandate violated RFRA, because it constituted facilitation of a sin. The legal issues at play in the courts were whether the opt-out process itself was a substantial burden, whether the governmental interest was compelling, and whether there was a less restrictive way of serving the same interest. All but one of the courts of appeals hearing these challenges held that the accommodation adequately protected plaintiffs' religious freedom. See, e.g., Little Sisters of the Poor v. Burwell, 794 F.3d 1151 (10th Cir. 2016) (holding that the accommodation scheme does not substantially burden their religious exercise under RFRA or infringe upon their First Amendment rights). Several of these challenges were consolidated and addressed by the Supreme Court in *Zubik v. Burwell,* but the Court expressed no view on the merits of the RFRA claims. Instead, there was a unanimous decision to remand the cases to afford the parties "an opportunity to arrive at an approach going forward that accommodates petitioners' religious exercise while at the same time ensuring that women covered by petitioners' health plans 'receive full and equal health coverage, including contraceptive coverage.' " 578 U.S. 403, 403 (2016). In October 2017, however, the Trump administration issued expansive new rules (described below) which appeared to relieve these entities of the procedural obligations to which they objected. The plaintiffs entered into agreements with the government to dismiss their cases because it looked like the new rules had settled the issue in their favor.

Trump Era Rules: Expanding Exemptions & Reducing Access

On October 13, 2017, two interim final rules were issued by the regulatory agencies charged with implementation of the ACA. Religious Exemptions and Accommodations for Coverage of Certain Preventive Services Under the Affordable Care Act, 82 Fed. Reg. 47792 (Oct. 13, 2017), and Moral Exemptions and Accommodations for Coverage of Certain Preventive Services Under the Affordable Care Act, 82 Fed. Reg. 47838 (Oct. 13, 2017). These rules expanded the scope of exemptions allowed in two important ways. First, the new rules allowed any objecting religious employers or universities, even those who expressly hire and recruit people from diverse religious and ideological backgrounds, to get an exemption from the law, as opposed to requiring them to follow an opt-out process provided as an accommodation. The difference between accommodation and exemption means the difference between an employee or student getting contraceptive coverage or not. The opt-out process was created precisely to accommodate religious objections while protecting access to coverage. Granting these entities exemptions means that covered individuals would not be offered the objected-to contraception by any means.

Second, the new rules allowed those who were not previously eligible for either an exemption or accommodation to now claim an exemption, as long as they have *moral* objections to the rule. This means the exemption is no longer specifically linked to religious exercise, which is the basis of the protections in RFRA. It also increased the number of entities that could deny contraceptive coverage for employees or students. The rules did preserve the opt-out process as an option; but it would not be required.

Legal Challenges to the Trump Administration Rules

Lawsuits challenging the new rules were filed immediately. A nationwide injunction was affirmed by the Third Circuit in *Pennsylvania v. President of United States*, 930 F.3d 543 (2019), but the decision was ultimately reversed by the Supreme Court in *Little Sisters of the Poor Saints Peter and Paul Home v. Pennsylvania*, 140 S.Ct. 2367 (2020). In a 7–2 decision by Justice Thomas, the Court held that the ACA provided statutory authority for the government to promulgate exemptions to the preventive services mandate, and it rejected the claim that the rules were procedurally invalid due to the agency's failure to issue a notice of proposed rulemaking.

Although the Court permitted the rules to take effect, outstanding questions remained. First, the Supreme Court did not decide whether the scope of the exemptions were required or even permitted by RFRA. The Court also did not address the specific question of whether the rules were arbitrary and capricious in light of the federal government's failure to require an accommodation process that would also protect women's health and access to contraceptive care. At least four Justices—Ginsburg, Bryer, Kagan, and Sotomayor—seemed amendable to this as a basis for future challenge. Justices Alito and Gorsuch, on the other hand, made clear they were ready to go further than the majority to find that RFRA actually compelled the religious exemption being challenged. Id. at 287 (Alito, J., concurring). But because the Court did not go so far as to hold that the Trump exemptions were required by RFRA, this seemed to leave room for the Biden administration to reconsider the rule and return to the accommodation-based approach of the Obama administration, something Biden has signaled a willingness to do.

NOTES AND QUESTIONS

1. *RFRA Challenge: Substantial Burden on Free Exercise of Religion.* In *Hobby Lobby*, the plaintiffs objected to two forms of intrauterine devices and two forms of emergency contraception forms on religious grounds. They believed that these medications and devices worked after fertilization and thus were abortifacients. They claimed that requiring them to include these forms of contraception in their health plans substantially burdened their free exercise of religion. The Court did not engage in the debate over where the line should be drawn between abortion and contraception, for purposes of evaluating plaintiffs' claim of a substantial burden. In religious freedom cases, courts have been careful not to weigh or evaluate the relevant doctrines of faith. As long as a belief is sincere, the Court will not second-guess the underlying basis for that belief.

2. *RFRA Challenge: Does the Contraceptive Mandate Serve a Compelling Interest?* RFRA provides that government action imposing a substantial burden on religious exercise must serve a "compelling government interest." The majority did not, however, discuss whether it viewed the government's interest as compelling. It merely assumed such an interest was met for purposes of its analysis and found the mandate unconstitutional on a different basis.

Addressing potential unintended pregnancies is widely recognized as an important issue, but would a majority of the Court consider it compelling? Consider that around the time of this case, statistics showed that half of all pregnancies were unintended, a million women who do not want to become pregnant have unprotected sex each day, and 25,000 women become pregnant through sexual assault each year. See Gardiner Harris, F.D.A. Approves 5-Day Emergency Contraception, N.Y. Times (Aug. 13, 2010). Studies have also shown that cost-sharing can act as a barrier to women getting birth control, or the most effective form of birth control.

What about regulators' initial assertions that the mandate serves compelling public health and gender equity goals? How much weight should these be given in assessing the strength of the interest? Should subsequent regulators be bound by a prior administration's conclusions that the contraceptive mandate serves compelling public health and gender equity goals? Although the Obama administration had consistently asserted the mandate served a compelling interest in regulatory action and litigation, the Trump administration seemed to retreat from this position. *See Little Sisters,* 140 S.Ct. at 2392 (Alito, J., concurring) ("In *Hobby Lobby*, the Government asserted and we assumed for the sake of argument that the Government had a compelling interest in 'ensuring that all women have access to all FDA-approved contraceptives without cost sharing'. [ ] Now, the Government concedes that it lacks a compelling interest in providing such access, [ ] and this time the Government is correct.") *But see id.* at 2399 (Kagan, J., concurring) (noting that "[r]ather than dispute HRSA's prior finding that the mandate is 'necessary for women's health and well-being,' the Departments left that determination in place .... [thus] commit[ing] themselves to minimizing the impact on contraceptive coverage, even as they sought to protect employers with continuing religious objections.").

3.   Little Sisters *vs.* Hobby Lobby. In *Hobby Lobby*, the Court's holding emphasized the fact that the government had already created an opt-out program that could accommodate religious objections without impeding access to contraception. In other words, employees would not be harmed by granting employers' relief because of the opt-out accommodation, which helped ensure prescription contraception coverage. *Little Sisters*, on the other hand, did not address the potential harm to employees from the blanket exemption created by the Trump-era rules. Is a blanket exemption that fails to try to minimize harm to those denied coverage consistent with the reasoning in *Hobby Lobby*? Should government have to require objecting entities to follow the opt-out process to preserve access, instead of simply giving them an exemption?

2.   Provider Conscientious Refusals to State Mandates

As explained in Section IV.B. above, states have enacted their own mandates to ensure access to contraceptive care—whether through private insurance mandates or requirements that certain types of health care providers must provide contraceptive care. Just as some organizations object to funding certain forms of contraception, health care professionals and entities, including hospitals and pharmacies, have raised religious objections to dispensing or even giving information about certain forms contraception. But religious challenges to state mandates tend not to be successful. First, RFRA only applies to federal law. State law challenges are subject to the Free Exercise Clause of the federal constitution, which unlike RFRA, does not require exemptions from neutral laws of general applicability. See, e.g., Storman's, Inc. v. Selecky, 794 F.3d 1064 (9th Cir. 2015), *cert. denied* 136 S.Ct. 2433 (2016) (holding that the Free Exercise Clause provided for only rational basis review of a state's requirement that a pharmacist dispense emergency contraception). Although state mandates have also been challenged under state constitutional free exercise protections, the results tend to be the same. See, e.g., Catholic Charities v. Superior Court, 85 P.3d 67 (Cal. 2004) (rejecting a state free exercise challenge to the state's Women's Contraceptive Equity Act). Twelve states have statutory conscience protections that allow health care providers to refuse to provide contraception-related services.

NOTE: THE ETHICS OF CONSCIENTIOUS REFUSALS

The materials in the first four sections of this chapter demonstrate that questions about the regulation of access to abortion and contraception can trigger sharply divided views shaped by deeply held beliefs. Such disputes implicate ethical questions about how to balance protections for religious and moral objections to certain procedures with patients' health care needs and gender equity concerns. Claims for legal protections of providers and other entities that refuse to provide or pay for certain care on religious or moral grounds are grounded in claims about the importance of conscience. Consider how one bioethicist and physician has explained the role of conscience in such conflicts:

> Conscientious persons strive to preserve moral integrity. This requires that their external behavior be congruent with their conscience's internal dictates about what they take to be morally right and feel compelled to do. In our morally diverse world, conscientious persons may come into conflict with each other and with society's moral values. . . .

> Any society purporting to serve the good of its members is therefore obliged to protect the exercise of conscience and conscientious objection. However, this involves a serious dilemma for any pluralist, democratic, liberal, or constitutional state. On the one hand, such a society is committed to tolerance to religious diversity, freedom of individual choice, and "neutrality" with respect to religious belief. On the other hand, optimizing freedom of conscience for some individuals may often limit the legal rights, social entitlements, and moral beliefs of others.

Edmund D. Pellegrino, The Physician's Conscience, Conscience Clauses, and Religious Belief: A Catholic Perspective, 30 Fordham Urban L.J. 221, 221 (2002). In this essay, Dr. Pellegrino explored the ethical implications of religious health care professionals and institutions objecting to procedures considered standard in the medical community. He questioned whether conscience clauses can provide adequate protection for religious and moral beliefs, such as where they are written narrowly to only apply to certain kinds of providers. In addressing the conflict between patients' needs and physicians' conscience, he stated that physicians should make their positions publicly known so that patients know in advance of a crisis that what they want may not be care the physician is willing to provide. At the same time, he concluded that "[t]he only viable course for the religious physician is to maintain fidelity to moral integrity and dictates of conscience" by refusing to perform a morally objectionable procedure or treatment, even in emergencies or remote areas where the choice of physicians is limited. For some religious professionals, this would include refusing to refer a patient to another practitioner willing to perform the disputed treatment, because to do so would be cooperation in an illicit act, making the person a moral accomplice.

Compare this approach to the ethics opinion on conscientious refusals in reproductive health published by the American College of Obstetrics and Gynecology (ACOG). ACOG has offered several recommendations, which it believes "maximize respect for health care professionals' consciences without compromising the health and well-being of the women they serve":

1. In the provision of reproductive services, the patient's well-being must be paramount. Any conscientious refusal that conflicts with a patient's well-being should be accommodated only if the primary duty to the patient can be fulfilled.

2. Health care providers must impart accurate and unbiased information so that patients can make informed decisions about their health care. . . .

3. Where conscience implores physicians to deviate from standard practices, including abortion, sterilization and provision of contraceptives, they must provide potential patients with accurate and prior notice of their personal moral commitments. . . .

4. Physicians and other health care professionals have the duty to refer patients in a timely manner to other providers if they do not feel that they can in conscience provide the standard reproductive services that their patients request.

5. In an emergency in which referral is not possible or might negatively affect a patient's physical or mental health, providers have an obligation to provide medically indicated and requested care regardless of the provider's personal moral objections.

6. In resource poor areas, access to safe and legal reproductive services should be maintained. . . .

7. Lawmakers should advance policies that balance protection of providers' consciences with the critical goal of ensuring timely, effective, evidence-based, and safe access to all women seeking reproductive services.

The Limits of Conscientious Refusal in Reproductive Medicine. ACOG Committee Opinion No. 385. American College of Obstetricians and Gynecologists. Obstet Gynecol 2007; 110:1203–8 (reaffirmed 2016).

V. STERILIZATION

Voluntary sterilization, like most forms of contraception, is generally considered less controversial than abortion. Nonetheless, as with abortion and contraception, some patients have had difficulty accessing sterilization procedures because of insurance exclusions and provider refusals to treat. Recall that the Church Amendments (discussed in the Note on Federal Conscience Protections at the end of the Section III of this chapter) provided that receipt of federal funding could not be used to require recipients to provide abortions or *sterilizations*. Some consider sterilization to be a sin because it separates sexual activity from a procreative purpose. In fact, the Church Amendments were a response to an injunction initially issued by a Montana district court requiring a religiously-affiliated hospital to perform a tubal ligation despite religious objections, based in part on the fact it received federal funding. The injunction was eventually dissolved, however, based on the Church Amendments. See Taylor v. St. Vincent Hospital, 369 F. Supp. 948 (D. Mont. 1973), aff'd, 523 F. 2d 75 (9th Cir. 1975). Today, eighteen states allow health care providers to refuse to provide sterilization services. Many of the issues discussed under the sections on abortion and contraception are thus relevant to sterilization as well.

Sterilization practices also implicate reproductive justice concerns often neglected in debates over access to contraception and abortion. While much of the "reproductive rights" movement has focused on barriers to preventing or terminating pregnancy, some individuals experience *barriers to choosing procreation*. There is a long and troubling history of forced and coerced sterilization of certain groups, which complicates discussions about sterilization as a reproductive health issue.

Forced and Coerced Sterilization

Discussions about sterilization often begin with the development of the eugenics movement in the late 19th century. Among the groups targeted for sterilization were people believed to have intellectual disabilities, as well as those engaging in criminal behavior or behavior viewed as immoral, reflecting low intelligence, and/or reflecting poor character. Despite lack of evidence of a link between such characteristics and heritability, some believed society could "purify" its gene pool by sterilizing those who would "pollute" it. The aim of the eugenics movement was articulated in a now infamous decision by Justice Holmes in Buck v. Bell, 274 U.S. 200 (1927). That case dealt with an attempt by the State of Virginia to sterilize Carrie Buck, who had been committed to the State Colony for Epileptics and the Feeble Minded. This action was opposed on the grounds that the statute authorizing sterilization violated the Fourteenth Amendment by denying Ms. Buck due process of law and the equal protection of the law. Justice Holmes responded:

> Carrie Buck is a feeble minded white woman who was committed to the State Colony above mentioned in due form. She is the daughter of a feeble minded mother in the same institution and the mother of an illegitimate feeble minded child . . . .

> [The lower court found] "that Carrie Buck is the probable potential parent of socially inadequate offspring, likewise afflicted, that she may be sterilized without detriment to her general health and that her welfare and that of society will be promoted by her sterilization." . . . We have seen more than once that the public welfare may call upon the best citizens for their lives. It would be strange if it could not call upon those who already sapped the strength of the state for these

lesser sacrifices, often not felt to be such by those concerned, in order to prevent our being swamped with incompetence. It is better for all the world if instead of waiting to execute degenerate offspring for crime, or to let them starve for their imbecility, society can prevent those who are manifestly unfit from continuing their kind. The principle that sustains compulsory vaccination is broad enough to cover cutting the fallopian tubes. [ ] Three generations of imbeciles are enough.

* * *

The Supreme Court addressed eugenic sterilizations once more, in Skinner v. Oklahoma, 316 U.S. 535 (1942). The Court determined that the equal protection clause prohibited Oklahoma from enforcing a statute requiring sterilization of persons convicted of repeated criminal acts only for crimes within special categories. White collar crimes were exempted from these categories, and the Supreme Court's holding was based on the state's irrational distinction between blue collar (sterilizable) and white collar (unsterilizable) crimes. The Court was asked to, but did not, overrule *Buck v. Bell.*

The Supreme Court has never overturned the decision in *Buck v. Bell*, although it is of questionable precedential value today. Society's perception of people with disabilities has undergone significant change and, especially after the Nazi experience, for which *Buck v. Bell* was cited as precedent, arguments based upon eugenics are now held in low regard. In fact, when Carrie Buck was discovered in the Appalachian hills in 1980, she was found to be mentally competent and extremely disappointed that throughout her life she was unable to bear another child. Subsequent research determined that Carrie Buck's illegitimate daughter Vivian, who was conceived as the result of a rape by a relative of Carrie Buck's foster parents, was an honor roll student. At the trial that, on appeal, gave rise to Justice Holmes' regrettable opinion, Carrie Buck was poorly represented by a lawyer who did not develop the evidence or the legal arguments necessary to overcome the statute or its application to his client. For an engaging account of this case by a scholar who met Ms. Buck in the 1980s, and for pictures of Ms. Buck and her daughter, see Paul Lombardo, Facing Carrie Buck, 33 Hastings Ctr. Rep. (March–April) 14 (2003).

The Governor of Virginia apologized for the sterilization of Carrie Buck on the seventy-fifth anniversary of the Supreme Court decision in 2002. But Oklahoma and Virginia were not alone: thirty-two states have had sterilization programs, many supported with federal funding, that have targeted individuals with developmental disabilities, those of low socioeconomic status, racial and ethnic minorities, unmarried women, and prisoners. And although eugenics laws grew out of favor after WWII, forced and involuntary or coerced sterilizations have continued, targeting immigrants, people of color, and the poor.

In the 1970s, two high-profile lawsuits raised public awareness of continuing widespread sterilization abuses. One of the cases, Relf v. Weinberger, 372 F. Supp. 1196 (D.D.C. 1974), is perhaps best known because of the shocking violations committed against the Relf sister plaintiffs—the forced sterilization of two Black sisters only 12 and 14 years old by a federally-funded family planning clinic in Alabama. Id. at 1199. But the D.C. district court's findings revealed much broader abuses:

> Over the last few years, an estimated 100,000 to 150,000 low-income persons have been sterilized annually under federally funded programs. . . . Although Congress has been insistent that all family planning programs function on a purely voluntary basis, there is uncontroverted evidence in the record that minors and other incompetents have been sterilized with federal funds and that an indefinite number of poor people have been improperly coerced into accepting a sterilization operation under the threat that various federally supported welfare benefits would be withdrawn unless they submitted to irreversible sterilization. Patients receiving Medicaid assistance at childbirth are evidently the most frequent targets of this pressure, as the experience [ ] of . . . . Mrs. Waters [who] was actually refused medical assistance by her attending physician unless she submitted to a tubal ligation after the birth.

Id. Similar abuses were revealed in a complaint brought against Los Angeles County's USC Medical Center around the same time, in *Madrigal v. Quilligan*. Several women brought a lawsuit alleging systematic coerced sterilizations, including medical personnel obtaining "consent" for sterilization while the plaintiffs were in labor and, in some cases heavily medicated. Patients signed consent forms either because they thought were consenting to a cesarean section or they believed the procedure they were undergoing was reversible. The plaintiffs were Mexican-American, with limited English proficiency (LEP), and they resided in a community with inadequate medical resources. Even before the lawsuit, a young resident working in the maternity ward tried to sound the alarm to journalists, civil rights organizations and government officials about the widespread abuses he witnessed. See Maya Manian, The Story of *Madrigal v. Quilligan*: Coerced Sterilization of Mexican-American Women, in Reproductive Rights and Justice Stories (Melissa Murray, Kate Shaw, & Reva Siegel eds., 2019).

These cases revealed the disturbing paradox that "the influx of federal family planning funds in the 1960s and 1970s both increased access to reproductive health care and increased abusive sterilization practices." Id. Sterilization efforts occurred throughout the U.S. and targeted Black,

Mexican-American, Native American, Puerto Rican, and poor white women. See Lisa C. Ikemoto, Infertile by Force and Federal Complicity: The Story of *Relf v. Weinberger*, *in* Women and the Law Stories 196 (Elizabeth M. Schneider & Stephanie M. Wildman eds., 2011). Although the *Madrigal* plaintiffs ultimately lost their suit, both cases helped spur reform efforts at the state and federal levels, including requirements to ensure access to relevant bilingual medical information. For a discussion of national origin discrimination and the meaningful access requirement for LEP individuals, see Chapter 2, Section IV.A.

Since 1975, it has been a federal crime for government officials and others receiving federal financial assistance to "coerce[ ] or endeavor[ ] to coerce any person to undergo an abortion or sterilization procedure by threatening such person with the loss of, or disqualification for the receipt of, any benefit or service" under a federal program. Violation is punishable by $1000, up to one year in jail, or both. 42 U.S.C.A. § 300a–8. Nonetheless, this disturbing legacy persists. For example, California has recently come under scrutiny for coerced and unlawful sterilization of its women prisoners. See California State Auditor, Sterilization of Female Inmates: Report 2013–120 (Jun. 2014) (revealing ethical and legal lapses by corrections officials and medical personnel, including failing to follow informed consent protocols requiring a physician to certify that the required waiting period has been satisfied, that the patient appears mentally competent and understands the lasting effects of sterilization, and that the patient has been given one final opportunity to change her mind). In response, the Governor of California signed a bill increasing restrictions on sterilization in state prisons. The law prohibits "sterilization for the purpose of birth control" and prohibits sterilization by any means except when "required for the immediate preservation of the individual's life in an emergency medical situation" or when "medically necessary, as determined by contemporary standards of evidence-based medicine, to treat a diagnosed condition. . . ." Cal. Penal Code § 3440(a) & (b). Additional protections, such as heightened informed consent requirements, must also be met. If a sterilization is performed, the law requires psychological consultation and medical follow up. And the law requires the state to publish data related to the number of sterilizations it performs, disaggregated by race, age, medical justification, and method of sterilization. Id. Even more recently, there have been allegations of women in immigration detention centers being sterilized without their consent. See Caitlin Dickerson, Inquiry Ordered Into Claims Immigrants Had Unwanted Gynecology Procedures, N.Y. Times, Sep. 16, 2020.

Forced Sterilization or Right to Care?

More complex legal and ethical challenges have arisen where guardians or family members of permanently disabled individuals who lack capacity to make their own medical decisions seek permission to authorize sterilization based on the belief that it would be in that person's best interest, as in the following case below.

CONSERVATORSHIP OF VALERIE N.

Supreme Court of California, 1985.
707 P.2d 760.

GRODIN, JUSTICE.

* * *

I.

Valerie was born on July 13, 1955, apparently a victim of Downs Syndrome as a result of which she is severely retarded. Her IQ is estimated to be 30. She is now 29 years old. She lives with her mother and stepfather. Although she has no comprehension of the nature of these proceedings, she has expressed her wish to continue to have her parents care for her. Her parents' long range plan for Valerie is that she will move to a residential home should they become mentally or physically unable to care for her. She has received therapy and training for behavior modification which was not successful in eliminating her aggressive sexual advances toward men. Her parents are attempting to prepare her for the time when they can no longer care for her, and to broaden her social activities as an aspect of this preparation. They have concluded that other methods of birth control are inadequate in Valerie's case.

On September 5, 1980, appellants [Valerie's mother and stepfather] filed their petition to be named conservators of Valerie's person in the Santa Clara County Superior Court pursuant to section 1820. In the same petition they sought the additional power to authorize "a Salpingectomy or any other operation that will permanently sterilize" Valerie. The petition was supported by the declaration of Valerie's personal physician who stated that the tubal ligation procedure is "advisable and medically appropriate."

[A]ppellants submitted a declaration by a physician who had treated Valerie from the time she was 10 years old. He stated that in his opinion a tubal litigation procedure was "advisable and medically appropriate in that a potential pregnancy would cause psychiatric harm to Valerie." A second declaration [ ] by a licensed marriage, family and child counselor having a master's degree in developmental psychology, was also submitted. This declarant had worked with Valerie on a weekly basis for a year during 1977–1978. She believed that a tubal ligation was "an appropriate means of guarding against pregnancy," and had observed that Valerie acted "affectionately" toward adult men and made "inappropriate" sexual advances toward them. This declarant was of the opinion that because Valerie's parents had found it necessary to be overly restrictive in order to avoid a possible pregnancy which would have "severe psychologically damaging consequences" to Valerie, close monitoring had severely hampered Valerie's ability to form social relationships. She also believed that the level of Valerie's retardation meant that no alternative birth control methods were available that would ensure against pregnancy.

Valerie's mother testified that Valerie had not been sexually active, apart from masturbation, because she had been closely supervised. She was aggressive and affectionate toward boys. On the street she approached men, hugged and kissed them, climbed on them, and wanted to sit on their laps. Valerie had been given birth control pills in her early teens, but she rejected them and became ill. Her doctor then recommended the tubal ligation. Valerie was unable to apply other methods of birth control such as a diaphragm, and would not cooperate in a pelvic examination for an intrauterine device which the witness believed was unsafe in any event.

* * *

The parties agree that [the state statutory scheme] bars nontherapeutic sterilization of conservatees.

III. *Constitutional Rights of the Developmentally Disabled*

Our conclusion regarding the present legislative scheme requires that we confront appellants' contention that the scheme is unconstitutional. Both appellants and counsel for Valerie pose the constitutional question in terms of the right of procreative choice. Appellants argue that [the statute] deprives Valerie of that right by precluding the only means of contraception realistically available to her, while counsel for Valerie contends that the legislation furthers that right by protecting her against sterilization forced upon her by the will of others. The sad but irrefutable truth, however, is that Valerie is not now nor will she ever be competent to choose between bearing or not bearing children, or among methods of contraception. The question is whether she has a constitutional right to have these decisions made for her, in this case by her parents as conservators, in order to protect her interests in living the fullest and most rewarding life of which she is capable. At present her conservators may, on Valerie's behalf, elect that she not bear or rear children. As means of avoiding the severe psychological harm which assertedly would result from pregnancy, they may choose abortion should she become pregnant; they may arrange for any child Valerie might bear to be removed from her custody; and they may impose on her other methods of contraception, including isolation from members of the opposite sex. They are precluded from making, and Valerie from obtaining the advantage of, the one choice that may be best for her, and which is available to all women competent to choose—contraception through sterilization. We conclude that the present legislative scheme, which absolutely precludes the sterilization option, impermissibly deprives developmentally disabled persons of privacy and liberty interests protected by the Fourteenth Amendment to the United States Constitution [ ].

The right to marriage and procreation are now recognized as fundamental, constitutionally protected interests. [ ] So too, is the right of a woman to choose not to bear children, and to implement that choice by use of contraceptive devices or medication, and, subject to reasonable restrictions, to terminate a pregnancy. These rights are aspects of the right of privacy which exists within the penumbra of the First Amendment to the United States Constitution [ ]. They are also within the concept of liberty protected against arbitrary restrictions by the Fourteenth Amendment.

Although the Supreme Court has not considered the precise question of the right to contraception in the context of an assertion that the right includes sterilization, that sterilization is encompassed within the right to privacy has been acknowledged in this state. . . .

[T]he Legislature has denied incompetent women the procreative choice that is recognized as a fundamental, constitutionally protected right of all other adult women. We realize that election of the method of contraception to be utilized, or indeed whether to choose contraception at all, cannot realistically be deemed a "choice" available to an incompetent since any election must of necessity be made on behalf of the incompetent by others. The interests of the incompetent which mandate recognition of procreative choice as an aspect of the fundamental right to privacy and liberty do not differ from the interests of women able to give voluntary consent to this procedure, however. That these interests include the individual's right to personal growth and development is implicit in decisions of both the United States Supreme Court and this court.

* * *

An incompetent developmentally disabled woman has no less interest in a satisfying or fulfilling life free from the burdens of an unwanted pregnancy than does her competent sister. Her interest in maximizing her opportunities for such a life through habilitation is recognized and given statutory protection by [other state laws that have recognized that many developmentally disabled persons lead self-sufficient, fulfilling lives, and become loving, competent, and caring marriage partners and parents, as well as laws creating social supports to assist these individuals in achieving their maximum developmental potential.] If the state withholds from her the only safe and reliable method of contraception suitable to her condition, it necessarily limits her opportunity for habilitation and thereby her freedom to pursue a fulfilling life. Therefore, whether approached as an infringement of the right of privacy under the First Amendment or the privacy right that is found within the liberty protected by the Fourteenth Amendment, and whether analyzed under due process or equal protection principles, the issue is whether withholding the option of sterilization as a method of contraception to this class of women is constitutionally permissible. Because the rights involved are fundamental the permissibility of the restriction must be justified by a "compelling state interest," and may be no broader than necessary to protect that interest. [ ]

* * *

The state interest [ ] must be in precluding the option of sterilization because it is in most cases an irreversible procedure. Necessarily implicit in the interest asserted by the state is an assumption that the conservatee may at some future time elect to bear children. While the prohibition of sterilization may be a reasonable means by which to protect some conservatees' right to procreative choice, here it sweeps too broadly for it extends to individuals who cannot make that choice and will not be able to do so in the future. [ ]

Respondent argues that the ban is, nonetheless, necessary because past experience demonstrates that when the power to authorize sterilization of incompetents has been conferred on the judiciary it has been subject to abuse. Again, however, the rationale fails since less restrictive alternatives to total prohibition are available in statutory and procedural safeguards as yet untried in this state. [ ]

NOTES AND QUESTIONS

1.   *Lessons of the Past.* One justice in the *Valerie N.* case wrote a particularly strong dissent:

> Today's holding will permit the state, through the legal fiction of substituted consent, to deprive many women permanently of the right to conceive and bear children. The majority run roughshod over this fundamental constitutional right in a misguided attempt to guarantee a procreative choice for one they assume has never been capable of choice and never will be. . . .

> The majority opinion opens the door to abusive sterilization practices which will serve the convenience of conservators, parents, and service providers rather than incompetent conservatees. The ugly history of sterilization abuse against developmentally disabled persons in the name of seemingly enlightened social policies counsels a different choice.

Do you share the dissent's concern? Is there a process we could use to ensure that decisions are truly being made in the person's best interest, as opposed to what is more convenient for parents or conservators? How do we ensure that discriminatory or uninformed assumptions about people with developmental disabilities are not seeping into the decision-making process and leading to decisions that devalue their autonomy and right to equal treatment?

Despite finding the sterilization ban unconstitutional, the majority ultimately affirmed the trial court's denial of the parents' request to authorize sterilization, based on insufficient evidence that sterilization was necessary and that no less intrusive means were available. The affirmance was without prejudice, though, which meant the plaintiffs could renew their application for the power to consent to Valerie's sterilization once they had supporting evidence.

2.   *Judicial Safeguards Against Abuse.* The majority opinion went on to discuss the kind of procedural safeguards that have been used in other states to avoid abuse in these circumstances. The framework it referenced (and commonly used by other states) was established by the case In re Guardianship of Hayes, 608 P.2d 635 (Wash. 1980):

> The decision [to authorize sterilization] can only be made in a superior court proceeding in which (1) the incompetent individual is represented by a disinterested guardian ad litem, (2) the court has received independent advice based upon a comprehensive medical, psychological, and social evaluation of the individual, and (3) to the greatest extent possible, the court has elicited and taken into account the view of the incompetent individual.

> Within this framework, the judge must first find [ ] the individual is (1) incapable of making his or her own decision about sterilization, and (2) unlikely to develop sufficiently to make an informed judgment about sterilization in the foreseeable future.

> Next it must be proved [ ] that there is a need for contraception. The judge must find that the individual is (1) physically capable of procreation, and (2) likely to engage in sexual activity at the present or in the near future under circumstances likely to result in pregnancy, and must find in addition that (3) the nature and extent of the individual's disability, as determined by empirical evidence and not solely on the basis of standardized tests, renders him or her permanently incapable of caring for a child, even with reasonable assistance.

> Finally, there must be no alternatives to sterilization. The judge must find that (1) all less drastic contraceptive methods, including supervision, education and training, have been proved unworkable or inapplicable, and (2) the proposed method of sterilization entails the least invasion of the body of the individual. In addition, it must be shown [ ] that (3) the current state of scientific and medical knowledge does not suggest either (a) that a reversible sterilization procedure or other less drastic contraceptive method will shortly be available or (b) that science is on the threshold of an advance in the treatment of the individual's disability.

The court also noted that the standard of proof for the above requirements was "clear, cogent and convincing evidence." Do you think this framework adequately safeguards against abuse? Some have read the *Hayes* requirements as effectively removing the possibility of the sterilization of the developmentally disabled. Can you imagine a case that would meet these requirements?

VI. DECISION-MAKING DURING PREGNANCY

The topics covered so far in this chapter have dealt with either government restrictions on women's right to prevent or terminate pregnancy, or government's interference with one's ability to procreate. But when a woman is pregnant and wants to carry the fetus to term, can the government compel medical intervention when the woman either does not or cannot consent, on the grounds that such intervention is necessary to protect fetal life? Traditionally, such situations have been framed as maternal-fetal conflicts, and the *In re A.C.* excerpt below presents a particularly extreme example—where the mother is incapacitated and both the mother's and fetus's lives appear to be at stake. Such disputes are more accurately understood, however, as reflecting a physician-patient conflict over medical decision-making during pregnancy. During pregnancy, physicians perceive dual duties to care for both the fetus and the pregnant patient. Where physicians believe that certain behavior by a patient could endanger fetal health, some physicians may view the patient's failure to follow medical advice as "noncompliance" that must be overridden, including through legal action, if necessary. Consider whether the majority and dissent's contrasting approaches in *In re A.C.* are helpful for resolving the more commonly occurring physician-patient conflicts described in the following notes.

IN RE A.C.

District of Columbia Court of Appeals, 1990.
573 A.2d 1235.

TERRY, ASSOCIATE JUDGE.

* * *

We are confronted here with two profoundly difficult and complex issues. First, we must determine who has the right to decide the course of medical treatment for a patient who, although near death, is pregnant with a viable fetus. Second, we must establish how that decision should be made if the patient cannot make it for herself—more specifically, how a court should proceed when faced with a pregnant patient, *in extremis,* who is apparently incapable of making an informed decision regarding medical care for herself and her fetus. We hold that in virtually all cases the question of what is to be done is to be decided by the patient—the pregnant woman—on behalf of herself and the fetus. If the patient is incompetent or otherwise unable to give an informed consent to a proposed course of medical treatment, then her decision must be ascertained through the procedure known as substituted judgment. . . .

I

This case came before the trial court when George Washington University Hospital petitioned the emergency judge in chambers for declaratory relief as to how it should treat its patient, A.C., who was close to death from cancer and was twenty-six and one-half weeks pregnant with a viable fetus. After a hearing lasting approximately three hours, which was held at the hospital (though not in A.C.'s room), the court ordered that a caesarean section be performed on A.C. to deliver the fetus. . . . The caesarean was performed, and a baby girl, L.M.C., was delivered. Tragically, the child died within two and one-half hours, and the mother died two days later.

* * *

II

A.C. was first diagnosed as suffering from cancer at the age of thirteen. In the ensuing years she underwent major surgery several times, together with multiple radiation treatments and chemotherapy. A.C. married when she was twenty-seven, during a period of remission, and soon thereafter she became pregnant. She was excited about her pregnancy and very much wanted the child. Because of her medical history, she was referred in her fifteenth week of pregnancy to the high-risk pregnancy clinic at George Washington University Hospital.

On Tuesday, June 9, 1987, when A.C. was approximately twenty-five weeks pregnant, she went to the hospital for a scheduled check-up. Because she was experiencing pain in her back and shortness of breath, an x-ray was taken, revealing an apparently inoperable tumor which nearly filled her right lung. On Thursday, June 11, A.C. was admitted to the hospital as a patient. By Friday her condition had temporarily improved, and when asked if she really wanted to have her baby, she replied that she did.

Over the weekend, A.C.'s condition worsened considerably. Accordingly, on Monday, June 15, members of the medical staff treating A.C. assembled, along with her family, in A.C.'s room. The doctors then informed her that her illness was terminal, and A.C. agreed to palliative treatment designed to extend her life until at least her twenty-eighth week of pregnancy. The "potential outcome [for] the fetus," according to the doctors, would be much better at twenty-eight weeks than at twenty-six weeks if it were necessary to "intervene." A.C. knew that the palliative treatment she had chosen presented some increased risk to the fetus, but she opted for this course both to prolong her life for at least another two weeks and to maintain her own comfort. When asked if she still wanted to have the baby, A.C. was somewhat equivocal, saying "something to the effect of 'I don't know, I think so.' " As the day moved toward evening, A.C.'s condition grew still worse, and at about 7:00 or 8:00 p.m. she consented to intubation to facilitate her breathing.

The next morning, June 16, the trial court convened a hearing at the hospital in response to the hospital's request for a declaratory judgment. The court appointed counsel for both A.C. and the fetus, and the District of Columbia was permitted to intervene for the fetus as *parens patriae*. . . .

* * *

There was no evidence before the court showing that A.C. consented to, or even contemplated, a caesarean section before her twenty-eighth week of pregnancy. There was, in fact, considerable dispute as to whether she would have consented to an immediate caesarean delivery at the time the hearing was held. A.C.'s mother opposed surgical intervention, testifying that A.C. wanted "to live long enough to hold that baby" and that she expected to do so, "even though she knew she was terminal." Dr. Hamner [a treating obstetrician] testified that, given A.C.'s medical problems, he did not think she would have chosen to deliver a child with a substantial degree of impairment. . . .

After hearing this testimony and the arguments of counsel, the trial court made oral findings of fact. It found, first, that A.C. would probably die, according to uncontroverted medical testimony, "within the next twenty-four to forty-eight hours;" second, that A.C. was "pregnant with a twenty-six and a half week viable fetus who, based upon uncontroverted medical testimony, has approximately a fifty to sixty percent chance to survive if a caesarean section is performed as soon as possible;" third, that because the fetus was viable, "the state has [an] important and legitimate interest in protecting the potentiality of human life;" and fourth, that there had been some testimony that the operation "may very well hasten the death of [A.C.]," but that there had also been testimony that delay would greatly increase the risk to the fetus and that "the prognosis is not great for the fetus to be delivered post-mortem . . . ." Most significantly, the court found:

> The court is of the view that it does not clearly know what [A.C.'s] present views are with respect to the issue of whether or not the child should live or die. She's presently unconscious. As late as Friday of last week, she wanted the baby to live. As late as yesterday, she did not know for sure.

Having made these findings of fact and conclusions of law, . . . the court ordered that a caesarean section be performed to deliver A.C.'s child.

The court's decision was then relayed to A.C., who had regained consciousness. [When the court reconvened later in the day, Dr. Hamner testified that A.C. then consented to the procedure.] When the court suggested moving the hearing to A.C.'s bedside, Dr. Hamner discouraged the court from doing so, but he and Dr. Weingold, together with A.C.'s mother and husband, went to A.C.'s room to confirm her consent to the procedure. What happened then was recounted to the court a few minutes later:

* * *

Dr. Weingold: She does not make sounds because of the tube in her windpipe. She nods and she mouths words. One can see what she's saying rather readily. She asked whether she would survive the operation. She asked [Dr.] Hamner if he would perform the operation. He told her he would only perform it if she authorized it but it would be done in any case. She understood that. She then seemed to pause for a few moments and then very clearly mouthed words several times, *I don't want it done, I don't want it done.* Quite clear to me.

I would obviously state the obvious and that is this is an environment in which, from my perspective as a physician, this would not be an informed consent one way or the other. She's under tremendous stress with the family on both sides, but I'm satisfied that I heard clearly what she said. . . .

Dr. Weingold later qualified his opinion as to A.C.'s ability to give an informed consent, stating that he thought the environment for an informed consent was non-existent because A.C. was in intensive care, flanked by a weeping husband and mother. He added:

I think she's in contact with reality, clearly understood who Dr. Hamner was. Because of her attachment to him [she] wanted him to perform the surgery. Understood he would not unless she consented and did not consent.

That is, in my mind, very clear evidence that she is responding, understanding, and is capable of making such decisions. . . .

After hearing this new evidence, the court found that it was "still not clear what her intent is" and again ordered that a caesarean section be performed. . . . The operation took place, but the baby lived for only a few hours, and A.C. succumbed to cancer two days later.

* * *

IV

* * *

A.   *Informed Consent and Bodily Integrity*

* * *

. . . [O]ur analysis of this case begins with the tenet common to all medical treatment cases: that any person has the right to make an informed choice, if competent to do so, to accept or forgo medical treatment. . . .

In the same vein, courts do not compel one person to permit a significant intrusion upon his or her bodily integrity for the benefit of another person's health. See, e.g., McFall v. Shimp, 10 Pa.D. & C.3d 90 (Allegheny County Ct.1978). In *McFall* the court refused to order Shimp to donate bone marrow which was necessary to save the life of his cousin, McFall:

> The common law has consistently held to a rule which provides that one human being is under no legal compulsion to give aid or to take action to save another human being or to rescue. . . .

. . . Even though Shimp's refusal would mean death for McFall, the court would not order Shimp to allow his body to be invaded. It has been suggested that fetal cases are different because a woman who "has chosen to lend her body to bring [a] child into the world" has an enhanced duty to assure the welfare of the fetus, sufficient even to require her to undergo caesarean surgery. [ ] Surely, however, a fetus cannot have rights in this respect superior to those of a person who has already been born.[8]

* * *

There are two additional arguments against overriding A.C.'s objections to caesarean surgery. First, as the American Public Health Association cogently states in its *amicus curiae* brief:

> Rather than protecting the health of women and children, court-ordered caesareans erode the element of trust that permits a pregnant woman to communicate to her physician—without fear of reprisal—all information relevant to her proper diagnosis and treatment. An even more serious consequence of court-ordered intervention is that it drives women at high risk of complications during pregnancy and childbirth out of the health care system to avoid coerced treatment.

Second, and even more compellingly, any judicial proceeding in a case such as this will ordinarily take place—like the one before us here—under time constraints so pressing that it is difficult or impossible for the mother to communicate adequately with counsel, or for counsel to organize an effective factual and legal presentation in defense of her liberty and privacy interests and bodily integrity. . . .

* * *

---

[8]   There are also practical consequences to consider. What if A.C. had refused to comply with a court order that she submit to a caesarean? Enforcement could be accomplished only through physical force or its equivalent. A.C. would have to be fastened with restraints to the operating table, or perhaps involuntarily rendered unconscious by forcibly injecting her with an anesthetic, and then subjected to unwanted major surgery. Such actions would surely give one pause in a civilized society, especially when A.C. had done no wrong.

B. Substituted Judgment

. . . Sometimes, however, as our analysis presupposes here, a once competent patient will be unable to render an informed decision. In such a case, we hold that the court must make a substituted judgment on behalf of the patient, based on all the evidence. This means that the duty of the court, "as surrogate for the incompetent, is to determine as best it can what choice that individual, if competent, would make with respect to medical procedures." [ ]

* * *

We have found no reported opinion applying the substituted judgment procedure to the case of an incompetent pregnant patient whose own life may be shortened by a caesarean section, and whose unborn child's chances of survival may hang on the court's decision. Despite this precedential void, we conclude that substituted judgment is the best procedure to follow in such a case because it most clearly respects the right of the patient to bodily integrity. . . .

* * *

Because it is the patient's decisional rights which the substituted judgment inquiry seeks to protect, courts are in accord that the greatest weight should be given to the previously expressed wishes of the patient. This includes prior statements, either written or oral, even though the treatment alternatives at hand may not have been addressed. . . .

Courts in substituted judgment cases have also acknowledged the importance of probing the patient's value system as an aid in discerning what the patient would choose. We agree with this approach. [Eds. Note: The court then discussed the ways in which it could determine the substituted judgment of the patient. For a fuller discussion of this issue, see Chapter 5, Section IV.]

C. The Trial Court's Ruling

. . . . The [trial] court did not . . . make a finding as to what A.C. would have chosen to do if she were competent. Instead, the court undertook to balance the state's and [the fetus's] interests in surgical intervention against A.C.'s perceived interest in not having the caesarean performed.

* * *

What a trial court must do in a case such as this is to determine, if possible, whether the patient is capable of making an informed decision about the course of her medical treatment. If she is, and if she makes such a decision, her wishes will control in virtually all cases. If the court finds that the patient is incapable of making an informed consent (and thus incompetent), then the court must make a substituted judgment. . . .

Having said that, we go no further. We need not decide whether, or in what circumstances, the state's interests can ever prevail over the interests of a pregnant patient. . . . Indeed, some may doubt that there could ever be a situation extraordinary or compelling enough to justify a massive intrusion into a person's body, such as a caesarean section, against that person's will. Whether such a situation may someday present itself is a question that we need not strive to answer here. . . .

* * *

BELSON, ASSOCIATE JUDGE, concurring in part and dissenting in part:

* * *

I think it appropriate . . . to state my disagreement with the very limited view the majority opinion takes of the circumstances in which the interests of a viable unborn child can afford such compelling reasons. The state's interest in preserving human life and the viable unborn child's interest in survival are entitled, I think, to more weight than I find them assigned by the majority when it states that "in virtually all cases the decision of the patient . . . will control." I would hold that in those instances, fortunately rare, in which the viable unborn child's interest in living and the state's parallel interest in protecting human life come into conflict with the mother's decision to forgo a procedure such as a caesarean section, a balancing should be struck in which the unborn child's and the state's interest are entitled to substantial weight.

* * *

The balancing test should be applied in instances in which women become pregnant and carry an unborn child to the point of viability. This is not an unreasonable classification because, I submit, a woman who carries a child to viability is in fact a member of a unique category of persons. Her circumstances differ fundamentally from those of other potential patients for medical procedures that will aid another person, for example, a potential donor of bone marrow for transplant. This is so because she has undertaken to bear another human being, and has carried an unborn child to viability. Another unique feature of the situation we address arises from the singular nature of the dependency of the unborn child upon the mother. A woman carrying a viable unborn child is not in the same category as a relative, friend, or stranger called upon to donate bone marrow or an organ for transplant. Rather, the expectant mother has placed herself in a special class of persons who are bringing another person into existence, and upon whom that other person's life is totally dependent. Also, uniquely, the viable unborn child is literally captive within the mother's body. No other potential beneficiary of a surgical procedure on another is in that position.

* * *

I next address the sensitive question of how to balance the competing rights and interests of the viable unborn child and the state against those of the rare expectant mother who elects not to have a caesarean section necessary to save the life of her child. The indisputable view that a woman carrying a viable child has an extremely strong interest in her own life, health, bodily integrity, privacy, and religious beliefs necessarily requires that her election be given correspondingly great weight in the balancing process. In a case, however, where the court in an exercise of a substituted judgment has concluded that the patient would probably opt against a caesarean section, the court should vary the weight to be given this factor in proportion to the confidence the court has in the accuracy of its conclusion. Thus, in a case where the indicia of the incompetent patient's judgment are equivocal, the court should accord this factor correspondingly less weight. The appropriate weight to be given other factors will have to be worked out by the development of law in this area, and cannot be prescribed in a single court opinion. Some considerations obviously merit special attention in the balancing process. One such consideration is any danger to the mother's life or health, physical or mental, including the relatively small but still significant danger that necessarily inheres in any caesarean delivery, and including especially any danger that exceeds that level. The mother's religious beliefs as they relate to the operation would appear to deserve inclusion in the balancing process.

On the other side of the analysis, it is appropriate to look to the relative likelihood of the unborn child's survival. . . . The child's interest in being born with as little impairment as possible should also be considered. This may weigh in favor of a delivery sooner rather than later. The most important factor on this side of the scale, however, is life itself, because the viable unborn child that dies because of the mother's refusal to have a caesarean delivery is deprived, entirely and irrevocably, of the life on which the child was about to embark.

* * *

NOTES AND QUESTIONS

1.   *Balancing of Interests?* As the opinions in *A.C.* suggest, decisions made by or on behalf of the mother to best serve her health or other interests may conflict with what a doctor believes is important for protection of the fetus; this conflict forces courts to answer two questions. First, should the court balance the interests of the fetus (or the interests of the state in protecting the fetus) with the interests of the mother? If the answer to that question is "no," as it was for the majority in *A.C.,* the issue becomes the comparatively simple one of determining the wishes of the mother. See In re Baby Boy Doe, 632 N.E.2d 326 (Ill. App. 1994), where the court refused to balance the interests of a fetus (or the interests of the state in protecting a fetus) against the interests

of a competent pregnant woman who refused a caesarean section, even though doctors believed her decision put the life of her fetus at risk.

If there is a decision to balance the interests, however, as the dissenting opinion suggests, the court must also face the question of what standards to apply. Are the interests of the fetus or the state as strong as the interests of the mother? Can the relative strengths of these different interests vary from case to case, depending upon the stage of development of the fetus, the consequences of the decision to be made, or other factors? How might those interests be balanced if the woman objects to recommended medical care on religious grounds? See, e.g., Raleigh Fitkin-Paul Morgan Memorial Hospital v. Anderson, 201 A.2d 537 (N.J. 1964) (ordering blood transfusions to save an "unborn child" over the objections of a woman who refused the transfusion due to her religious convictions as a Jehovah's Witness); In re Application of Jamaica Hospital, 491 N.Y.S.2d 898 (Sup.Ct. 1985) (ordering blood transfusion to save fetus prior to viability); Jefferson v. Griffin Spalding County Hospital Authority, 274 S.E.2d 457 (Ga. 1981) (ordering a caesarean section against the religiously motivated wishes of the mother when physicians argued that failure to do so would result in the death of both the mother and the fetus, noting the "duty of the state to protect a living, unborn human being from meeting [ ] death before being given the opportunity to live.").

2.   *Living Wills & Pregnancy. In re A.C.* illustrates how courts use the process of substituted judgment to try to determine what a patient would have wanted, when that patient is no longer capable of making an informed decision about care. (This is discussed further in Chapter 5). Living wills and advance directives are important tools that allow patients to make their health care wishes known in the event of incapacity, and they are expressly permitted by almost every state. Yet, in most states, the law may not allow a living will to be used to refuse treatment during pregnancy, and in some states a living will is not available to pregnant patients at all. Such pregnancy exemptions have been getting increased attention since the 2014 report about a pregnant woman in Texas, Marlise Munoz, who was kept alive using life support for two months after being declared brain dead, against her family's wishes. For a critique of such exemptions and a discussion of how living wills can be used to better capture patients' preferences in the event of pregnancy, see Elizabeth Villarreal, Pregnancy and Living Wills: A Behavioral Economic Analysis, 128 Yale L. J. (2019), online form at https://www.yalelawjournal.org/forum/.

3.   *Assessing Risk to Fetal Health.* One reason that many commentators have opposed judicial intervention to require pregnant women to undergo medical care for the presumed benefit of their fetuses is the fact that medical diagnosis in this area can be wrong. Doctors seem willing to testify that intervention is necessary to save the fetus even when the prognosis following nonintervention is quite uncertain. There is a series of cases suggesting that when doctors testify that a caesarean is necessary for the health of the fetus, if the mother "escapes" and attempts a normal vaginal birth, there is a good chance the child will be born without complication. J. Fletcher, Drawing Moral

Lines in Fetal Therapy, 29 Clin. Obstetrics & Gynecology 595 (1986). Fletcher summarizes his review of six cases:

> First, inaccuracies and possible misdiagnosis appear to be involved in half of the cases . . . since babies were born healthy after vaginal delivery. If precedents flow from examples flawed by faulty assumptions or mistaken evaluations, errors may be replicated. These outcomes also should remind all concerned of the possibility of misdiagnosis before fetal therapy. Forced fetal therapy on the basis of misdiagnosis would constitute an ethical megadisaster.

29 Clin. Obstetrics and Gynecology at 599. See also In re Baby Boy Doe, supra (mother delivered a "normal and healthy" baby despite predictions of disaster). In fact, using legal action to override a pregnant patient's refusal to undergo a cesarean section or other medical intervention not only implicates autonomy principles, but also seems contrary to evolving evidence of the overuse of cesarean sections as increasing the risk of poor outcomes. See Stephanie Teleki, Birthing a Movement to Reduce Unnecessary C-Sections: An Update from California, Health Aff. Blog (Oct. 31, 2017). How closely should the court scrutinize medical assertions about the risks and benefits of procedures for fetal and maternal health?

4.   *Using the Criminal System to Compel Medical Treatment.* Medical treatment has also been coerced by threatening criminal punishment or confinement. In 1989, the Medical University of South Carolina, along with other state and local law enforcement and social service agencies, hatched a plan to deal with what they saw as the pernicious effects of drug use by pregnant women. The program targeted women who met certain guidelines, for example, those with little or incomplete prenatal care. This meant the women who were targeted were also low-income and dependent on public health care providers, such as the Medical University of South Carolina. The women were also predominantly African American.

The women were tested for cocaine use when they arrived at the hospital to deliver their babies. Those who tested positive were given a "choice" between enrolling in a substance abuse program and being arrested for their drug use. Those who chose the treatment option were not prosecuted unless they failed to adhere to the treatment plan or failed a subsequent drug test; the others were reported to, and subsequently arrested by, the police. No other hospital in Charleston chose to participate in the program.

Ten of the arrested mothers brought suit against the hospital and others. In 2001 the Supreme Court, in a 6–3 opinion, determined that the Charleston process constituted an illegal warrantless search and seizure prohibited by the Fourth Amendment. Ferguson v. City of Charleston, 532 U.S. 67 (2001). Justice Stevens, for the Court, pointed out that the "direct and primary purpose" of the test was to aid the police in enforcing a criminal law, and that the Fourth Amendment was implicated even if the ultimate goal of the hospital was to encourage women to get treatment. Justice Kennedy, concurring, pointed out that the case did not prohibit a state from punishing women who put their fetuses at risk, and Justice Scalia, dissenting, said that the majority "proves once again that no good deed goes unpunished."

Indeed, criminal prosecutions for maternal substance use, based on asserted harm to the fetus, have persisted. For example, two years after the *Ferguson* decision, a woman in South Carolina was convicted of homicide by child abuse for using cocaine during pregnancy after she suffered a stillbirth, and she was sentenced to 20 years imprisonment. State v. McKnight, 576 S.E.2d 168 (S.C. 2003). See also Ex Parte Ankrom, 152 So.3d 397 (Ala. 2013) (upholding convictions of two women for chemical endangerment of child because they ingested illegal substances during pregnancy). Most state courts have rejected the use of traditional child endangerment or abuse laws for such prosecutions. See, e.g., State v. Wade, 232 S.W.3d 663 (Mo. App. 2007); State v. Martinez, 137 P.3d 1195 (N.M. App. 2006); State v. Aiwohi, 123 P.3d 1210 (Haw. 2005); Reinesto v. Superior Court, 894 P.2d 733 (Ariz. App. 1995); Johnson v. State, 602 So.2d 1288 (Fla. 1992) (reversing a conviction for delivery of a controlled substance from a mother to her child after birth, but before the umbilical cord was severed). Nonetheless, thirty-eight states have enacted statutes specifically prohibiting feticide or harm to unborn children, and some of these have been used as authority for arresting pregnant women. Prosecutors have brought cases and secured convictions despite weak or no evidence that the pregnant woman's behavior *actually* caused harm. See Lynn M. Paltrow and Jeanne Flavin, Arrests of and Forced Interventions on Pregnant Women in the United States, 1973–2005: Implications for Women's Legal Status and Public Health, 38 J. of Health Pol., Pol'y, & L. 299 (2013). Even states that reject such criminal prosecutions may be willing to allow other government action against women who ingest drugs or alcohol during pregnancy. For example, in State v. Louk, 786 S.E.2d 219 (W. Va. 2016), the Supreme Court of Appeals of West Virginia held that an unborn fetus was not a "child" within the meaning of a statute imposing criminal penalty for child neglect resulting in death. But just one year later, the same court found that the presence of illegal drugs in a child's system at birth, in light of mother's admitted use of illegal drugs during pregnancy, was sufficient evidence that the child was abused and/or neglected for purposes of a civil petition by the state Department of Health and Human Resources. In re ALCM, 801 S.E.2d 260 (W. Va. 2017).

5.   *Civil Confinement.* The remedy of civil commitment or court ordered protective custody may be used as an alternative to criminal law to prevent pregnant mothers from engaging in certain behavior. Some states have promulgated statutes that provide explicitly that women who put their fetuses at risk may be restrained from doing so by the court. Is the use of the civil commitment or protective custody remedy any less of an infringement on women's liberty than the use of the criminal sanction?

6.   *Connection Between the State's Action and Fetal Health.* A balancing test to determine whether the state can compel intervention depends upon the state's ability to show that its action is justified by its interest in protecting fetal health. In the case of patients' drug use, however, this assertion has been

challenged by health care providers and public health officials. They agree that fetal exposure to drugs and alcohol is an important public health concern because of the risk it poses to the child's development, but they argue that a treatment-focused approach that is supportive and respects patient autonomy, rather than a punitive approach, is more effective for the health of the child and woman. See American Society of Addiction Medicine, Substance Use, Misuse, and Use Disorders During and Following Pregnancy, with an Emphasis on Opioids (Jan. 18, 2017) (noting "the importance of working with a pregnant woman to facilitate her quitting or at least reducing substance use during pregnancy, and engaging in addiction-related treatment if necessary," but "oppos[ing] criminalizing and other punitive approaches to substance use during pregnancy as they turn women away from prenatal care, thus compromising maternal and fetal well-being;" also asserting that "[p]unishing pregnant women impedes proper medical care and the promotion of public health"); American College of Obstetricians and Gynecologists Committee on Health Care for Underserved Women, Substance Abuse Reporting and Pregnancy: The Role of Obstetrician-Gynecologist, 117 Obstetrics Gynecology 200 (2011) (reaffirmed 2014) ("incarceration and the threat of incarceration have proved to be ineffective in reducing the incidence of alcohol or drug abuse" and providers are "encouraged to work with state legislators to retract legislation that punishes women for substance abuse during pregnancy"); American Academy of Pediatrics (AAP), Policy Statement: A Public Health Response to Opioid Use in Pregnancy, 139 Pediatrics 1 (Mar. 2017) (noting that punitive measures "are not in the best interest of the health of the mother-infant dyad," and recommending treatment programs that provide "nonjudgmental, trauma-informed services").

A study of drug treatment programs has shown that many pregnant women have trouble accessing them voluntarily, either because the programs will not take them or because of limited slots. See Rebecca Stone, Pregnant Women and Substance Use: Fear, Stigma, and Barriers to Care, 3 Health and Justice (2015). From a public health perspective, it seems counterproductive for states to use threats of criminal enforcement or civil confinement to force women into treatment, given the dearth of resources for women who want to access such programs voluntarily. In addition, some have argued that the state's poor treatment of pregnant women once confined actually undermines the state's purported interests in maternal and fetal health. See Paltrow & Flavin, supra. Examples of this include shackling women during childbirth and preventing incarcerated women from caring for their newborns in ways that are important for child health. See Priscilla Ocen, Punishing Pregnancy: Race, Incarceration and the Shackling of Pregnant Prisoners, 100 Cal. L. Rev. 1239 (2012).

Forced medical interventions, in combination with criminal or civil confinement, have even been used in cases where there was no evidence of conduct that created a risk of harm to the fetus. Consider an example described in a recent news article: A women who was 14 weeks pregnant voluntarily disclosed at a prenatal checkup that she had been addicted to Percocet the year before. She was not still using drugs and only mentioned it to her doctor because she thought her doctor should know. A urine test confirmed that she was not using drugs, but her doctor and a social worker insisted that she start an anti-addiction drug anyway. The patient refused, and some weeks later county sheriffs came to her home and took her to court, where her doctor accused her of endangering her fetus and she was ordered to attend a mandatory 78-day stay at a drug treatment facility or risk going to jail. Erik Eckholm, Case Explores Rights of Fetus Versus Mother, N.Y. Times (Oct. 23, 2013).

7. *Punishing Pregnancy and Other Status?* Criminal and civil actions against pregnant women in the name of protecting fetal health have led some scholars to suggest such programs constitute a status-based system of punishment implicating reproductive justice concerns. See, e.g., Priscilla A. Ocen, Birthing Injustice: Pregnancy as a Status Offense, 85 Geo. Wash. L. Rev. 1163 (2017). If a mother is imprisoned for the term of her pregnancy *in order to protect her fetus,* is she being punished for the crime she committed, or for her pregnancy? What if the behavior the woman is punished for would otherwise be lawful if she weren't pregnant? Consider that what constitutes an illness may be a contested issue, especially in the case of substance use disorder (SUD). There is a tension between the various characteristics of addiction that cause society to simultaneously classify it as behavior deserving of punishment and as an illness deserving treatment. Much has been made about the growing understanding of, and empathy for, people who struggle with substance use disorder, especially with the national spotlight on opioid addiction. How are the pregnant women who suffer from SUD treated in this public discourse?

Consider also that the testing and surveillance of pregnant women for drug use effectively targets *poor* pregnant women who depend on public resources, as well as those living in disproportionately minority communities that tend to be over policed by the criminal justice system. South Carolina did not have a similar program targeting mothers who delivered their babies at private hospitals in affluent communities. The study by Lynn Paltrow and Jeanne Flavin, supra, documented over four hundred cases of medical and criminal interventions targeting pregnant women, finding that arrests or deprivations of physical liberty disproportionately impacted minorities and low-income individuals. This is despite evidence that drug use is comparable among whites and non-whites. Some scholars have argued that uneven criminal enforcement against certain groups reflects troubling assumptions that parenting by these women—namely poor women and racial or ethnic minorities—is "inherently risky"—a view that harkens back to earlier sterilization programs targeting these same groups. See, e.g., Khiara Bridges, The Poverty of Privacy Rights (2017); Michele Goodwin, Fetal Protection Laws: Moral Panic and the New Constitutional Battlefront, 102 Calif. L. Rev. 781 (2014).

The role of poverty in these cases presents another complicating factor. There is growing recognition of the outsized role of social determinants of health: food insecurity, housing insecurity, exposure to violence, trauma, environmental hazards, discrimination, and other poverty-related factors can contribute to poor physical and mental health, which in turn pose risks to maternal and fetal health. See, e.g., Z. Mahmoodi et al., Association of Maternal Working Condition with Low Birth Weight: The Social Determinants of Health Approach, 5 Ann. Med. Health Sci. Res. 385 (2015); N.N. Sarkar et al., The Impact of Intimate Partner Violence on Women's Reproductive Health and Pregnancy Outcome, 28 J. of Obstetrics and Gynaecology 266 (2008). How do we determine whether poor fetal outcomes are due to a specific act like drug or alcohol use, to some other behavioral, social, or environmental factor, or to some combination of factors? If causation cannot be proved, is it fair to use criminal or civil sanctions against pregnant women to prevent certain behavior? Consider the case of McKnight v. State, 661 S.E.2d 354, 358 n.10 (S.C. 2008). The state of South Carolina charged Regina McKnight, a twenty-one-year-old Black woman, with homicide by child abuse based on the theory that her use of cocaine caused her stillbirth. After only fifteen minutes of deliberation, the jury found her guilty and she was sentenced to twelve years in prison. Her conviction was eventually overturned due to ineffective assistance counsel because the state relied on outdated research as to the effect of drugs on pregnancy, and her own counsel failed to present "recent studies showing that cocaine is no more harmful to a fetus than nicotine use, poor nutrition, or other conditions commonly associated with the urban poor." Id. at 358 n.2. How effective will such government action be as health policy, if it only addresses substance use and only does so temporarily with confinement or imprisonment, but does not address the other social factors shaping fetal and child health?

8. *Pregnancy, Employment & Sex Discrimination.* As noted above, environmental hazards may pose a risk to fetal health, and some employment environments that are safe for workers may not be safe for workers' fetuses. Even in cases where there may be some risk to fetal health, the question of whether such employment is "good" or "bad" for the fetus may not be simple to answer. For example, employees may not have a choice of where to work, and the job may provide financial resources essential for securing basic needs for the child's well-being. Even if one has a choice, certain jobs may provide greater financial security, independence, and personal fulfillment to the parent, promoting greater psychological health and enhancing the family's resources in ways that may have a significant positive effect on pregnancy outcomes and long-term well-being.

Employers have sought to exclude pregnant women (or even women who may become pregnant) from jobs that could put fetuses at risk—either because the employers wanted to protect the fetuses, or because the employers wanted to protect themselves from adverse publicity, increased health insurance costs and potential tort liability. Should an employer be able to refuse to assign people to certain jobs based on their capacity to get pregnant because of the possible risk of future harm? This issue has received attention almost exclusively where the employee is a woman (i.e., a potential mother) rather than a man (i.e., a potential father) even though occupational hazards to men may affect the health of the fetus as well.

The U.S. Supreme Court considered this issue under Title VII of the Civil Rights Act of 1964 in International Union, United Automobile Workers v. Johnson Controls, Inc., 499 U.S. 187 (1991). Johnson Controls excluded women of child-bearing age from its battery manufacturing operation unless they had been sterilized, because workers in that process sometimes developed very high levels of lead in their blood. The Supreme Court determined that this policy constituted disparate treatment by sex, under Title VII as amended by the Pregnancy Discrimination Act, which could be justified only if sex was a bona fide occupational qualification. The Court concluded:

> It is no more appropriate for the courts than it is for individual employers to decide whether a woman's reproductive role is more important to herself and her family than her economic role. Congress has left this choice to the woman as hers to make.

VII. ASSISTED REPRODUCTIVE TECHNOLOGIES (ART)

A. INTRODUCTION

1. The Process of Human Reproduction

Those seeking medical help in facilitating reproduction do so because they want a child. The birth of a child requires the growth of a fetus in a uterus. This, in turn, requires the implantation of a fertilized ovum (also called an egg or an oocyte) in the uterine wall. At least until the cloning of human beings is perfected, the ovum can implant in the uterine wall only if it has been fertilized by a sperm, a process which generally takes place in the fallopian tube.

Within a day of an ovum being fertilized by a sperm, it has divided twice, into a two- and then a four-celled "embryo" (sometimes called a "preembryo") and it is here that we must pause to consider the impact that our language has on our conceptualization and therefore our law. The terminology of early pregnancy is rife with alternative phraseology. For example, a fertilized ovum/egg/oocyte is also called a "zygote," or a cluster of dividing "blastomeres" which eventually divide into a "blastocyst" or hollow cellular ball. Some will argue that because there is no differentiated living matter generated in forming the blastocyst (i.e., there are no fetal or placental cells yet), this division of cell material is not a constructive development of living substance. It is a process which does nothing to distribute particular developmental qualities to particular parts of a resulting fetus. On the other hand, it is the preliminary stage which makes it possible for the blastocyst cell ball to implant and use materials from the uterine wall to grow and develop into a fetus. Obviously, the fertilization of the ovum by the sperm begins the embryonic process, although some use the term "embryo" only to apply to later stages of development.

At any rate, whether it is called an embryo, a preembryo or something else, the fertilized ovum continues its migration through the fallopian tube toward the uterus. The fertilized and subdividing ovum arrives at the lower end of the fallopian tube, hesitating for about two days before being expelled into the uterus. This delay allows the uterine lining to build up for successful implantation. The implantation of this cell cluster is a complicated biological process that takes several days.

Even when all other conditions are met, not all fertilized ova actually do implant and not all implanted ova survive until birth. About 50% of fertilized ova are expelled from the body before pregnancy is discovered. Some of these spontaneous expulsions occur before implantation, and some occur during the first few days after the commencement of the implantation process.

There are several reasons that those who wish to raise children may not conceive them through this process. First, some are unable to participate in coital sexual relations, or choose not to participate in such relations. Others are infertile—that is, normal coitus does not result in a fertilized ovum that implants in the uterine wall and grows into a fetus and then a child. Although estimates vary, about twelve to fifteen percent of American couples are unable to conceive after a year of having unprotected sex. There are a variety of reasons for infertility. One may not be able to produce sperm capable of fertilizing the ovum. Others may be unable to produce ova, or they may have fallopian tubes that cannot adequately accept ova to be fertilized or do not permit the travel of fertilized ova to the uterus. In some cases, the "shell" around the ovum may not admit any sperm at all. In other cases, the ovum is successfully fertilized but the uterus is incapable of allowing for implantation of the fertilized ovum or of maintaining the implanted embryo through the pregnancy. In all of these cases, some form of intervention may allow those who could not become parents through unassisted coitus to achieve parenthood nonetheless.

2. ART Interventions

The range of problems that prevent conception and childbearing give rise to a wide variety of alternative interventions to facilitate reproduction:

- AIH (artificial insemination-homologous): artificial insemination with the sperm of the person who is intended to be a biological and nurturing co-parent, often the husband/spouse;

- AID (artificial insemination-donor): artificial insemination with the sperm of someone who is presumed to have no continuing relationship with the child;

- IVF (*in vitro* fertilization): fertilizing an ovum outside of the uterus and then implanting it;

- embryo or egg transfer;

- genetic surrogacy: a person who is not an intended parent agrees to become pregnant through assisted reproduction using that person's own gamete; and

- gestational surrogacy: a person who is not an intended parent agrees to become pregnant through assisted reproduction using a different person's gametes.

In many cases ART requires the involvement of a host of other people. A sperm source (who is not necessarily a sperm donor—but is usually a sperm vendor) is necessary to address the inability to produce physiologically adequate sperm; an ovum source is necessary where the problem is the inability to provide a potentially fertile ovum; and a uterus source (originally called a "surrogate," now more generally described as a "gestational surrogate") is necessary where the intended parent(s) is(are) unable or unwilling to carry the fertilized ovum through the pregnancy. The interventions may also vary in complexity, depending on the specific infertility program. Medical treatment of an ovum may allow it to be fertilized, and treatment of a fertilized ovum may allow it to be implanted. And sometimes a combination of these needs requires that there be a series of interventions to produce a pregnancy and a child.

3. The Role of the Law

The law is often required to help parties sort out the formal relationships of those who undertake processes designed to facilitate reproduction. Legal rights and responsibilities arising out of ART may include obligations of support, decision-making authority, inheritance, contractual duties, and so on. There has been an increase in the variety of non-traditional legal relationships that have developed as a consequence of medical and non-medical reproductive techniques. It is possible to have a child who is the product of sperm from one source (whose sperm may be mixed with sperm from several sources), an ovum extracted from one source, implanted in another, and carried for the benefit of yet another set of parents who intend to nurture the child from the time of its birth.

This implicates social and cultural norms about family, which have changed over time, due in part to the ability of ARTs to facilitate nontraditional family structures. In the face of this change, the law has consistently been forced to answer questions about who *ought* to be treated as the legal parents. What rights, if any, should others involved in the process have? In addition, certain forms of ART result in the production of embryos that may never be used for procreative purposes. For what other purposes can they be used? What legal principles should guide our regulation of these embryos? And how might different theories of personhood or definitions of the beginning of life be relevant to these legal principles? The remainder of this section considers these questions.

B. ART & PARENTAGE LAW

In the Interest of K.M.H.

Supreme Court of Kansas, 2007.
169 P.3d 1025.

BEIER, J. This appeal from a consolidated child in need of care (CINC) case and a paternity action arises out of an artificial insemination leading to the birth of twins K.M.H. and K.C.H. We are called upon to decide the existence and extent of the parental rights of the known sperm donor, who alleges he had an agreement with the children's mother to act as the twins' father.

The twins' mother filed a CINC petition to establish that the donor had no parental rights under Kansas law. The donor sued for determination of his paternity. The district court sustained the mother's motion to dismiss, ruling that [the Kansas Parentage Act] was controlling and constitutional. That statute provides:

> "The donor of semen provided to a licensed physician for use in artificial insemination of a woman other than the donor's wife is treated in law as if he were not the birth father of a child thereby conceived, unless agreed to in writing by the donor and the woman." K.S.A. 38–1114(f).

Factual and Procedural Background

. . . The mother, S.H., is an unmarried female lawyer who wanted to become a parent through artificial insemination from a known donor. She was a friend of the donor, D.H., an unmarried male nonlawyer, who agreed to provide sperm for the insemination. Both S.H. and D.H. are Kansas residents, and their oral arrangements for the donation occurred in Kansas, but S.H. underwent two inseminations with D.H.'s sperm in Missouri. [After the second, she became pregnant and ultimately gave birth to twins.]

* * *

There was no formal written contract between S.H. and D.H. concerning the donation of sperm, the artificial insemination, or the expectations of the parties with regard to D.H.'s parental rights or lack thereof.

* * *

The district judge ruled that Kansas law governed, that K.S.A. 38–1114(f) was constitutional and applicable .... The [trial court judge concluded] as a matter of law that D.H. had no legal rights or responsibilities regarding K.M.H. and K.C.H.

Constitutionality of K.S.A. 38–1114(f)

In his brief, D.H. makes a general allegation that K.S.A. 38–1114(f) offends the [equal protection and due process clauses of the] Constitution. [At oral argument D.H.] acknowledged that he no longer challenges the statute as unconstitutional on its face; rather, he argues it cannot be constitutionally applied to him, as a known sperm donor who alleges he had an oral agreement with the twins' mother that granted him parental rights. . . .

* * *

Given the relative newness of the medical procedure of artificial insemination, and thus the newness of K.S.A. 38–1114(f)'s attempt to regulate the relationships arising from it, it is not surprising that the issue raised by D.H. is one of first impression, not only in Kansas but nationally. We therefore begin our discussion of the constitutionality of the statute by surveying the landscape of various states' laws governing the rights of sperm donors for artificial insemination. This landscape and its ongoing evolution provide helpful context for our analysis of K.S.A. 38–1114(f).

The majority of states that have enacted statutes concerning artificial insemination state that the husband of a married woman bears all rights and obligations of paternity as to any child conceived by artificial insemination, whether the sperm used was his own or a donor's.[ ] The 1973 Uniform Parentage Act . . . provided the model for many of the state artificial insemination statutes . . . . Section 5 of the original uniform Act provided:

> (a) If, under the supervision of a licensed physician and with the consent of her husband, a wife is inseminated artificially with semen donated by a man not her husband, the husband is treated in law as if he were the natural father of a child thereby conceived. The husband's consent must be in writing and signed by him and his wife. The physician shall certify their signatures and the date of the insemination, and file the husband's consent with the [State Department of Health], where it shall be kept confidential and in a sealed file. However, the physician's failure to do so does not affect the father and child relationship. All papers and records pertaining to the insemination, whether part of the permanent record of a court or of a file held by the supervising physician or elsewhere, are subject to inspection only upon an order of the court for good cause shown.

(b) The donor of semen provided to a licensed physician for use in artificial insemination of a *married* woman other than the donor's wife is treated in law as if he were not the natural father of a child thereby conceived." (Emphasis added.) Uniform Parentage Act (1973) § 5; 9B U.L.A. at 407–08.

The wording of this original Act and statutes that imitated it did not address the determination of a sperm donor's paternity when an *unmarried* woman conceived a child through artificial insemination. . . .

* * *

Certain states [ ] either anticipated the need for their original statutes to govern the relationship of a sperm donor to the child of an unmarried recipient as well as a married recipient or modified their original Uniform Act-patterned statutes to remove the word "married" from the § 5 (b) language. This meant these states' statutes contained complete bars to paternity for any sperm donor not married to the recipient, regardless of whether the recipient was married to someone else and regardless of whether the donor was known or anonymous. An example of such a provision reads: "The donor of semen provided to a licensed physician for use in artificial insemination of a woman other than the donor's wife is treated in law as if he were not the natural father of a child thereby conceived." See, e.g., Cal. Fam. Code § 7613(b) (West 2004) [ ].

* * *

In 1985, Kansas became one of the states that adopted portions of the Uniform Parentage Act of 1973 regarding presumptions of paternity, but it did not adopt any provision relating to artificial insemination. [ ]

In 1994, Kansas amended its statute to incorporate the 1973 Uniform Act's § 5(b) . . . . [ ] It did not differentiate between known and unknown or anonymous donors, but it did make two notable changes in the uniform language.

As discussed above, although the 1973 Uniform Act governed the paternity of children born only to married women as a result of artificial insemination with donor sperm, the version adopted by Kansas omitted the word "married." [ ] This drafting decision demonstrates the legislature's intent that the bar to donor paternity apply regardless of whether the recipient was married or unmarried.

The other alteration in the 1973 Uniform Act's language is directly at issue here. The Kansas Legislature provided that a sperm donor and recipient could choose to opt out of the donor paternity bar by written agreement. [ ] . . . .

This second drafting decision is critical and sets this case apart from [the approach taken by other states]. . . .

Ultimately, in view of the requirement that we accept as true D.H.'s evidence supporting existence of an oral agreement, we are faced with a very precise question: Does our statute's requirement that any opt-out agreement between an unmarried mother and a known sperm donor be "in writing" result in an equal protection or due process violation? . . .

Equal Protection

K.S.A. 38–1114(f) draws a gender-based line between a necessarily female sperm recipient and a necessarily male sperm donor for an artificial insemination. By operation of the statute, the female is a potential parent or actual parent under all circumstances; by operation of the same statute, the male will never be a potential parent or actual parent unless there is a written agreement to that effect with the female. [T]he male's ability to insist on father status effectively disappears once he donates sperm. Until that point, he can unilaterally refuse to participate unless a written agreement on his terms exists. After donation, the male cannot force the fatherhood issue. The female can unilaterally decide if and when to use the donation for artificial insemination and can unilaterally deny any wish of the male for parental rights by refusing to enter into a written agreement.

The guiding principle of equal protection analysis is that similarly situated individuals should be treated alike. . . . In order to pass muster under the federal and state equal protection provisions, a classification that treats otherwise similarly situated individuals differently based solely on the individuals' genders must substantially further a legitimate legislative purpose; the government's objective must be important, and the classification substantially related to achievement of it. [ ]

Given the biological differences between females and males and the immutable role those differences play in conceiving and bearing a child, regardless of whether conception is achieved through sexual intercourse or artificial insemination, we are skeptical that S.H. and D.H. are truly similarly situated. However, assuming for purposes of argument that they are, we perceive several legitimate legislative purposes or important governmental objectives underlying K.S.A. 38–1114(f).

. . . . K.S.A. 38–1114(f) envisions that both married and unmarried women may become parents without engaging in sexual intercourse, either because of personal choice or because a husband or partner is infertile, impotent, or ill. It encourages men who are able and willing to donate sperm to such women by protecting the men from later unwanted claims for support from the mothers or the children. It protects women recipients as well, preventing potential claims of donors to parental rights and responsibilities, in the absence of an agreement. Its requirement that any such agreement be in writing enhances predictability, clarity, and enforceability. Although the timing of entry into a written agreement is not set out explicitly, the design of the statute implicitly encourages early resolution of the elemental question of whether a donor will have parental rights. Effectively, the parties must decide whether they will enter into a written agreement before any donation is made, while there is still balanced bargaining power on both sides of the parenting equation.

In our view, the statute's gender classification substantially furthers and is thus substantially related to these legitimate legislative purposes and important governmental objectives. K.S.A. 38–1114(f) establishes the clear default positions of parties to artificial insemination. If these parties desire an arrangement different from the statutory norm, they are free to provide for it, as long as they do so in writing. . . .

Due Process

* * *

We simply are not persuaded that the requirement of a writing transforms what is an otherwise constitutional statute into one that violates D.H.'s substantive due process rights. Although we agree . . . that one goal of the Kansas Parentage Act as a whole is to encourage fathers to voluntarily acknowledge paternity and child support obligations, the obvious impact of the plain language of this particular provision in the Act is to prevent the creation of parental status where it is not desired or expected. To a certain extent, D.H. . . . evidently misunderstand[s] the statute's mechanism. It ensures no *attachment* of parental rights to sperm donors in the absence of a written agreement to the contrary; it does not *cut off* rights that have already arisen and attached.

* * *

We also reject the argument from D.H. . . . that the statute inevitably makes the female the sole arbiter of whether a male can be a father to a child his sperm helps to conceive. This may be true, as we discussed above, once a donation is made, a recipient who becomes pregnant through artificial insemination using that donation can refuse to enter into an agreement to provide for donor paternity. This does not make the requirement of written agreement unconstitutional. Indeed, it is consistent with United States Supreme Court precedent making even a married pregnant woman the sole arbiter, regardless of her husband's wishes, of whether she continues a pregnancy to term. . . .

All of this being said, we cannot close our discussion of the constitutionality of K.S.A. 38–1114(f) without observing that all that is constitutional is not necessarily wise. We are mindful of, and moved by, the Center's advocacy for public policy to maximize the chance of the availability of two parents—and two parents' resources—to Kansas children. We are also aware of continued evolution in regulation of artificial insemination in this and other countries. In particular, Britain and The Netherlands now ban anonymous sperm donations, near-perfect analogs to donations from known donors who will have no role beyond facilitating artificial insemination. These shifts formally recognize the understandable desires of at least some children conceived through artificial insemination to know the males from whom they have received half of their genes. The Human Fertilisation and Embryology Authority Act of 1990, as amended by Disclosure of Donor Information, Regulations 2004 No. 1511 (requiring, effective April 2005, British donors' identities to be made available to donor-conceived children when children become 18); Netherlands Embryos Bill, Article 3 Dutch Ministry of Health, Welfare, and Sport (2004) www. minvws.nl/en (effective June 2004, child born using donated sperm has right to obtain information about biological father at age 16). As one such child recently wrote,

> "[t]hose of us created with donated sperm won't stay bubbly babies forever. We're all going to grow into adults, and form opinions about the decision to bring us into the world in a way that deprives us of the basic right to know where we came from, what our history is and who both our parents are."

[ ] We sympathize. However, weighing of the interests of all involved in these procedures as well as the public policies that are furthered by favoring one or another in certain circumstances, is the charge of the Kansas Legislature, not of this court.

* * *

Equity

[D.H.] asserts that the district court must be reversed because S.H. has "unclean hands." In essence, he argues that he, a nonlawyer, was tricked by lawyer S.H., who failed to inform him of the statute and failed to explain how the absence of independent legal advice or a written agreement could affect his legal rights. He asserts that he asked S.H. about whether he needed a lawyer or whether they should put their arrangement in writing and was told neither was necessary. This behavior, he alleges, may have constituted a violation of S.H.'s ethical duties as a licensed lawyer.

* * *

Generally speaking, mere ignorance of the law is no excuse for failing to abide by it. [ ] There may be a case in the future in which a donor can prove that the existence of K.S.A. 38–1114(f) was concealed, or that he was fraudulently induced *not* to obtain independent legal advice or *not* to enter into a written agreement to ensure creation and preservation of his parental rights to a child conceived through artificial insemination. This is not such a case.

Affirmed.

CAPLINGER, J., dissenting: I respectfully disagree with the majority's analysis of the constitutionality of K.S.A. 38–1114(f) as applied to D.H. I would hold the statute unconstitutional as applied to D.H. for the reason that it violates his fundamental right to parent his children without due process of law.

* * *

Requirement that donor take affirmative action to protect his parental rights

In concluding that the opt-out provision in K.S.A. 38–1114(f) satisfies due process requirements, the majority states that D.H.'s "own *inaction* before donating his sperm" left him unable to meet the statute's requirements of a written agreement. [ ] Therein lies the constitutional problem with the statute. Fundamental rights must be actively waived, rather than passively lost due to inaction.

Initially, before analyzing this issue, I would note that the terminology employed by the majority, i.e., that D.H. failed to "opt out" of the statute, is a misnomer. In effect, the statute requires a known sperm donor, regardless of any agreement or understanding the donor may have as to his role in parenting a child conceived from his sperm, to *opt in* to parenthood or forever waive his right to parent. . . .

* * *

Effect of "ignorance of the law" on an individual's fundamental right to parent

Nor can I agree with the majority's conclusion that D.H.'s ignorance of the statute's writing requirement has no effect on the statute's application. . . .

* * *

I would urge the majority to consider the *complete* rationale of *Lehr*: "When an unwed father demonstrates a full commitment to the responsibilities of parenthood by 'com[ing] forward to participate in the rearing of his child,' his interest in personal contact with his child acquires substantial protection under the Due Process Clause." [ ]

That is the scenario with which this court is faced. A putative father has come forward to participate in the rearing of his children, emotionally and financially; consequently, his interest in doing so is entitled to full protection under the Due Process Clause. Instead of being given this protection and an opportunity to prove that he intended to actively parent his children, D.H. has been subjected to the workings of a statute of which he was unaware, that required him to "opt in" to fatherhood before ever donating his sperm, or be forever barred from parenting his children.

. . . Because the rights to parent are fundamental, those rights may be waived only through an intentional, free, and meaningful choice. Here, the record indicates D.H. was not even aware of K.S.A. 38–1114(f), much less its requirement that he must enter into a written agreement formalizing his intent to parent his child before he provided his sperm to S.H. . . .

* * *

HILL, J., dissenting: I must respectfully join with Judge Caplinger in her dissent. I too agree that as applied in this case, K.S.A. 38–1114(f) is unconstitutional when applied to a known donor.

But I raise my hand and ask a different question. Who speaks for the children in these proceedings? As applied by the majority in this case, this generative statute of frauds slices away half of their heritage. A man who was once considered a "putative father" in the initial child in need of care proceeding is now branded a mere "semen donor." The majority offers the children sympathy. But is this in their best interests? The trial court never got to the point of deciding the best interests of the children because it was convinced that such a consideration was barred by the operation of K.S.A. 38–1114(f) to a known donor.

None of the elaborate and meticulous safeguards our Kansas laws afford parents *and children* in proceedings before our courts when confronted with questions of parentage have been extended to these children. . . .

I agree with the Ohio Court of Common Pleas when it said:

"A father's voluntary assumption of fiscal responsibility for his child should be endorsed as a socially responsible action. A statute which absolutely extinguishes a father's efforts to assert the rights and responsibility of being a father, in a case with such facts as those *sub judice,* runs contrary to due process safeguards. [ ]" [ ]

I think the same can be said about our statute.

NOTE: THE 2017 UNIFORM PARENTAGE ACT

Most states adopted some version of the 1973 Uniform Parentage Act (UPA). In 2000, the original (1973) UPA and the Uniform Status of Children of Assisted Conception Act were combined and rewritten in the form of a new (2000) UPA, which was amended in 2002. The 2002 UPA was designed to address some of the uncertainty around parentage in assisted conception cases, but those portions of the statute proved controversial and were not embraced by many states. In addition, the 2002 statute continued to reflect certain gendered assumptions about parentage, such as the assumption that children would be born to couples with one man and one woman, which impacted legal presumptions and rights determining parentage.

The Uniform Law Commission (ULC) determined that the UPA needed updating in light of the growth of non-traditional family structures facilitated by ART, as well as the equity implications of the Supreme Court case *Obergefell v. Hodges*, 576 U.S. 644 (2015), which held laws barring marriage of same sex couples unconstitutional. This ruling was widely understood as rendering vulnerable parentage laws' differential treatment of same-sex couples versus different-sex couples, and raising questions about the constitutionality of gendered provisions of prior versions of the UPA. See also, *Pavan v. Smith*, 137 S.Ct. 2075 (2017) (invalidating an Arkansas law denying married same-sex couples the same recognition on their children's birth certificates granted to married different-sex couples). To address these inequities, and to modernize parentage laws in other ways, the Uniform Parentage Act was updated in 2017. Select provisions are reproduced below.

UNIFORM PARENTAGE ACT (2017)

* * *

[ARTICLE 1—GENERAL PROVISIONS]

§ 102.   DEFINITIONS.

(4) "Assisted reproduction" means a method of causing pregnancy other than sexual intercourse. The term includes:

> (A) intrauterine or intracervical insemination;
>
> (B) donation of gametes;
>
> (C) donation of embryos;
>
> (D) in-vitro fertilization and transfer of embryos; and
>
> (E) intracytoplasmic sperm injection.

* * *

(9) "Donor" means an individual who provides gametes intended for use in assisted reproduction, whether or not for consideration. The term does not include:

> (A) a woman who gives birth to a child by means of assisted reproduction[, except as otherwise provided in [Article] 8]; or
>
> (B) a parent under Article 7 [or an intended parent under Article 8].

* * *

(10) "Gamete" means sperm, egg, or any part of a sperm or egg.

* * *

(13) "Intended parent" means an individual, married or unmarried, who manifests an intent to be legally bound as a parent of a child conceived by assisted reproduction.

* * *

[ARTICLE 2—PARENT-CHILD RELATIONSHIP]

§ 201.   ESTABLISHMENT OF PARENT-CHILD RELATIONSHIP.

A parent-child relationship is established between an individual and a child if:

(1)   the individual gives birth to the child[, except as otherwise provided in [Article] 8, concerning surrogacy agreements];

(2)   there is a presumption under Section 204 of the individual's parentage of the child, unless the presumption is overcome in a judicial proceeding or a valid denial of parentage is made [pursuant to the rules established elsewhere in this Act];

(3)   the individual is adjudicated a parent of the child . . . ;

(4)   the individual adopts the child;

(5)   the individual acknowledges parentage of the child under [the section of the Act establishing the process for voluntary acknowledgment of parentage], unless [such] acknowledgment is rescinded [ ] or successfully challenged [under other provisions of this Act];

(6)   the individual's parentage of the child is established under [Article] 7 [concerning assisted reproduction] [;or

(7)   the individual's parentage of the child is established under [Article] 8]

* * *

§ 202.   NO DISCRIMINATION BASED ON MARITAL STATUS. A parent-child relationship extends equally to every child and parent, regardless of the marital status of the parent.

* * *

§ 204.   PRESUMPTION OF PARENTAGE.

(a)   An individual is presumed to be a parent of a child if:

(1)   except as otherwise provided [in this act]:

(A)   the individual and the woman who gave birth to the child are married to each other and the child is born during the marriage . . . ;

(B)   the individual and the woman who gave birth to the child were married to each other and the child is born not later than 300 days after the marriage is terminated by death, [divorce, dissolution, annulment, or declaration of invalidity, or after a decree of separation or separate maintenance] . . . ; or

(C) the individual and the woman who gave birth to the child married each other after the birth of the child . . . , the individual at any time asserted parentage of the child, and [the assertion is in a [state birth] record or on the child's birth certificate]; or

(2) the individual resided in the same household with the child for the first two years of the life of the child, including any period of temporary absence, and openly held out the child as the individual's child. . . .

* * *

[ARTICLE 7—ASSISTED REPRODUCTION]

§ 701.   SCOPE OF [ARTICLE].

This [article] does not apply to the birth of a child conceived by means of sexual intercourse[, or assisted reproduction under a surrogacy agreement under [Article] 8]. [Eds. Note: For a discussion of article 8, see the Note on Gestational Surrogacy and the UPA after *Johnson v. Calvert* below.]

§ 702.   PARENTAL STATUS OF DONOR.

A donor is not a parent of a child conceived by means of assisted reproduction.

§ 703.   PARENTAGE OF CHILD OF ASSISTED REPRODUCTION.

An individual who consents under Section 704 to assisted reproduction by a woman with the intent to be a parent of a child conceived by the assisted reproduction is a parent of the child.

§ 704.   CONSENT TO ASSISTED REPRODUCTION.

(a) Except as otherwise provided in subsection (b), the consent described in Section 703 must be in a record signed by a woman giving birth to a child conceived by assisted reproduction and an individual who intends to be a parent of the child.

(b) Failure to consent in a record as required by subsection (a), before, on, or after birth of the child, does not preclude the court from finding consent to parentage if: (1) the woman or the individual proves by clear-and-convincing evidence the existence of an express agreement entered into before conception that the individual and the woman intended they both would be parents of the child; or (2) the woman and the individual for the first two years of the child's life, including any period of temporary absence, resided together in the same household with the child and both openly held out the child as the individual's child . . . .

* * *

NOTES AND QUESTIONS

1.   *UPA 1973 vs. UPA 2017. K.M.H.* applied the Kansas version of the UPA. Would the outcome have been different under the 1973 UPA (excerpts cited in *K.M.H.*)? What about the 2017 UPA? Are there children who would have no fathers under the 1973 Act or the Kansas statute but who would have a father under the 2017 UPA? If we reject genetic essentialism (i.e., that the source of the genetic contribution is the parent), where should we look to decide if a sperm source ought to be the legal parent? Should we look to the intent of the parties? Should we look to the best interest of the child, as the dissent in *K.M.H.* suggests?

2.   *Parentage Under Common Law.* The 1973 version of the UPA applies only if the woman who is inseminated is a wife (i.e., is married), if she proceeds with the consent of her husband and the procedure is carried out under the supervision of a licensed physician. If any of these conditions is not met, the statute is inapplicable. See, e.g., Marriage of Witbeck-Wildhagen, 667 N.E.2d 122 (Ill. App. Ct. 1996) (lack of husband's consent to wife's insemination precluded statutory or equitable basis for recognition of paternity); Patton v. Vanterpool, 806 S.E.2d 493 (Ga. 2017) (statute creating an irrebuttable presumption of paternity with respect to children born during marriage who were conceived through artificial insemination did not apply to children conceived by IVF therapy). The third requirement may not be met because artificial insemination need not be done as a medical procedure by a physician. A syringe can be used to place "fresh" semen in the cervical canal or uterus directly into the person who intends to become pregnant, which can be done successfully at home by anyone who understands the underlying biological principles. Why should the involvement of a health care professional matter? When the statute does not apply, the common law has had to define the rights and interests of all of the parties. In a part of the opinion not reproduced in this casebook, the *K.M.H.* case noted that some common law cases have treated the source of the sperm to be the father, with all of the rights (including visitation) and obligations (including child support) that come with such a designation; others have not.

3.   *State Statutes & Donor Parentage.* The *K.M.H.* opinion noted that in adopting their own versions of the 1973 UPA, some states added explicit bars to donor paternity, regardless of whether the recipient was married to someone else and regardless of whether the donor was known or anonymous. Some courts have found that such a bar, while constitutional on its face, could be an unconstitutional violation of a donor's due process rights under certain conditions. See, e.g., McIntyre v. Crouch, 780 P.2d 239 (Or. Ct. App. 1989) (statute unconstitutional as applied if the donor can prove that he had an agreement with the mother to share the rights and responsibilities of parenthood); accord C.O. v. W.S., 64 Ohio Misc.2d 9 (1994). See also, L.F. v. Breit, 736 S.E. 2d 711 (Va. 2013) (noting that the assisted conception statute

did not contemplate situations where unmarried donors have long-term relationships, as well as biological ties, where the donor and birth mother have entered into a written agreement of paternity, and where the donor has voluntarily assumed parental responsibilities; and finding that a bar to donor paternity in this case would violate the donor's due process rights). Other courts have found that a man subject to a donor paternity bar may be able to prove parentage under a different statutory provision. See Jason P. v. Danielle S. 226 Cal.App.4th 167, 796–797 (2014) (holding that the donor paternity bar in Cal. Fam. Code § 7613(b)(West 2004) does not prevent donor from attempting to qualify as a presumed parent under § 7611(d) based on evidence the donor received the child into the donor's home and held out the child out as the donor's natural child).

Although earlier versions of the UPA focused on donor *paternity*, the 2017 Act focuses more broadly on donors' *parentage* rights, in ways that eschew earlier gender assumptions. For example, Section 703 in the 2017 UPA (reproduced above) is titled "*Parentage* of Child of Assisted Reproduction" and addresses the rights of "an individual" who intends to parent; but in the 2002 version, this section was titled "*Paternity* of Child Assisted Reproduction." Section 705 in the 2002 UPA addresses limitations on *husband's dispute of paternity*, while the corresponding section in the 2017 UPA refers to limitations on *spouse's dispute of parentage*. Such changes reflect an acknowledgement that children are not necessarily born to couples with one man and one woman, as well as the increased use of egg donations by those not intending to parent. As of this writing, a number of states have enacted reforms that provide greater legal protection for children's relationship to their unmarried and LGBTQ parents, and recognize legal parentage for children born through various forms of assisted reproduction. This includes eliminating much gendered terminology, as well as eliminating stigmatizing terms, like "illegitimate", that have long been used to classify and determine the rights of children born to unmarried parents. A growing number of states are adopting or considering reforms consistent with the 2017 UPA. See the Note on ART & Changing Norms later in this chapter.

4.   *Objections to Artificial Insemination.* Virtually no legal questions have arisen surrounding the use of the sperm of a husband to inseminate his wife. Such a process may be used within a marriage to allow for processing of the semen to overcome low sperm count. In these cases, there is no question about the identity of the mother and the father of the child. While religious objections have been raised to masturbation (which is required) and to any process that alters the "natural" procreative process, artificial insemination-homologous (AIH) of a married woman with sperm from her husband has been well accepted as a social and medical matter. Artificial insemination by a donor (AID), however, has generated greater objections based on religious beliefs for some and more general social mores for others. For example, in the case of married couples who need to use a donor, some believe that AID constitutes adultery. Some have also expressed concern that the availability of AID will undermine traditional assumptions about appropriate family structure. For

others, however, the possibility that ART facilitates new and different family combinations is viewed as empowering for individuals who want to have children and as reshaping society's assumptions about what constitutes a strong and moral family unit in positive ways.

Other objections are focused on the anonymity of the process, in light of the fact that donation of sperm often occurs anonymously in the U.S. Some are concerned that the uncertainty of the screening of the sperm source in AID can provide for uncertainty in the health of the child, and that anonymity of the sperm source may have unforeseen emotional consequences. There is also an argument that AID is inappropriate because if there is no limit on the number of times one person may donate sperm there may be a genetically unacceptably large number of children who unknowingly have the same genetic parent, which may lead to the dilution of the genetic pool of some communities.

5. *Donor Anonymity.* The debate over whether donor anonymity should be protected has historically centered around sperm donation. The question of whether a prospective mother (and, ultimately, the child) should know the identity of the sperm source was seen as having important legal and practical implications for donation. An anonymous (or, at least, unknown) sperm source would have a more difficult time establishing parentage than one who is known or identifiable, which may encourage prospective mothers to choose unknown sperm sources. That, in turn, may encourage women seeking artificial insemination to seek the help of cryobanks, run as medical institutions with medical control over the processing of the gametes. In the case of both sperm and egg donations, those who seek anonymous donors typically still want to know something about those donors. Gamete banks provide profiles of their various donors, and some issue catalogs of these profiles allowing their consumers to choose the profile that seems most appropriate. A "donor catalog" may provide information on race/ethnicity, hair color and texture, eye color, height, weight, blood type, skin tone, years of education, and occupation (or major in college). In addition to relevant health information, what kind of information, if any, about the gamete source ought to be provided to the intended parents?

6. *Obligations Concerning Donor Identity.* UPA 2017 requires gamete banks and fertility clinics ("ART clinics") to collect and store certain identifying information and medical history of donors. UPA 2017 also allows donors to maintain anonymity by signing a declaration stating that the donor does not agree to have the donor's identity disclosed. Even in these cases, though, upon request from the child, the ART clinics must make a good-faith effort to notify the donor, who may then elect to withdraw the earlier nondisclosure declaration. The clinic must also make a good-faith effort to provide nonidentifying medical history if requested. The American legal preference for unknown or anonymous donors is not reflected elsewhere in the world, and, especially in Europe, the trend is toward a legal requirement that the source be made known to the child when the child becomes an adult. Which approach is better—the American protection for anonymity, or the preference elsewhere for release of this information? In the United States there has been an increase

in the number of "directed donations" of gametes—i.e., donations made by identifiable persons for use by identified persons.

At the very least, should a child be informed about being the product of ART, and that there may be a gamete source other than the parents? The most recent recommendations by the American Society of Reproductive Medicine (ASRM) encourage disclosure:

- Disclosure to donor-conceived persons of the use of donor gametes or embryos in their conception is strongly encouraged, while ultimately the choice of recipient parents.

- Counseling and informed consent about disclosure and information sharing are essential for donors and recipients.

- [ART and donation programs and facilities] should expect inquiries from donor-conceived persons about their genetic background and should develop written policies to respond to these inquiries.

- [ART and donation programs and facilities] should gather, maintain, and permanently store medical and genetic information about donors.

- Donors and recipient parents should be informed in advance about how and when ART programs; sperm, oocyte, and embryo banks; and oocyte and embryo donation programs will release donor information to recipients and offspring. Donors and recipients also should be counseled that later changes in the law may affect any agreements.

Informing Offspring of Their Conception by Gamete or Embryo Donation: An Ethics Committee Opinion. American Society for Reproductive Medicine, Fertil. Steril. 2018; 109:601–5.

7. *Timing.* Cryopreservation—the ability to freeze and thaw with viability retention—is beneficial for many reasons. For example, cryopreservation permits testing for the presence of any latent contaminant (such as HIV), and it provides greater flexibility in family planning. This flexibility can have legal and ethical implications, however, because it gives rise to the possibility that the parent of a child conceived through ART could have died years before that child's birth. The Social Security Administration faced this issue in a number of cases where claimants have sought survivor's benefits from fathers who had died before they were born (or even conceived). Their eligibility depends on whether they are recognized as heirs by the states where they reside. In some states a child may be an heir to a parent who died pre-conception; in some states a child may not be. The Uniform Status of Children of Assisted Conception Act explicitly provided that one who died before implantation of the genetic contribution was not a parent of the "resulting child." In the 2017 version of the Uniform Parentage Act, Section 708 provides that "if an individual who consented . . . to assisted reproduction by a woman who agreed to give birth to a child dies before a transfer of gametes or embryos, the deceased individual is only a parent if: (1) either[ ] the individual consented in a record that if assisted reproduction were to occur

after the death of the individual, the individual would be a parent of the child [or such intent is proved by clear-and-convincing evidence]; and (2) either [ ] the embryo is in utero not later [36] months after the individual's death or [ ] the child is born not later than [45] months after the individual's death." See also Section C below, discussing the legal and ethical implications of IVF and the disposition of frozen gametes.

JOHNSON V. CALVERT
Supreme Court of California, 1993.
851 P.2d 776.

PANELLI, J.

* * *

Mark and Crispina Calvert are a married couple who desired to have a child. Crispina was forced to undergo a hysterectomy in 1984. Her ovaries remained capable of producing eggs, however, and the couple eventually considered surrogacy. In 1989 Anna Johnson heard about Crispina's plight from a coworker and offered to serve as a surrogate for the Calverts.

On January 15, 1990, Mark, Crispina, and Anna signed a contract providing that an embryo created by the sperm of Mark and the egg of Crispina would be implanted in Anna and the child born would be taken into Mark and Crispina's home "as their child." Anna agreed she would relinquish "all parental rights" to the child in favor of Mark and Crispina. In return, Mark and Crispina would pay Anna $10,000 in a series of installments, the last to be paid six weeks after the child's birth. Mark and Crispina were also to pay for a $200,000 life insurance policy on Anna's life.

The zygote was implanted on January 19, 1990. Less than a month later, an ultrasound test confirmed Anna was pregnant.

Unfortunately, relations deteriorated between the two sides. Mark learned that Anna had not disclosed she had suffered several stillbirths and miscarriages. Anna felt Mark and Crispina did not do enough to obtain the required insurance policy. She also felt abandoned during an onset of premature labor in June.

In July 1990, Anna sent Mark and Crispina a letter demanding the balance of the payments due her or else she would refuse to give up the child. The following month, Mark and Crispina responded with a lawsuit, seeking a declaration they were the legal parents of the unborn child. Anna filed her own action to be declared the mother of the child, and the two cases were eventually consolidated. The parties agreed to an independent guardian ad litem for the purposes of the suit.

The child was born on September 19, 1990, and blood samples were obtained from both Anna and the child for analysis. The blood test results excluded Anna as the genetic mother. The parties agreed to a court order providing that the child would remain with Mark and Crispina on a temporary basis with visits by Anna.

Discussion

*Determining Maternity Under the Uniform
Parentage Act [of 1973]*

. . . . [W]e are left with the undisputed evidence that Anna, not Crispina, gave birth to the child and that Crispina, not Anna, is genetically related to him. Both women thus have adduced evidence of a mother and child relationship as contemplated by the [Uniform Parentage] Act [1973].[ ] Yet for any child California law recognizes only one natural mother, despite advances in reproductive technology rendering a different outcome biologically possible.

We decline to accept the contention of amicus curiae . . . that we should find the child has two mothers. Even though rising divorce rates have made multiple parent arrangements common in our society, we see no compelling reason to recognize such a situation here. The Calverts are the genetic and intending parents of their son and have provided him, by all accounts, with a stable, intact, and nurturing home. To recognize parental rights in a third party with whom the Calvert family has had little contact since shortly after the child's birth would diminish Crispina's role as mother.

We see no clear legislative preference in [the statutory law] as between blood testing evidence and proof of having given birth.

* * *

Because two women each have presented acceptable proof of maternity, we do not believe this case can be decided without enquiring into the parties' intentions as manifested in the surrogacy agreement. Mark and Crispina are a couple who desired to have a child of their own genes but are physically unable to do so without the help of reproductive technology. They affirmatively intended the birth of the child, and took the steps necessary to effect in vitro fertilization. But for their acted-on intention, the child would not exist. Anna agreed to facilitate the procreation of Mark's and Crispina's child. The parties' aim was to bring Mark's and Crispina's child into the world, not for Mark and Crispina to donate a zygote to Anna. Crispina from the outset intended to be the child's mother. Although the gestative function Anna performed was necessary to bring about the child's birth, it is safe to say that Anna would not have been given the opportunity to gestate or deliver the child had she, prior to implantation of the zygote, manifested her own intent to be the child's mother. No reason appears why Anna's later change of heart should vitiate the determination that Crispina is the child's natural mother.

We conclude that although the Act recognizes both genetic consanguinity and giving birth as means of establishing a mother and child relationship, when the two means do not coincide in one woman, she who intended to procreate the child—that is, she who intended to bring about the birth of a child that she intended to raise as her own—is the natural mother under California law.[10]

* * *

Anna urges that surrogacy contracts violate several social policies. Relying on her contention that she is the child's legal, natural mother, she cites the public policy embodied in [the] Penal Code [ ], prohibiting the payment for consent to adoption of a child. She argues further that the policies underlying the adoption laws of this state are violated by the surrogacy contract because it in effect constitutes a prebirth waiver of her parental rights.

We disagree. Gestational surrogacy differs in crucial respects from adoption and so is not subject to the adoption statutes. The parties voluntarily agreed to participate in in vitro fertilization and related medical procedures before the child was conceived; at the time when Anna entered into the contract, therefore, she was not vulnerable to financial inducements to part with her own expected offspring. As discussed above, Anna was not the genetic mother of the child. The payments to Anna under the contract were meant to compensate her for her services in gestating the fetus and undergoing labor, rather than for giving up "parental" rights to the child. Payments were due both during the pregnancy and after the child's birth.

* * *

Finally, Anna and some commentators have expressed concern that surrogacy contracts tend to exploit or dehumanize women, especially women of lower economic status. Anna's objections center around the psychological harm she asserts may result from the gestator's relinquishing the child to whom she has given birth. Some have also

---

[10] Thus, under our analysis, in a true "egg donation" situation, where a woman gestates and gives birth to a child formed from the egg of another woman with the intent to raise the child as her own, the birth mother is the natural mother under California law.

The dissent would decide parentage based on the best interests of the child. Such an approach raises the repugnant specter of governmental interference in matters implicating our most fundamental notions of privacy, and confuses concepts of parentage and custody. Logically, the determination of parentage must precede, and should not be dictated by, eventual custody decisions. The implicit assumption of the dissent is that a recognition of the genetic intending mother as the natural mother may sometimes harm the child. This assumption overlooks California's dependency laws, which are designed to protect all children irrespective of the manner of birth or conception. Moreover, the best interests standard poorly serves the child in the present situation: it fosters instability during litigation and, if applied to recognize the gestator as the natural mother, results in a split of custody between the natural father and the gestator, an outcome not likely to benefit the child. Further, it may be argued that, by voluntarily contracting away any rights to the child, the gestator has, in effect, conceded the best interests of the child are not with her.

cautioned that the practice of surrogacy may encourage society to view children as commodities, subject to trade at their parents' will.

We are unpersuaded that gestational surrogacy arrangements are so likely to cause the untoward results Anna cites as to demand their invalidation on public policy grounds. Although common sense suggests that women of lesser means serve as surrogate mothers more often than do wealthy women, there has been no proof that surrogacy contracts exploit poor women to any greater degree than economic necessity in general exploits them by inducing them to accept lower-paid or otherwise undesirable employment. We are likewise unpersuaded by the claim that surrogacy will foster the attitude that children are mere commodities; no evidence is offered to support it.

The argument that a woman cannot knowingly and intelligently agree to gestate and deliver a baby for intending parents carries overtones of the reasoning that for centuries prevented women from attaining equal economic rights and professional status under the law. To resurrect this view is both to foreclose a personal and economic choice on the part of the surrogate mother, and to deny intending parents what may be their only means of procreating a child of their own genes.

* * *

The judgment of the Court of Appeal is affirmed.

[ARABIAN, J., concurred with the majority's Uniform Parentage Act analysis, but would leave the issue of whether surrogacy contracts could be consistent with public policy to the legislature.]

KENNARD, J., dissenting.

When a woman who wants to have a child provides her fertilized ovum to another woman who carries it through pregnancy and gives birth to a child, who is the child's legal mother? Unlike the majority, I do not agree that the determinative consideration should be the intent to have the child that originated with the woman who contributed the ovum. In my view, the woman who provided the fertilized ovum and the woman who gave birth to the child both have substantial claims to legal motherhood. Pregnancy entails a unique commitment, both psychological and emotional, to an unborn child. No less substantial, however, is the contribution of the woman from whose egg the child developed and without whose desire the child would not exist.

For each child, California law accords the legal rights and responsibilities of parenthood to only one "natural mother." When, as here, the female reproductive role is divided between two women, California law requires courts to make a decision as to which woman is the child's natural mother, but provides no standards by which to make that decision. The majority's resort to "intent" to break the "tie" between the genetic and gestational mothers is unsupported by statute, and, in the absence of appropriate protections in the law to guard against abuse of surrogacy arrangements, it is ill-advised. To determine who is the legal mother of a child born of a gestational surrogacy arrangement, I would apply the standard most protective of child welfare—the best interests of the child.

* * *

Analysis of the Majority's "Intent" Test

* * *

The majority offers four arguments in support of its conclusion to rely on the intent of the genetic mother as the exclusive determinant for deciding who is the natural mother of a child born of gestational surrogacy. Careful examination, however, demonstrates that none of the arguments mandates the majority's conclusion.

The first argument that the majority uses in support of its conclusion that the intent of the genetic mother to bear a child should be dispositive of the question of motherhood is "but-for" causation. Specifically, the majority relies on a commentator who writes that in a gestational surrogacy arrangement, " 'the child would not have been born but for the efforts of the intended parents.' " [ ] [But the resort to the "but-for" test derived from tort law is unprecedented and unjustified here.]

* * *

Behind the majority's reliance on "but-for" causation as justification for its intent test is a second, closely related argument. The majority draws its second rationale from a student note: " 'The mental concept of the child is a controlling factor of its creation, and the originators of that concept merit full credit as conceivers.' " [ ]

* * *

[This concept is taken from the law of intellectual property.] The problem with this argument, of course, is that children are not property. Unlike songs or inventions, rights in children cannot be sold for consideration, or made freely available to the general public.

Next, the majority offers as its third rationale the notion that bargained-for expectations support its conclusion regarding the dispositive significance of the genetic mother's intent. Specifically, the majority states that " 'intentions that are voluntarily chosen, deliberate, express and bargained-for ought presumptively to determine legal parenthood.' "

* * * But the courts will not compel performance of all contract obligations. [ ] The unsuitability of applying the notion that, because contract intentions are "voluntarily chosen, deliberate, express and bargained-for," their performance ought to be compelled by the courts is even more clear when the concept of specific performance is used to determine the course of the life of a child. Just as children are not the intellectual property of their parents, neither are they the personal property of anyone, and their delivery cannot be ordered as a contract remedy on the same terms that a court would, for example, order a breaching party to deliver a truckload of nuts and bolts.

* * *

The majority's final argument in support of using the intent of the genetic mother as the exclusive determinant of the outcome in gestational surrogacy cases is that preferring the intending mother serves the child's interests, which are " '[u]nlikely to run contrary to those of adults who choose to bring [the child] into being.' " [ ]

I agree with the majority that the best interests of the child is an important goal . . . . The problem with the majority's rule of intent is that application of this inflexible rule will not serve the child's best interests in every case.

* * *

The Best Interests of the Child

* * *

In the absence of legislation that is designed to address the unique problems of gestational surrogacy, this court should look not to tort, property or contract law, but to family law, as the governing paradigm and source of a rule of decision. The allocation of parental rights and responsibilities necessarily impacts the welfare of a minor child. And in issues of child welfare, the standard that courts frequently apply is the best interests of the child. [ ] This "best interests" standard serves to assure that in the judicial resolution of disputes affecting a child's well-being, protection of the minor child is the foremost consideration. Consequently, I would apply "the best interests of the child" standard to determine who can best assume the social and legal responsibilities of motherhood for a child born of a gestational surrogacy arrangement.

* * *

Factors that are pertinent to good parenting, and thus that are in a child's best interests, include the ability to nurture the child physically and psychologically [ ] and to provide ethical and intellectual guidance. [ ] Also crucial to a child's best interests is the "well recognized right" of every child "to stability and continuity." [ ] The intent of the genetic mother to procreate a child is certainly relevant to the question of the child's best interests; alone, however, it should not be dispositive.

* * *

In this opinion, I do not purport to offer a perfect solution to the difficult questions posed by gestational surrogacy; perhaps there can be no perfect solution. But in the absence of legislation specifically designed to address the complex issues of gestational surrogacy and to protect against potential abuses, I cannot join the majority's uncritical validation of gestational surrogacy.

I would reverse the judgment of the Court of Appeal, and remand the case to the trial court for a determination of disputed parentage on the basis of the best interests of the child.

NOTES AND QUESTIONS

1. *Objections to Surrogacy.* Five years earlier, in In the Matter of Baby M, 109 N.J. 396 (1988), the Supreme Court of New Jersey invalidated a surrogacy contract under facts similar to *Johnson v. Calvert*. In both cases, the surrogate and the commissioning parents developed palpable animosity for each other and the surrogate sought to invalidate the surrogacy contract. The court in *Baby M* invalidated the contract as inconsistent with the prohibition on the use of money in adoption, requirements concerning parental fitness in custody determinations, and public policy. The *Baby M* decision reflected many of the objections raised to surrogacy agreements at that time. These objections generally fall into three categories—those related to the contracting parties; those related to the child; and those related to the effect of the process on society as a whole.

The first argument generally advanced against surrogacy is that it exploits women who are willing to give or rent their bodies as vessels to carry other people's children. Could the inducement of payment for pregnancy cause a woman to consent to something that otherwise would be an unthinkable intrusion upon her body? Will the development of commercial surrogacy lead to a class of poor women who will become child-bearers for wealthy women who do not want to undergo pregnancy? Does the relationship between ethnicity and the distribution of wealth in this society mean it is likely that we will develop separate childbearing races and child-raising races? For an interesting discussion of American women who "outsource" pregnancy to women in India, where the service is far cheaper than it is in the United States, see Judith Warner, Outsourced Wombs (Domestic Disturbances), N.Y. Times (Jan. 3, 2008).

Feminists have been divided on this issue. Some are deeply offended by the use of the woman's body in this way; others believe this is misguided paternalism that leads courts (and others) to conclude that women are incapable of deciding for themselves whether they should enter surrogacy contracts. There are echoes of this paternalism concern in *Johnson*, when Justice Panelli points out that "there is no proof that surrogacy contracts exploit poor women to any greater degree than economic necessity in general exploits them by inducing them to accept lower-paid or otherwise undesirable employment." Some economists argue that making surrogacy contracts unenforceable will merely lower the amount that is paid to surrogates. Thus, they suggest, making surrogacy contracts unenforceable is just another in a long history of allegedly protectionist regulations that restrict what a woman may choose to do with her body. See Debra Satz, Feminist Perspectives on Reproduction and the Family, Stanford Encyclopedia of Philosophy (rev. 2013).

Another argument is that surrogacy contracts ought to be prohibited or discouraged because they advance only the best interests of the contracting parties, not the best interest of the child. Normally in a custody dispute between those with claims as parents, the court will look to the best interest of the child in determining the appropriate placement. Enforcing a surrogacy contract seems inconsistent with this principle. How does Justice Kennard say this concern should be addressed in the dissent? In addition, some believe that children who find out that they were carried by a surrogate will be injured by that discovery; others argue that surrogacy contracts render children instruments for the use of parents, not ends in themselves, and that this is necessarily harmful for children.

Finally, some believe that the fabric of society as a whole is weakened by surrogacy arrangements, at least commercial ones. In *Baby M*, Chief Justice Wilentz compared the surrogacy contract to the sale of a child and noted that "[a]lmost every evil that prompted the prohibition of the payment of money in connection with adoptions exists here." Surrogacy contracts are often condemned on the ground that they constitute baby selling, which is illegal in every state. There is confusion over the purpose of statutes that prohibit baby selling, though: they may be intended to protect parents from financial inducements to give up their children, or they may be intended to protect children from being reduced to the status of an ordinary commodity. Should surrogacy arrangements be governed by baby selling statutes? For exploration of the debate on baby selling and its logical consequences, see E. Landes & R. Posner, The Economics of the Baby Shortage, 7 J. Legal. Stud. 323 (1978); J.R.S. Prichard, A Market for Babies?, 34 U. Toronto L.J. 341 (1984); and Michele Goodwin, The Free-Market Approach to Adoption: The Value of a Baby, 26 B.C. Third World L.J. 61 (2006).

Although surrogacy remains hotly debated in parts of the U.S., there is increased acceptance. This is likely due, in part, to growing public awareness of the benefits of surrogacy in overcoming infertility or other reproductive barriers for those who want children. This may also be due to the increased visibility of surrogates willing to publicly share their experiences and explain their choices in their own voice. In New York, for example, the Protecting Modern Families Coalition, which describes itself as "a coalition of advocates from LGBTQ+, women's, religious, and infertility organizations" worked with state lawmakers to overhaul New York's parentage law through the Child-Parent Security Act enacted in 2020. This Act legalized surrogacy and modernized rules establishing parentage for children conceived through ART. A number of surrogates wrote a letter supporting the Act, challenging what they characterized as inaccurate assumptions about surrogacy based on outdated practices or troubling but unrepresentative examples of unethical practices in some developing countries, as well as assumptions about women's lack of knowledge or agency. In describing their reason for choosing surrogacy, they explain:

> All of us are lucky to have families of our own. We know the incomparable joy that comes from raising children, and also know—from our experiences with intended parents—the deep pain that comes when you are told that parenthood isn't possible for you.

> The decision to serve as a surrogate is not one that any of us entered into lightly. We each researched the science and the legal guidelines, learned about the experiences of other surrogates, and thought deeply about the commitment and process involved. For us, it came down to giving the gift of life to people who want nothing more than to raise children of their own.

Surrogates Issue Letter in Support of Parent-Child Security Act, Jan. 23, 2020, at familyequality.org/press-releases/.

2. *The Role of Parentage Law.* Compare the different approaches taken by the majority and dissent in *Johnson.* What do they reveal about the judges' understanding of the role of parentage law and the interests it should advance? As the *Johnson* and *Baby M* cases illustrate, courts have taken different approaches to these questions. Compare McDonald v. McDonald, 196 A.D.2d 7 (N.Y. App. Div. 1994) (a "true 'egg donation'" case, in which the court determined that a woman who gestated a child produced through the fertilization of another woman's ovum with her husband's sperm was to be considered the mother of the child that resulted, relying heavily on the reasoning of *Johnson v. Calvert*) with Belsito v. Clark, 67 Ohio Misc.2d 54 (1994) (rejecting *Johnson's* reasoning and holding that parentage was to be determined by genetic contribution, not by the intent-to-procreate of the parties to the original surrogacy agreement). But see S.N. v. M.B., 935 N.E.2d 463 (Ohio Ct. App. 2010), *rev. dismissed* 931 N.E.2d 126 (Ohio 2010) (relying on intent to enforce a surrogacy contract).

Uncertainty about the enforceability of surrogacy arrangements have motivated amendments to the original UPA. The 2002 UPA included an article recognizing and regulating surrogacy contracts, but these provisions were not popular with states. Almost half of the states had not addressed surrogacy by statute, and states that did tended to craft different approaches. In response, further updates were made in the 2017 UPA, with the goal of making them more consistent with current surrogacy practice and recently adopted statutes. Consider how the 2017 UPA addresses surrogacy in the excerpts below.

UNIFORM PARENTAGE ACT (2017)
[ARTICLE 8—SURROGACY AGREEMENT]

§ 801.   DEFINITIONS. In this [article]:

(1) "Genetic surrogate" means a woman who is not an intended parent and who agrees to become pregnant through assisted reproduction

using her own gamete, under a genetic surrogacy agreement as provided in this [article].

(2) "Gestational surrogate" means a woman who is not an intended parent and who agrees to become pregnant through assisted reproduction using gametes that are not her own, under a gestational surrogacy agreement as provided in this [article].

(3) "Surrogacy agreement" means an agreement between one or more intended parents and a woman who is not an intended parent in which the woman agrees to become pregnant through assisted reproduction and which provides that each intended parent is a parent of a child conceived under the agreement. Unless otherwise specified, the term refers to both a gestational surrogacy agreement and a genetic surrogacy agreement.

§ 802.   ELIGIBILITY TO ENTER GESTATIONAL OR GENETIC SURROGACY AGREEMENT.

(a) To execute an agreement to act as a gestational or genetic surrogate, a woman must: (1) have attained 21 years of age; (2) previously have given birth to at least one child; (3) complete a medical evaluation related to the surrogacy arrangement by a licensed medical doctor; (4) complete a mental-health consultation by a licensed mental-health professional; and (5) have independent legal representation of her choice throughout the surrogacy arrangement regarding the terms of the surrogacy agreement and the potential legal consequences of the agreement.

(b) To execute a surrogacy agreement, each intended parent, whether or not genetically related to the child, must: (1) have attained 21 years of age; (2) complete a medical evaluation related to the surrogacy arrangement by a licensed medical doctor; (3) complete a mental-health consultation by a licensed mental health professional; and (4) have independent legal representation of the intended parent's choice throughout the surrogacy arrangement regarding the terms of the surrogacy agreement and the potential legal consequences of the agreement.

* * *

§ 804.   REQUIREMENTS OF GESTATIONAL OR GENETIC SURROGACY AGREEMENT: CONTENT.

(a) A surrogacy agreement must comply with the following requirements:

(1) A surrogate agrees to attempt to become pregnant by means of assisted reproduction.

(2) Except as otherwise provided [in this Act], the surrogate and the surrogate's spouse or former spouse, if any, have no claim to parentage of a child conceived by assisted reproduction under the agreement.

(3) The surrogate's spouse, if any, must acknowledge and agree to comply with the obligations imposed on the surrogate by the agreement.

(4) Except as otherwise provided [in this Act], the intended parent or, if there are two intended parents, each one [ ] will be the exclusive parent or parents of the child, regardless of number of children born or gender or mental or physical condition of each child.

(5) Except as otherwise provided [in this Act], [the intended parents] immediately on birth will assume responsibility for the financial support of the child, regardless of number of children born or gender or mental or physical condition of each child.

* * *

(7) The agreement must permit the surrogate to make all health and welfare decisions regarding herself and her pregnancy. This [Act] does not enlarge or diminish the surrogate's right to terminate her pregnancy.

(8) The agreement must include information about each party's right under this [article] to terminate the surrogacy agreement.

(b) A surrogacy agreement may provide for: (1) payment of consideration and reasonable expenses; and (2) reimbursement of specific expenses if the agreement is terminated under this [Article]. . . .

* * *

§ 808. TERMINATION OF GESTATIONAL SURROGACY AGREEMENT.

(a) A party to a gestational surrogacy agreement may terminate the agreement, at any time before an embryo transfer, by giving notice of termination in a record to all other parties. If an embryo transfer does not result in a pregnancy, a party may terminate the agreement at any time before a subsequent embryo transfer.

(b) Unless a gestational surrogacy agreement provides otherwise, on termination of the agreement under subsection (a), the parties are released from the agreement, except that each intended parent remains responsible for expenses that are reimbursable under the agreement and incurred by the gestational surrogate through the date of termination.

(c) Except in a case involving fraud, neither a gestational surrogate nor the surrogate's spouse or former spouse, if any, is liable to the intended parent or parents for a penalty or liquidated damages, for terminating a gestational surrogacy agreement under this section.

§ 809.   PARENTAGE   UNDER   GESTATIONAL   SURROGACY AGREEMENT.

(a)   Except as otherwise provided in [this Act], on birth of a child conceived by assisted reproduction under a gestational surrogacy agreement, each intended parent is, by operation of law, a parent of the child.

(b)   Except as otherwise provided in [this Act], neither a gestational surrogate nor the surrogate's spouse or former spouse, if any, is a parent of the child.

* * *

(d)   Except as otherwise provided in [this Act], if, due to a clinical or laboratory error, a child conceived by assisted reproduction under a gestational surrogacy agreement is not genetically related to an intended parent or a donor who donated to the intended parent or parents, each intended parent, and not the gestational surrogate and the surrogate's spouse or former spouse, if any, is a parent of the child, subject to any other claim of parentage.

* * *

§ 812.   EFFECT OF GESTATIONAL SURROGACY AGREEMENT.

(a)   A gestational surrogacy agreement that [complies with §§ 802–804] is enforceable.

(b)   If a child was conceived by assisted reproduction under a gestational surrogacy agreement that does not comply [with the above requirements], the court shall determine the rights and duties of the parties to the agreement consistent with the intent of the parties at the time of execution of the agreement. . . .

* * *

§ 813.   REQUIREMENT TO VALIDATE GENETIC SURROGACY AGREEMENT.

(a)   Except as otherwise provided in Section 816, to be enforceable, a genetic surrogacy agreement must be validated by [designate court] [prior to commencement of the assisted reproduction].

§ 814.   TERMINATION   OF   GENETIC   SURROGACY AGREEMENT.

(a) A party to a genetic surrogacy agreement may terminate the agreement as follows:

   * * *

   (2)   A genetic surrogate . . . may withdraw consent to the agreement any time before 72 hours after the birth of a child conceived by assisted reproduction under the agreement. . . .

[Eds. Note: The other provisions of this section follow the rules for termination of gestational surrogacy in § 808.]

§ 815. PARENTAGE UNDER VALIDATED GENETIC SURROGACY.

(a) Unless a genetic surrogate exercises the right under Section 814 to terminate a genetic surrogacy agreement, each intended parent is a parent of a child conceived by assisted reproduction under an agreement validated under Section 813.

* * *

§ 816. EFFECT OF NONVALIDATED GENETIC SURROGACY AGREEMENT.

* * *

(b) If all parties agree, a court may validate a genetic surrogacy agreement after assisted reproduction has occurred but before the birth of a child conceived by assisted reproduction. . . .

(d) If . . . a genetic surrogate [withdraws] her consent to the agreement [after 72 hours after the birth of child], the court shall adjudicate parentage of the child based on the best interest of the child, taking into account . . . the intent of the parties . . . .

* * *

NOTES AND QUESTIONS

1. *Relaxing Surrogacy Rules.* Is the legal standard in Article 8 of the Uniform Parentage Act (2017) different from the test applied by the court in *Johnson v. Calvert*? Would the case have been decided differently under the 2017 UPA? As noted above, one of the goals of the drafters of the 2017 UPA was to update the surrogacy provisions to make them more consistent with current practice and law, including liberalizing the rules governing gestational surrogacy agreements. For example, the 2002 UPA requires procedures similar to those for adoption: parties to a surrogacy agreement must obtain court approval of the agreement before seeking medical procedures relating to the agreement, and the court's approval must be based, in part, on a home study by the relevant child-welfare agency (unless waived by the court). The 2017 version eliminates the requirement of pre-pregnancy court validation for gestational surrogacy agreements, but retains this requirement for genetic surrogacy contracts. The 2017 UPA also eliminates the home study requirement. The 2017 version contains other procedural protections for surrogacy agreements, such as requiring the intended legal parents to pay for independent legal representation for the surrogate, requiring the surrogate's spouse to be a party to the agreement, and requiring execution of the agreement before any medical procedure is performed. Importantly, the 2017

UPA provides guidance for how parentage may be determined in cases where the surrogacy agreement does not comply with the law.

2.   *Gestational vs. Genetic Surrogacy.* Another difference between the 2017 UPA and the prior version is how it regulates genetic versus gestational surrogacy agreements. The 2017 UPA permits both, but it imposes additional safeguards or requirements on genetic surrogacy agreements. As noted above, court approval is required for genetic surrogacy contracts only. In addition, while gestational surrogacy agreements are binding once successful transfer has occurred, genetic surrogates are allowed to withdraw their consent any time up to 72 hours after birth. The comments explain that "[t]his differentiation . . . is intended to reflect both the factual differences between the two types of surrogacy as well as the reality that policy makers view these two forms of surrogacy as being quite different." What are those differences?

3.   *Healthcare Decision-Making by Surrogates.* Recall the earlier discussion about decision-making during pregnancy in Section VI. What are the implications of this for intended parents who may want to require or prohibit certain behavior by surrogates? Modern surrogacy laws may address this potential conflict in various ways. How does the UPA 2017 address health care decision-making during a surrogate's pregnancy? Compare this to the approach in the Connecticut Parentage Act:

(8)   The surrogacy agreement shall not infringe on the rights of the person acting as surrogate to make all health and welfare decisions regarding the person, the person's body and the person's pregnancy throughout the duration of the surrogacy arrangement, including during attempts to become pregnant, pregnancy, delivery and post-partum. The surrogacy agreement shall not infringe upon the right of the person acting as surrogate to autonomy in medical decision making by, including, but not limited to, requiring the person acting as surrogate to undergo a scheduled, nonmedically indicated caesarean section or to undergo multiple embryo transfer. Except as otherwise provided by law, any written or oral agreement purporting to waive or limit the rights described in this subdivision are void as against public policy.

NOTE: ART & CHANGING NORMS

As noted above, the technology for facilitating reproduction has challenged social and legal decisions based on traditional assumptions of family models. It has also highlighted justice concerns arising out of the disparate treatment of certain groups under parentage laws and when trying to access to fertility care.

1.   *Application of Parentage Laws to Same Sex Couples.* Same sex couples have had difficulty establishing parentage for a couple of reasons. First, as the previous cases demonstrate, parentage laws have reflected certain heteronormative assumptions about who could be considered parents. Recall that in *Johnson v. Calvert*, the court noted that "for any child California law recognizes only one natural mother, despite advances in reproductive

technology rendering a different outcome biologically possible." Consider the implications for a same sex lesbian couple in which one person contributes the ovum and the other carries the child to term. A Florida court confronted this issue in T.M.H. v. D.M.T., 79 So.3d 787 (Fla. App. 2011). In *T.M.H.*, the ovum source and the birth mother planned to raise the child together, paid for the infertility treatment together, and did raise the child together after the child's birth. When the relationship went bad, though, the birth mother absconded with the child to Australia. The genetic mother sought an order that she was a parent of the child and entitled to parental rights. The birth mother disagreed and argued that under the relevant Florida statute an egg donor could not be mother at all:

> The donor of any egg, sperm, or preembryo, other than the commissioning couple or a father who has executed a preplanned adoption agreement [ ], shall relinquish all maternal or paternal rights and obligations with respect to the donation or the resulting children. . . .

Fla. Stat. sec. 742.14. The trial court judge reluctantly agreed that the ovum source could not be a parent under the statute, but the decision was appealed. The State Supreme Court ultimately held the application of the statute to preclude assertion of parental rights unconstitutional for two reasons. First, it deprived the biological mother of parental rights where she was an intended parent and actually established a parental relationship with the child. Second, to the extent the statutory exception to relinquishment of rights provided for a "commissioning couple" was applied to opposite sex, but not same sex, couples, this violated state and federal equal protection law. D.M.T. v. T.M.H., 129 So.3d 320 (Fla. 2013). Same sex couples have also faced barriers as a result of parentage assumptions that rely on marriage status, in states where same sex couples have been denied the right to marry. See, e.g., Miller-Jenkins v. Miller-Jenkins, 912 A.2d 951 (Vt. 2006).

Such discrimination appears to be unconstitutional after Obergefell v. Hodges, 576 U.S. 644 (2015). In a five-to-four decision, the Court held that denying the right to marry to same sex couples violated the Fourteenth Amendment, and that a state cannot treat married couples differently because of the sexual identity or preference of the spouses. The Court also expressly noted that the right to marry protects not only romantic relationships, but "safeguards children and families" as well. *Obergefell* is understood as establishing the parenting rights of same sex couples. For example, before *Obergefell* a dozen states limited the opportunity of same sex couples to adopt or foster children. In 2016, Mississippi, the last state to maintain such a statutory or regulatory limitation, saw it overturned by judicial decision in Campaign for S. Equal. v. Mississippi Dep't. of Human Servs., 175 F. Supp. 3d 691 (S.D. Miss. 2016):

> In sum, the majority opinion [in Obergefell] foreclosed litigation over laws interfering with the right to marry and "rights and responsibilities intertwined with marriage." [ ] It also seems highly unlikely that the same court that held a state cannot ban gay marriage because it would deny

benefits—expressly including the right to adopt—would then conclude that married gay couples can be denied that very same benefit.

Nonetheless, according to one scholar's comprehensive survey of contemporary regulation of parental recognition in the context of ART, even after *Obergefell* those departing from traditional norms of gender and sexuality have continued to have their parent-child relationships discounted. See Douglas NeJaime, The Nature of Parenthood, 126 Yale L.J. 2260 (2017). Indeed, the 2017 revisions to the UPA were in part an attempt to remedy apparent constitutional deficiencies of state parentage laws in light of *Obergefell*.

2.   *State Recognition of Multiple Parents*. Parentage law has undergone a significant shift since the court's observation in *Johnson v. Calvert* that a child could not have two mothers under California law. Since 2014, California law has allowed legal recognition that a child has more than two parents, but only where such a finding "is necessary to protect the child from the detriment of being separated from one of his or her parents." Cal. Fam. Code § 7612(c) provides:

> In determining detriment to the child, the court shall consider all relevant factors, including, but not limited to, the harm of removing the child from a stable placement with a parent who has fulfilled the child's physical needs and the child's psychological needs for care and affection, and who has assumed that role for a substantial period of time. A finding of detriment to the child does not require a finding of unfitness of any of the parents or persons with a claim to parentage.

Families with more than two parents are permitted but not preferred, it seems, under the judicial considerations required by the statute. A handful of other states have also recognized more-than-two-parent families, either by statute or by judicial determination. See, e.g., Nevada Assembly Bill 115 (enacted Jun. 8, 2021.

3.   *Refusal to Provide Fertility Treatment*. Despite greater social and legal recognition of non-traditional family structures, some physicians and hospitals refuse to provide the full range of infertility services to single people or those who may not conform to traditional norms of gender and sexuality. See, e.g., North Coast Women's Care Medical Group v. Superior Court, 44 Cal. 4th 1145, 189 P.3d 959 (2008) (plaintiff who was a lesbian alleged that a health care provider refused to provide her with necessary infertility treatment because of her sexual orientation in violation of state antidiscrimination law). Some states have laws that prohibit such discrimination, such as California's Unruh Civil Rights Act, which prohibits discrimination on the basis of sex, marital status, and sexual orientation, among other categories, by public accommodations. Cal. Civ. Code § 51. Although not all states contain similar antidiscrimination protections, recall from Chapter 2, Section IV.B. that the ACA's Section 1557 prohibition against sex discrimination likely includes discrimination on the basis of sexual orientation and gender identity.

Provider refusal to treat raises a broader question about the gatekeeping role of physicians generally in the area of ART. May physicians refuse to help someone have a child because that person is unmarried (and with no partner), considered psychologically immature, welfare dependent, has a history of a serious genetic disorder, or is HIV positive? Historically, such rejections were commonplace. Office of Technology Assessment, Artificial Insemination Practice in the United States (1988). A 2005 survey of assisted reproductive technology programs revealed that ART providers screened out people for a whole host of reasons. For example, among those surveyed: 81% were very or extremely likely to turn away a couple where the man had been physically abusive to his existing child; 59%—a woman who was HIV+; 55%—a woman who had severe diabetes with a 10% chance that pregnancy would lead to her death; 48%—a gay couple wanting to use a surrogate (yet only 17% were likely to turn away a lesbian couple wanting to use a sperm donor); 38%—a couple dependent on welfare; 20%—a woman with no partner. Andrea D. Gurmankin et al., Screening Practices and Beliefs of Assisted Reproductive Technology Programs, 83 Fertility and Sterility 61 (2005). Are some of these categories of refusals more troubling than others on ethical grounds? Do the underlying reasons for refusal matter? Should ART practitioners be able to consider parental fitness at all? If so, how would fitness be determined and might this trigger the kind of reproductive justice concerns identified earlier in this chapter. For example, there is evidence that some providers assume unfitness among people with certain kinds of disabilities. See Kimberly M. Mitcherson, Disabling Dreams of Parenthood: The Fertility Industry, Anti-discrimination, and Parents with Disabilities, 27 Law & Ineq. 311 (2009). Would any of the above refusals implicate antidiscrimination law, such as the Americans with Disabilities Act or Section 1557 of the ACA? (See Chapter 2).

The Ethics Committee of the American Society of Reproductive Medicine (ASRM) has issued several relevant ethics opinions concerning provider refusals to treat. See Ethics Committee Opinions and Webinars at the ASRM website at ASRM.org. In one opinion, the Committee explains that "fertility programs may withhold services from prospective patients on the basis of well-grounded reasons that those patients will be unable to provide minimally adequate or safe care for offspring." ASRM, Child-Rearing Ability and the Provision of Fertility Services, 108 Fertil. & Steril., 944 (Dec. 2017). But it goes on to advise programs to develop written policies and procedures for making such determinations, and cautions that "[b]ecause of the importance of reproduction, judgments to deny treatment should be made only when there is a strong and substantial basis for doing so." In addition to reproductive interests, this Opinion identifies important antidiscrimination principles codified in federal and state laws that limit providers' ability to refuse treatment. This opinion, and a series of others, advise that providers should not deny care solely on the basis of disability, HIV status, sexual orientation, marital status, or gender identity. See, e.g., Human Immunodeficiency Virus (HIV) and Infertility Treatment 104 Fertil. & Steril. e1 (Jul. 2015) (emphasizing that HIV is a manageable disease, and that certain ART and other medical treatments have proven highly effective in avoiding newborn infection in the case of HIV+ parents); Access to Fertility Treatment by Gays, Lesbians, and Unmarried Persons, 100 Fertil. & Steril. 1514 (Dec. 2013) (finding that no reasonable claim can be made that children reproduced outside of heterosexual marriage are harmed and taking the position that "[m]oral objection . . . itself is not an acceptable basis for limiting childrearing or reproduction" by single individuals or same-sex couples); Access to fertility services by transgender persons: an Ethics Committee opinion, 104 Fertil Steril 1111 (Nov. 2015) (providing similar guidance).

C. *IN VITRO* FERTILIZATION, EGG TRANSFER, AND EMBRYO TRANSFER

In vitro (literally, "in glass") fertilization is a highly technical medical intervention. In the normal *in vitro* case, an ovum that is ready to be released from the ovary is identified through laparoscopy or ultrasound and removed from the ovary by surgery or aspiration through a hollow needle. The ovum is placed in a container with the appropriate amount of semen containing fertile sperm. The fertilized ovum is then placed in the woman's uterus, where it is permitted to implant and develop. But in the course of this process, circumstances may change that affect the parties' intent to use the embryos for their original procreative purposes. How do courts decide what happens to the embryos in these instances?

DAVIS V. DAVIS
Supreme Court of Tennessee, 1992.
842 S.W.2d 588.

DAUGHTREY, J.

This appeal presents a question of first impression, involving the disposition of the cryogenically-preserved product of in vitro fertilization (IVF), commonly referred to in the popular press and the legal journals as "frozen embryos." The case began as a divorce action, filed by the appellee, Junior Lewis Davis, against his then wife, appellant Mary Sue Davis. The parties were able to agree upon all terms of dissolution, except one: who was to have "custody" of the seven "frozen embryos" stored in a Knoxville fertility clinic that had attempted to assist the Davises in achieving a much-wanted pregnancy during a happier period in their relationship.

I.   Introduction

Mary Sue Davis originally asked for control of the "frozen embryos" with the intent to have them transferred to her own uterus, in a post-divorce effort to become pregnant. Junior Davis objected, saying that he preferred to leave the embryos in their frozen state until he decided whether or not he wanted to become a parent outside the bounds of marriage.

Based on its determination that the embryos were "human beings" from the moment of fertilization, the trial court awarded "custody" to Mary Sue Davis and directed that she "be permitted the opportunity to bring these children to term through implantation." The Court of Appeals reversed, finding that Junior Davis has a "constitutionally protected right not to beget a child where no pregnancy has taken place" and holding that "there is no compelling state interest to justify[ ] ordering implantation against the will of either party." The Court of Appeals further held that "the parties share an interest in the seven fertilized ova" and remanded the case to the trial court for entry of an order vesting them with "joint control . . . and equal voice over their disposition."

Mary Sue Davis then sought review in this Court, contesting the validity of the constitutional basis for the Court of Appeals decision. We granted review, not because we disagree with the basic legal analysis utilized by the intermediate court, but because of the obvious importance of the case in terms of the development of law regarding the new reproductive technologies, and because the decision of the Court of Appeals does not give adequate guidance to the trial court in the event the parties cannot agree.

We note, in this latter regard, that their positions have already shifted: both have remarried and Mary Sue Davis (now Mary Sue Stowe) has moved out of state. She no longer wishes to utilize the "frozen embryos" herself, but wants authority to donate them to a childless couple. Junior Davis is adamantly opposed to such donation and would prefer to see the "frozen embryos" discarded. The result is, once again, an impasse, but the parties' current legal position does have an effect on the probable outcome of the case, as discussed below.

If we have no statutory authority or common law precedents to guide us [in this case], we do have the benefit of extensive comment and analysis in the legal journals. In those articles, medical-legal scholars and ethicists have proposed various models for the disposition of "frozen embryos" when unanticipated contingencies arise, such as divorce, death of one or both of the parties, financial reversals, or simple disenchantment with the IVF process. Those models range from a rule requiring, at one extreme, that all embryos be used by the gamete-providers or donated for uterine transfer, and, at the other extreme, that any unused embryos be automatically discarded. Other formulations would vest control in the female gamete-provider—in every case, because of her greater physical and emotional contribution to the IVF process, or perhaps only in the event that she wishes to use them herself. There are also two "implied contract" models: one would infer from enrollment in an IVF program that the IVF clinic has authority to decide in the event of an impasse whether to donate, discard, or use the "frozen embryos" for research; the other would infer from the parties' participation in the creation of the embryos that they had made an irrevocable commitment to reproduction and would require transfer either to the female provider or to a donee. There are also the so-called "equity models": one would avoid the conflict altogether by dividing the "frozen embryos" equally between the parties, to do with as they wish; the other would award veto power to the party wishing to avoid parenthood, whether it be the female or the male progenitor.

Each of these possible models has the virtue of ease of application. Adoption of any of them would establish a bright-line test that would dispose of disputes like the one we have before us in a clear and predictable manner. As appealing as that possibility might seem, we conclude that given the relevant principles of constitutional law, the existing public policy of Tennessee with regard to unborn life, the current state of scientific knowledge giving rise to the emerging reproductive technologies, and the ethical considerations that have developed in response to that scientific knowledge, there can be no easy answer to the question we now face. We conclude, instead, that we must weigh the interests of each party to the dispute, in terms of the facts and analysis set out below, in order to resolve that dispute in a fair and responsible manner.

* * *

IV. The "Person" vs. "Property" Dichotomy

One of the fundamental issues the inquiry poses is whether the preembryos in this case should be considered "persons" or "property" in the contemplation of the law. The Court of Appeals held, correctly, that they cannot be considered "persons" under Tennessee law . . . .

Nor do preembryos enjoy protection as "persons" under federal law. . . .

Left undisturbed, the trial court's ruling would have afforded preembryos the legal status of "persons" and vested them with legally cognizable interests separate from those of their progenitors. Such a decision would doubtless have had the effect of outlawing IVF programs in the state of Tennessee. But in setting aside the trial court's judgment, the Court of Appeals, at least by implication, may have swung too far in the opposite direction.

To our way of thinking, the most helpful discussion on this point is found not in the minuscule number of legal opinions that have involved "frozen embryos," but in the ethical standards set by The American Fertility Society, as follows:

> Three major ethical positions have been articulated in the debate over preembryo status. At one extreme is the view of the preembryo as a human subject after fertilization, which requires that it be accorded the rights of a person. This position entails an obligation to provide an opportunity for implantation to occur and tends to ban any action before transfer that might harm the preembryo or that is not

immediately therapeutic, such as freezing and some preembryo research.

At the opposite extreme is the view that the preembryo has a status no different from any other human tissue. With the consent of those who have decision-making authority over the preembryo, no limits should be imposed on actions taken with preembryos.

A third view—one that is most widely held—takes an intermediate position between the other two. It holds that the preembryo deserves respect greater than that accorded to human tissue but not the respect accorded to actual persons. The preembryo is due greater respect than other human tissue because of its potential to become a person and because of its symbolic meaning for many people. Yet, it should not be treated as a person, because it has not yet developed the features of personhood, is not yet established as developmentally individual, and may never realize its biologic potential.

* * *

In its report, the Ethics Committee then calls upon those in charge of IVF programs to establish policies in keeping with the "special respect" due preembryos and suggests:

Within the limits set by institutional policies, decision-making authority regarding preembryos should reside with the persons who have provided the gametes. . . . As a matter of law, it is reasonable to assume that the gamete providers have primary decision-making authority regarding preembryos in the absence of specific legislation on the subject. A person's liberty to procreate or to avoid procreation is directly involved in most decisions involving preembryos.[ ]

We conclude that preembryos are not, strictly speaking, either "persons" or "property," but occupy an interim category that entitles them to special respect because of their potential for human life. It follows that any interest that Mary Sue Davis and Junior Davis have in the preembryos in this case is not a true property interest. However, they do have an interest in the nature of ownership, to the extent that they have decision-making authority concerning disposition of the preembryos, within the scope of policy set by law.

V.   The Enforceability of Contract

* * *

We believe, as a starting point, that an agreement regarding disposition of any untransferred preembryos in the event of contingencies (such as the death of one or more of the parties, divorce, financial reversals, or abandonment of the program) should be presumed valid and should be enforced as between the progenitors. This conclusion is in keeping with the proposition that the progenitors, having provided the gametic material giving rise to the preembryos, retain decision-making authority as to their disposition.

At the same time, we recognize that life is not static, and that human emotions run particularly high when a married couple is attempting to overcome infertility problems. It follows that the parties' initial "informed consent" to IVF procedures will often not be truly informed because of the near impossibility of anticipating, emotionally and psychologically, all the turns that events may take as the IVF process unfolds. Providing that the initial agreements may later be modified by agreement will, we think, protect the parties against some of the risks they face in this regard. But, in the absence of such agreed modification, we conclude that their prior agreements should be considered binding.

* * *

[In this case,] we are . . . left with this situation: there was initially no agreement between the parties concerning disposition of the preembryos under the circumstances of this case; there has been no agreement since; and there is no formula in the Court of Appeals opinion for determining the outcome if the parties cannot reach an agreement in the future.

* * *

VI. The Right of Procreational Autonomy

Although an understanding of the legal status of preembryos is necessary in order to determine the enforceability of agreements about their disposition, asking whether or not they constitute "property" is not an altogether helpful question. As the appellee points out in his brief, "[as] two or eight cell tiny lumps of complex protein, the embryos have no [intrinsic] value to either party." Their value lies in the "potential to become, after implantation, growth and birth, children." Thus, the essential dispute here is not where or how or how long to store the preembryos, but whether the parties will become parents. The Court of Appeals held in effect that they will become parents if they both agree to become parents. The Court did not say what will happen if they fail to agree. We conclude that the answer to this dilemma turns on the parties' exercise of their constitutional right to privacy.

* * *

[The Court found there was a right to individual privacy in both Federal and Tennessee State law, and that this right to individual privacy encompassed the right of procreational autonomy.]

For the purposes of this litigation it is sufficient to note that, whatever its ultimate constitutional boundaries, the right of procreational autonomy is composed of two rights of equal significance—the right to procreate and the right to avoid procreation. Undoubtedly, both are subject to protections and limitations.

The equivalence of and inherent tension between these two interests are nowhere more evident than in the context of in vitro fertilization. None of the concerns about a woman's bodily integrity that have previously precluded men from controlling abortion decisions is applicable here. We are not unmindful of the fact that the trauma (including both emotional stress and physical discomfort) to which women are subjected in the IVF process is more severe than is the impact of the procedure on men. In this sense, it is fair to say that women contribute more to the IVF process than men. Their experience, however, must be viewed in light of the joys of parenthood that is desired or the relative anguish of a lifetime of unwanted parenthood. As they stand on the brink of potential parenthood, Mary Sue Davis and Junior Lewis Davis must be seen as entirely equivalent gamete-providers.

It is further evident that, however far the protection of procreational autonomy extends, the existence of the right itself dictates that decisional authority rests in the gamete-providers alone, at least to the extent that their decisions have an impact upon their individual reproductive status. . . . [N]o other person or entity has an interest sufficient to permit interference with the gamete-providers' decision to continue or terminate the IVF process, because no one else bears the consequences of these decisions in the way that the gamete-providers do.

Further, at least with respect to Tennessee's public policy and its constitutional right of privacy, the state's interest in potential human life is insufficient to justify an infringement on the gamete-providers' procreational autonomy.

Certainly, if the state's interests do not become sufficiently compelling in the abortion context until the end of the first trimester, after very significant developmental stages have passed, then surely there is no state interest in these preembryos which could suffice to overcome the interests of the gamete-providers. The [Tennessee] abortion statute reveals that the increase in the state's interest is marked by each successive developmental stage such that, toward the end of a pregnancy, this interest is so compelling that abortion is almost strictly forbidden. This scheme supports the conclusion that the state's interest in the potential life embodied by these four-to eight-cell preembryos (which may or may not be able to achieve implantation in a uterine wall and which, if implanted, may or may not begin to develop into fetuses, subject to possible miscarriage) is at best slight. When weighed against the interests of the individuals and the burdens inherent in parenthood, the state's interest in the potential life of these preembryos is not sufficient to justify any infringement upon the freedom of these individuals to make their own decisions as to whether to allow a process to continue that may result in such a dramatic change in their lives as becoming parents. The unique nature of this case requires us to note that the interests of these parties in parenthood are different in scope than the parental interests considered in other cases. Previously, courts have dealt with the childbearing and child-rearing aspects of parenthood. Abortion cases have dealt with gestational parenthood. In this case, the Court must deal with the question of genetic parenthood. We conclude, moreover, that an interest in avoiding genetic parenthood can be significant enough to trigger the protections afforded to all other aspects of parenthood. The technological fact that someone unknown to these parties could gestate these preembryos does not alter the fact that these parties, the gamete-providers, would become parents in that event, at least in the genetic sense. The profound impact this would have on them supports their right to sole decisional authority as to whether the process of attempting to gestate these preembryos should continue. This brings us directly to the question of how to resolve the dispute that arises when one party wishes to continue the IVF process and the other does not.

VII.   Balancing the Parties' Interests

Resolving disputes over conflicting interests of constitutional import is a task familiar to the courts. One way of resolving these disputes is to consider the positions of the parties, the significance of their interests, and the relative burdens that will be imposed by differing resolutions. In this case, the issue centers on the two aspects of procreational autonomy—the right to procreate and the right to avoid procreation. We start by considering the burdens imposed on the parties by solutions that would have the effect of disallowing the exercise of individual procreational autonomy with respect to these particular preembryos.

Beginning with the burden imposed on Junior Davis, we note that the consequences are obvious. Any disposition which results in the gestation of the preembryos would impose unwanted parenthood on him, with all of its possible financial and psychological consequences. The impact that this unwanted parenthood would have on Junior Davis can only be understood by considering his particular circumstances, as revealed in the record.

Junior Davis testified that he was the fifth youngest of six children. When he was five years old, his parents divorced, his mother had a nervous break-down, and he and three of his brothers went to live at a home for boys run by the Lutheran Church. Another brother was taken in by an aunt, and his sister stayed with their mother. From that day forward, he had monthly visits with his mother but saw his father only three more times before he died in 1976. Junior Davis testified that, as a boy, he had severe problems caused by separation from his parents. He said that it was especially hard to leave his mother after each monthly visit. He clearly feels that he has suffered because of his lack of opportunity to establish a relationship with his parents and particularly because of the absence of his father.

In light of his boyhood experiences, Junior Davis is vehemently opposed to fathering a child that would not live with both parents. Regardless of whether he or Mary Sue had custody, he feels that the child's bond with the non-custodial parent would not be satisfactory. He testified very clearly that his concern was for the psychological obstacles a child in such a situation would face, as well as the burdens it would impose on him. Likewise, he is opposed to donation because the recipient couple might divorce, leaving the child (which he definitely would consider his own) in a single-parent setting.

Balanced against Junior Davis's interest in avoiding parenthood is Mary Sue Davis's interest in donating the preembryos to another couple for implantation. Refusal to permit donation of the preembryos would impose on her the burden of knowing that the lengthy IVF procedures she underwent were futile, and that the preembryos to which she contributed genetic material would never become children. While this is not an insubstantial emotional burden, we can only conclude that Mary Sue Davis's interest in donation is not as significant as the interest Junior Davis has in avoiding parenthood. If she were allowed to donate these preembryos, he would face a lifetime of either wondering about his parental status or knowing about his parental status but having no control over it. He testified quite clearly that if these preembryos were brought to term he would fight for custody of his child or children. Donation, if a child came of it, would rob him twice—his procreational autonomy would be defeated and his relationship with his offspring would be prohibited.

The case would be closer if Mary Sue Davis were seeking to use the preembryos herself, but only if she could not achieve parenthood by any other reasonable means. We recognize the trauma that Mary Sue has already experienced and the additional discomfort to which she would be subjected if she opts to attempt IVF again. Still, she would have a reasonable opportunity, through IVF, to try once again to achieve parenthood in all its aspects—genetic, gestational, bearing, and rearing.

Further, we note that if Mary Sue Davis were unable to undergo another round of IVF, or opted not to try, she could still achieve the child-rearing aspects of parenthood through adoption. The fact that she and Junior Davis pursued adoption indicates that, at least at one time, she was willing to forego genetic parenthood and would have been satisfied by the child-rearing aspects of parenthood alone.

VIII.   Conclusion

In summary, we hold that disputes involving the disposition of preembryos produced by in vitro fertilization should be resolved, first, by looking to the preferences of the progenitors. If their wishes cannot be ascertained, or if there is dispute, then their prior agreement concerning disposition should be carried out. If no prior agreement exists, then the relative interests of the parties in using or not using the preembryos must be weighed. Ordinarily, the party wishing to avoid procreation should prevail, assuming that the other party has a reasonable possibility of achieving parenthood by means other than use of the preembryos in question. If no other reasonable alternatives exist, then the argument in favor of using the preembryos to achieve pregnancy should be considered. However, if the party seeking control of the preembryos intends merely to donate them to another couple, the objecting party obviously has the greater interest and should prevail.

But the rule does not contemplate the creation of an automatic veto, and in affirming the judgment of the Court of Appeals, we would not wish to be interpreted as so holding.

For the reasons set out above, the judgment of the Court of Appeals is affirmed, in the appellee's favor. This ruling means that the Knoxville Fertility Clinic is free to follow its normal procedure in dealing with unused preembryos, as long as that procedure is not in conflict with this opinion.

NOTES AND QUESTIONS

1. Davis *Aftermath*. The Knoxville Fertility Clinic could not "follow its normal procedure in dealing with unused pre-embryos" because its normal procedure was to provide those pre-embryos to infertile couples—a procedure that would have been in conflict with the opinion. On rehearing directed to this issue, the Tennessee court looked to the report of the American Fertility Society's Ethics Committee, which authorizes the pre-embryos to be donated for research purposes (which would require the consent of both of the Davises) or to be discarded. See Davis v. Davis, Order on Petitions to Rehear (Sup.Ct.Tenn., Nov. 23, 1992).

2. *The Relevance of Abortion Law.* The court suggests that the state interest in assuring that the embryo is implanted and allowed to develop cannot be sufficiently compelling to overcome a gamete-provider's right to avoid parenthood because, by analogy, "the state's interest does not become sufficiently compelling in the abortion context until the end of the first trimester, after very significant developmental stages have passed . . . ." Is the analogy to the abortion context a persuasive one? In the same year in which this case was decided, the U.S. Supreme Court reconsidered the value of drawing trimester lines and pointed out that the "State has legitimate interests from the outset of the pregnancy in protecting . . . the life of a fetus that may become a child." Planned Parenthood of Southeastern Pennsylvania v. Casey, *supra* Section III.B. Could the interest of the state in protecting the fetus extend back to before a fertilized ovum is even introduced into the mother's body, back to the pre-embryo in the laboratory? As a constitutional matter, could a state have an interest in protecting fertilized ova that would

be sufficiently compelling to overcome the interests of one or both of the gamete providers?

3.   *Property vs. Custody Theories.* The court describes several different rules that could be applied in disputes over the status and control of extracorporeal fertilized ova or pre-embryos. One of the alternatives is to treat the pre-embryos as personal property and apply the common law of property to them; another would be to treat those pre-embryos as if they were children entitled to protection of the state law. Each view has some legal support. See York v. Jones, 717 F. Supp. 421 (E.D. Va. 1989) (treating frozen embryos prepared in the course of infertility treatment as property that would be subject to the ordinary law of bailment). See also La. Rev. Stat. § 9:131 (applying the "best interest of the in vitro fertilized ovum" standard).

4.   *Enforceability of Contract.* Several courts have followed a path close to that taken in *Davis.* See Kass v. Kass, 696 N.E.2d 174 (N.Y. 1998) (enforcing informed consent document declaring that the clinic would use any frozen pre-embryos for research in the event that the progenitors could not agree on their disposition); Roman v. Roman, 193 S.W.3d 40 (Tex. Civ. App. 2006, rev. den. 2007) (enforcing the prior agreement allowing the clinic to destroy the embryos); Dahl v. Angle, 194 P.3d 834 (Or. Ct. App. 2008) (enforcing prior agreement).

Two courts to confront this issue expressly rejected the notion that the prior agreement of the progenitors should always govern the disposition of frozen embryos. See A.Z. v. B.Z., 725 N.E.2d 1051 (Mass. 2000) (refusing to enforce agreement that provided that if the intended parents were to become separated, the frozen embryos then available would be given to one of the parents for implantation, because it forced the former husband to become a parent against his will and was thus contrary to public policy); J.B. v. M.B., 783 A.2d 707 (N.J. 2001) (holding the parties would not normally be bound by a prior IVF agreement that would now require one of those parties to become a parent against that person's will).

Notably, in *Davis, Kass, Roman,* and *Dahl,* the enforceable prior agreements provided for the destruction of frozen embryos rather than their implantation; but in *A.Z.* and *J.B.,* the agreements would have resulted in the creation of a child (and, of course, parents) against one of the parties' wishes. How important might that distinction have been to explain the different outcomes? Compare A.Z. and J.B., supra, with Szafranski v. Dunston, 34 N.E.3d 1132 (Ill. App. Ct. 2015) (enforcing prior agreement allowing woman to use cryopreserved pre-embryos created with ex-boyfriend, despite his desire to avoid becoming a parent). See also, Judith Daar, Litowitz v. Litowitz: Feuding over Frozen Embryos and Forecasting the Future of Reproductive Medicine in Sandra Johnson, et al. (eds), Health Law and Bioethics Cases in Context (2009), describing the backstory of an embryo disposition case that involved a battle over which of several agreements signed by the parties should prevail and providing a catalog of frozen embryo cases.

5.   *Other Relevant Factors?* In the United States, clinics conducting *in vitro* fertilization generally try to remove several eggs so that at least a few will be fertilized *in vitro* and can be replaced in the uterus. In some cases there is no fertilization or inadequate fertilization *in vitro,* and, thus, no return to the uterus, although fertilization occurs in most of the ova in which it is attempted. ART success rates can vary based on treatment factors, such as number of embryos transferred and type of ART procedure, as well as patient characteristics, such as age, infertility diagnosis, and history of prior births or miscarriages. For recent data on ART success rates, see CDC, 2018 Assisted Reproductive Technology: Fertility Clinic Success Rates Report 24 (2020).

Because the chances of success are substantially increased if several fertilized ova are returned to the uterus, and because of the high cost and physical burden of repeating the egg retrieval procedure, women undergoing *in vitro* fertilization are generally given drugs to increase the number of ova that become ripe and ready for release and fertilization in one cycle. While this "superovulation" increases the chances of successful fertilization and implantation, inducing ovulation itself may have adverse side effects. In addition, the simultaneous placement of multiple fertilized ova results in a higher rate of multiple births than in the rest of the population, increasing health risks for the woman and child. Given the difference in the burden of retrieval processes, should the ovum source (who went through an invasive surgical procedure, following sometimes debilitating hormone therapy) have greater say in the disposition of the frozen embryo than the sperm source (who went into the next room with a cup)? Would this approach violate principles of equal protection?

Most of the cases suggest that the result in a frozen embryo disposition case might be different if the party who wished to bring the frozen embryo to life were otherwise infertile, and would have no other way to become a parent. In fact, this seemed to be an important factor in the court's decision in Szafranski v. Dunston, supra Note 4. Justice Verniero, concurring in the *J.B.* case, also described in Note 4, pointed out that "the same principles that compel the outcome in this case would permit an infertile party to assert his or her right to use a preembryo against the objections of the other party, if such use were the only means of procreation. In that instance, the balance arguably would weigh in favor of the infertile party absent countervailing factors of greater weight." What might count as "countervailing factors of greater weight"? The *J.B.* majority didn't seem convinced the result would be different if the one seeking to use the embryo were otherwise infertile, saying, "[w]e express no opinion in respect of a case in which a party who has become infertile seeks use of stored preembryos against the wishes of his or her partner, noting only that the possibility of adoption also may be a consideration, among others, in the court's assessment." J.B., supra, at 720.

6.   *Poor Quality Care & Other Harms.* Reproductive technologies, such as *in vitro* fertilization and embryo cryopreservation, can give rise to a host of different kinds of liability. Of course, the physician and medical institution would be liable as in an ordinary medical malpractice case for negligently

performing the medical procedure. Would the duty of due care extend to the fetus as well as the parents? Negligently freezing an ovum or embryo, or negligently thawing it and thus destroying or damaging it, would also subject the appropriate professionals to malpractice liability. For an example of a case based on both intentional and negligent infliction of emotional distress against a provider who destroyed tissue that was cryogenically frozen to allow for later conception, see Witt v. Yale-New Haven Hosp, 977 A.2d 779 (Conn. Super. Ct. 2008). But see Miller v. American Fertility Group of Illinois, 897 N.E.2d 837 (Ill. App. Ct. 2009) (no wrongful death action allowed on behalf of mistakenly destroyed pre-embryo before its implantation in the mother).

The value of cryogenically preserved fertilized pre-embryos (or ova) may also make them targets for theft; clinic employees (or others) may be tempted to provide excess genetic material to those who need it without seeking consent from the original progenitors. Indeed, during the late 1990s that kind of redirection of ova apparently became fairly common at the University of California-Irvine's fertility clinic, where at least fifteen children were born as a result of the use of ova from women who had not consented to their transfer. It also creates the opportunity for mistakes to occur. Consider the case of Prato-Morrison v. Doe, 103 Cal.App.4th 222 (2002). Donna Prato-Morrison and Robert Morrison engaged the services of a fertility clinic without success. When the clinic later became the target of an investigation into its widespread misuse of genetic materials, the Morrisons questioned whether another family who used the fertility clinic (Judith and Jacob Doe) innocently received Donna Morrison's genetic material. The Morrisons filed a complaint in which they asked the court to determine whether they were the genetic parents of the Does' twin daughters (at the time, almost 14 years old); they sought to compel blood tests and obtain the right to visit the twins (who knew nothing about the claim). In rejecting the Morrisons' claims, the court held that plaintiffs failed to introduce admissible evidence that defendant wife had received eggs supplied by plaintiff wife, and that plaintiffs' rights as alleged biological parents did not overcome defendants' rights as presumed parents, so as to entitle plaintiffs to discover whether the twins were born as a result of the theft of their genetic materials. Finally, the court went even further to hold that even had plaintiffs presented proof of a genetic link to the twins sufficient to establish their standing to pursue a parentage action, it would not be in the best interests of the twins to have plaintiffs intrude into their lives, or to be subjected to the blood tests and mental health evaluation suggested by plaintiffs.

But in Robert B. v. Susan B., 109 Cal.App.4th 1109 (2003), the husband of a couple subject to an unintentional mix-up was able to establish parentage. In that case, a husband and wife brought a parentage action against a single woman who had given birth to the genetic brother of a child born to the couple, as the result of the placement of the wrong embryo in the single woman's uterus by a fertility clinic. The husband and wife had sought ova from an anonymous donor for fertilization with the husband's sperm, and the single woman had sought an embryo created from anonymously donated ova and sperm; but after receiving the results of genetic testing of the husband, the single woman learned that she had received ova from an anonymous donor fertilized by the husband's sperm. The Court of Appeal affirmed the trial court's determination that the husband was the child's father, but that the single woman was the child's mother. The husband's wife had no standing because she was biologically unrelated to the child.

7.   *Selective Reduction.* One consequence of the implantation of several embryos fertilized *in vitro* is that many of the embryos may continue to develop *in vivo*. While most multiple births pose no substantial problems to the mother or the neonatal siblings, sometimes the life or health of the mother (or some of the developing fetuses) can be preserved only if the number of fetuses that the mother is carrying is reduced. Is the case of such selective reduction governed by the more general law applying to abortion, described earlier in the chapter? Are there any ethical or legal concerns present in cases of selective reduction that are not present in other situations that might give rise to abortion? What legal or ethical considerations, if any, should guide the choice of which of two or more fetuses should be aborted to save the life or health of the mother or the well-being of the potential siblings?

8.   *Genetic Screening.* One consequence of the ability to freeze an embryo is that it permits evaluation of the genetic structure of the embryo before it is thawed and implanted. This allows for pre-implantation genetic screening or diagnosis (PIGD), which allows for a determination of several genetic attributes of the pre-embryo. This screening (for a wide range of conditions or attributes) or diagnosis (for a particular genetic disease) is increasingly available, as a practical matter, for more conditions, and for a lower cost. For an exploration of the social and ethical questions raised by such advances, see Jaime S. King, And Genetic Testing for All . . . The Coming Revolution in Non-Invasive Prenatal Genetic Testing, 42 Rutgers L.J. 599 (2011), and Erik Parens & Adrienne Asch, The Disability Rights Critique of Prenatal Genetic Testing: Reflections and Recommendations, 29 The Hastings Ctr Rpt S1 (1999).

9.   *ART Regulation.* Other than questions of parentage, ART is still relatively unregulated. Why are state legislatures, many of which have been active in restricting access to abortion, not banning IVF, especially to the extent that it may involve selective reduction? Why might state legislatures and the federal government be less interested in setting standards for care in this area as compared to other areas, such as nursing home care?

D. REPRODUCTIVE CLONING AND STEM CELL RESEARCH

1. Reproductive Cloning

In 1997 the world was stunned by the revelation that a Scottish laboratory had cloned a mammal—to be precise, a Finn Dorset sheep. Scientists achieved this by removing the nucleus from a sheep ovum, replacing it with the genetic material from the sheep to be cloned, and then reimplanting the genetically changed ovum back into the surrogate mother sheep. In fact, the one cloned sheep was the only success in almost 300 attempts. Other mammals were cloned using the same process, but scientists warned that the likelihood that this technology would be available for human beings remained years away. Despite this, the bioethics community began a world-wide discussion on the consequences of cloning as a form of human reproduction.

Is it possible to talk about the creation of human beings through cloning in the same terms as we talk about the creation of human beings through other processes of assisted reproduction? How are the questions raised by the existence of cloned human beings truly different from the questions raised by human beings created by other technically driven medical processes? Cloning requires in vitro manipulation (like in vitro fertilization) and then implantation in a womb (like surrogacy). Is it different from either of those two forms of assisted reproduction? Is cloning really a form of reproduction at all, or is it something different—perhaps, merely replication? Who counts as the parents of the cloned individual? Does it make sense to speak of a "parent" or must we limit ourselves to speaking of the cloned person's "antecedent" or "precursor" or "genetic source"?

For generations, science fiction has raised fears of totalitarian regimes that can clone human beings for enslavement for the benefit of the regime. Should this present a genuine worry to this society? Could cloning be appropriated by the state in support of a totalitarian eugenics policy? A related worry is that a society could treat a newly cloned person as nothing more than the continuation of the source of its genetic material—or as a body parts source for the person who provided the genetic material. But wouldn't a cloned person have all of the rights of any other person? As some bioethicists argue, if our society were to treat cloned human beings as enslaved persons or spare parts warehouses, the existence of cloning would be the least of our problems.

Would cloned people have the same relationship to each other that identical twins maintain? Identical twins have a common genetic make-up, as a clone and his source would, yet we treat those twins as separate human beings with separate lives and we have identified no social dislocation that arises out of their existence. Why should we be more worried about cloned human beings than about identical twins? In fact, because a cloned person and his source would be different ages, they would be likely to be raised in different environments, and thus to be less similar than identical twins. On the other hand, the number of identical twins is naturally limited, and there are very few identical triplets. We could, through cloning, create a much larger number of identical "siblings."

In any case, is there any justifiable reason for cloning a full human being? A few have been suggested. Suppose a child were dying of a disease that could be treated with a bone marrow transplant from a genetically identical donor, but such a donor could not be found. Might it be acceptable to clone the patient, and then, independently, determine if the newly replicated person ought to be a donor for his "brother"? (See the discussion of "savior siblings" in Note on Organ Transplantation from Living Donors at Chapter 4, Section I.C.1.) Alternatively, suppose a couple has an infertility problem which makes it impossible for either of them to become a parent in the ordinary course of events. Should they be able to opt for cloning so that their child will be related to one of them? What if only one of the parents is infertile, and they want a child that does not require the potentially interfering involvement of a third party?

The issue has become more complicated because human reproduction is not the only goal of cloning. Cloning of human stem cells may also be useful in developing therapies for medical conditions that are now untreatable. Many view cloning with the intent of creating a new human being—*reproductive cloning*—as posing far more ethical and legal difficulty than cloning human stem cells with the intent to research or create a medical treatment—*therapeutic cloning*.

Both forms of cloning generally require somatic cell nuclear transfer, a process in which the nucleus of an unfertilized human ovum is removed and replaced with the nucleus from an adult cell. The resulting ovum cell, which has the full complement of human genetic material, is then stimulated so that it will divide and form a pre-embryo. In *therapeutic cloning*, some cells from the pre-embryo are removed for purposes of medical research or treatment. These stem cells have the potential to develop into almost any human cells, and they may be useful in repairing almost any human tissue, but once those cells are removed, the pre-embryo cannot develop into a human being. In *reproductive cloning*, the pre-embryo would be placed in a uterus to develop into a human being.

The National Bioethics Advisory Commission published a comprehensive and thoughtful report on the cloning of human beings within months of the revelation that it had been successfully performed on sheep. The report includes a useful description of the relevant scientific procedures as well as a fully annotated analysis of the ethical, theological, legal and policy issues, and it was careful to distinguish between reproductive cloning (which it considered medically risky) and therapeutic cloning (which, it thought, held great scientific promise).

RECOMMENDATIONS OF THE NATIONAL BIOETHICS ADVISORY COMMISSION (NBAC) WITH REGARD TO CLONING

(1997).

* * *

As somatic cell nuclear transfer cloning could represent a means of human reproduction for some people, limitations on that choice must be made only when the societal benefits of prohibition clearly outweigh the value of maintaining the private nature of such highly personal decisions. Especially in light of some arguably compelling cases for attempting to clone a human being using somatic cell nuclear transfer, the ethics of policy making must strike a balance between the values society wishes to reflect and issues of privacy and the freedom of individual choice.

* * *

. . . . At present, the use of this technique to create a child would be a premature experiment that exposes the developing child to unacceptable risks. This in itself is sufficient to justify a prohibition on cloning human beings at this time, even if such efforts were to be characterized as the exercise of a fundamental right to attempt to procreate. More speculative psychological harms to the child, and effects on the moral, religious, and cultural values of society may be enough to justify continued prohibitions in the future, but more time is needed for discussion and evaluation of these concerns.

* * *

Within this overall framework the Commission came to the following conclusions and recommendations:

I. The Commission concludes that at this time it is morally unacceptable for anyone in the public or private sector, whether in a research or clinical setting, to attempt to create a child using somatic cell nuclear transfer cloning. We have reached a consensus on this point because current scientific information indicates that this technique is not safe to use in humans at this time. . . .

II. Federal legislation should be enacted to prohibit anyone from attempting, whether in a research or clinical setting, to create a child through somatic cell nuclear transfer cloning. It is critical, however, that such legislation include a sunset clause to ensure that Congress will review the issue after a specified time period (three to five years) in order to decide whether the prohibition continues to be needed. . . .

III. Any regulatory or legislative actions undertaken to effect the foregoing prohibition on creating a child by somatic cell nuclear transfer should be carefully written so as not to interfere with other important areas of scientific research. In particular, no new regulations are required regarding the cloning of human DNA sequences and cell lines, since neither activity raises the scientific and ethical issues that arise from the attempt to create children through somatic cell nuclear transfer, and these fields of research have already provided important scientific and biomedical advances.

* * *

This was not the last federal bioethics panel to consider the issue. The President's Council on Bioethics (which effectively replaced NBAC, and then itself was subsequently replaced by the Presidential Commission for the Study of Bioethical Issues) issued a series of reports, including Human Cloning and Human Dignity (2002), Monitoring Stem Cell Research (2004), and Reproduction and Responsibility (2004), which supported a ban on reproductive cloning and both a moratorium and then careful regulation of all other cloning. Although disagreements between those who opposed reproductive cloning and those who supported therapeutic cloning (and disagreements among those on each side, too) made it difficult for Congress to legislate with regard to reproductive cloning, many state legislatures have done so. Several have outlawed reproductive cloning, some have outlawed cloning of any kind, and some have outlawed reproductive cloning while at the same time authorizing state funding for stem cell research (which is thought to be both scientifically and economically promising as described further below). See National Conference of State Legislatures (NCSL), Embryonic and Fetal Research Laws, at ncsl.org (last updated Jan. 1, 2016).

2.   Stem Cell Research

Because of the scientific promise shown by stem cell research, many people who support legislation banning reproductive cloning vigorously oppose placing any limitation upon therapeutic cloning believed necessary for effective stem cell research. The ethical issues which have given rise to the legal argument are a bit more subtle than one might expect.

JUDITH JOHNSON AND ERIN WILLIAMS, CRS REPORT FOR CONGRESS: STEM CELL RESEARCH (2006)

* * *

Stem cell research is controversial not because of its goals, but rather because of the means of obtaining some of the cells. Research involving most types of stem cells, such as those derived from adult tissues and umbilical cord blood, is uncontroversial, except when its effectiveness as an alternative to embryonic stem cells is debated. The crux of the debate centers around embryonic stem cells, which enable research that may facilitate the development of medical treatments and cures, but which require the destruction of an embryo to derive. In addition, because cloning is one method of producing embryos for research, the ethical issues surrounding cloning are also relevant.

* * *

Most positions on embryonic stem cell research rest at least in part on the relative moral weight accorded to embryos and that accorded to the prospect of saving, prolonging, or improving others' lives. . . .

Some groups explore the moral standing of human embryos, and also consider the "duty to relieve the pain and suffering of others." Others take the position that embryos do not have the same moral status as persons. They acknowledge that embryos are genetically human, but hold that they do not have the same moral relevance because they lack specific capacities, including consciousness, reasoning and sentience. They also argue that viewing embryos as persons would "rule out all fertility treatments that involve the creation and discarding of excess embryos," and further assert that we do not have the same "moral or religious" response to the natural loss of embryos (through miscarriage) that we do to the death of infants. . . . They conclude that performing research to benefit persons justifies the destruction of embryos. Acceptance of the notion that the destruction of embryos can be justified in some circumstances forms the basis of pro-stem cell research opinions, and is usually modified with some combination of the distinctions and limitations that follow.

* * *

Some proponents of embryonic stem cell research base their support on the question of whether an embryo is viable. The relevance of the viability distinction rests on the premise that it is morally preferable for embryos that will not grow or develop beyond a certain stage and/or those that would otherwise be discarded to be used for the purpose of alleviating human suffering.

* * *

A separate distinction that often leads to the same conclusions as viability is the purpose for which embryos are created. This distinction draws an ethical line based upon the intent of the people creating embryos. In the view of some, it is permissible to create an embryo for reproductive purposes (such as IVF), but impermissible to create one with the intention of destroying it for research.

* * *

A further distinction has been drawn based upon the timing of the creation of embryonic stem cell lines. Here, the premise is that it is unacceptable to induce the destruction of embryos for the creation of new lines. However, in cases in which embryos have already been destroyed and the lines already exist, it is morally preferable to use those lines for research to improve the human condition. This was one central distinction drawn in the 2001 Bush policy, which limited the use of federal funding to research on lines derived on or before the date of the policy. Supporters of the Bush policy on both sides of the issue favor this distinction as a compromise.

* * *

One factual distinction that has been used to support competing ethical viewpoints is the efficacy of alternatives to embryonic stem cell research. The promise of stem cell therapies derived from adult tissue and umbilical cord blood have buttressed opposition to embryonic stem cell research. . . . These opponents argue that therapies and cures can be developed without the morally undesirable destruction of embryos. However, not all scientists agree that adult stem cells hold as much potential as embryonic stem cells. Most supporters of embryonic stem cell research believe that it is the quickest and, perhaps in some cases, the only path that will yield results. . . .

* * *

Some division over the support for and opposition to embryonic stem cell research focuses on the question of whether the use of federal funding is appropriate. Those who oppose federal funding argue that the government should not be associated with embryo destruction. They point out that embryo destruction violates the "deeply held moral beliefs of some citizens," and suggest that "funding alternative research is morally preferable." Proponents of federal funding argue that it is immoral to discourage life-saving research by withholding federal funding. They point out that consensus support is not required for many federal spending policies, as it "does not violate democratic principles or infringe on the rights of dissent of those in the minority." They argue that the efforts of both federally supported and privately supported researchers are necessary to keep the United States at the forefront of what they believe is a very important, cutting edge area of science. Furthermore, supporters believe that the oversight that comes with federal dollars will result in better and more ethically controlled research in the field.

* * *

42 U.S.C. § 289g

(Dickey-Wicker Amendment, 2006).

(a) None of the funds made available [in appropriations for the Departments of Labor, Education, and Health and Human Services] may be used for—

(1) the creation of a human embryo or embryos for research purposes; or

(2) research in which a human embryo or embryos are destroyed, discarded, or knowingly subjected to risk of injury or death greater than that allowed for research on fetuses in utero . . . .

(b) For purposes of this section, the term "human embryo or embryos" includes any organism, not protected as a human subject . . . that is derived by fertilization, parthenogenesis, cloning, or any other means from one or more human gametes or human diploid cells.

When the first Dickey-Wicker Amendment was passed, scientists had no access to stem cells derived from fertilized ova ("embryonic stem cells"), but in 1998 a laboratory at the University of Wisconsin stabilized such stem cells so that they could be used in research. In 1999 the Department of Health and Human Services determined that the Dickey-Wicker Amendment did not apply to embryonic stem cell research because the stem cells were not embryos under the statute. Despite this, no federal research grants were made for stem cell research during the Clinton administration. In 2001, President Bush issued an executive order providing that federal funding could be made available for research involving embryonic stem cell lines that were already available for research, but that no new stem cell lines could be used in federally-funded research. He reasoned that the damage had already been done to the embryos that were destroyed to create the stem cell lines that were already available, but that federal funding agencies should act to assure that no new embryos would be destroyed for use in federally funded research.

In 2009, President Obama greatly expanded the class of embryonic stem cells that could be used in federally-funded research. He determined that the Dickey-Wicker Amendment permitted the use of embryonic stem cells in federally funded research, but that it did not permit federal funds to be used in deriving the stem cell from the embryo. That is, private resources would have to be expended to obtain the embryonic stem cells, an act which requires destroying the embryo, but federal funds could be used for the subsequent research upon the stem cells derived from the embryo. The Obama regulations provided that the stem cells used in federally funded research could only come from leftover fertilized ova created by infertility laboratories for patients engaged in infertility treatment, and only when the progenitors approved of the use of their cells for this purpose in writing. See 74 Fed.Reg. 32, 170–32, 175 (July 7, 2009). Two scientists tried, unsuccessfully, to challenge these regulations as a violation of the Dickey-Wicker Amendment. Sherley v. Sebelius, 689 F.3d 776 (D.C. Cir. 2012), cert. den. 568 U.S. 1087 (2013). Near the end of the Obama administration, the 21st Century Cures Act was signed into law, authorizing $30 million for regenerative medicine research but limiting its application to research with adult stem cells. Since then, the federal approach to stem cell research has continued to shift with the political tides: the Trump administration undertook several steps to undermine much of the progress made during the Obama administration, and the election of President Biden has renewed hope for a return to a more supportive approach.

State legislatures have produced far more legislation on the use of stem cells than has Congress. Some states have sought to outlaw all cloning, and thus most stem cell research, while others have encouraged stem cell research, including embryonic stem cell research, subject to certain safeguards. For example, California, which is one of a dozen states to outlaw reproductive cloning, is also one of several states enacting laws to facilitate stem cell research for therapeutic purposes. Through a ballot initiative formally called the Stem Cell Research and Cures Act, California established the California Institute of Regenerative Medicine (CIRM), originally funded by the authorization of $3 billion in state bonds, to stimulate the scientific and economic benefits many believe will follow in the wake of this research. CIRM is a state agency authorized to make grants and loans for stem cell research and facilities, and to establish related regulatory standards and oversight bodies. States encouraging stem cell research have included important protections against abuses, such as opposing reproductive cloning and the sale of embryos, and ensuring fully informed consent for the donation of gametes or embryos. States have also created an additional layer of specialized oversight, on top of the normal oversight required under federal human research subject regulation. (See Chapter 7 for more on research regulation generally). For a thoughtful overview of the various state approaches to stem cell research and important lessons to be gleaned, see Nefi D. Acosta & Sidney H. Golub, The New Federalism: State Policies Regarding Embryonic Stem Cell Research, 44 J. L. Med. & Ethics 419 (2016).

CHAPTER 4

ORGAN TRANSPLANTATION AND THE DETERMINATION OF DEATH

■ ■ ■

I. ORGAN TRANSPLANTATION

A. RATIONING SCARCE ORGANS

As with other limited or shared health care resources, decision-making regarding the retrieval and allocation of scarce human organs brings issues of equity and justice to the forefront. The COVID-19 pandemic brought renewed attention to the equitable allocation of limited medical resources such as ventilators and hospital beds, as well as the potential for allocation decisions to exacerbate pre-existing health inequities. Questions about equitable access to health care resources also figure prominently in debates over the Affordable Care Act (ACA). How can these decisions be made fairly and equitably? Is it ethical, or good public policy, to ration medical care by ability to pay? And, in the case of scarce resources such as human organs or certain technologies, if not by ability to pay, then how?

Many argue that health care is distinct from ordinary consumer goods—such as iPads and cars—that we distribute entirely by ability to pay. The argument that health care is distinct rests on the dignity of persons (and what is required for human flourishing); the moral duty to relieve human suffering; and the communal benefits of fostering good health (in terms of public health and a robust economy). Counterarguments in favor of treating health care, even life-saving medical treatment, as an ordinary consumer good have been prominent during the debates over the ACA. Some argue that placing responsibility on the individual to pay for medical care incentivizes work, savings, and fitness, whereas guaranteeing health care rewards irresponsibility.

Most arguments in favor of providing health care to those who cannot pay reference a "basic" package of health care services. What are the relevant considerations in deciding whether particular medical treatments, such as organ transplants, fit within the basic package? Short-term or long-term survival rates? Quality of life post-transplant? And if so, how should quality be measured? Global budgets that trade off some health care interventions (e.g., transplants vs. preventive care) in order to spread limited funds most effectively? Compare Gil Siegal & Richard Bonnie,

Closing the Organ Gap: A Reciprocity-based Social Contract Approach, 34 J. L. Med. & Ethics 415 (2006), calling organ transplantation a "remarkable achievement" that saves lives, improves quality of life, and reduces costs (for kidney transplants as compared to hemodialysis, for example) with Albert R. Jonsen, The God Squad and the Origins of Transplantation Ethics and Policy, 35 J. L. Med. & Ethics 238 (2007), arguing that the looming problem behind transplantation policy is the "incessant . . . demand that life must be salvaged at all costs."

What other factors should be considered? The supply of transplantable organs simply does not satisfy the need, and rationing follows. A rather famous example of explicit rationing occurred in the 1960s when the Admissions and Policy Committee (sometimes known as the "God Squad") of the Seattle Artificial Kidney Center decided which patients would receive the lifesaving "artificial kidney" (i.e., hemodialysis). The God Squad was reported to have selected patients partly on the basis of age, gender, marital status, number of dependents, emotional stability, education, and occupation, among other characteristics. Media exposure of the work of the God Squad contributed to the decision to have Medicare cover dialysis for all individuals with end-stage renal disease (ESRD), even those not otherwise eligible for Medicare; this decision was followed by one to have Medicare cover kidney transplants as well. Richard A. Rettig, Special Treatment—The Story of Medicare's ESRD Entitlement, 364 NEJM 596 (2011).

Of course, efforts have been made to increase the supply of transplantable organs, within limits. The use of brain death as the medical and legal standard for determination of death originally emerged in part in response to the need for transplantable human organs. More recent developments, i.e., donation after cardiac or circulatory death, discussed at Section II.C, further illustrate how policy decisions and legal norms can produce direct consequences for the availability and distribution of health resources, even where those norms appear unrelated to such concerns.

Ability to pay actually is one factor that determines who gets an organ transplant. A candidate generally is not placed on the active waiting list unless there is a source of payment for the transplant and the expensive lifelong post-transplant care. Even after the ACA, private insurance, Medicaid, and Medicare will not cover all medically possible transplants or necessary aftercare.

While ability to pay is the gateway to being listed for a transplant, one cannot buy an organ from a willing seller. Instead, we use other criteria to ration scarce organs. Consider the following problem.

PROBLEM: SELECTING AN ORGAN TRANSPLANT RECIPIENT

For the last few weeks, two patients have been under treatment at General Hospital for acute liver failure. One, James Patterson, is a 65-year-old retired CEO of a major computer software company. He has two children, a 40-year-old daughter with a child of her own and a 24-year-old son. Patterson is active in his church and is a financial supporter of the local university athletics department. He was exposed to hepatitis B, which has destroyed most of his liver. The second patient is Antonia Friedman, a 30-year-old attorney with two children, ages 2 and 4. She is an active member of the local city council and has contributed generously in the past year to the hospital's building fund. Friedman is in recovery from alcohol use disorder and has been through rehabilitation programs several times. This time, though, she has abstained from alcohol for six months and so meets common minimum criteria for liver transplant applied to individuals with alcohol use disorder. Within the last week, both patients have taken a turn for the worse, and both will die within the next few weeks if they do not receive a liver transplant.

A few hours ago, a patient was admitted to the hospital with massive head trauma caused by an automobile accident. The patient is brain dead but is being kept on support systems to preserve his organs for transplantation. The liver is undamaged. Tissue matching shows that it is an equally acceptable organ for either Patterson or Friedman. Circumstances make it impossible for the liver to be divided between the two patients. Patterson's physician was the first to list his patient on the transplant list, which he did three days earlier than did Friedman's physician. Who should receive the single available organ?

Is this decision essentially a medical decision to be made by the physician? Would you prefer that a committee decide? Who should be on the committee? Should your hospital establish a procedure in which patients are notified that listing is being considered and given an opportunity to advocate for themselves? Should the hospital keep a record documenting the reasons a patient wasn't listed or selected? Should the government set the criteria for deciding who gets the organ? See Section I.B, below.

Would you prefer to use a lottery system in which a recipient is selected from among all qualified candidates by drawing lots? What advantages and disadvantages would a pure "first come-first served" policy present? Should time on the waiting list make any difference? What if you knew that Patterson's doctor was more aggressive in listing patients than was Friedman's? What if you knew that other doctors at that hospital simply refuse to list individuals in recovery from alcohol use disorder for a transplant? Is considering Friedman's alcohol use disorder inappropriate or is it morally required? See, e.g., Robert S. Brown, Transplantation for Alcoholic Hepatitis—Time to Rethink the 6-Month "Rule," 365 NEJM 1836 (2011), discussing a study demonstrating the effectiveness of transplantation, in terms of survival and post-transplant abstinence from alcohol, even without a 6-month pre-transplant abstinence period; Erin Minelli & Bryan Liang, Transplant

Candidates and Substance Use: Adopting Rational Health Policy for Resource Allocation, 44 U. Mich. J. L. Reform 667 (2011).

How important is the risk of graft failure, i.e., that the organ itself will fail due to rejection? Will that consideration always outweigh other concerns? Consider that under current national kidney allocation policies, patients who have previously received a kidney that failed within a short time after transplant receive priority even though the presensitization caused by the earlier transplant increases the risk of graft failure. In addition, a person who has donated specific organs as a living donor receives priority in receiving an organ should the need arise in the future.

Projected lifespan of the patient is not the same as the risk of graft failure. Should younger patients always have precedence over older patients in the same medical situation? See Alan B. Leichtman et al., Improving the Allocation System for Deceased-Donor Kidneys, 364 NEJM 1287 (2011); Benjamin E. Hippen et al., Risk, Prognosis, and Unintended Consequences in Kidney Allocation, 364 NEJM 1285 (2011). Aside from age, are there any characteristics that are more important than others? Or more suspect? Should the post-transplant coverage provided by each patient's insurance or, alternatively, the net worth of each patient be considered?

B. REGULATING ORGAN DISTRIBUTION

The National Organ Transplant Act of 1984 (NOTA) requires the Department of Health and Human Services (HHS) to establish an Organ Procurement and Transplantation Network (OPTN) to organize the retrieval, distribution, and transplantation of human organs. OPTN is the primary vehicle for federal regulation of the organ transplant system in the U.S. (42 C.F.R. § 121). The Centers for Medicare & Medicaid Services also sets quality and policy standards as conditions for federal payments to transplant hospitals and other facilities involved in organ retrieval and transplantation.

HHS contracts with the United Network for Organ Sharing (UNOS), a private nonprofit organization, for management of the federal OPTN. OPTN/UNOS manages procedures and standards for listing patients, retrieval and testing of organs, distribution to transplant hospitals, and selection of recipients, among other issues. OPTN/UNOS policies are being constantly revised and can be found on the OPTN and UNOS websites.

Access to the Transplant Waiting List

Individual doctors and hospitals decide who gets on the transplant waiting list and when. Patient listing practices vary substantially among transplant centers around the country. Psychosocial factors, including anticipated compliance with medical recommendations, absence of a reliable support system, availability of alternative therapies, and the cause of the original organ failure influence physician listing decisions. See

Denise M. Dudzinski, Shifting to Other Justice Issues: Examining Listing Practices, 4 Am. J. Bioethics 35 (2004) (reviewing studies of listing practices that conclude that standards and procedures tend to be informal and disadvantage racial minorities and women); Geert Verleden et al., Recipient Selection Processes and Listing for Lung Transplantation, 9 J. Thoracic Disease 3372 (2017); C.K. Argo et al., Regional Variability in Symptom-Based MELD Exceptions: A Response to Organ Shortage? 11 Am J. Transplantation 2353 (2011). For a compelling narrative of the effect of race on listing decisions, see Vanessa Grubbs, Good for Harvest, Bad for Planting, 26 Health Aff. 232 (2007).

Assumptions, biases, or lack of knowledge about living with disability compromise access to the transplant waiting list for people with disabilities. A 2019 report found that disabilities unrelated to a person's need for an organ transplant generally have little or no impact on the likelihood that the transplant will be successful. However, many organ transplant centers maintain policies that categorially exclude or disfavor placing people with HIV, psychiatric disabilities, or intellectual or developmental disabilities on the organ transplant waiting list. Nat'l Council on Disability, Organ Transplant Discrimination Against People with Disabilities (2019). Assumptions that the lives of people with disabilities are of poorer quality than those of people without disabilities, or that people with certain disabilities may be unable to comply with post-operative care, also compromise evaluation of psychosocial factors. Id. Recent federal enforcement activity may suggest an increased focus on disability antidiscrimination laws in this context. See, e.g., Dept. of Justice, Massachusetts General Hospital Enters Agreement with U.S. Attorney's Office to Better Ensure Equal Access for Individuals with Disabilities (Aug. 7, 2020); Office of Civil Rights, U.S. Dept. of Health and Human Servs., OCR Resolved Disability Complaint of Individual Who Was Denied the Opportunity for Health Transplant List Placement (Feb. 12, 2019). See also Chapter 2 for discussion of federal laws that prohibit discrimination based on disability.

Hospitals doing transplants have an incentive to engage in listing practices that increase volume. Federal investigations revealed that physicians at some transplant hospitals exaggerated their patients' medical condition to move them up higher on the list so the hospital would be able to perform more transplants. The charged hospitals settled with the government without admitting guilt, with one facility paying $1 million. Nara Schoenberg, A Man of Principle, Chicago Trib. (Jan. 25, 2004). Several studies have found that transplant centers located in high-competition areas listed a higher percentage of patients for transplant than areas with less competition. See, e.g., P.S. Cho et al., Competitive Market Analysis of Transplant Centers and Discrepancy of Wait-Listing of Recipients for Kidney Transplantation, 6 Int'l J. Organ Transplant Med.

141 (2015); E. Haavi Morreim, Another Kind of End-run: Status Upgrades, 5 Am. J. Bioethics 11 (2005).

Patients seeking transplants have an incentive to seek out the list with the shortest waiting time. In fact, median waiting times for particular organs have varied significantly among geographic regions despite a federal regulation requiring that organs be distributed "over as broad a geographic region as feasible" consistent with assuring that organs are not damaged. The reasons for these variations are many: all organs are limited as to the distance they can safely be transported, so distribution favors closer organs over more distant organs, and some organs (such as hearts) are particularly vulnerable in transport; some regions have higher rates of organ donation while others have higher rates of medical need; and some localities have hospitals with more aggressive listing practices, thereby drawing in a higher share of available organs. David L. Weimer, Public and Private Regulation of Organ Transplantation: Liver Allocation and the Final Rule, 32 J. Health Polit. Pol'y & L. 9 (2007). Another potential cause of persisting regional variation may be that transplant centers in different regions vary in their assessment of and requests for symptom-based exceptions (SBE) to move patients up in line. See Argo et al., above, pointing out that Medicaid patients are least likely to receive SBEs. It is not surprising that efforts to reduce geographic disparities have provoked debate. See Lara C. Pullen, Lawsuits Drive Transplant Community Debate Over Liver Allocation, 19(5) Am. J. Transplantation 1251 (Apr. 24, 2019).

The persistent difference in regional wait times for organs encourages the practice of "multiple listing," in which patients with the means to travel at a moment's notice get listed in more than one region. Several websites provide transplant candidates with the wait time for specific organs in specific localities, along with the travel time required to get there from their home. OPTN policy allows multiple listing, but the AMA holds that physicians should "refrain" from multiple listing and instead place candidates on a "single waiting list for each type of organ." AMA Code of Medical Ethics Opinion E–6.2.1(g).

Selecting the Recipient

Federal regulations require that OPTN organ allocation policies be "equitable," "seek to achieve the best use of donated organs," "avoid wasting organs," and rely on "objective and measurable medical criteria." 42 C.F.R. § 121.8. Indeed, OPTN policies governing selection of a recipient from among listed patients use "objective and measurable medical criteria," quantifying medical urgency, measuring how good a match the organ and recipient are, and so on. But these apparently scientific criteria have measurable effects on access to organs for specific populations (minority, ethnic, age) and most obviously for specific individuals. Kidney allocation policies provide an illustration of apparently neutral criteria that have had an adverse racial impact. For years, UNOS kidney allocation policies used very granular levels of "antigen matching" to select recipients from the transplant list even though the difference in graft failure was miniscule as between each level. Before UNOS revised these criteria, whites received double the number of kidney transplants as compared to African Americans. The two populations were equally represented on the transplant waiting list (even though African Americans suffer a higher rate of total kidney failure) but in raw numbers more whites donated organs. OPTN, Annual Report 2003. Revised matching criteria have significantly increased the proportion of organs allocated to African Americans. A. Hart et al., OPTN/SRTR 2015 Annual Data Report: Early Effects of the New Kidney Allocation System, 17 Am. J. Transplantation 543 (2017).

Outside claims of discrimination, should courts get involved in deciding whether an individual patient receives an available organ? In Murnaghan v. HHS, Civ. Action No. 13-3083 (E.D. Pa. Jan. 5 and 13, 2013), the judge enjoined application of the OPTN "Under 12 Rule," so that Sarah Murnaghan, a 10-year-old girl facing imminent death, could have access to an adult lung rather than waiting for a pediatric lung. OPTN policy at the time allocated pediatric lungs to children and adult lungs to individuals over the age of 12 unless no appropriate adult recipient was on the list, in which case the lungs would be allocated to a younger child. The judge's order called the policy discriminatory, arbitrary, and capricious. The order required OPTN to consider Sarah as a candidate without applying the "Under 12 Rule."

The Secretary of HHS refused to set aside the policy on her own for Sarah's case, commenting that making a decision on "who lives and who dies" without following policies that applied to all was inappropriate. She noted that 40 other children in Pennsylvania were awaiting lung transplants at that time. Paige Cunningham, Sibelius Won't Intervene on Transplant, Politico (June 5, 2013). See also Daniel Bruggebrew, The Administrative Law Implications of Quasi-Governmental Organ Allocation, 41 Seton Hall Legis. J. 1 (2016). Sarah's first lung transplant, received within several days of the judge's order, failed, but a second transplant has sustained her since that time.

C. LEGAL FRAMEWORK FOR THE PROCUREMENT OF ORGANS AND TISSUE

NOTA defers to state law on most questions related to organ donation, and most states have adopted the Uniform Anatomical Gift Act (UAGA), with some few variations, to govern donation of organs that will be retrieved post-mortem. Donation of non-life-sustaining organs by living donors is governed by familiar informed-consent principles and by policies issued by OPTN. Legal restraints on the collection of human organs for transplantation contribute to the current shortage because they require that there be consent for organ retrieval and that organs be donated instead of purchased. Some provisions of the UAGA are excerpted below.

1. The Uniform Anatomical Gift Act (2009) (UAGA)

§ 4. Who May Make Anatomical Gift Before Donor's Death.

(1) [T]he donor, if the donor is an adult or if the donor is a minor and is:

   (A) emancipated; or

   (B) authorized under state law to apply for a driver's license . . . :

(2) an agent of the donor, unless the power or attorney for health care or other record prohibits the agent from making an anatomical gift;

(3) a parent of the donor, if the donor is an unemancipated minor; or

(4) the donor's guardian.

§ 5. Manner of Making Anatomical Gift Before Donor's Death.

[This section provides that a donor may make an anatomical gift by indicating so on the driver's license, in a will, on a donor card, through a donor registry, or by any means of communication if terminally ill or fatally injured.]

§ 8. Preclusive Effect of Anatomical Gift, Amendment, or Revocation.

(a) Except as otherwise provided in subsection (g) . . . , in the absence of an express, contrary indication by the donor, a person other than the donor is barred from making, amending, or revoking an anatomical gift of a donor's body or part if the donor made an anatomical gift of the donor's body or part under Section 5. . . .

(g) If a donor who is an unemancipated minor dies, a parent of the donor who is reasonably available may revoke or amend an anatomical gift of the donor's body or part.

(h) If an unemancipated minor who signed a refusal dies, a parent of the minor who is reasonably available may revoke the minor's refusal.

§ 9. Who May Make Anatomical Gift of Decedent's Body or Part.

(a) [A]n anatomical gift of a decedent's body or part for purpose of transplantation, therapy, research, or education may be made by any member of the following classes of persons who is reasonably available, in the order of priority listed:

   (1) an agent [appointed by the decedent prior to death];

   (2) the spouse of the decedent;

   (3) adult children of the decedent;

(4) parents of the decedent;

(5) adult siblings of the decedent;

(6) adult grandchildren of the decedent;

(7) grandparents of the decedent;

(8) an adult who exhibited special care and concern for the decedent;

(9) the persons who were acting as the [guardians] of the person of the decedent at the time of death; and

(10) any other person having the authority to dispose of the decedent's body.

§ 14. Rights and Duties of Procurement Organization and Others.

(a) When a hospital refers an individual at or near death to a procurement organization, the organization shall make a reasonable search of the records of the [state department of motor vehicles] and any donor registry . . . to ascertain whether the individual has made an anatomical gift.

* * *

(c) When a hospital refers an individual at or near death to a procurement organization, the organization may conduct any reasonable examination necessary to ensure the medical suitability of a part that is or could be the subject of an anatomical gift . . . from a donor or a prospective donor. During the examination period, measures necessary to ensure the medical suitability of the part may not be withdrawn unless the hospital or procurement organization knows that the individual expressed a contrary intent.

* * *

(f) Upon the death of a minor who was a donor or had signed a refusal, unless a procurement organization knows the minor is emancipated, the procurement organization shall conduct a reasonable search for the parents of the minor and provide the parents with an opportunity to revoke or amend the anatomical gift or revoke the refusal.

(g) Upon referral by a hospital . . . , a procurement organization shall make a reasonable search for any person listed in Section 9 having priority to make an anatomical gift on behalf of a prospective donor. . . .

* * *

(i) Neither the physician who attends the decedent at death nor the physician who determines the time of the decedent's death may participate in the procedures for removing or transplanting a part from the decedent.

§ 18. Immunity.

(a) A person that acts in accordance with this [act] or with the applicable anatomical gift law of another state, or attempts in good faith to do so, is not liable for the act in a civil action, criminal prosecution, or administrative proceeding.

* * *

PROBLEM: ORGAN DONATION AND THE UAGA

Laurel Singer, age 16, her sister Rebecca, age 22, and Rebecca's friend Peter Klaus, also age 22, were brought to General Hospital after an extremely serious car accident. Although they were alive when treated by the emergency medical technicians at the scene, Laurel, Rebecca, and Peter have died. Physicians have initiated standard procedures to maintain respiration and circulation in their bodies and preserve their organs for transplantation, including administering certain drugs and inserting a catheter into an artery to provide perfusion (delivery of blood to the organs). Time is critical as organs deteriorate quickly. The Singer sisters' parents have arrived at the hospital, but Peter's parents are on vacation somewhere in Asia. Peter's friends and the hospital are continuing efforts to contact his parents but without success.

Laurel and Rebecca each signed the organ donor card on the back of their driver's licenses; however, their parents told the hospital to discontinue the organ maintenance procedures and objected to removal of their organs. Peter didn't have a driver's license, but his friends all agree that he was clear about his intentions to be an organ donor.

May surgeons remove Laurel's, Rebecca's, and Peter's transplantable organs? Under the UAGA, what must the hospital do in order to proceed? Do the Singer parents have the right to stop the removal of their daughters' organs? Should the hospital abide by their wishes? If Peter's parents arrive after the surgery has been completed and object to the removal of their son's organs, would they have any action against the physicians or the hospital? Why would Peter not have signed a donor card or signed up with his state's online donor registry?

NOTES AND QUESTIONS

1. *Consent Not Required.* The UAGA speaks in terms of a "gift" made by the decedent prior to death or by authorized decision-makers after the death. It does not require that there be informed consent for organ donation. What information would be required for the donor or surrogate to give informed consent for the retrieval of organs? Should they be informed of the exact medical procedures that will be performed to preserve the organs prior to death, as they would be if they were consenting to medical treatment; the criteria that will be used to determine that the patient has died; or the nature of the procedures that will be used to remove the organs (or bones or skin)? Why is informed consent not required? See Ana Iltis, Organ Donation, Brain

Death and the Family: Valid Informed Consent, 43 J.L. Med. & Ethics 369 (2015); Leili Monfared & Lois Shepherd, Organ Procurement Now: Does the U.S. Still "Opt In?", 2017 U. Ill. L. Rev. 1003. See Section I.C.2, below, as to whether the donor's permission ought to be required at all.

2. *Immunity Provision.* Although the UAGA provides immunity for "good faith" attempts at compliance (§ 18, above), courts have interpreted good faith as a question of fact, which allows the case to go to the jury. See, for example, Cooper v. Louisiana Organ Procurement, 146 So. 3d 908 (La. Ct. App. 2014); Siegel v. LifeCenter Organ Donor Network, 969 N.E.2d 1271 (Ohio Ct. App. 2011); Schembre v. Mid-America Transplant Assn., 135 S.W.3d 527 (Mo. Ct. App. 2004) (plaintiff claimed nurse was negligent in describing procedures used to harvest bones).

3. *Medicare and Medicaid Requirements.* Hospitals receiving Medicare or Medicaid are required to contact the appropriate organ procurement organization whenever they have a patient whose death is imminent or who has died. The procurement organization is then to evaluate the viability of organs and tissue and speak with the family about the option of donation. 42 C.F.R. § 482.45. Under the UAGA, interventions to preserve the organs are continued without permission unless the patient had earlier expressed a "contrary intent." See UAGA § 14, above, and Section II.C, below.

NOTE: ORGAN TRANSPLANTATION FROM LIVING DONORS

Individuals who have decisional capacity may donate non-life-sustaining organs for transplantation; such donors have supplied over 5,000 donations annually. Most frequently, the recipients are family members or friends but, increasingly, strangers in need. The possibility of donation from a stranger has led patients and families to campaign publicly to recruit a donor. Several organizations offer matching services as well.

A willing donor's organ may not be a match for the intended recipient, however. Transplant centers have overcome the problem of incompatibility by arranging for kidney paired donations (KPD). In KPD, a mismatched donor and recipient are matched with another pair in the same situation and the kidneys are exchanged, with the surgeries occurring simultaneously. Some transplant centers arrange kidney chains that result in a long series of transplantation surgeries as kidneys incompatible with a designated recipient proceed down a chain of possibly dozens of persons. Evelyn Tenenbaum, Bartering for a Compatible Kidney Using Your Incompatible, Live Kidney Donor: Legal and Ethical Issues Related to Kidney Chains, 42 Am. J.L. & Med. 129 (2016); Kimberly Krawiec et al., Contract Development in a Matching Market: The Case of Kidney Exchange, 80 L. & Contemp. Probs. 11 (2017).

Concerns over safety and outcomes for living donors are significant. The Renal and Lung Living Donors Evaluation Study (RELIVE), a multiyear study of long-term outcomes for living lung and kidney donors, produced data on several measures including long-term health effects, emotional well-being, and satisfaction. See, e.g., C.R. Gross et al., Health Related Quality of Life in

Kidney Donors from the Last Five Decades: Results from the RELIVE Study, 13 Am. J. Transplantation 2924 (Nov. 2013), reporting health status roughly equivalent to that of the general population but identifying pre-surgical conditions that predict negative health outcomes for donors.

Legal issues relating to consent can arise in any intervivos transplant (see Green v. Buell, 215 So. 3d 715 (La. Ct. App. 2017), but they are more common when the donor is legally incapacitated. The principles relied on in such cases are the same as those that govern other medical treatment decisions for incompetent patients: they rely on either the donor's best interests or substituted judgment criterion for decision-making. See Chapter 5. See also Macey Henderson and Jed Gross, Living Organ Donation and Informed Consent in the United States: Strategies to Improve the Process, 45 J.L. Med. & Ethics 66 (2017); Michael T. Morley, Proxy Consent to Organ Donation by Incompetents, 111 Yale L.J. 1215 (2002). Ethical concerns relating to coercion of living donors are not easily accounted for in a legal framework, however. See, e.g., Nat'l Council on Disability, Organ Transplant Discrimination Against People with Disabilities (2019) (examining impact of organ procurement policies and practices on people with disabilities).

Advances in reproductive technologies, particularly preimplantation genetic diagnosis (see Chapter 3), create the capacity for parents of a child in need of a transplant to conceive and birth a child (a so-called "savior sibling") who is certain to be a match for the older sibling. The practice raises serious ethical and legal concerns at the intersection of reproductive rights and organ transplants. See, e.g., Zachary E. Shapiro, Savior Siblings in the United States: Ethical Conundrums, Legal and Regulatory Void, 24 Wash. & Lee Civil Rts. & Soc. Just. 419 (2018).

2.   Presumed Consent and Property Rights

NEWMAN V. SATHYAVAGLSWARAN

United States Court of Appeals, Ninth Circuit, 2002.
287 F.3d 786.

FISHER, CIRCUIT JUDGE.

Parents, whose deceased children's corneas were removed by the Los Angeles County Coroner's office without notice or consent, brought this 42 U.S.C. § 1983 action alleging a taking of their property without due process of law. The complaint was dismissed by the district court for a failure to state a claim upon which relief could be granted. We must decide whether the longstanding recognition in the law of California, paralleled by our national common law, that next of kin have the exclusive right to possess the bodies of their deceased family members creates a property interest, the deprivation of which must be accorded due process of law under the Fourteenth Amendment of the United States Constitution. We hold that it does. . . .

* * *

Robert Newman and Barbara Obarski (the parents) each had children, Richard Newman and Kenneth Obarski respectively, who died in Los Angeles County in October 1997. Following their deaths, the Office of the Coroner for the County of Los Angeles (the coroner) obtained possession of the bodies of the children and, under procedures adopted pursuant to California Government Code § 27492.47 as it then existed, removed the corneas from those bodies without the knowledge of the parents and without an attempt to notify them and request consent. The parents became aware of the coroner's actions in September 1999 and subsequently filed this § 1983 action alleging a deprivation of their property without due process of law in violation of the Fourteenth Amendment.

II.   Property Interests in Dead Bodies

[T]he Supreme Court repeatedly has affirmed that "the right of every individual to the possession and control of his own person, free from all restraint or interference of others," [ ] is "so rooted in the traditions and conscience of our people," [ ] as to be ranked as one of the fundamental liberties protected by the "substantive" component of the Due Process Clause. [ ] This liberty, the Court has "strongly suggested," extends to the personal decisions about "how to best protect dignity and independence at the end of life." [ ] The Court has not had occasion to address whether the rights of possession and control of one's own body, the most "sacred" and "carefully guarded" of all rights in the common law, [ ] are property interests protected by the Due Process Clause. Nor has it addressed what Due Process protections are applicable to the rights of next of kin to possess and control the bodies of their deceased relatives.

A.   History of Common Law Interests in Dead Bodies

* * *

Many early American courts adopted Blackstone's description of the common law, holding that "a dead body is not the subject of property right." [ ] The duty to protect the body by providing a burial was often described as flowing from the "universal . . . right of sepulture," rather than from a concept of property law. [ ] As cases involving unauthorized mutilation and disposition of bodies increased toward the end of the 19th century, paralleling the rise in demand for human cadavers in medical science and use of cremation as an alternative to burial, [ ] courts began to recognize an exclusive right of the next of kin to possess and control the disposition of the bodies of their dead relatives, the violation of which was actionable at law. Thus, in holding that a city council could not "seize upon existing private burial grounds, make them public, and exclude the proprietors from their management," the Supreme Court of Indiana commented that "the burial of the dead can [not] . . . be taken out of the hands of the relatives thereof" because "we lay down the proposition, that the bodies of the dead belong to the surviving relations, in the order of inheritance, as property, and that they have the right to dispose of them as such, within restrictions analogous to those by which the disposition of other property may be regulated." [ ] . . .

B. Interests in Dead Bodies in California Law

[The court traces the history of California law concerning the disposition of dead bodies. California courts referred to "quasi-property" rights in cases disputing the handling of cadavers, including cases in which the courts held that civil litigants had no right to demand an autopsy; held that next-of-kin could exclude the decedent's friends from the funeral; and permitted an action for retaining organs after autopsy.]

C. The Right to Transfer Body Parts

The first successful transplantation of a kidney in 1954 led to an expansion of the rights of next of kin to the bodies of the dead. In 1968, the National Conference of Commissioners on Uniform State Laws approved the Uniform Anatomical Gift Act (UAGA), adopted by California the same year, which grants next of kin the right to transfer the parts of bodies in their possession to others for medical or research purposes. [ ] The right to transfer is limited. . . .

In the 1970s and 1980s, medical science improvements and the related demand for transplant organs prompted governments to search for new ways to increase the supply of organs for donation. [ ] Many perceived as a hindrance to the supply of needed organs the rule implicit in the UAGA that donations could be effected only if consent was received from the decedent or next of kin. [ ] In response, some states passed "presumed consent" laws that allow the taking and transfer of body parts by a coroner without the consent of next of kin as long as no objection to the removal is known. [ ] California Government Code § 27491.47, enacted in 1983, was such a law.

III. Due Process Analysis

"[T]o provide California non-profit eye banks with an adequate supply of corneal tissue," S. Com. Rep. SB 21 (Cal. 1983), § 27491.47(a) authorized the coroner to "remove and release or authorize the removal and release of corneal eye tissue from a body within the coroner's custody" without any effort to notify and obtain the consent of next of kin "if . . . [t]he coroner has no knowledge of objection to the removal." The law also provided that the coroner or any person acting upon his or her request "shall [not] incur civil liability for such removal in an action brought by any person who did not

object prior to the removal . . . nor be subject to criminal prosecution."
§ 27491.47(b).[11]

* * *

In two decisions the Sixth Circuit, the only federal circuit to address
the issue until now, held that the interests of next of kin in dead bodies
recognized in Michigan and Ohio allowed next of kin to bring § 1983 actions
challenging implementation of cornea removal statutes similar to
California's. [ ] . . .

The supreme courts of Florida and Georgia, however, have held that
similar legal interests of next of kin in the possession of the body of a
deceased family member, recognized as "quasi property" rights in each
state, are "not . . . of constitutional dimension." *Georgia Lions Eye Bank,
Inc. v. Lavant,* 255 Ga. 60, 335 S.E.2d 127, 128 (1985); *State v. Powell,* 497
So.2d 1188, 1191 (Fla. 1986) (commenting that "[a]ll authorities generally
agree that the next of kin have no property right in the remains of a
decedent"). The Florida Supreme Court recently rejected the broad
implications of the reasoning in *Powell* distinguishing that decision as
turning on a balance between the public health interest in cornea donation
and the " 'infinitesimally small intrusion' " of their removal. *Crocker v.
Pleasant,* 778 So.2d 978, 985, 988 (Fla. 2001) (allowing a § 1983 action to
go forward for interference with the right of next of kin to possess the body
of their son because "in Florida there is a legitimate claim of entitlement
by the next of kin to possession of the remains of a decedent for burial or
other lawful disposition").

We agree with the reasoning of the Sixth Circuit and believe that
reasoning is applicable here. Under traditional common law principles,
serving a duty to protect the dignity of the human body in its final
disposition that is deeply rooted in our legal history and social traditions,
the parents had exclusive and legitimate claims of entitlement to possess,
control, dispose and prevent the violation of the corneas and other parts of
the bodies of their deceased children. With California's adoption of the
UAGA, [ ], it statutorily recognized other important rights of the parents
in relation to the bodies of their deceased children—the right to transfer
body parts and refuse to allow their transfer. These are all important
components of the group of rights by which property is defined, each of
which carried with it the power to exclude others from its exercise,
"traditionally . . . one of the most treasured strands in an owner's bundle
of property rights." [ ]

* * *

---

[11] For body parts other than corneas, California adopted the 1987 version of the UAGA
authorizing transfer when no knowledge of objection is known and after "[a] reasonable effort has
been made to locate and inform [next of kin] of their option to make, or object to making, an
anatomical gift." Cal. Health & Safety Code § 7151.5(a)(2).

Nor does the fact that California forbids the trade of body parts for profit mean that next of kin lack a property interest in them. The Supreme Court has "never held that a physical item is not 'property' simply because it lacks a positive economic or market value." [ ]

We need not . . . decide whether California has transgressed basic property principles [ ] because that statute did not extinguish California's legal recognition of the property interests of the parents to the corneas of their deceased children. It allowed the removal of corneas only if "the coroner has no knowledge of objection," a provision that implicitly acknowledges the ongoing property interests of next of kin.

* * *

. . . The property rights that California affords to next of kin to the body of their deceased relatives serve the premium value our society has historically placed on protecting the dignity of the human body in its final disposition. California infringed the dignity of the bodies of the children when it extracted the corneas from those bodies without the consent of the parents. The process of law was due the parents for this deprivation of their rights.

* * *

The scope of the process of law that was due the parents is not a question that we can answer based on the pleadings alone. This question must be addressed in future proceedings.

* * *

We do not hold that California lacks significant interests in obtaining corneas or other organs of the deceased in order to contribute to the lives of the living. Courts are required to evaluate carefully the state's interests in deciding what process must be due the holders of property interests for their deprivation. [ ] An interest so central to the state's core police powers as improving the health of its citizens is certainly one that must be considered seriously in determining what process the parents were due. [ ] But our Constitution requires the government to assert its interests and subject them to scrutiny when it invades the rights of its subjects. Accordingly, we reverse the district court's dismissal of the parents' complaint and remand for proceedings in which the government's justification for its deprivation of parents' interests may be fully aired and appropriately scrutinized.

NOTES AND QUESTIONS

1. *Satisfying Due Process Requirements.* As noted in *Newman*, the California statute departed from the UAGA provision of the time, which required a "reasonable effort" to contact persons authorized to permit donation

before tissue or organs could be removed by the coroner. The Ninth Circuit leaves open the possibility that the State could structure the coroner's removal of tissue in a way that would satisfy due process concerns. Would requiring that there be a reasonable attempt to contact next-of-kin satisfy due process requirements? Would the existence of an easily accessible system for registering objections (an opt-out system) satisfy the court? The UAGA provision was removed in 2006 in response to this litigation and now provides that no tissue or organ can be removed unless there is an anatomical gift (although in an uncommon scenario, the coroner can gift a decedent's tissue under the authority provided for in the UAGA § 9(a)(10), above).

2. *"Presumed Consent."* The California statute may be seen as relying on "presumed consent," long advocated as a tool for increasing the supply of organs and tissue for transplant. Presumed consent relies on the assumption, grounded in repeated surveys, that most people intend to donate when the opportunity arises but simply fail to do so. Presumed consent shifts the responsibility for objecting to the individual. Before the decision in *Newman*, a series of news articles demonstrated that the practice had a disproportionate impact on individuals from disadvantaged racial and ethnic groups. At the time, 80% of coroner autopsies were performed on African American or Latino individuals and only 16% on whites.

Is the assumption that the individual would consent if asked a persuasive claim in this situation? See Michele Goodwin, *Newman v. Sathyavaglswaran: Unbundling Property in the Dead*, in Health Law and Bioethics Cases in Context (Sandra Johnson et al., eds. 2009); Gil Siegel & Richard Bonnie, Closing the Organ Gap: A Reciprocity-Based Social Contract Approach, 34 J.L. Med. & Ethics 415 (2006). Almost all states have followed the lead of the 2006 UAGA in repealing presumed consent provisions. But see, e.g., Mo. Stat. § 58–770, which allows for the removal of the pituitary gland without attempting to notify the family when there is no known objection. See also David Orentlicher, Presumed Consent to Organ Donation: Its Rise and Fall in the United States, 61 Rutgers L. Rev. 295 (2009).

3. *Commercial Market.* Several cases have focused attention on the robust commercial market for human cadaveric tissue. In Adams v. King County, 192 P.3d 891 (Wash. 2008), for example, the Washington Supreme Court held that the adult decedent's parents might have a cause of action under Washington law where their son's brain was removed by the medical examiner and given to the Stanley Medical Research Institute (SMRI). SMRI funded a position in the examiner's office in exchange for the examiner providing brains to SMRI for research. The decedent had signed a donor card donating "any organ," but the court held that Washington's anatomical gift statute provided that only hospitals could receive undesignated gifts; the pre-mortem donation document did not apply. See also Geary v. Stanley Med. Res. Inst., 939 A.2d 86 (Me. 2008), in which multiple plaintiffs alleged that an agent who was paid $1,000 per brain falsified consent forms; Commonwealth v. Mastromarino, 2 A.3d 581 (Pa. Super. Ct. 2010), reviewing a prison sentence of 25–58 years for a tissue procurer who had taken cadaver tissue and bones

without consent, earning over $1 million from tissue banks. See also Michele Goodwin, Empires of the Flesh: Tissue and Organ Taboos, 60 Ala. L. Rev. 1219 (2009).

4.   *Beyond Gifts.* When claims relating to the bodies of deceased family members are raised outside the context of anatomical gifts, the cases are decided differently. See, e.g., Shelley v. County of San Joaquin, 996 F. Supp. 2d 921 (E.D. Cal. 2014) (distinguishing *Newman* as applying only to anatomical gifts and holding that loss of decedent's remains by the sheriff's office did not violate family's procedural constitutional rights). See also Chapter 7 on the ownership of tissue in research.

3.   Paying "Donors" for Organs and Tissue

FLYNN V. HOLDER

United States Court of Appeals, Ninth Circuit, 2012.
684 F.3d 852.

KLEINFELD, SENIOR CIRCUIT JUDGE.

* * *

The complaint challenges the constitutionality of the ban on compensation for human organs in the National Organ Transplant Act, as applied to bone marrow transplants. . . .

Some plaintiffs are parents of sick children who have diseases such as leukemia and a rare type of anemia, which can be fatal without bone marrow transplants. Another plaintiff is a physician and medical school professor, and an expert in bone marrow transplantation. He says that at least one out of five of his patients dies because no matching bone marrow donor can be found, and many others have complications when scarcity of matching donors compels him to use imperfectly matched donors. One plaintiff is a parent of mixed race children, for whom sufficiently matched donors are especially scarce, because mixed race persons typically have the rarest marrow cell types. . . .

Another plaintiff is a California nonprofit corporation that seeks to operate a program incentivizing bone marrow donations. The corporation proposes to offer $3,000 awards in the form of scholarships, housing allowances, or gifts to charities selected by donors, initially to minority and mixed race donors of bone marrow cells, who are likely to have the rarest marrow cell type. The corporation, MoreMarrowDonors.org, alleges that it cannot launch this program because the National Organ Transplant Act criminalizes payment of compensation for organs, and classifies bone marrow as an organ.

* * *

Until about twenty years ago, bone marrow was extracted from donors' bones by "aspiration." Long needles, thick enough to suck out the soft, fatty marrow, were inserted into the cavities of the anesthetized donor's hip bones. These are large bones with big central cavities full of marrow. Aspiration is a painful, unpleasant procedure for the donor. It requires hospitalization and general or local anesthesia, and involves commensurate risks.

. . . With [a] new technique, now used for at least two-thirds of bone marrow transplants, none of the soft, fatty marrow is actually donated. Patients who need bone marrow transplants do not need everything that the soft, fatty substance from bone cavities contains, just some of the marrow's "hematopoietic stem cells." These stem cells are seeds from which white blood cells, red blood cells, and platelets grow.

These are not the embryonic stem cells often the subject of controversy. Those stem cells, taken from human embryos, are "pluripotent," that is, they can turn into any kind of cell—brain, blood, retina, toenail, whatever. The stem cells at issue in this case are "hematopoietic stem cells." . . . Hematopoietic stem cells turn into blood cells and nothing else. Humans and other large mammals produce these blood stem cells constantly in vast numbers, because our blood cells die within a few months and need continual replacement. . . . [N]ew ones are made in the bone marrow, as long as we live.

[S]ome blood stem cells flow into and circulate in the bloodstream before they mature [into blood cells]. These are called "peripheral" blood stem cells. . . . The new bone marrow donation technique, developed during the past twenty years, is called "peripheral blood stem cell apheresis." . . . This procedure begins with five days of injections of a medication called a "granulocyte colony-stimulating factor" into the donor's blood. The medication accelerates blood stem cell production in the marrow, so that more stem cells go into the bloodstream. . . .

The main difference between an ordinary blood donation and apheresis is that instead of just filling up a plastic bag with whole blood, the donor sits for some hours in a recliner while the blood passes through the apheresis machine. This same apheresis technique is sometimes used for purposes other than bone marrow donations, such as when the machine is set up to collect plasma or platelets, rather than stem cells, from a donor's blood. When it is used for these other purposes, the identical technique is called a "blood donation" or "blood plasma donation." When used to separate out and collect hematopoietic stem cells from the donor's bloodstream, apheresis is called "peripheral blood stem cell apheresis" or a "bone marrow donation."

* * *

[E]ven with [the National Marrow Donor Program registry, funded by the federal government], good [marrow] matches often cannot be found. And even when a good match is found in the registry, tracking down the potential donor from what may be an outdated address may be impossible to accomplish in time to save the patient's life—assuming the potential donor is willing to go through with the process when found.

The plaintiff nonprofit proposes to mitigate this matching problem by using a financial incentive [which plaintiff concedes constitutes "valuable consideration" under NOTA]. The idea is that the financial incentive will induce more potential donors to sign up, stay in touch so that they can be located when necessary, and go through with the donations. . . .

Plaintiffs argue that the National Organ Transplant Act, as applied to MoreMarrowDonors.org's planned pilot program, violates the Equal Protection Clause. They claim that blood stem cell harvesting is not materially different from blood, sperm, and egg harvesting, which are not included under the statutory or regulatory definitions of "human organ." Like donors of blood and sperm, a bone marrow donor undergoing apheresis suffers no permanent harm, experiences no significant risk, and quickly regenerates what is donated. Plaintiffs also argue that any rational basis that Congress had when it passed the statute no longer exists with respect to the pilot program, because of the subsequent development of the apheresis method. Plaintiffs seek declaratory and injunctive relief so that MoreMarrowDonors.org can proceed with the initiative.

* * *

. . . [NOTA] makes it a felony "to knowingly acquire, receive, or otherwise transfer any human organ for valuable consideration for use in human transplantation." And it defines the term "human organ" to include "bone marrow." Ergo, the statute expressly prohibits compensating bone marrow donors. According to the government's brief, Congress took the view that "human body parts should not be viewed as commodities," [citing a Senate Report] and had several policy reasons for disallowing compensation to donors, which suffice to serve as a rational basis for the prohibition.

[T]he government argues that because it is much harder to find a match for patients who need bone marrow transplants than for patients who need blood transfusions, exploitative market forces could be triggered if bone marrow could be bought. The government also asserts that peripheral blood stem cell apheresis poses greater health risks for the donor than blood donations do, because of the side effects of the medicine used to increase stem cell secretion. [The court holds that the question of risk is one for trial or summary judgment rather than a motion to dismiss, but notes that "there are significant risks from egg donations, but the government concedes that the statute allows them."]

. . . To the extent that plaintiffs challenge the constitutionality of the compensation ban on bone marrow donation by the old aspiration method . . . the challenge must fail.

. . . It is irrelevant that the legislative history indicates that Congress viewed certain types of regenerable tissue, such as blood, as falling outside the statutory definition of "human organ." It may be that senators themselves, their staffs, or lobbyists for blood banks argued for an exception for body substances that can regenerate, and persuaded committee staffers to put that reason in the legislative history. But the statute does not say that compensation is permitted for organs or body parts that regenerate and prohibited for those that do not. Nor is the statute consistent with such a construction. The statute defines the liver "or any subpart thereof" as an organ for which compensation is prohibited. The drafters doubtless knew that a partial resection of a liver can yield a donation that will save the recipient's life, and that the donor's liver will grow back. So the statute does expressly prohibit compensation for at least one explicitly denoted "human organ" that will regenerate.

. . . Congress may have been concerned that if donors could be paid, rich patients or the medical industry might induce poor people to sell their organs, even when the transplant would create excessive medical risk, pain, or disability for the donor. Or, looking from the other end, Congress might have been concerned that every last cent could be extracted from sick patients needful of transplants, by well-matched potential donors making "your money or your life" offers. The existing commerce in organs extracted by force or fraud by organ thieves might be stimulated by paying for donations. Compensation to donors might also degrade the quality of the organ supply, by inducing potential donors to lie about their medical histories in order to make their organs marketable. Plaintiffs argue that a $3,000 housing subsidy, scholarship, or charitable donation is too small an amount to create a risk of any of these evils, but for a lot of people that could amount to three to six months' rent.

Congress may have had philosophical as well as policy reasons for prohibiting compensation. People tend to have an instinctive revulsion at denial of bodily integrity, particularly removal of flesh from a human being for use by another, and most particularly "commodification" of such conduct, that is, the sale of one's bodily tissue. While there is reportedly a large international market for the buying and selling of human organs, in the United States, such a market is criminal and the commerce is generally seen as revolting. Leon Kass [observes]:

> [A]lthough we allow no commerce in organs, transplant surgeons and hospitals are making handsome profits from the organ-trading business, and even the not-for-profit transplant registries and procurement agencies glean for their employees a middleman's

livelihood. Why . . . should everyone be making money from this business except the person whose organ makes it possible? Could it be that [the] real uneasiness [lies] with organ donation or with transplantation itself, for if not, what would be objectionable about its turning a profit? [Leon R. Kass, *Life, Liberty and the Defense of Dignity: The Challenge for Bioethics* 177 (2002).]

. . . Kass points to the idea of "psychophysical unity, a position that regards a human being as largely, if not wholly, self-identical with his enlivened body," so that, as [Immanuel] Kant put it, to " 'dispose of oneself as a mere means to some end of one's own liking is to degrade the humanity in one's person.' " . . .

[T]here are strong arguments for contrary views. But these policy and philosophical choices are for Congress to make, not us. The distinctions made by Congress must have a rational basis, but do not need to fit perfectly with that rational basis, and the basis need merely be rational, not persuasive to all. Here, Congress made a distinction between body material that is compensable and body material that is not. The distinction has a rational basis, so the prohibition on compensation for bone marrow donations by the aspiration method does not violate the Equal Protection Clause.

The focus, though, of plaintiffs' arguments is compensation for "bone marrow donations" by the peripheral blood stem cell apheresis method. For this, we need not answer any constitutional question, because the statute contains no prohibition. Such donations of cells drawn from blood flowing through the veins may sometimes anachronistically be called "bone marrow donations," but none of the soft, fatty marrow is donated, just cells found outside the marrow, outside the bones, flowing through the veins.

Congress could not have had an intent to address the apheresis method when it passed the statute, because the method did not exist at that time. We must construe the words of the statute to see what they imply about extraction of hematopoietic stem cells by this method. . . .

Since payment for blood donations has long been common, the silence in [NOTA] on compensating blood donors is loud. . . . The government concedes that the common practice of compensating blood donors is not prohibited by the statute.

The government argues that hematopoietic stem cells in the veins should be treated as "bone marrow" because "bone marrow" is a statutory organ, and the statute prohibits compensation not only for donation of an organ, but also "any subpart thereof." . . .

We reject this argument. . . . If the government's argument that what comes from the marrow is a subpart of the marrow were correct, then the statute would prohibit compensating blood donors. . . .

[E]very blood draw includes some hematopoietic stem cells. . . . Once the stem cells are in the bloodstream, they are a "subpart" of the blood, not the bone marrow. The word "subpart" refers to the organ from which the material is taken, not the organ in which it was created. . . .

* * *

We construe "bone marrow" to mean the soft, fatty substance in bone cavities, as opposed to blood, which means the red liquid that flows through the blood vessels. The statute does not prohibit compensation for donations of blood and the substances in it, which include peripheral blood stem cells. The Secretary of Health and Human Services has not exercised regulatory authority to define blood or peripheral blood stem cells as organs. We therefore need not decide whether prohibiting compensation for such donations would be unconstitutional.

NOTES AND QUESTIONS

1. *Tissue vs. Organs.* Flynn relies on the specific type of tissue involved (as well as the method for removing it from the body) to hold that payment is not prohibited under NOTA. Should tissue, such as corneas, be treated differently than solid organs for purposes of payment? For purposes of consent? See *Newman* and notes following, above. See also Emily Largent, NOTA: Not a Good Act for Tissues to Follow, 19 Quinnipiac Health L.J. 179 (2016), arguing that tissue generally should be subject to a different regulatory regime than are organs but that the post-retrieval market should be regulated.

2. *Payment for Organs?* The AMA recommends that payment for organs be studied with the following limitations: incentives are modest and set at the lowest level that can reasonably be expected to increase donations, and payment should be limited to "voluntary donation" of cadaveric organs only with "an explicit prohibition against the selling of organs." AMA Code of Medical Ethics Opinion E–6.1.3. Do these limits respond persuasively to objections to payment for human body parts or not? How would you separate the act of payment from the act of selling? See, e.g., Glenn Cohen, Regulating the Organ Market: Normative Foundations for Market Regulation, 77 L. & Contemp. Probs. 71 (2014); Radhika Rao, A Small Price to Pay: Incentivizing Cadaveric Organ Donation with Posthumous Payments, 18 Minn. J.L. Sci. & Tech. 273 (2017); Jake Linford, The Kidney Donor Scholarship Act: How College Scholarships Can Provide Financial Incentives for Kidney Donation While Preserving Altruistic Meaning, 2 St. Louis U.J. Health L. & Pol'y 265 (2009); Michele Goodwin, Black Markets: The Supply and Demand of Body Parts (2006) (arguing in favor of payment). But see Institute of Medicine, Organ Donation: Opportunities for Action (2006) (rejecting market exchanges).

3. *Other Financial Incentives.* The incentives proposed in *Flynn* did not include payments of cash to the donor. Does the form the financial transfer takes make a difference to legal, ethical, or public policy issues in providing financial incentives for donors? NOTA provides that "valuable consideration"

does not include "reasonable payments associated with the removal, transportation, implantation, processing, preservation, quality control, and storage of a human organ or the expenses of travel, housing, and lost wages incurred by the donor of a human organ in connection with the donation . . .". 42 U.S.C.A. § 274e(c)(2). (The UAGA, which unlike NOTA applies only to cadaveric organs and tissue, has a similar provision. § 16.) "Reasonable payments" allows profit for the firms involved in transplantation. See discussion following *Newman*, above.

Some states compensate donors within these restrictions. Pennsylvania, for example, has established a public trust fund for families of decedents who donate organs. The statute limits payments to $3,000 per donor (although the program currently pays only $300) to be used to cover funeral expenses and incidental expenses (lodging, food) borne by the family in relation to the donation. 20 Pa. Cons. Stat. § 8622. Other states have state income tax incentives allowing living donors to deduct expenses related to donation, including lost wages. Wis. Stat. Ann. § 71.05(10)(i). The federal Organ Donation and Recovery Improvement Act of 2004 provides grants to procurement organizations (and other groups) for reimbursing living donors for travel and certain living expenses. 42 U.S.C. § 274f. NOTA was amended in 2007 (42 U.S.C. § 274e) to specifically exclude paired exchanges from the Act's prohibition of the payment of "valuable consideration" for organ donation even though each organ is donated for the purpose of procuring another organ for a designated recipient. See Note on Living Donors, above.

4. *Transplant Tourism.* Organ procurement and transplantation operate on a global scale. Persons of means in need of transplants can go where the organs are. While there had once been significant concern that transplant tourism was bringing foreign nationals to the U.S. to receive one of "our" scarce organs, the situation has shifted dramatically. Now a primary concern is U.S. citizens who avoid the waiting list here by travelling abroad to receive transplants. There are concerns for public safety and quality as these patients return to the U.S. for post-transplant care. See, e.g., Anthony J. Polcari et al., Transplant Tourism—A Dangerous Journey?, 25 Clinical Transplantation 633 (2011).

The most dramatic concern, however, is that transplant tourism may have created a global black market in which organs are bought and sold or retrieved from persons under even more extreme circumstances. See, e.g., Robert Ainley, Organ Transploitation: A Model Law Approach to Combat Human Trafficking and Transplant Tourism, 13 Or. R. Int'l L. 427 (2011). But see F. Ambagtsheer et al., On Patients Who Purchase Organ Transplants Abroad, 10 Am J. Transplantation 2800 (2016), arguing that data do not support claims of this abuse.

II. THE DETERMINATION OF DEATH

A. THE UNIFORM DETERMINATION OF DEATH ACT

Determining that a person has died used to be simple. If the heart and lungs had stopped functioning, eventually other less observable processes stopped as well. The cessation of cardiopulmonary function signaled death without the need for testing whether other processes, such as neurological activity, continued.

The first well-accepted definition of death to include what has become known as "brain death" came in 1968 from the Ad Hoc Committee of the Harvard Medical School to Examine the Definition of Brain Death. The Committee identified two developments that required an expansion of the traditional heart-lung definition of death. First, the Committee pointed out that "resuscitative and support measures" had improved and were being used more frequently, though often unsuccessfully, leaving patients with serious brain injuries and burdening patients, families, and society in unacceptable ways. These technological advances could "restore 'life' as judged by the ancient standards of persistent respiration and continuing heart beat" even when "there is not the remotest possibility of an individual recovering consciousness following massive brain damage." Second, the Committee explained that the uncertainty that "brain death" was real death was hindering the ability to recover life-sustaining organs that would be used for transplantation. The full report of the Committee is available at A Definition of Irreversible Coma: Report of the Ad Hoc Committee of the Harvard Medical School to Examine the Definition of Brain Death, 205 JAMA 337 (1968).

The Committee listed characteristics of brain death (which the Committee called "irreversible coma") as unreceptivity and unresponsivity, no autonomous movements or breathing, no reflexes, and flat electroencephalogram. The Committee recommended that death be declared before the respirator is turned off ("in our judgment it will provide a greater degree of legal protection to those involved"); that the physician in charge consult with others before the declaration of death is made; that the physician (rather than the family) determine death; and that the decision to declare death be made by physicians who are not involved "in any later effort to transplant organs or tissue from the deceased individual."

The Ad Hoc Committee suggested that "no statutory change in the law should be necessary [to accommodate brain death] since the law treats this question essentially as one of fact to be determined by physicians." Nevertheless, brain death statutes began appearing shortly after the publication of the Committee's report. In 1980, the Uniform Determination of Death Act (UDDA) ultimately was developed with the cooperation of both the AMA and the ABA.

UNIFORM DETERMINATION OF DEATH ACT (1980)

§ 1. An individual who has sustained either (1) irreversible cessation of circulatory and respiratory functions, or (2) irreversible cessation of all functions of the entire brain, including the brain stem, is dead. A determination of death must be made in accordance with accepted medical standards.

PROBLEM: WHEN DOES DEATH OCCUR?

Alberto Arcturus was found lying face down by the side of the road, apparently after being run down by a hit-and-run driver. A passing motorist called the local emergency medical services, and an ambulance arrived on the scene about 15 minutes after the call. The two paramedics found that Mr. Arcturus was not breathing and had no pulse. They also discovered that a substantial portion of his head (including his forehead and forebrain) was crushed. One paramedic looked at the other and said, "He's dead; let's call the morgue." The second, less experienced paramedic insisted on trying to resuscitate Mr. Arcturus, as was required by the emergency medical services manual for paramedics. They placed him in the ambulance and administered cardiopulmonary resuscitation throughout the fifteen-minute ride to the nearest hospital emergency room.

At the emergency room, physicians confirmed that Mr. Arcturus did not breathe spontaneously and had no spontaneous cardiac activity. One doctor told the charge nurse that he was dead and that "dead on arrival" should be marked on his chart. A young intern balked at this, however, because the hospital emergency room protocol required more before a brain-injured patient could be declared dead. The physicians then administered drugs and used paddles that sent an electric current through Mr. Arcturus's chest in an effort to start his heart. After some time, they managed to get a weak pulse, and they placed Mr. Arcturus on a ventilator and moved him to the intensive care unit. A neurology consult revealed that Mr. Arcturus's neocortex was completely and irreversibly destroyed—most of it was literally gone, left on the highway— although his brain stem remained intact. With the help of a ventilator, Mr. Arcturus's body could continue functioning indefinitely. If the ventilator were to be removed, Mr. Arcturus's heart and lung function would cease within some minutes as repeated trials proved that he could not breathe on his own.

After considerable investigation, police have captured the hit-and-run driver. They want to know whether to charge him with vehicular homicide or something else. The doctors and Mr. Arcturus's family want to know whether he is dead or alive. Further, if he is dead, the family wants to know when he died—at the roadside, in the hospital emergency room, in the intensive care unit, or somewhere else. Mr. Arcturus's health insurer is involved because insurance ordinarily won't cover medical services to a dead patient. Finally, the regional Organ Procurement Organization has been notified that there is a potential donor, and the transplant team at the hospital wants to know if Mr. Arcturus's healthy organs (including his kidneys and liver) are available for transplantation to save the life of patients on the transplant waiting list.

Is Mr. Arcturus dead? If so, when did he die?

B. VARIATION IN MEDICAL AND LEGAL STANDARDS

Although the Ad Hoc Committee advocated the adoption of neurological determination of death over 50 years ago and "brain death" has been adopted as a legal standard for death by all of the states for several decades, controversy has persisted. Cases in which families refuse to accept that their loved one is dead garner headlines because they touch a nerve. These cases are often dismissed as instances where laypersons are in denial, do not trust physicians and hospitals, are incapable of understanding that the condition of the body supported by mechanical ventilation is not evidence of life, or are relying on religious exceptionalism.

However, a deeper look reveals more. There are fissures in the medical community itself in terms of the standards and procedures used to determine brain death and, for some, even whether the legal standard requiring the cessation of *all* brain activity aligns with current medical practice in determining death. See Robert D. Truog et al., Understanding Brain Death, 323(21) JAMA 2139 (May 1, 2020); Thaddeus Pope, Brain Death Forsaken: Growing Conflict and New Legal Challenges, 37 J. Leg. Med. 265 (2017). See also Rachel Aviv, The Death Debate, New Yorker 30 (Feb. 5, 2018), describing the complexities arising in the dispute over the status of Jahi McMath, a 13-year-old girl declared brain dead by the hospital after surgery. The Nevada Supreme Court considered a similar situation in *Hailu*, below.

IN RE GUARDIANSHIP OF HAILU
Supreme Court of Nevada, 2015.
361 P.3d 524.

PICKERING, J.:

"For legal and medical purposes, a person is dead if the person has sustained an irreversible cessation of . . . [a]ll functions of the person's entire brain, including his or her brain stem." NRS 451.007(1). The determination of death "must be made in accordance with accepted medical standards." NRS 451.007(2). Here, we are asked to decide whether the American Association of Neurology [AAN] guidelines are considered "accepted medical standards" that satisfy the definition of brain death in NRS 451.007. We conclude that the district court failed to properly consider whether the [AAN] guidelines adequately measure all functions of the entire brain, including the brain stem, . . . and are considered accepted medical standards by states that have adopted the Uniform Determination of Death Act. Accordingly, we reverse the district court's order denying a petition for temporary restraining order and remand.

FACTS

Medical history

On April 1, 2015, 20-year-old university student Aden Hailu went to St. Mary's Regional Medical Center (St. Mary's) after experiencing abdominal pain. Medical staff could not determine the cause of her pain and decided to perform an exploratory laparotomy and remove her appendix. During the laparotomy, Hailu's blood pressure was low and she suffered "severe, catastrophic anoxic, or lack of brain oxygen damage," and she never woke up. After her surgery, Hailu was transferred to the St. Mary's Intensive Care Unit (ICU), under the care of Dr. Anthony Floreani. Within the first two weeks of April, three different electroencephalogram (EEG) tests were conducted, all of which showed brain functioning.

On April 13, 2015, Dr. Aaron Heide, the Director of Neurology and Stroke at St. Mary's, first examined Hailu. Dr. Heide concluded that Hailu was not brain dead at that time but was "rapidly declining." To make that determination, Dr. Heide conducted an examination of Hailu's neurological functions; her left eye was minimally responsive, she was chewing on the ventilator tube, and she moved her arms with stimulation. The next day, April 14, 2015, Hailu did not exhibit these same indicia of neurological functioning.

On May 28, 2015, St. Mary's performed an apnea test, which involved taking Hailu off ventilation support for ten minutes to see if she could breathe on her own; Hailu failed the apnea test, leading St. Mary's to conclude that "[t]his test result confirms Brain Death unequivocally." Based on Hailu's condition, Dr. Jeffrey Bacon wrote the following in his notes: "Awaiting administration and hospital lawyers for direction re care—withdrawal of Ventilator support indicated NOW in my opinion as brain death unequivocally confirmed." On June 2, 2015, St. Mary's notified Hailu's father and guardian, Fanuel Gebreyes, that it intended to discontinue Hailu's ventilator and other life support. Gebreyes opposed taking Hailu off life support and sought judicial relief.

Procedural history

Gebreyes filed an emergency motion for [a] temporary restraining order to enjoin St. Mary's from removing Hailu from life-sustaining services. On June 18, 2015, the district court held a hearing on the matter. The parties stipulated that St. Mary's would continue life-sustaining services until July 2, 2015, at 5:00 p.m. to allow Gebreyes to have an independent neurologist examine Hailu. They further stipulated that if, after the independent examination, Gebreyes wished St. Mary's to continue life support, he would need to request it through guardianship court. However, "if on July 2, 2015, it is determined that Aden Hailu is legally and clinically deceased, the hospital shall proceed as they see fit." Based on the stipulation, the district court dismissed the complaint for a temporary restraining order.

For reasons unknown, Gebreyes was unable to obtain the services of a neurologist before the stipulated July 2, 2015, deadline. Consequently, on July 1, 2015, Gebreyes filed an "Emergency Petition for Order Authorizing Medical Care, Restraining Order and Permanent Injunction." In the petition, he alleged that the doctors at St. Mary's had prematurely determined that Hailu had experienced brain death and sought to prevent the hospital from removing Hailu from the ventilator. St. Mary's opposed the emergency petition on July 2, 2015, and the district court held a hearing that same day.

At the July 2, 2015, hearing, the district court heard from four witnesses. First, Gebreyes testified that he wanted Hailu to get a tracheostomy and feeding tube to prepare her for transport; he hoped to take her home or relocate her to Las Vegas, where he resides. When asked why he did not obtain the services of another doctor to perform the tracheostomy, he stated that it was something he thought St. Mary's had to do because Hailu is at St. Mary's. Second, Gebreyes obtained the services of Dr. Paul Byrne—a known opponent of brain-death declarations who is unlicensed in Nevada—to testify that Hailu is still alive. Dr. Byrne complained that Hailu was never treated for thyroid problems and testified that this treatment will help her improve.

Third, Dr. Aaron Heide testified on behalf of St. Mary's. Dr. Heide applied the [AAN] guidelines to Hailu to determine if she was brain dead. He testified that the AAN guidelines are the accepted medical standard in Nevada. The AAN guidelines call for three determinations: (1) whether there is a coma and unresponsiveness; (2) whether there is brainstem activity (determined by conducting a clinical examination of reflexes, eyes, ears, etc.); and (3) whether the patient can breathe on her own (determined by conducting an apnea test). Although another doctor conducted the apnea test one month after Dr. Heide's last examination of Hailu, Dr. Heide believed that Hailu "had zero percent chance of any form of functional neurological outcome." Further, Dr. Heide also administered a Transcranial Doppler test, which is a test that measures blood flow to the brain. While there was still some blood flow to Hailu's brain, the lack of blood flow was consistent with brain death.

Last, Helen Lidholm, the CEO of St. Mary's, testified that the hospital is in favor of allowing Hailu to be transported to Las Vegas, where her father lives. Lidholm stated that St. Mary's "could make that happen" as long as Gebreyes arranges the proper medical equipment and transportation for Hailu and ensures a transfer location that can care for her. St. Mary's would allow the family to retain the services of any neurologist to come in and test Hailu as long as the physician is licensed in the State of Nevada; St. Mary's also offered to pay for the physician's examination fee. On cross-examination, Lidholm clarified that if the family has a licensed neurologist examine Hailu and determine that she is still alive, the physician can then order treatment for Hailu. Gebreyes said that he never received this offer before the hearing.

After Gebreyes said that he wanted to take advantage of the opportunity to bring in his own neurologist, the parties stipulated to extend the hearing until July 21, 2015, to give Gebreyes time to retain the services of a neurologist. . . . [T]he district court stated that Gebreyes needs a neurological expert because the matter involves "primarily neurological issues." . . .

On July 21, 2015, Gebreyes presented a plan to transport Hailu to Las Vegas based on the testimony of two physicians. First, Gebreyes called Dr. Brian Callister to testify. Dr. Callister is not a neurologist, but specializes in internal medicine and hospitalist medicine. He examined Hailu the day of his testimony and reviewed her medical records. Based on his examination of Hailu and review of her records, Dr. Callister testified: "I believe that her status is quite grim. I think that her chance of survival, her chance of awakening from her current state is a long shot. However, I do not think that the chance is zero." Dr. Callister stated that all three EEG tests did show brainwaves, albeit abnormal and slow. In Dr. Callister's opinion, the EEG tests are "something that should give you just enough pause to say you can't say with certainty that her chances are zero." Although Dr. Callister admitted that under the AAN guidelines Hailu's condition looks irreversible, Dr. Callister pointed to other factors that demonstrate improvement is a possibility. As examples, Dr. Callister cites Hailu's young age, her health, her skin, her ability to make urine and pass bowel movements, and the fact that the general functioning of the rest of her body is good. . . .

. . . [O]n cross-examination, Dr. Callister conceded that under "a strict definition" of the AAN guidelines, Hailu "would meet their category [of brain death]." On redirect, Dr. Callister concluded that "there's enough variables and enough questions based on the condition of her physical body, the EEG's and the fact that no further neurological testing has been done in several months, and the fact that no outside third party neurologist has looked at her that I would have pause."

[Gebreyes also called a doctor who was willing to perform the tracheostomy but not to take Hailu into his care for other purposes.]

* * *

Next, St. Mary's called Dr. Anthony Floreani to testify. Dr. Floreani took care of Hailu in the ICU since the night following her surgery. Dr. Floreani is a pulmonary doctor, not a neurologist. Dr. Floreani agreed with the conclusions of Dr. Heide that Hailu is brain dead. . . . Dr. Floreani testified that the St. Mary's doctors did the tests "by the book exactly how you should do it." Based on all of the evidence from the July 2 hearing and the July 21 hearing, the district court ruled in favor of St. Mary's. The district court stated that a restraining order should not be granted because the medical evidence from Dr. Heide and Dr. Floreani suggested that the AAN guidelines were followed, and thus, "medical standards were met, the outcome and criteria were satisfied in terms of the statute, the [AAN] protocol was followed, the outcome of the various three step tests under the [AAN] protocol all direct certification of death, and I agree." . . .

DISCUSSION

* * *

The district court focused exclusively on whether St. Mary's physicians satisfied the AAN guidelines, without discussing whether the AAN guidelines satisfy NRS 451.007. Although St. Mary's presented testimony that the AAN guidelines are the accepted medical standard in Nevada— albeit a simple "yes" to the question of whether the AAN guidelines are the accepted medical standard in Nevada—the district court and St. Mary's failed to demonstrate that the AAN guidelines are considered "accepted medical standards" that are applied uniformly throughout states that have enacted the UDDA . . . . The district court did not reach this issue at all, while St. Mary's has only cited one source to support its argument that the AAN guidelines are the nationally accepted medical standard.

St. Mary's cites the New Jersey Law Revision Commission's Report relating to the Declaration of Death Act. However, the report actually supports the opposite conclusion for which St. Mary's argues. . . . *See* N.J. Law Revision Comm'n, Final Report Relating to New Jersey Declaration of Death Act, at 14 (Jan. 18, 2013). [T]he report cited to multiple studies suggesting that "the AAN guidelines need more research" and "there is still a great variety of practice in U.S. hospitals" even though the AAN guidelines were published in 1995. Despite recognizing the AAN as guidelines "upon which most hospitals and physicians rely," the report concluded that the AAN guidelines were not so broadly adopted and utilized as to have become *the* accepted medical standard for determining brain death. Based on the foregoing, and the record before us, we are not convinced that the AAN guidelines are considered the accepted medical standard that can be applied in a way to make Nevada's Determination of Death Act uniform with states that have adopted it, as the UDDA requires. . . .

* * *

It appears from a layperson's review of the Harvard criteria versus the AAN guidelines that the AAN guidelines incorporated many of the clinical tests used in the Harvard criteria. [ ] However, the AAN guidelines do not require confirmatory/ancillary testing, such as EEGs. Although the AAN guidelines state that ancillary testing should be ordered "only if clinical examination cannot be fully performed due to patient factors, or if apnea testing is inconclusive or aborted," the AAN's own study recognized that a decade after publication of the guidelines, 84 percent of brain death determinations still included EEG testing. [ ]

* * *

[T]he briefing and record before us do not answer two key questions. First, the briefing and testimony do not establish whether the AAN guidelines are considered accepted medical standards among states that have enacted the UDDA. . . . Second, whatever their medical acceptance generally, the briefing and testimony do not establish whether the AAN guidelines adequately measure the extraordinarily broad standard laid out by NRS 451.007, which requires, before brain death can be declared under the UDDA, an "irreversible cessation" of "[a]*ll functions* of the person's *entire* brain, including his or her brain stem." [ ] Emphasis added. Though courts defer to the medical community to determine the applicable criteria to measure brain functioning, it is the duty of the law to establish the applicable standard that said criteria must meet. . . .

* * *

We concur: HARDESTY, C.J., and PARRAGUIRRE, DOUGLAS, CHERRY, SAITTA, and GIBBONS, JJ.

NOTES AND QUESTIONS

1.   *Appearances Are Deceiving?* The position that dead is dead (whether measured by the irreversible cessation of cardiopulmonary function or total brain function) forms the foundation for legal standards and medical practice determining whether death has occurred. It certainly can be difficult for laypersons to accept that someone who has been declared brain dead, currently often referred to as "neurological determination of death," is dead. The person who is being supported by a ventilator but who is declared dead by neurological criteria still looks so alive—warm to the touch and with a steady rhythm of breath due to the ventilator. Even health care professionals may experience cognitive dissonance in "caring for" persons who are brain dead while preparing the body for retrieval of organs for transplantation or maintaining the body on support while family matters are settled. Robert D. Truog, Brain Death—Too Flawed to Endure, Too Ingrained to Abandon, 35 J.L. Med. & Ethics 273 (2007). It should not be surprising, then, that conflicts between families and hospitals regarding the determination of death by neurological criteria occur and sometimes reach the courts.

2. *Avoiding Conflicts.* Cases like Hailu always raise questions about whether there were steps that could have been taken to avoid the conflict and lead to acceptance of the determination of death. Do you see any opportunities in Hailu? Both Aden Hailu and Jahi McMath, the subject of the New Yorker piece cited above, were young women of color. What impact might the gender and race of the patients or their next of kin make in this context? See Chapter 2. Both Ms. Hailu and Ms. McMath suffered severe injuries during surgery. As is routine, the declaration of death in each case was made by physicians affiliated with that hospital. Should hospitals establish different default procedures when the patient has been injured during the course of treatment in the hospital? If a patient is not declared to be dead, life-sustaining treatment may still be withdrawn. See Chapter 5. Is it possible that approaching the situation from that direction could produce a different result in some cases? The Ad Hoc Committee (see Section II.A, above) rejected involving the family in the determination of death and allocated the process entirely to the physician. Does a hard-edge boundary protect families or does it create alienation in the process?

3. *Defining Accepted Medical Standards.* The court is careful to note that it is not concluding that the AAN criteria are not the generally accepted medical standard for determining death, but only that the issue was not addressed in the court below. In fact, there is significant variability in the protocols doctors and hospitals use to determine whether brain death has occurred. See David M. Greer et al., Variability of Brain Death Policies in the U.S., 73 JAMA Neurology 213 (2016); David M. Greer et al., Determination of Brain Death/Death by Neurologic Criteria, The World Brain Death Project, 324(11) JAMA 1078 (Aug. 3, 2020).

After the Law Revision Commission's report cited in the court's opinion, the New Jersey legislature amended the state's determination of death statute to require that the determination of brain death be "based upon the exercise of the physician's best medical judgment and in accordance with currently accepted medical standards that are based upon nationally recognized sources of practice guidelines, including, but not limited to, those adopted by the [AAN]." After *Hailu*, the Nevada legislature went even further by requiring that the determination of brain death in adults "must be made" in compliance with the 2010 AAN Guidelines and subsequent revisions adopted by the AAN. Nev. Rev. Stat. § 451.007(2)(b)(1). See also AAN standards at James A. Russell et al., Brain Death, The Determination of Brain Death, and Member Guidance for Brain Death Accommodation Requests: AAN Position Statement, 92(5) Neurology 228 (Jan 2019). Should legislatures select the specific medical protocol to be used, or is that a question for medicine? Should the determination of death be treated differently than other medical decisions in that regard?

4. *Do Medical Standards Align with Legal Requirements?* The court implies that the brain death statute requires that it must be determined that there is absolutely no function in the brain at all. In fact, some extremely minimal activity continues in the brain (such as the secretion of hormones

regulating salts and fluids) that can continue certain bodily functions in some patients who are determined to be brain dead under current medical standards. See, e.g., Seema K. Shah, Piercing the Veil: The Limits of Brain Death as a Legal Fiction, 48 Mich. J.L. Reform 301 (2015). See also President's Council on Bioethics, Controversies in the Determination of Death (2008), recognizing the persistence of some activity but recommending continuation of the current standards as applied, although shifting the term to "total brain failure." Must state statutes be revised to account for the persistence of some brain activity if the whole brain death standard is to be retained? See *T.A.C.P.*, below. Or should the statutes remain the same, and the interpretation of "all functions of the entire brain" be understood to tolerate the incidental activity that remains?

5. *Accommodating Objections.* Should state statutes provide for individual choice as between cardiopulmonary or neurological determination of death? New Jersey formally adopted an exemption from the application of the brain death standard for determination of death for patients who reject brain death on specific grounds:

> The death of an individual shall not be declared upon the basis of neurological criteria . . . when the licensed physician authorized to declare death, has reason to believe, on the basis of information in the individual's available medical records, or information provided by a member of the individual's family or any other person knowledgeable about the individual's personal religious beliefs that such a declaration would violate the personal religious beliefs of the individual. In these cases, death shall be declared, and the time of death fixed, solely upon the basis of cardio-respiratory criteria. . . . N.J. Stat. § 26:6A–5.

A few other states require reasonable accommodation, rather than an exemption, for families that object to the brain death standard for determination of death. The New York State Department of Health requires that hospitals provide reasonable accommodations for those with "religious or moral objections" to brain death, but explicitly does not require accommodation for persons with objections based "solely on the psychological denial that death has occurred or on an alleged inadequacy of the brain death determination." The Department's guidelines advise sensitivity in the latter case. N.Y. State Dept. of Health, Guidelines for Determining Brain Death (2011). Would either the New Jersey or New York provisions apply to the situation in *Hailu*? Compare Cal. Health & Safety Code § 1254.4, encompassing a broader range of objections but providing only for a reasonable time to allow the family to adjust to the diagnosis and, in the case of cultural or religious objections, some reasonable efforts to accommodate those "religious or cultural practices or concerns."

6. *Consent Required?* After remand in *Hailu*, the hospital asked the trial court for permission to perform additional testing, including an electroencephalogram (EEG), over the father's objection, and the court allowed two EEGs to be performed. It's not clear whether that testing took place.

Shortly before the date of the hearing on the issue of the remand, Aden Hailu suffered cardiopulmonary arrest. The death certificate listed the date of death as January 4, 2016, the date of the arrest. Siobhan McAndres, The Contested Death of Aden Hailu, Reno-Gazette-Journal (Mar. 25, 2016). Should surrogate consent be required before a physician performs brain death diagnostic tests? One practical consequence of requiring consent (or honoring family objections) is to enable families to avoid determination of death using the brain death standard. See Pope, Brain Death and the Law: Hard Cases and Legal Challenges, above.

7. *Revising the UDDA.* There have been many suggestions for revisions to the UDDA, often centered on the issues raised by *Hailu* and the notes above. See, e.g., Ariane Lewis et al., Determination of Death by Neurologic Criteria in the United States: The Case for Revising the Uniform Determination of Death Act, 47(4 supp.) J. Law Med. Ethics 9 (Dec. 2019); D. Alan Shewmon, Statement in Support of Revising the Uniform Determination of Death Act and in Opposition to a Proposed Reform, J. Med. & Phil. (2021) (advance article). In 2021, the Uniform Law Commission convened a committee to amend or revise the UDDA. What revisions, if any, would you suggest and why?

IN RE T.A.C.P.

Supreme Court of Florida, 1992.
609 So. 2d 588.

KOGAN, JUSTICE.

We have for review an order of the trial court certified by the Fourth District Court of Appeal as touching on a matter of great public importance requiring immediate resolution by this Court. We frame the issue as follows:

Is an anencephalic newborn considered "dead" for purposes of organ donation solely by reason of its congenital deformity?

* * *

I. Facts

At or about the eighth month of pregnancy, the parents of the child T.A.C.P. were informed that she would be born with anencephaly. . . .

In this case, T.A.C.P. actually survived only a few days after birth. The medical evidence in the record shows that the child T.A.C.P. was incapable of developing any sort of cognitive process, may have been unable to feel pain or experience sensation due to the absence of the upper brain, and at least for part of the time was placed on a mechanical ventilator to assist her breathing. . . .

On the advice of physicians, the parents continued the pregnancy to term and agreed that the mother would undergo caesarean section during birth. The parents agreed to the caesarean procedure with the express hope that the infant's organs would be less damaged and could be used for transplant in other sick children. Although T.A.C.P. had no hope of life herself, the parents both testified in court that they wanted to use this opportunity to give life to others. However, when the parents requested that T.A.C.P. be declared legally dead for this purpose, her health care providers refused out of concern that they thereby might incur civil or criminal liability. . . .

II. The Medical Nature of Anencephaly

Although appellate courts appear never to have confronted the issue, there already is an impressive body of published medical scholarship on anencephaly. From our review of this material, we find that anencephaly is a variable but fairly well defined medical condition. Experts in the field have written that anencephaly is the most common severe birth defect of the central nervous system seen in the United States, although it apparently has existed throughout human history.

A statement by the Medical Task Force on Anencephaly ("Task Force") printed in the New England Journal of Medicine generally described "anencephaly" as "a congenital absence of major portions of the brain, skull, and scalp, with its genesis in the first month of gestation." David A. Stumpf et al., *The Infant with Anencephaly*, 322 New Eng.J.Med. 669, 669 (1990). The large opening in the skull accompanied by the absence or severe congenital disruption of the cerebral hemispheres is the characteristic feature of the condition. *Id.*

The Task Force defined anencephaly as diagnosable only when all of the following four criteria are present:

> (1) A large portion of the skull is absent. (2) The scalp, which extends to the margin of the bone, is absent over the skull defect. (3) Hemorrhagic, fibrotic tissue is exposed because of defects in the skull and scalp. (4) Recognizable cerebral hemispheres are absent.

* * *

Thus, it is clear that anencephaly is distinguishable from some other congenital conditions because its extremity renders it uniformly lethal. . . .

The Task Force stated that most reported anencephalic children die within the first few days after birth, with survival any longer being rare. . . . The Task Force reported, however, that these survival rates are confounded somewhat by the variable degrees of medical care afforded to anencephalics. Some such infants may be given considerable life support while others may be given much less care.

The Task Force reported that the medical consequences of anencephaly can be established with some certainty. All anencephalics by definition are permanently unconscious because they lack the cerebral cortex necessary for conscious thought. Their condition thus is quite similar to that of persons in a persistent vegetative state. Where the brain stem is functioning, as it was here, spontaneous breathing and heartbeat can occur. In addition, such infants may show spontaneous movements of the extremities, "startle" reflexes, and pupils that respond to light. Some may show feeding reflexes, may cough, hiccup, or exhibit eye movements, and may produce facial expressions.

* * *

There appears to be general agreement that anencephalics usually have ceased to be suitable organ donors by the time they meet all the criteria for "whole brain death," i.e., the complete absence of brainstem function. . . . There also is no doubt that a need exists for infant organs for transplantation. Nationally, between thirty and fifty percent of children under two years of age who need transplants die while waiting for organs to become available. . . .

III. Legal Definitions of "Death" & "Life"

[The court considers Florida's statute:]

> For legal and medical purposes, *where respiratory and circulatory functions are maintained by artificial means of support* so as to preclude a determination that these functions have ceased, the occurrence of death *may* be determined where there is the irreversible cessation of the functioning of the entire brain, including the brain stem, determined in accordance with this section.

§ 382.009(1), Fla.Stat. (1991) (emphasis added). A later subsection goes on to declare:

> Except for a diagnosis of brain death, the standard set forth in this section is not the exclusive standard for determining death or for the withdrawal of life-support systems.

§ 382.009(4), Fla.Stat. (1991). This language is highly significant for two reasons.

First, the statute does not purport to codify the common law standard applied in some other jurisdictions, as does the uniform act. The use of the permissive word "may" in the statute in tandem with the savings clause of section 382.009(4) buttresses the conclusion that the legislature envisioned other ways of defining "death." Second, the statutory framers clearly did not intend to apply the statute's language to the anencephalic infant not being kept alive by life support. To the contrary, the framers expressly limited the statute to that situation in which "respiratory and circulatory functions are maintained by artificial means of support."

* * *

IV. Common Law & Policy

* * *

The question remaining is whether there is good reason in public policy for this Court to create an additional common law standard applicable [only] to anencephalics. Alterations of the common law, while rarely entertained or allowed, are within this Court's prerogative. . . . We believe, for example, that our adoption of the cardiopulmonary definition of death today is required by public necessity and, in any event, merely formalizes what has been the common practice in this state for well over a century.

Such is not the case with petitioners' request. Our review of the medical, ethical, and legal literature on anencephaly discloses absolutely no consensus that public necessity or fundamental rights will be better served by granting this request.

We are not persuaded that a public necessity exists to justify this action, in light of the other factors in this case—although we acknowledge much ambivalence about this particular question. We have been deeply touched by the altruism and unquestioned motives of the parents of T.A.C.P. The parents have shown great humanity, compassion, and concern for others. The problem we as a Court must face, however, is that the medical literature shows unresolved controversy over the extent to which anencephalic organs can or should be used in transplants.

* * *

We express no opinion today about who is right and who is wrong on these issues—if any "right" or "wrong" can be found here. The salient point is that no consensus exists as to: (a) the utility of organ transplants of the type at issue here; (b) the ethical issues involved; or (c) the legal and constitutional problems implicated.

V. Conclusions

Accordingly, we find no basis to expand the common law to equate anencephaly with death. We acknowledge the possibility that some infants' lives might be saved by using organs from anencephalics who do not meet the traditional definition of "death" we reaffirm today. But weighed against this is the utter lack of consensus, and the questions about the overall utility of such organ donations. The scales clearly tip in favor of not extending the common law in this instance.

To summarize: We hold that Florida common law recognizes the cardiopulmonary definition of death as stated above; and Florida statutes create a "whole-brain death" exception applicable whenever cardiopulmonary function is being maintained artificially. There are no other legal standards for determining death under present Florida law.

Because no Florida statute applies to the present case, the determination of death in this instance must be judged against the common law cardiopulmonary standard. The evidence shows that T.A.C.P.'s heart was beating and she was breathing at the times in question. Accordingly, she was not dead under Florida law, and no donation of her organs would have been legal.

NOTES AND QUESTIONS

1.   *Variation Among the States.* Despite the UDDA, there is significant variation among state statutes recognizing irreversible cessation of all brain function as a legal standard for the determination of death. These differences can be significant, as illustrated in the court's discussion of the Florida statute in *T.A.C.P.* Several states have accepted brain death for the determination of death by case law rather than by statute. In departing from the UDDA, some states adopted definitions for "irreversible," which the UDDA leaves undefined; some provide that brain function is relevant for the determination of death *only* where cardiopulmonary function is maintained by mechanical means. Some statutes provide that a person *may* (instead of *shall*) be considered dead in the presence of irreversible cessation of cardiopulmonary or all brain function. Jason Goldsmith, Wanted! Dead and/or Alive: Choosing Among the Not-So-Uniform Statutory Definitions of Death, 61 U. Miami L. Rev. 871 (2007).

2.   *Implications for Others.* If the Florida court had declared that anencephaly was acceptable as a legal standard for the determination of death, what implications would that have for medical treatment of individuals with the same medical condition? Shortly after the *T.A.C.P.* opinion was issued, a mother successfully fought a hospital refusal to repeatedly resuscitate her infant who had anencephaly, with the court ultimately holding that a federal statute requiring that hospitals provide emergency care covered the child. In re Baby K, 16 F. 3d 590 (4th Cir. 1994). The hospital in *Baby K* claimed that continued treatment was futile, not that Baby K met a legal standard for death. See Chapter 5. Some states already allow individual choice in the application of the neurological determination of death. See discussion after *Hailu*, above. What are the implications of allowing individualized standards for the determination of death?

3.   *"Whole" vs. "Higher" Brain Death.* Some have argued that the legal standard for death should be expanded to embrace the irreversible cessation of so-called "higher brain" functions related to cognition, identity, and personality. Proponents argue that neocortical function is essential to being human; that without it, humans cannot function as a whole; and that counting brain function in the absence of neocortical function as life is a base and animalistic view of the nature of being human. Ben Sarbey, Definition of Death: Brain Death and What Matters in a Person, 3 J. L. & Biosciences 745 (2016). Objections to this expansion range from practical to metaphysical concerns, including that death would be pronounced while the body is

breathing on its own; diagnosis of the condition continues to be challenging; newer knowledge about higher brain injuries continues to emerge; declaring a person to be dead when there is still brain function violates the sanctity of human life; and reliance on cognitive function as a measure of personhood leads inevitably to a slippery slope and devaluation of persons with cognitive disabilities. See, e.g., James Bernat et al., Defining Death in Theory and Practice, 12 Hastings Ctr. Rep. 5 (1982).

C. CARDIOPULMONARY DEATH REVIVED IN ORGAN RETRIEVAL

The adoption of the neurological determination of death, or brain death, was designed in part to increase the retrieval of life-sustaining organs for transplant. In the face of continuing shortages of organs for transplant, an Institute of Medicine report (Organ Donation: Opportunities for Action 2006) advocated the widespread adoption of the then-emerging practice of donation after cardiopulmonary (or circulatory or cardiac) death (DCD), also known as donation after circulatory determination of death (DCDD). (The literature and hospital policies have used a variety of terms over time to describe this process, including "non-heart-beating donors," "donors without a heartbeat," or "asystolic donors.") The volume of organs retrieved through DCD has increased dramatically. In 2020, there were 3,223 donors through DCD, representing 25.6% of donated cadaveric organs, a 18.6% increase over the 2019 total.

In a way, DCD appears to return to the older, common law standard as it relies on the cessation of cardiopulmonary function as the indicator that death has occurred. Although DCD at first glance appears to bring us full circle, the procedures differ quite substantially from the procedures used to determine cardiopulmonary death in earlier times and in contexts other than organ donation. The practice of DCD is now a mainstream practice, but it continues to raise significant ethical and legal controversies, including challenges to legal standards for the determination of death and the continued viability of the "dead donor rule," i.e., the rule that a person's life-sustaining organs may be removed only after death. See, e.g., Robert D. Truog, Defining Death: Getting it Wrong for All the Right Reasons, 93 Tex. L. Rev. 1885 (2015).

PROBLEM: CENTRAL CITY HOSPITAL'S PROPOSAL FOR DONATION AFTER CIRCULATORY DEATH

Central City Hospital (CCH) is a large teaching hospital. The hospital has decided to increase the volume of its organ procurement and transplant program in an effort to develop programs that produce a substantial net revenue stream, to meet the need of desperate patients, to realize the improved outcomes that a higher volume of transplants produces, and to develop a source of pride for hospital staff. CCH hired a new medical director for its transplant program, Dr. Joshua Niblet, who had been directing a very successful program at a hospital in another state.

Until now, all cadaveric organs donated at CCH have come from patients who have been declared brain dead while they are on mechanical cardiopulmonary support. Dr. Niblet is recommending that the hospital engage in organ recovery in controlled DCD. "Controlled DCD" occurs when "death is anticipated and occurs after medical supportive therapy is withdrawn [at the request of the patient, the patient's surrogate, or as directed by an advance directive], usually in an operating room, or hospital intensive care unit, or within a closely monitored time frame." IOM Report (2006). "Uncontrolled DCD" (or unexpected, unplanned, or unanticipated cardiac death) occurs when cardiopulmonary function has stopped spontaneously and usually unexpectedly. The IOM report recognizes that organ retrieval in uncontrolled cardiac death presents special challenges. Dr. Niblet is not seeking implementation of uncontrolled DCD at this time.

As a newly minted health lawyer and newly appointed assistant hospital counsel, you have been assigned to work with Dr. Niblet to help him institute the new program. You are being asked to assist in drafting the substance of the hospital's policies for DCD. Identify the issues you see in the following plan and how you would resolve them. Assume that your state has adopted the UDDA without revision. As a guide for your work, use the following federal regulation, as well as the materials that follow the regulation:

42 C.F.R. § 486.344(f)

If an [Organ Procurement Organization] recovers organs from donors after cardiac death, the OPO must have protocols that address the following: (1) Criteria for evaluating patients for donation after cardiac death; (2) Withdrawal of support, including the relationship between the time of consent to donation and the withdrawal of support; (3) Use of medications and interventions not related to withdrawal of support; (4) Involvement of family members prior to organ recovery; (5) Criteria for declaration of death and the time period that must elapse prior to organ recovery.

Potential DCD Donors

The federal policy governing organ transplantation currently provides that potential DCD donors include "patients who have died, or whose death is imminent, whose medical treatment no longer offers medical benefit to the patient as determined by the patient, the patient's authorized surrogate, or the patient's advance directive, if applicable." OPTN Policy 2.15. See Chapter 5 on treatment decision-making at the end of life. OPTN policy specifies that the decision to withdraw life-sustaining treatment be made before organ donation is discussed. If the patient who is the potential donor is making the donation decision, the patient's capacity to make that decision must be assessed. If the patient is incompetent but had earlier executed a document indicating intent to donate organs, that document (which typically does not include specific directions concerning DCD) controls. Failing those first two options, the individual identified in state law decides whether to donate organs. See Section I.C., above.

Declaration of Death

The determination of death must be made in compliance with state law. Once the decision to withdraw treatment is made, medical interventions to preserve the organs are begun, and the patient is taken to the operating room. Mechanical support sustaining cardiac function is removed, and the patient is observed to see whether the heart will begin to function on its own, before death is declared and the surgeon makes the incision and removes the organs. Although an uncommon occurrence, the hospital must be ready to treat the patient if the heart begins beating spontaneously.

Hospital protocols concerning how long to watch for spontaneous breathing vary significantly, but typically range from two to five minutes. The OPTN policy does not adopt a specific wait time. It is generally recognized that measurable brain activity continues for a significant time, perhaps up to 15 minutes after cessation of cardiopulmonary function. See Sam David Shemie and Dale Gardner, Circulatory Arrest, Brain Arrest and Death Determination, 5(15) Frontiers in Cardiovascular Med. 1 (2018); James DuBois, Is Organ Procurement Causing the Death of Patients?, 18 Issues L. & Med. 21 (2002); Jerry Menikoff, The Importance of Being Dead: Non-Heart-Beating Organ Donation, 18 Issues L. & Med. 3 (2002).

Dr. Niblet wants to set the shortest time possible, as the organs begin to deteriorate quickly once circulation stops.

Medical Interventions Prior to DCD

Medications to thin the blood, expand the blood vessels, cool down the body, and preserve the organs are administered before the patient is declared dead. Where the patient does not have decisional capacity, the decision to withdraw treatment is made by a surrogate or pursuant to the patient's advance directive (such as a living will or durable power of attorney). Most advance directives do not include instructions concerning medical treatment provided for pre-mortem organ preservation in DCD and instead only specify that medical treatment is to be stopped if certain medical conditions are met.

The Uniform Anatomical Gift Act provides that during the time in which a potential donor is being evaluated, "measures necessary to ensure the medical suitability of the [organs] may not be withdrawn unless the hospital . . . knows that the individual expressed a contrary intent." § 14(c). As to whether the typical advance directive evidences contrary intent, the Act provides that interventions required for successful organ recovery continue until the apparent conflict between the two instructions (the advance directive and the document for organ donation) is resolved, unless the interventions to preserve the organs are contraindicated for end-of-life care. § 21. See Richard Bonnie, Stephanie Wright, & Kelly Dineen, Legal Authority to Preserve

Organs in Cases of Uncontrolled Cardiac Death, 36 J.L. Med. & Ethics 741 (2008).

Some protocols use extracorporeal membrane oxygenation (ECMO) to provide oxygenated blood to preserve organs after the death. Because there is a concern that oxygenated blood flowing through the heart after mechanical support has been withdrawn would be evidence of spontaneous cardiac function, physicians may block blood flow above the diaphragm so that only the abdominal organs receive oxygenated blood. See A.L. Dalle Ave et al., Ethical Issues in the Use of ECMO in Controlled Donation after Circulatory Determination of Death, 16 Am J. Transplantation 2293 (Aug. 2016). The demand for heart transplantation, however, is driving increasing retrievals from DCD donors. Aravinda Page et al., Heart Transplantation from Donation after Circulatory Determined Death, 7 Annals of Cardiothoracic Surgery 75 (2018).

Dr. Niblet wants full discretion to employ the most effective organ preservation interventions as needed.

Questions for Hospital Policy Development

When patients or families are considering decisions relating to treatment at the end of life, when should the possibility of organ donation be raised? Who should present the option?

If the patient no longer has the capacity to make decisions but has executed a document for donation of organs upon death by signing up with the state's donor registry, as signified on the driver's license, or by signing a donor card, does that indicate consent to donation after DCD? These documents ordinarily do not provide information concerning DCD or measures that must be taken to preserve the organs. If the patient lacks decisional capacity, would you treat the organ donation document as determinative, or would you require that the consent of the surviving next-of-kin be secured?

Under the UAGA, pre-mortem documents of gift are binding on family members and others (see Section I.C, above). Would you recommend that the hospital secure informed consent from the surrogate for the specific interventions required to preserve organs, or would you structure the consent to donate organs to include permission to perform "all procedures necessary to retrieve the organs"? If the patient left an advance directive declining all medical treatment, would that preclude donation upon DCD?

What would you recommend as to wait time for the CCH policy? How does the wait time relate to whether the donor is dead under the UDDA? Could the precise language of your state's determination of death statute make a difference to your legal analysis? See *T.A.C.P.* and the notes following, above. Would you prefer that the state legislature amend its statute to specifically allow DCD, or are you satisfied that it falls within the UDDA?

CHAPTER 5

LIFE AND DEATH DECISIONS

■ ■ ■

I. INTRODUCTION

Concerns about the ethical, legal, and medical propriety of discontinuing what is now generally referred to as "life-sustaining treatment" have played a leading role in the development of law and bioethics over the last few decades. These concerns continue to capture our attention due to several factors: medical advances that have increased our ability to prolong life, debates over the "brain death" standard for determination of death (discussed in Chapter 4), increased access to medical aid in dying (discussed in Chapter 6), and attention to the consequences of individual and structural inequities in health care.

While the law is regularly invoked to resolve disputes in this area, a number of the questions brought before courts and legislatures arguably are outside the competence of the law. For example, consider what might constitute a "terminal illness," a concept that some have considered relevant in bioethical decision-making. While there surely is an element of medical judgment in a determination that a patient is "terminally ill," more than that is involved. Classifying a patient as "terminally ill" considers a combination of the patient's medical condition and anticipated life expectancy, as well as the social, ethical, and legal consequences of the classification. If such a classification triggers a provision in an advance directive, or if it allows a physician to participate in aid in dying, for example, the classification might be treated differently than if it triggers the patient's relocation from one room to another within a hospital.

If a determination of terminal illness is not solely a medical decision, but rather a hybrid of medical, ethical, social, political, and legal decisions, where is the locus of appropriate decision-making? Should the decision be made by health care professionals alone? By a patient and the patient's family? By a hospital committee? By some external committee? By a court-appointed guardian? By the court itself? By the state legislature? By Congress? Should the decision be based upon principles and rules that emerge from medicine, ethics, religion, litigation, or legislative social policy making? Finally, what substantive principles ought to govern the decision-maker? And what safeguards should be present?

The difficulties in allocating decision-making authority and developing appropriate substantive principles are not limited to the "terminal illness" classification. They extend to such defining terms as "life-sustaining treatment," "death prolonging," "extraordinary means," "heroic efforts," "intractable pain," "suffering," "persistent vegetative state," "futile," and even breathing and feeding.

For almost every question that arises within bioethics, we must make two determinations: (1) a procedural determination—who should be authorized to resolve the problem, and (2) a substantive determination—what substantive principles should apply. As one might expect, the substantive questions are often hidden in apparently procedural inquiries.

As these issues have come before the courts over the past few decades, the courts have looked to both the traditions of the common law and the traditions of ethics and medicine to discover principles for decision-making. In turn, judicial decisions have often formed the basis for new ethical and medical approaches. In fact, the debate over appropriate ethical policy in determining when life-sustaining treatment should be initiated or discontinued now involves lawyers as much as bioethics scholars, and the public debates on these issues have centered on the judicial resolution of cases and the promulgation of statutes as much as on any other source of formal principles. The law is not merely looking to ethics for potential methods of analysis, it is supplanting (some would say usurping) ethics in debate on these issues.

The law is not developed only through rationally justified and formally articulated judicial opinions. The law also comes out of political compromises in legislative action and public perceptions of well-publicized bioethical cases. The increasingly overt political nature of some questions—such as questions about the definition, meaning, and quality of life—are both profoundly personal and deeply related to larger concerns of equity and justice. Medical advances have increased our ability to prolong life. Is society spending too much to extend the life of people who are very ill or at the end of their lives? Forty years ago, Governor Lamm of Colorado was considered outrageous when he suggested that the elderly may have an obligation to die; today the social consequences of health care decisions have become a matter of real concern to those who realize that all of a family's assets easily could be consumed by a final illness.

While some are concerned that they will receive too much care at the end of life, others are concerned that they will receive too little. Many disability advocates, for example, have expressed well-founded concerns that life and death decisions by and for people perceived as ill or impaired reflect long-standing assumptions and negative beliefs about the value of life with a disability, sometimes with life-ending results (discussed in Chapter 2). If end-of-life care is a limited health care resource, how can we ensure that it is distributed equitably, regardless of age, disability, race or ethnicity, gender, or other characteristic?

The starting point for case law concerning health care decision-making is generally the law of informed consent. The law of informed consent is bolstered by state administrative regulations (often from the state health department or its equivalent), state statutes (for example, patients' rights provisions in some states), state constitutional provisions (including the sometimes state-protected rights of privacy and dignity), federal statutes and regulations (like HIPAA and those assuring the privacy and security of medical records, and those governing research involving human subjects), and the U.S. Constitution. But another starting point for end-of-life treatment decisions is drawn from cases involving persons who are incapacitated and cannot make decisions for themselves. In such cases, the courts have often referenced the traditional law of guardianship in which the state assumes a parens patriae role to protect the incompetent person. While at times the two approaches work in harmony, at other times they lead to conflicting results.

This chapter begins with a discussion of the constitutional foundation upon which health care decision-making is based. This chapter addresses life and death decisions, but the principles that arise out of these cases are extended to other health care contexts. In the *Cruzan* case, which you will be reading in this chapter, and in the *Glucksberg* case, which you will be reading in the next chapter, the U.S. Supreme Court was called upon to apply the U.S. Constitution to end-of-life care, but the principles established in those cases have ramifications throughout the health care system. The Constitution may set the basic ground rules for recognizing health care decision-making authority, but it leaves a great deal of room for state regulation.

After examining the constitutional framework, this chapter examines the authority of an adult who has decisional capacity—a competent adult— to make his or her own health care decisions. We then move on to review the decision-making rules for adults who lack that capacity. We review advance directives such as individual instructions (previously called living wills) and durable powers of attorney, tools used for decision-making for a now-incapacitated individual. The chapter concludes with a discussion of decision-making surrounding treatment deemed futile.

In reading this chapter, consider what role individuals, families, hospitals, institutional ethics committees, health care professionals, courts, legislatures, advocacy organizations, and others ought to play in dealing with the ethical, medical, legal, social, and political questions that often arise out of our new-found technical ability to maintain life. Also consider what safeguards should be available to ensure that bias, unequal treatment, and discrimination do not find their way into decisions by each of these actors regarding withholding or withdrawing life-sustaining care.

As you review the way courts and others have considered individual cases, attempt to distill reasoned principles from their judgments. Is your analysis a procedural or substantive one? Who ought to be involved in the decision-making at each point? What principles are relevant to your analysis at each point?

II. MAKING HEALTH CARE DECISIONS ABOUT DEATH AND DYING: THE CONSTITUTIONAL FOUNDATION

CRUZAN V. DIRECTOR, MISSOURI DEPARTMENT OF HEALTH

Supreme Court of the United States, 1990.
497 U.S. 261.

CHIEF JUSTICE REHNQUIST delivered the opinion of the Court.

Petitioner Nancy Beth Cruzan was rendered incompetent as a result of severe injuries sustained during an automobile accident. Co-petitioners Lester and Joyce Cruzan, Nancy's parents and co-guardians, sought a court order directing the withdrawal of their daughter's artificial feeding and hydration equipment after it became apparent that she had virtually no chance of recovering her cognitive faculties. The Supreme Court of Missouri held that because there was no clear and convincing evidence of Nancy's desire to have life-sustaining treatment withdrawn under such circumstances, her parents lacked authority to effectuate such a request. We granted certiorari and now affirm.

On the night of January 11, 1983, Nancy Cruzan lost control of her car as she traveled down Elm Road in Jasper County, Missouri. The vehicle overturned, and Cruzan was discovered lying face down in a ditch without detectable respiratory or cardiac function. Paramedics were able to restore her breathing and heartbeat at the accident site, and she was transported to a hospital in an unconscious state. An attending neurosurgeon diagnosed her as having sustained probable cerebral contusions compounded by significant anoxia (lack of oxygen). The Missouri trial court in this case found that permanent brain damage generally results after 6 minutes in an anoxic state; it was estimated that Cruzan was deprived of oxygen from 12 to 14 minutes. She remained in a coma for approximately three weeks and then progressed to an unconscious state in which she was able to orally ingest some nutrition. In order to ease feeding and further the recovery, surgeons implanted a gastrostomy feeding and hydration tube in Cruzan with the consent of her then husband. Subsequent rehabilitative efforts proved unavailing. She now lies in a Missouri state hospital in what is commonly referred to as a persistent vegetative state: generally, a condition in which a person exhibits motor reflexes but evinces no indications of significant cognitive function. The State of Missouri is bearing the cost of her care.

After it had become apparent that Nancy Cruzan had virtually no chance of regaining her mental faculties her parents asked hospital employees to terminate the artificial nutrition and hydration procedures. All agree that such a removal would cause her death. The employees refused to honor the request without court approval. The parents then sought and received authorization from the state trial court for termination. The court found that a person in Nancy's condition had a fundamental right under the State and Federal Constitutions to refuse or direct the withdrawal of "death prolonging procedures." The court also found that Nancy's "expressed thoughts at age twenty-five in somewhat serious conversation with a housemate friend that if sick or injured she would not wish to continue her life unless she could live at least halfway normally suggests that given her present condition she would not wish to continue on with her nutrition and hydration."

The Supreme Court of Missouri reversed by a divided vote. . . .

We granted certiorari to consider the question of whether Cruzan has a right under the United States Constitution which would require the hospital to withdraw life-sustaining treatment from her under these circumstances.

* * *

State courts have available to them for decision a number of sources— state constitutions, statutes, and common law—which are not available to us. In this Court, the question is simply and starkly whether the United States Constitution prohibits Missouri from choosing the rule of decision which it did. This is the first case in which we have been squarely presented with the issue of whether the United States Constitution grants what is in common parlance referred to as a "right to die."

* * *

The Fourteenth Amendment provides that no State shall "deprive any person of life, liberty, or property, without due process of law." The principle that a competent person has a constitutionally protected liberty interest in refusing unwanted medical treatment may be inferred from our prior decisions. . . .

But determining that a person has a "liberty interest" under the Due Process Clause does not end the inquiry;[7] "whether respondent's

---

[7]   Although many state courts have held that a right to refuse treatment is encompassed by a generalized constitutional right of privacy, we have never so held. We believe this issue is more

constitutional rights have been violated must be determined by balancing his liberty interests against the relevant state interests." [ ]

Petitioners insist that under the general holdings of our cases, the forced administration of life-sustaining medical treatment, and even of artificially-delivered food and water essential to life, would implicate a competent person's liberty interest. Although we think the logic of the cases discussed above would embrace such a liberty interest, the dramatic consequences involved in refusal of such treatment would inform the inquiry as to whether the deprivation of that interest is constitutionally permissible. But for purposes of this case, we assume that the United States Constitution would grant a competent person a constitutionally protected right to refuse lifesaving hydration and nutrition.

Petitioners go on to assert that an incompetent person should possess the same right in this respect as is possessed by a competent person. . . .

The difficulty with petitioners' claim is that in a sense it begs the question: an incompetent person is not able to make an informed and voluntary choice to exercise a hypothetical right to refuse treatment or any other right. Such a "right" must be exercised for her, if at all, by some sort of surrogate. Here, Missouri has in effect recognized that under certain circumstances a surrogate may act for the patient in electing to have hydration and nutrition withdrawn in such a way as to cause death, but it has established a procedural safeguard to assure that the action of the surrogate conforms as best it may to the wishes expressed by the patient while competent. Missouri requires that evidence of the incompetent's wishes as to the withdrawal of treatment be proved by clear and convincing evidence. The question, then, is whether the United States Constitution forbids the establishment of this procedural requirement by the State. We hold that it does not.

Whether or not Missouri's clear and convincing evidence requirement comports with the United States Constitution depends in part on what interests the State may properly seek to protect in this situation. Missouri relies on its interest in the protection and preservation of human life, and there can be no gainsaying this interest. As a general matter, the States—indeed, all civilized nations—demonstrate their commitment to life by treating homicide as serious crime. Moreover, the majority of States in this country have laws imposing criminal penalties on one who assists another to commit suicide. We do not think a State is required to remain neutral in the face of an informed and voluntary decision by a physically-able adult to starve to death.

But in the context presented here, a State has more particular interests at stake. The choice between life and death is a deeply personal

---

properly analyzed in terms of a Fourteenth Amendment liberty interest. See *Bowers v. Hardwick,* 478 U.S. 186, 194–195 (1986).

decision of obvious and overwhelming finality. We believe Missouri may legitimately seek to safeguard the personal element of this choice through the imposition of heightened evidentiary requirements. It cannot be disputed that the Due Process Clause protects an interest in life as well as an interest in refusing life-sustaining medical treatment. Not all incompetent patients will have loved ones available to serve as surrogate decisionmakers. . . . A State is entitled to guard against potential abuses in such situations. Similarly, a State is entitled to consider that a judicial proceeding to make a determination regarding an incompetent's wishes may very well not be an adversarial one, with the added guarantee of accurate factfinding that the adversary process brings with it. [ ] Finally, we think a State may properly decline to make judgments about the "quality" of life that a particular individual may enjoy, and simply assert an unqualified interest in the preservation of human life to be weighed against the constitutionally protected interests of the individual.

In our view, Missouri has permissibly sought to advance these interests through the adoption of a "clear and convincing" standard of proof to govern such proceedings.

* * *

We think it self-evident that the interests at stake in the instant proceedings are more substantial, both on an individual and societal level, than those involved in a run-of-the-mine civil dispute. But not only does the standard of proof reflect the importance of a particular adjudication, it also serves as "a societal judgment about how the risk of error should be distributed between the litigants." [ ] The more stringent the burden of proof a party must bear, the more that party bears the risk of an erroneous decision. We believe that Missouri may permissibly place an increased risk of an erroneous decision on those seeking to terminate an incompetent individual's life-sustaining treatment. An erroneous decision not to terminate results in a maintenance of the status quo; the possibility of subsequent developments such as advancements in medical science, the discovery of new evidence regarding the patient's intent, changes in the law, or simply the unexpected death of the patient despite the administration of life-sustaining treatment, at least create the potential that a wrong decision will eventually be corrected or its impact mitigated. An erroneous decision to withdraw life-sustaining treatment, however, is not susceptible of correction.

* * *

In sum, we conclude that a State may apply a clear and convincing evidence standard in proceedings where a guardian seeks to discontinue nutrition and hydration of a person diagnosed to be in a persistent vegetative state. . . .

The Supreme Court of Missouri held that in this case the testimony adduced at trial did not amount to clear and convincing proof of the patient's desire to have hydration and nutrition withdrawn. . . . The testimony adduced at trial consisted primarily of Nancy Cruzan's statements made to a housemate about a year before her accident that she would not want to live should she face life as a "vegetable," and other observations to the same effect. The observations did not deal in terms with withdrawal of medical treatment or of hydration and nutrition. We cannot say that the Supreme Court of Missouri committed constitutional error in reaching the conclusion that it did.

* * *

JUSTICE O'CONNOR, concurring.

I agree that a protected liberty interest in refusing unwanted medical treatment may be inferred from our prior decisions, and that the refusal of artificially delivered food and water is encompassed within that liberty interest. I write separately to clarify why I believe this to be so.

As the Court notes, the liberty interest in refusing medical treatment flows from decisions involving the State's invasions into the body. Because our notions of liberty are inextricably entwined with our idea of physical freedom and self-determination, the Court has often deemed state incursions into the body repugnant to the interests protected by the Due Process Clause. [ ] The State's imposition of medical treatment on an unwilling competent adult necessarily involves some form of restraint and intrusion. A seriously ill or dying patient whose wishes are not honored may feel a captive of the machinery required for life-sustaining measures or other medical interventions. Such forced treatment may burden that individual's liberty interests as much as any state coercion. [ ]

The State's artificial provision of nutrition and hydration implicates identical concerns. Artificial feeding cannot readily be distinguished from other forms of medical treatment. . . . Whether or not the techniques used to pass food and water into the patient's alimentary tract are termed "medical treatment," it is clear they all involve some degree of intrusion and restraint. Feeding a patient by means of a nasogastric tube requires a physician to pass a long flexible tube through the patient's nose, throat and esophagus and into the stomach. Because of the discomfort such a tube causes, "[m]any patients need to be restrained forcibly and their hands put into large mittens to prevent them from removing the tube." . . . A gastrostomy tube (as was used to provide food and water to Nancy Cruzan), or jejunostomy tube must be surgically implanted into the stomach or small intestine. . . . Requiring a competent adult to endure such procedures against her will burdens the patient's liberty, dignity, and freedom to determine the course of her own treatment. Accordingly, the liberty guaranteed by the Due Process Clause must protect, if it protects anything, an individual's deeply personal decision to reject medical treatment, including the artificial delivery of food and water.

I also write separately to emphasize that the Court does not today decide the issue whether a State must also give effect to the decisions of a surrogate decisionmaker. In my view, such a duty may well be constitutionally required to protect the patient's liberty interest in refusing medical treatment. Few individuals provide explicit oral or written instructions regarding their intent to refuse medical treatment should they become incompetent. States which decline to consider any evidence other than such instructions may frequently fail to honor a patient's intent. Such failures might be avoided if the State considered an equally probative source of evidence: the patient's appointment of a proxy to make health care decisions on her behalf.

* * *

Today's decision, holding only that the Constitution permits a State to require clear and convincing evidence of Nancy Cruzan's desire to have artificial hydration and nutrition withdrawn, does not preclude a future determination that the Constitution requires the States to implement the decisions of a patient's duly appointed surrogate. Nor does it prevent States from developing other approaches for protecting an incompetent individual's liberty interest in refusing medical treatment. . . . Today we decide only that one State's practice does not violate the Constitution; the more challenging task of crafting appropriate procedures for safeguarding incompetents' liberty interests is entrusted to the "laboratory" of the States, in the first instance.

JUSTICE SCALIA, concurring.

* * *

While I agree with the Court's analysis today, and therefore join in its opinion, I would have preferred that we announce, clearly and promptly, that the federal courts have no business in this field; that American law has always accorded the State the power to prevent, by force if necessary, suicide—including suicide by refusing to take appropriate measures necessary to preserve one's life; that the point at which life becomes "worthless," and the point at which the means necessary to preserve it become "extraordinary" or "inappropriate," are neither set forth in the Constitution nor known to the nine Justices of this Court any better than they are known to nine people picked at random from the Kansas City telephone directory; and hence, that even when it *is* demonstrated by clear and convincing evidence that a patient no longer wishes certain measures to be taken to preserve her life, it is up to the citizens of Missouri to decide, through their elected representatives, whether that wish will be honored. It is quite impossible (because the Constitution says nothing about the matter) that those citizens will decide upon a line less lawful than the one we would choose; and it is unlikely (because we know no more about "life-and-death" than they do) that they will decide upon a line less reasonable.

The text of the Due Process Clause does not protect individuals against deprivations of liberty *simpliciter*. It protects them against deprivations of liberty "without due process of law." To determine that such a deprivation would not occur if Nancy Cruzan were forced to take nourishment against her will, it is unnecessary to reopen the historically recurrent debate over whether "due process" includes substantive restrictions. [ ] It is at least true that no "substantive due process" claim can be maintained unless the claimant demonstrates that the State has deprived him of a right historically and traditionally protected against State interference. [ ] That cannot possibly be established here.

. . . "[T]here is no significant support for the claim that a right to suicide is so rooted in our tradition that it may be deemed 'fundamental' or 'implicit in the concept of ordered liberty.' "[ ]

Petitioners rely on three distinctions to separate Nancy Cruzan's case from ordinary suicide: (1) that she is permanently incapacitated and in pain; (2) that she would bring on her death not by any affirmative act but by merely declining treatment that provides nourishment; and (3) that preventing her from effectuating her presumed wish to die requires violation of her bodily integrity. None of these suffices.

[Scalia points out (1) that pain and incapacity have never constituted legal defenses to a charge of suicide, (2) that the distinction between "action" and "inaction" is logically and legally meaningless, and (3) that preventing suicide often (or always) requires the violation of bodily integrity, and it begs the question of whether the refusal of treatment is itself suicide.]

* * *

Are there, then, no reasonable and humane limits that ought not to be exceeded in requiring an individual to preserve his own life? There obviously are, but they are not set forth in the Due Process Clause. What assures us that those limits will not be exceeded is the same constitutional guarantee that is the source of most of our protection—what protects us, for example, from being assessed a tax of 100% of our income above the subsistence level, from being forbidden to drive cars, or from being required to send our children to school for 10 hours a day, none of which horribles is categorically prohibited by the Constitution. Our salvation is the Equal Protection Clause, which requires the democratic majority to accept for themselves and their loved ones what they impose on you and me. This Court need not, and has no authority to, inject itself into every field of human activity where irrationality and oppression may theoretically occur, and if it tries to do so it will destroy itself.

JUSTICE BRENNAN, with whom JUSTICE MARSHALL and JUSTICE BLACKMUN join, dissenting.

* * *

Today the Court, while tentatively accepting that there is some degree of constitutionally protected liberty interest in avoiding unwanted medical treatment, including life-sustaining medical treatment such as artificial nutrition and hydration, affirms the decision of the Missouri Supreme Court. The majority opinion, as I read it, would affirm that decision on the ground that a State may require "clear and convincing" evidence of Nancy Cruzan's prior decision to forgo life-sustaining treatment under circumstances such as hers in order to ensure that her actual wishes are honored. Because I believe that Nancy Cruzan has a fundamental right to be free of unwanted artificial nutrition and hydration, which right is not outweighed by any interests of the State, and because I find that the improperly biased procedural obstacles imposed by the Missouri Supreme Court impermissibly burden that right, I respectfully dissent. Nancy Cruzan is entitled to choose to die with dignity.

* * *

The right to be free from unwanted medical attention is a right to evaluate the potential benefit of treatment and its possible consequences according to one's own values and to make a personal decision whether to subject oneself to the intrusion. For a patient like Nancy Cruzan, the sole benefit of medical treatment is being kept metabolically alive. Neither artificial nutrition nor any other form of medical treatment available today can cure or in any way ameliorate her condition. Irreversibly vegetative patients are devoid of thought, emotion and sensation; they are permanently and completely unconscious. As the President's Commission concluded in approving the withdrawal of life support equipment from irreversibly vegetative patients:

> "[T]reatment ordinarily aims to benefit a patient through preserving life, relieving pain and suffering, protecting against disability, and returning maximally effective functioning. If a prognosis of permanent unconsciousness is correct, however, continued treatment cannot confer such benefits. Pain and suffering are absent, as are joy, satisfaction, and pleasure. Disability is total and no return to an even minimal level of social or human functioning is possible." [ ]

There are also affirmative reasons why someone like Nancy might choose to forgo artificial nutrition and hydration under these circumstances. Dying is personal. And it is profound. For many, the thought of an ignoble end, steeped in decay, is abhorrent. A quiet, proud death, bodily integrity intact, is a matter of extreme consequence. "In certain, thankfully rare, circumstances the burden of maintaining the corporeal existence degrades the very humanity it was meant to serve." . . .

Such conditions are, for many, humiliating to contemplate, as is visiting a prolonged and anguished vigil on one's parents, spouse, and children. A long, drawn-out death can have a debilitating effect on family members. [ ] For some, the idea of being remembered in their persistent vegetative states rather than as they were before their illness or accident may be very disturbing.

* * *

The only state interest asserted here is a general interest in the preservation of life. But the State has no legitimate general interest in someone's life, completely abstracted from the interest of the person living that life, that could outweigh the person's choice to avoid medical treatment. . . . [T]he State's general interest in life must accede to Nancy Cruzan's particularized and intense interest in self-determination in her choice of medical treatment. There is simply nothing legitimately within the State's purview to be gained by superseding her decision.

* * *

As the majority recognizes Missouri has a *parens patriae* interest in providing Nancy Cruzan, now incompetent, with as accurate as possible a determination of how she would exercise her rights under these circumstances. . . .

Accuracy, therefore, must be our touchstone. Missouri may constitutionally impose only those procedural requirements that serve to enhance the accuracy of a determination of Nancy Cruzan's wishes or are at least consistent with an accurate determination. The Missouri "safeguard" that the Court upholds today does not meet that standard. The determination needed in this context is whether the incompetent person would choose to live in a persistent vegetative state on life-support or to avoid this medical treatment. Missouri's rule of decision imposes a markedly asymmetrical evidentiary burden. Only evidence of specific statements of treatment choice made by the patient when competent is admissible to support a finding that the patient, now in a persistent vegetative state, would wish to avoid further medical treatment. Moreover, this evidence must be clear and convincing. No proof is required to support a finding that the incompetent person would wish to continue treatment.

Even more than its heightened evidentiary standard, the Missouri court's categorical exclusion of relevant evidence dispenses with any semblance of accurate factfinding. The court adverted to no evidence supporting its decision, but held that no clear and convincing, inherently reliable evidence had been presented to show that Nancy would want to avoid further treatment. . . . The court did not specifically define what kind of evidence it would consider clear and convincing, but its general discussion suggests that only a living will or equivalently formal directive from the patient when competent would meet this standard.

* * *

Finally, I cannot agree with the majority that where it is not possible to determine what choice an incompetent patient would make, a State's role as *parens patriae* permits the State automatically to make that choice itself. [ ] Under fair rules of evidence, it is improbable that a court could not determine what the patient's choice would be. Under the rule of decision adopted by Missouri and upheld today by this Court, such occasions might be numerous. But in neither case does it follow that it is constitutionally acceptable for the State invariably to assume the role of deciding for the patient. A State's legitimate interest in safeguarding a patient's choice cannot be furthered by simply appropriating it.

* * *

JUSTICE STEVENS, dissenting.

* * *

Choices about death touch the core of liberty. Our duty, and the concomitant freedom, to come to terms with the conditions of our own mortality are undoubtedly "so rooted in the traditions and conscience of our people as to be ranked as fundamental," [ ] and indeed are essential incidents of the unalienable rights to life and liberty endowed us by our Creator. [ ]

The more precise constitutional significance of death is difficult to describe; not much may be said with confidence about death unless it is said from faith, and that alone is reason enough to protect the freedom to conform choices about death to individual conscience. We may also, however, justly assume that death is not life's simple opposite, or its necessary terminus, but rather its completion. Our ethical tradition has long regarded an appreciation of mortality as essential to understanding life's significance. It may, in fact, be impossible to live for anything without being prepared to die for something. . . .

These considerations cast into stark relief the injustice, and unconstitutionality, of Missouri's treatment of Nancy Beth Cruzan. Nancy Cruzan's death, when it comes, cannot be an historic act of heroism; it will inevitably be the consequence of her tragic accident. But Nancy Cruzan's interest in life, no less than that of any other person, includes an interest in how she will be thought of after her death by those whose opinions mattered to her. There can be no doubt that her life made her dear to her family, and to others. How she dies will affect how that life is remembered.

The trial court's order authorizing Nancy's parents to cease their daughter's treatment would have permitted the family that cares for Nancy to bring to a close her tragedy and her death. Missouri's objection to that order subordinates Nancy's body, her family, and the lasting significance of her life to the State's own interests. The decision we review thereby interferes with constitutional interests of the highest order.

To be constitutionally permissible, Missouri's intrusion upon these fundamental liberties must, at a minimum, bear a reasonable relationship to a legitimate state end. [ ] Missouri asserts that its policy is related to a state interest in the protection of life. In my view, however, it is an effort to define life, rather than to protect it, that is the heart of Missouri's policy.

* * *

Life, particularly human life, is not commonly thought of as a merely physiological condition or function. Its sanctity is often thought to derive from the impossibility of any such reduction. When people speak of life, they often mean to describe the experiences that comprise a person's history, as when it is said that somebody "led a good life."[20] They may also mean to refer to the practical manifestation of the human spirit, a meaning captured by the familiar observation that somebody "added life" to an assembly. If there is a shared thread among the various opinions on this subject, it may be that life is an activity which is at once the matrix for and an integration of a person's interests. In any event, absent some theological abstraction, the idea of life is not conceived separately from the idea of a living person. Yet, it is by precisely such a separation that Missouri asserts an interest in Nancy Cruzan's life in opposition to Nancy Cruzan's own interests.

* * *

Only because Missouri has arrogated to itself the power to define life, and only because the Court permits this usurpation, are Nancy Cruzan's life and liberty put into disquieting conflict. If Nancy Cruzan's life were defined by reference to her own interests, so that her life expired when her biological existence ceased serving *any* of her own interests, then her constitutionally protected interest in freedom from unwanted treatment would not come into conflict with her constitutionally protected interest in life. Conversely, if there were *any* evidence that Nancy Cruzan herself defined life to encompass every form of biological persistence by a human being, so that the continuation of treatment would serve Nancy's own liberty, then once again there would be no conflict between life and liberty. The opposition of life and liberty in this case are thus not the result of Nancy Cruzan's tragic accident, but are instead the artificial consequence

---

[20] It is this sense of the word that explains its use to describe a biography: for example, Boswell's Life of Johnson or Beveridge's The Life of John Marshall. The reader of a book so titled would be surprised to find that it contained a compilation of biological data.

of Missouri's effort, and this Court's willingness, to abstract Nancy Cruzan's life from Nancy Cruzan's person.

* * *

The Cruzan family's continuing concern provides a concrete reminder that Nancy Cruzan's interests did not disappear with her vitality or her consciousness. However commendable may be the State's interest in human life, it cannot pursue that interest by appropriating Nancy Cruzan's life as a symbol for its own purposes. Lives do not exist in abstraction from persons, and to pretend otherwise is not to honor but to desecrate the State's responsibility for protecting life. A State that seeks to demonstrate its commitment to life may do so by aiding those who are actively struggling for life and health. In this endeavor, unfortunately, no State can lack for opportunities: there can be no need to make an example of tragic cases like that of Nancy Cruzan.

NOTES AND QUESTIONS

1.   *Subsequent Proceedings.* Subsequent to this judgment the Missouri trial court heard additional evidence, provided by Nancy Cruzan's friends and colleagues, that she had made explicit and unambiguous statements that demonstrated, clearly and convincingly, that she would not want to continue the treatment that she was receiving. Without opposition from the Attorney General of Missouri, the trial court authorized Ms. Cruzan's guardians to terminate her nutrition and hydration. See Sandra H. Johnson, Quinlan and Cruzan: Beyond the Symbols, in Health Law: Cases in Context (Sandra H. Johnson et al., eds. 2009).

2.   *Ethical Principles of Autonomy, Beneficence, and Social Justice.* There are many ethical approaches to health care decision-making. The approach that has most influenced judicial decision-making is one that relies on the identification and application of particular principles, often called the "principlist approach." Tom Beauchamp and James Childress, Principles of Biomedical Ethics (7th ed. 2013). There are three primary substantive principles that operate here: autonomy, beneficence, and social justice. Autonomy and beneficence have been the principles most often relied upon in bioethics case law, which has largely ignored questions of cost (and allocation and rationing) and health disparities that implicate social justice. The principle of autonomy declares that each person is in control of his own person, including his body and mind. This principle, in its purest form, presumes that no other person or social institution ought to overrule a person's choice, whether or not that choice is "right" from an external perspective. Essentially, it is a libertarian principle.

The principle of beneficence declares that what is best for each person should be done. The principle incorporates both the negative obligation of nonmaleficence ("primum non nocere"—"first of all, do no harm"—the foundation of the Hippocratic Oath) and the positive obligation to do that which is good. Thus, a physician is obliged to provide the highest quality of medical care for her patients. Similarly, a physician ought to treat a seriously ill newborn, or incapacitated adult, in a way that best serves that patient, whatever she may think her patient "wants" and whatever the parents of the patient may desire.

When a person does not desire what others determine to be in her best interests, the principles of autonomy and beneficence conflict. For example, if we consider the continued life of a healthy person to be in that person's interest, the values of autonomy and beneficence conflict when a healthy competent adult decides to take his own life. You can see that there can be serious conflict over what is in any individual's self-interest and that values other than autonomy and beneficence will come into play in some of these cases. Did the *Cruzan* opinions address autonomy and beneficence? Did they imply any other values?

3. *A "Right to Die"?* Does the opinion of the Court recognize a constitutional right to die? Many authoritative sources presumed that the opinion did recognize a constitutionally protected liberty interest in a competent person to refuse unwanted medical treatment. Indeed, the syllabus prepared for the Court says just that, and the case was hailed by the New York Times as the first to recognize a right to die. On the other hand, the Chief Justice's language does not support such a conclusion. While the majority agrees that "[t]he principle that a competent person has a constitutionally protected liberty interest in refusing unwanted medical treatment *may* be inferred from our prior decisions," (emphasis added) the Court never makes the inference itself. In fact, the opinion says explicitly that *"for purposes of this case,* we assume that the United States Constitution would grant a competent person a Constitutionally protected right to refuse life saving nutrition and hydration." (emphasis added).

Why is this assumption limited to the "purposes of this case"? Does the Court question (1) whether there is a constitutionally protected liberty interest in refusing unwanted medical treatment, (2) whether the right extends to life-sustaining treatment, or (3) whether it covers hydration and nutrition? It must have been difficult for the Chief Justice to craft an opinion that would be joined by a majority of the court. Justice Scalia clearly did not believe that there was any constitutional right implicated. If the Chief Justice were to formally recognize a constitutional right, he might have lost Justice Scalia's signature— and thus lost an opportunity for there to be any majority opinion. The dissents filed in this case are long and obviously heartfelt. Do the dissenters, all of whom would recognize a constitutionally protected right to refuse life-sustaining treatment, and Justice O'Connor, who would also do so, create a majority in support of this constitutional position?

4. *Heeding the Patient or Surrogate?* The majority opinion permits a state to limit its consideration to those wishes previously expressed by the patient and to ignore the decisions of another person acting on behalf of the patient. In fact, the Court explicitly does not address the question of whether

a state must defer to an appropriately nominated surrogate acting on behalf of the patient. On the other hand, the dissenting justices would recognize the decisions of a surrogate under appropriate circumstances, and Justice O'Connor suggests that the duty to give effect to those decisions "may well be constitutionally required." What is the constitutional status of surrogate decision-making after *Cruzan*?

5.   *Liberty, Not Privacy*. Note that none of the opinions refers to the "right of privacy," a term which had caused the Court such tremendous grief in the abortion context. The Chief Justice analyzes this issue in the more general terms of a Fourteenth Amendment liberty interest, and none of the counsel argued the case in terms of the right to privacy. Apparently, the Court just did not wish to entangle itself any further with the "P" word. See Chapter 3.

6.   *Characterizing* Cruzan. Seven years after *Cruzan* was decided, the Supreme Court again considered end-of-life medical decision-making in Washington v. Glucksberg (discussed in Chapter 5), which addressed the constitutional status of medically assisted dying. Chief Justice Rehnquist, writing for the Court, announced in *Glucksberg* that "We have . . . assumed, and strongly suggested, that the Due Process Clause protects the traditional right to refuse unwanted lifesaving medical treatment." Is that an accurate description of the holding in *Cruzan*? A few pages later, in the same opinion, Chief Justice Rehnquist describes the *Cruzan* case slightly differently: "[A]lthough Cruzan is often described as a 'right to die' case [ ], we were, in fact, more precise: we assumed that the Constitution granted competent persons a 'constitutionally protected right to refuse lifesaving hydration and nutrition.'" Is that a more accurate account of what the Court decided in *Cruzan*?

Justice O'Connor, concurring in *Glucksberg*, says that "there is no need to address the question whether suffering patients have a constitutionally cognizable interest in obtaining relief from the suffering that they may experience in the last days of their lives." This issue, according to Justice O'Connor, was decided by *Cruzan*. Is she right? Justice Stevens also commented on the *Cruzan* case in the course of his concurring opinion in *Glucksberg*. He explained that "Cruzan did give recognition . . . to the more specific interest in making decisions about how to confront an imminent death. . . . Cruzan makes it clear that some individuals who no longer have the option of deciding whether to live or to die because they are already on the threshold of death have a constitutionally protected interest [in deciding how they will die] that may outweigh the State's interest in preserving life at all costs." Is this an accurate description of *Cruzan*? If the Justices who participated in both the *Cruzan* and *Glucksberg* cases cannot agree on just what the case really means, how can your health law teacher expect you to do so?

7.   *Right to Try*. Constitutional arguments are not limited to those who want to forgo treatment; they can be asserted by seriously ill patients who want

access to treatment, too. In Abigail Alliance v. von Eschenbach, 495 F.3d 695 (D.C. Cir. 2007), the D.C. Circuit, *en banc*, addressed the argument that a terminally ill patient had a constitutional right to access drugs that had not yet been approved by the FDA. The Abigail Alliance, named for a 21-year-old student who died of cancer after being denied drugs in the earliest stages of testing, argued that *Cruzan* and *Glucksberg* paved the way for the recognition of a constitutional right to access these not-yet-approved treatments, at least when those treatments provided the only hope of survival for the patient. The Court found, 8–2, that there was no such right, reversing the decision of the original three judge panel, Abigail Alliance v. von Eschenbach, 445 F.3d 470 (D.C. Cir. 2006), in part because of the longstanding tradition of governmental drug safety efforts in our legal system. The majority and the spirited dissenters disagreed about the way to articulate the right that the Abigail Alliance sought to have recognized in this case. Was the right asserted the right "to access experimental and unproven drugs," as the majority suggested, or the right to "try to save one's life," as the dissent argued?

As you have seen in *Cruzan*, and as you will recognize in *Glucksberg*, below, the due process arguments often depend upon the precise articulation of the right sought to be protected. Do you think the *Cruzan* case provides support for the position that an otherwise terminally ill cancer patient has the right to access to drugs that have not yet been found to be safe or effective through the government's required administrative process? From a constitutional perspective, is the argument that a patient is entitled to forgo treatment any different from the argument that a patient is entitled to have access to treatment?

In any case, the notion that a terminally ill person for whom no established treatment is effective should be able to try other, unapproved, treatments is a powerful one. After *Abigail Alliance*, several states passed "Right to Try" laws to permit such otherwise unauthorized experimentation with unproven treatment. In 2018, the federal Right to Try Act created a process through which terminally ill persons may seek quicker access to certain investigational treatment. Trickett Wendler, Frank Mongiello, Jodan McLinn, and Matthew Bellina, Right to Try Act of 2017, Pub. L. No. 115–176, 132 Stat. 1372 (2018). For a discussion of the federal Act, see Jordan Paradise, Three Framings of "Faster" at the FDA and the Federal Right to Try, 11 Wake Forest J. of L. & Pol'y 53 (2020).

8.   *Should Cost Be Considered?* Except for a glancing reference by Justice Stevens in his dissent, the opinions do not consider the cost of providing care to Nancy Cruzan. Should the cost be relevant? Should the constitutional right (to liberty or to life) vary depending on who bears the cost, or how great the cost actually is to keep a patient alive under particular circumstances? Would your analysis of this case be any different if the costs were being paid by an insurance company, by Ms. Cruzan's parents, by a gofundme.com account, or by community fund-raising in Nancy Cruzan's neighborhood, rather than by the state of Missouri? Should the one who pays the bills get to do the health care decision-making?

Judge Blackmar's dissent to the Missouri Supreme Court's opinion in *Cruzan* addressed the inconsistency of requiring some patients to be kept alive, at great expense, while the state is unable (or unwilling) to provide adequate care to others who actually want that care. He pointed out:

> The absolutist position is also infirm because the state does not stand prepared to finance the preservation of life, without regard to the cost, in very many cases. In this particular case the state has Nancy in its possession, and is litigating its right to keep her. Yet, several years ago, a respected judge needed extraordinary treatment which the hospital in which he was a patient was not willing to furnish without a huge advance deposit and the state apparently had no desire to help out. Many people die because of the unavailability of heroic medical treatment. It simply cannot be said that the state's interest in preserving and prolonging life is absolute.

760 S.W.2d at 429. Judge Blackmar also pointed out, in a footnote, that "an absolutist would undoubtedly be offended by an inquiry as to whether the state, by prolonging Nancy's life at its own expense, is disabling itself from [providing] needed treatment to others who do not have such dire prognosis." 760 S.W.2d at 429 n. 4.

9. *Different State, Different Outcome?* The result of the *Cruzan* case (confirmed in *Glucksberg*) is that most law regarding health care decision-making has continued to be established on a state-by-state basis. Consider a patient who would have a right to forgo life-sustaining treatment that could be exercised by his family in California or New Jersey, but would not have that right (or would not have a right that could be exercised by his family) in Missouri or Michigan. Indeed, the conditions and extent of, and the restrictions and exceptions to, any right to forgo life-sustaining treatment might be different in each state. State policies will thus require different results in factually identical cases. Is there anything wrong with this?

What would happen if Nancy Cruzan's family had decided to move her to the Yale Medical Center "because of the more favorable medical facilities" there? Could they have moved her from Missouri to Connecticut, where removal of the gastrostomy tube clearly would be legally permitted, just for the purpose of removing the gastrostomy tube? If they could not, then Nancy Cruzan could have become a prisoner of a state that rejects her family's values—values that have been incorporated into official state policies in other jurisdictions. If they could move her, however, Missouri would have allowed the family to undercut the important policy objectives of the state law and imperil the very life the law was designed to protect. Would it violate any criminal statute to move someone across state lines for the purpose of avoiding the laws governing termination of life support in the first state? Could a state make such an action a crime?

In early 1991 the father and guardian of Christine Busalacchi sought to have his daughter moved from Missouri to Minnesota for medical consultation with a nationally known neurologist who had consulted on several leading cases that resulted in the withdrawal of life-sustaining treatment. Ms. Busalacchi, who had been living in the same nursing home that had housed Nancy Cruzan, was arguably in a persistent vegetative state. The State of Missouri sought (and obtained) an order forbidding the move because of the fear that her father wanted only to find some place where his daughter could die. A divided Missouri Court of Appeals determined that the trial court was required to commence a new hearing on whether the move could be justified by other medical objectives. In deciding the case, the majority made it clear that "... we will not permit [the] guardian to forum shop in an effort to control whether Christine lives or dies." The dissent argued that "Minnesota is not a medical or ethical wasteland.... There is a parochial arrogance in suggesting, as the state does, that only in Missouri can Christine's medical, physical, and legal well being be protected and only here will her best interests be considered." Matter of Busalacchi, 1991 WL 26851 (Mo. App.1991). Ultimately, the State decided not to pursue the case, and Busalacchi died in Missouri. See also Mack v. Mack, 329 Md. 188, 618 A.2d 744 (1993) (Maryland Court denies full faith and credit to Florida judgment appointing the Florida-resident wife of a Maryland patient in persistent vegetative state as guardian so that patient could be moved to Florida, where life-sustaining treatment could be withdrawn).

10. *Adoption of Heightened Standard.* Missouri is not the only state to adopt the strict standard approved by the majority and decried by the dissenters. See, e.g., In re Martin, 450 Mich. 204, 538 N.W.2d 399 (1995); Couture v. Couture, 549 N.E.2d 571 (Ohio Ct. App. 1989). Arguably, the same standard has also been adopted in California, Conservatorship of Wendland, 26 Cal. 4th 519, 110 Cal. Rptr. 2d 412, 28 P.3d 151 (2001), at least when a patient is in a minimally conscious state. The strict Missouri "clear and convincing evidence" rule was also applied in New York, see In re Westchester County Med. Ctr. on Behalf of O'Connor, 72 N.Y.2d 517, 534 N.Y.S.2d 886, 531 N.E.2d 607 (1988), until the Legislature adopted the Family Health Care Decisions Act, N.Y. Pub. Health Sec. 2994–d, in 2010.

NOTE: STATE LAW BASES FOR A "RIGHT TO DIE"

Although some courts have found the right to forgo life-sustaining treatment in the U.S. Constitution, *Cruzan*'s interpretation of the Fourteenth Amendment has encouraged state courts to look for other bases for this right, too. State courts find this right in state common law, state statutes, or state constitutions.

The vast majority of state courts recognizing a right to refuse life-sustaining treatment have found that right in state common law, usually in the law of informed consent, often applied even after the patient has lost the capacity to make decisions. As the Chief Justice recognized in *Cruzan*, the informed consent doctrine has become firmly entrenched in American tort law and "... the logical corollary of the doctrine of informed consent is that the patient generally possesses the right not to consent, that is, to refuse treatment...." Once a court finds a common law right, it is not necessary to determine whether the right is also conferred by statute or by the U.S. or a state constitution. See, e.g., In re Storar, 52 N.Y.2d 363, 438 N.Y.S.2d 266, 420 N.E.2d 64 (1981).

Some courts, however, bolster their common law basis for a "right to die" with references to the state and federal Constitutions. See In the Matter of Tavel, 661 A.2d 1061 (Del. 1995). While the New Jersey court initially recognized a constitutional "right to die" in In re Quinlan, 70 N.J. 10, 355 A.2d 647, 664 (N.J. 1976), it later recognized that the constitutional determination was unnecessary and retrenched:

> While the right of privacy might apply in a case such as this, we need not decide that since the right to decline medical treatment is, in any event, embraced within the common law right to self determination. In re Conroy, 98 N.J. 321, 486 A.2d 1209, 1223 (1985).

Some courts have found the right to refuse life-sustaining treatment in state statutes. Generally, courts that rely on a statutory "right to die" also find a consistent common law right. See, e.g., McConnell v. Beverly Enterprises-Connecticut, 209 Conn. 692, 553 A.2d 596, 601–602 (1989). The Illinois Supreme Court, for example, explicitly rejected state and federal constitutional justifications for a "right to die" because of the existence of both state common law and state statutory remedies. In re Estate of Longeway, 133 Ill. 2d 33, 139 Ill. Dec. 780, 549 N.E.2d 292, 297 (1989).

Several state courts have found the "right to die" in their state constitutions. A decision based on the state constitution may be the strongest kind of support such a right can ever find, because it is not subject to review by the U.S. Supreme Court (absent an improbable argument that a state-created right would itself violate the U.S. Constitution) and it is not subject to change by the state legislature (except through the generally cumbersome state constitutional amendment process). Relevant state constitutional provisions take different forms. For example, the Florida Constitution provides that "[e]very natural person has the right to be let alone and free from governmental intrusion into his private life except as otherwise provided herein. . . . " Fla. Const., art. 1, section 23. The Arizona Constitution provides that "[n]o person shall be disturbed in his private affairs or his home invaded, without authority of law." Arizona Const., art. 2, section 8. Both of these constitutional provisions have given rise to state-court-recognized rights to forgo life-sustaining treatment. See In re Guardianship of Barry, 445 So. 2d 365 (Fla. App.1984) and Rasmussen v. Fleming, 154 Ariz. 207, 741 P.2d 674 (1987). See also DeGrella v. Elston, 858 S.W.2d 698 (Ky. 1993) and Lenz v. L.E. Phillips Career Dev. Ctr., 167 Wis. 2d 53, 482 N.W.2d 60 (1992). The California Court of Appeal also found that such a right for competent patients could be found in the California Constitution. See Bouvia v. Superior Court, 179 Cal. App. 3d 1127, 225 Cal. Rptr. 297 (1986), reprinted below following the next problem. In 2009 the Montana Supreme Court hinted that the Montana Constitution's "dignity" clause might be interpreted to give competent terminally ill individuals a right to seek and receive a prescription for a lethal dose in some circumstances. It seems likely that the same substantive constitutional source of law would guarantee the right to remove life-sustaining medical care. See Baxter v. Montana, 354 Mont. 234, 224 P.3d 1211 (2009).

III. HEALTH CARE DECISION-MAKING BY ADULTS WITH DECISIONAL CAPACITY

PROBLEM: THE CHRISTIAN SCIENTIST IN THE EMERGENCY ROOM

Shortly after Ms. Elizabeth Boroff was hit by a drunk driver who went through a red light and directly into her Volkswagen bus, she found herself being attended by paramedics and loaded into an ambulance for a trip to the Big County General Hospital emergency room. Although she was briefly unconscious at the scene of the accident, and although she suffered a very substantial blood loss, several broken bones and a partially crushed skull, she had regained consciousness by her arrival at the hospital. The doctors explained to her that her life was at risk and that she needed a blood transfusion and brain surgery immediately. She explained that she was a Christian Scientist, that she believed in the healing power of prayer, that she rejected medical care, and that she wished to be discharged immediately so that she could consult a Christian Science healer.

A quick conference of emergency room staff revealed a consensus that failure to relieve the pressure caused by her intracranial bleed would result in loss of consciousness within a few hours, and, possibly, her death. When this information was provided to her, she remained unmoved. The hospital staff asked her to identify her next of kin, and she explained that she was a widow with no living relatives except for her seven minor children, ages 1 through 9. Further inquiries revealed that she was the sole support for these children, that she had no life insurance, that she had an elementary school education, and that she had been employed as a clerk since her husband, a self-employed maintenance man, was himself killed in an automobile accident a year ago.

Uncertain of what to do, the emergency room staff called you, the hospital legal counsel, for advice. What advice should you give? Should they discharge Ms. Boroff, as she requests? Should you commence a legal action to keep her in the hospital and institute treatment? If you were to file a legal action, what relief would you seek, and what would be the substantive basis of your claim?

BOUVIA v. SUPERIOR COURT

California Court of Appeal, Second District, 1986.
179 Cal.App.3d 1127, 225 Cal.Rptr. 297.

BEACH, ASSOCIATE JUSTICE.

Petitioner, Elizabeth Bouvia, a patient in a public hospital, seeks the removal from her body of a nasogastric tube inserted and maintained against her will and without her consent by physicians who so placed it for the purpose of keeping her alive through involuntary forced feeding.

* * *

Petitioner is a 28-year-old woman. Since birth she has been afflicted with and suffered from severe cerebral palsy. She is quadriplegic. She is now a patient at a public hospital maintained by one of the real parties in interest, the County of Los Angeles. Other parties are physicians, nurses and the medical and support staff employed by the County of Los Angeles. Petitioner's physical handicaps of palsy and quadriplegia have progressed to the point where she is completely bedridden. Except for a few fingers of one hand and some slight head and facial movements, she is immobile. She is physically helpless and wholly unable to care for herself. . . . She suffers also from degenerative and severely crippling arthritis. She is in continual pain. . . .

She is intelligent, very mentally competent. She earned a college degree. She was married but her husband has left her. She suffered a miscarriage. She lived with her parents until her father told her that they could no longer care for her. She has stayed intermittently with friends and at public facilities. A search for a permanent place to live where she might receive the constant care which she needs has been unsuccessful. She is without financial means to support herself and, therefore, must accept public assistance for medical and other care.

She has on several occasions expressed the desire to die. In 1983 she sought the right to be cared for in a public hospital in Riverside County while she intentionally "starved herself to death." A court in that county denied her judicial assistance to accomplish that goal. . . . Thereafter, friends took her to several different facilities, both public and private, arriving finally at her present location. Efforts by . . . social workers to find her an apartment of her own with publicly paid live-in help or regular visiting nurses to care for her, or some other suitable facility have proved fruitless.

Petitioner must be spoon fed in order to eat. Her present medical and dietary staff have determined that she is not consuming a sufficient amount of nutrients. Petitioner stops eating when she feels she cannot orally swallow more, without nausea and vomiting. As she cannot now retain solids, she is fed soft liquid-like food. Because of her previously announced resolve to starve herself, the medical staff feared her weight loss might reach a life-threatening level. Her weight since admission to real parties' facility seems to hover between 65 and 70 pounds. Accordingly, they inserted the subject tube against her will and contrary to her express written instructions.[2]

Petitioner's counsel argue that her weight loss was not such as to be life threatening and therefore the tube is unnecessary. However, the trial court found to the contrary as a matter of fact, a finding which we must accept. Nonetheless, the point is immaterial, for, as we will explain, a patient has the right to refuse any medical treatment or medical service, even when such treatment is labeled "furnishing nourishment and hydration." This right exists even if its exercise creates a "life threatening condition."

The Right to Refuse Medical Treatment

"[A] person of adult years and in sound mind has the right, in the exercise of control over his own body, to determine whether or not to submit to lawful medical treatment." [ ] It follows that such a patient has the right to refuse *any* medical treatment, even that which may save or prolong her life. [ ]

* * *

A recent Presidential Commission for the Study of Ethical Problems in Medicine and Biomedical and Behavioral Research concluded in part: "The voluntary choice of a competent and informed patient should determine whether or not life-sustaining therapy will be undertaken, just as such choices provide the basis for other decisions about medical treatment. Health care institutions and professionals should try to enhance patients' abilities to make decisions on their own behalf and to promote understanding of the available treatment options .... Health care professionals serve patients best by maintaining a presumption in favor of sustaining life, while recognizing that competent patients are entitled to choose to forgo any treatments, including those that sustain life."

* * *

The American Hospital Association Policy and Statement of Patients' Choices of Treatment Options, approved by the American Hospital Association in February of 1985 discusses the value of a collaborative relationship between the patient and the physician and states in pertinent part: "Whenever possible, however, the authority to determine the course of treatment, if any, should rest with the patient" and "the right to choose

---

[2] Her instructions were dictated to her lawyers, written by them and signed by her by means of her making a feeble "x" on the paper with a pen which she held in her mouth.

treatment includes the right to refuse a specific treatment *or all treatment*
. . . ."

* * *

Significant also is the statement adopted on March 15, 1986, by the Council on Ethical and Judicial Affairs of the American Medical Association. It is entitled "Withholding or Withdrawing Life Prolonging Medical Treatment." In pertinent part, it declares: "The social commitment of the physician is to sustain life and relieve suffering. Where the performance of one duty conflicts with the other, the choice of the patient, or his family or legal representative if the patient is incompetent to act in his own behalf, should prevail."

* * *

It is indisputable that petitioner is mentally competent. She is not comatose. She is quite intelligent, alert and understands the risks involved.

The Claimed Exceptions to the Patient's Right to Choose Are Inapplicable

. . . The real parties in interest, a county hospital, its physicians and administrators, urge that the interests of the State should prevail over the rights of Elizabeth Bouvia to refuse treatment. Advanced by real parties under this argument are the State's interests in (1) preserving life, (2) preventing suicide, (3) protecting innocent third parties, and (4) maintaining the ethical standards of the medical profession, including the right of physicians to effectively render necessary and appropriate medical service and to refuse treatment to an uncooperative and disruptive patient. Included, whether as part of the above or as separate and additional arguments, are what real parties assert as distinctive facts not present in other cases, i.e., (1) petitioner is a patient in a public facility, thereby making the State a party to the result of her conduct, (2) she is not comatose, nor incurably, nor terminally ill, nor in a vegetative state, all conditions which have justified the termination of life-support system in other instances, (3) she has asked for medical treatment, therefore, she cannot accept a part of it while cutting off the part that would be effective, and (4) she is, in truth, trying to starve herself to death and the State will not be a party to a suicide.

* * *

At bench the trial court concluded that with sufficient feeding petitioner could live an additional 15 to 20 years; therefore, the preservation of petitioner's life for that period outweighed her right to decide. In so holding the trial court mistakenly attached undue importance to the *amount of time* possibly available to petitioner, and failed to give equal weight and consideration for the *quality* of that life; an equal, if not more significant, consideration.

All decisions permitting cessation of medical treatment or life-support procedures to some degree hastened the arrival of death. In part, at least, this was permitted because the quality of life during the time remaining in those cases had been terribly diminished. In Elizabeth Bouvia's view, the quality of her life has been diminished to the point of hopelessness, uselessness, unenjoyability and frustration. She, as the patient, lying helplessly in bed, unable to care for herself, may consider her existence meaningless. She cannot be faulted for so concluding. If her right to choose may not be exercised because there remains to her, in the opinion of a court, a physician or some committee, a certain arbitrary number of years, months, or days, her right will have lost its value and meaning.

Who shall say what the minimum amount of available life must be? Does it matter if it be 15 to 20 years, 15 to 20 months, or 15 to 20 days, if such life has been physically destroyed and its quality, dignity and purpose gone? As in all matters lines must be drawn at some point, somewhere, but that decision must ultimately belong to the one whose life is in issue.

Here Elizabeth Bouvia's decision to forgo medical treatment or life-support through a mechanical means belongs to her. It is not a medical decision for her physicians to make. Neither is it a legal question whose soundness is to be resolved by lawyers or judges. It is not a conditional right subject to approval by ethics committees or courts of law. It is a moral and philosophical decision that, being a competent adult, is hers alone.

* * *

Here, if force fed, petitioner faces 15 to 20 years of a painful existence, endurable only by the constant administrations of morphine. Her condition is irreversible. There is no cure for her palsy or arthritis. Petitioner would have to be fed, cleaned, turned, bedded, toileted by others for 15 to 20 years! Although alert, bright, sensitive, perhaps even brave and feisty, she must lie immobile, unable to exist except through physical acts of others. Her mind and spirit may be free to take great flights but she herself is imprisoned and must lie physically helpless subject to the ignominy, embarrassment, humiliation and dehumanizing aspects created by her helplessness. We do not believe it is the policy of this State that all and every life must be preserved against the will of the sufferer. It is incongruous, if not monstrous, for medical practitioners to assert their right to preserve a life that someone else must live, or, more accurately, endure, for "15 to 20 years." We cannot conceive it to be the policy of this State to inflict such an ordeal upon anyone.

. . . Being competent she has the right to live out the remainder of her natural life in dignity and peace. It is precisely the aim and purpose of the many decisions upholding the withdrawal of life-support systems to accord and provide as large a measure of dignity, respect and comfort as possible to every patient for the remainder of his days, whatever be their number. This goal is not to hasten death, though its earlier arrival may be an expected and understood likelihood.

* * *

Moreover, the trial court seriously erred by basing its decision on the "motives" behind Elizabeth Bouvia's decision to exercise her rights. If a right exists, it matters not what "motivates" its exercise. We find nothing in the law to suggest the right to refuse medical treatment may be exercised only if the patient's *motives* meet someone else's approval. It certainly is not illegal or immoral to prefer a natural, albeit sooner, death than a drugged life attached to a mechanical device.

* * *

We do not purport to establish what will constitute proper medical practice in all other cases or even other aspects of the care to be provided petitioner. We hold only that her right to refuse medical treatment even of the life-sustaining variety, entitles her to the immediate removal of the nasogastric tube that has been involuntarily inserted into her body. The hospital and medical staff are still free to perform a substantial, if not the greater part of their duty, i.e., that of trying to alleviate Bouvia's pain and suffering.

Petitioner is without means to go to a private hospital and, apparently, real parties' hospital as a public facility was required to accept her. Having done so it may not deny her relief from pain and suffering merely because she has chosen to exercise her fundamental right to protect what little privacy remains to her.

Personal dignity is a part of one's right of privacy. . . .

NOTES AND QUESTIONS

1.   *Reliance on Earlier Case.* The *Bouvia* court depended, in large part, upon Bartling v. Superior Court, 163 Cal. App. 3d 186, 209 Cal. Rptr. 220 (1984), the first case to confirm a competent patient's right to make decisions to forgo life-sustaining treatment. Mr. Bartling was a competent adult suffering from depression (the original reason for his hospitalization), a tumor on his lung, and emphysema. He had a living will, a separate declaration asking that treatment be discontinued, and a durable power of attorney appointing his wife to make his health care decisions. He and his wife continuously asked that the ventilator that was preserving his life be removed, and he, his wife, and his daughter all executed documents releasing the hospital from any liability claims arising out of honoring Mr. Bartling's request. Still, the hospital, which was a Christian hospital established and

operated on pro-life principles, opposed discontinuation of Mr. Bartling's ventilator on ethical grounds. The California Court of Appeal found that the trial court should have granted Mr. Bartling's request for an injunction against the hospital, concluding that, "if the right to patient self-determination as to his own medical treatment means anything at all, it must be paramount to the interests of the patient's hospital and doctors. The right of a competent adult to refuse medical treatment is a constitutionally guaranteed right which must not be abridged."

2.   *The Disability Critique.* The court describes the quality of Elizabeth Bouvia's life in startling terms. Are they also offensive terms? Does the court describe her life as useless, meaningless, and embarrassing solely because that was her view of her own life, or does it reflect the court's assumptions? Many disability rights advocates have critiziced *Bouvia* as a case which denigrated the quality and value of life with disability, and in which a disabled person was driven to desperate measures for lack of needed opportunities and supports to continue to live in the community. See, e.g., Paul K. Longmore, Elizabeth Bouvia, Assisted Suicide and Social Prejudice, 3 Issues L. & Med. 141 (1987).

Disability rights advocates have become key players in controversies over end-of-life decision-making, generally seeking to assure that those who seek to justify ending life by terminating life-sustaining treatment or by medically assisted dying have not equated the worth or dignity of an individual with the absence of disability or need for support. See Nat'l Council on Disability, Medical Futility and Disability Bias (2019); Mary Crossley, Ending-Life Decisions: Some Disability Perspectives, 33 Ga. S. Univ. L. Rev. 893 (2017). For a wonderful attempt to reconcile the principles of bioethics (and especially the principle of autonomy) with the principles of protection (and beneficence) that underlie the disability rights movement, see Alicia Ouellette, Bioethics and Disability: Toward a Disability-Conscious Bioethics (2011). For a discussion of racial disparities in end-of-life care, see Barbara Noah, The Role of Race in End-of-Life Care, 15 J. Health Care L. & Pol'y 349 (2012).

3.   *Conscience-Based Objections.* Do you agree that the hospital had an obligation to accept Ms. Bouvia and provide her with medical relief from her pain and suffering, even though the physicians and hospital found her conduct immoral and her request an abuse of the medical profession? Is the obligation anything more than to provide adequate end-of-life care, even when the patient refuses a particular course of treatment? Compare Brophy v. New England Sinai Hospital, Inc., 398 Mass. 417, 497 N.E.2d 626 (1986), where the Massachusetts Supreme Judicial Court found that a patient in a persistent vegetative state could, through his family, deny consent to feeding through a gastric tube, but that the hospital need not remove or clamp the tube if it found it to be contrary to the ethical dictates of the medical profession. The *Brophy* decision required that the family move the patient to another medical institution more receptive to his apparent desire for his feeding tube to be removed.

The New Jersey Supreme Court took a middle ground in In re Jobes, 108 N.J. 394, 529 A.2d 434, 450 (1987):

> The trial court held that the nursing home could refuse to participate in the withdrawal of the j-tube by keeping Mrs. Jobes connected to it until she is transferred out of that facility. Under the circumstances of this case, we disagree, and we reverse that portion of the trial court's order.
>
> Mrs. Jobes' family had no reason to believe that they were surrendering the right to choose among medical alternatives when they placed her in the nursing home. [ ] The nursing home apparently did not inform Mrs. Jobes' family about its policy toward artificial feeding until May of 1985 when they requested that the j-tube be withdrawn. In fact there is no indication that this policy has ever been formalized. Under these circumstances Mrs. Jobes and her family were entitled to rely on the nursing home's willingness to defer to their choice among courses of medical treatment. . . .
>
> We do not decide the case in which a nursing home gave notice of its policy not to participate in the withdrawal or withholding of artificial feeding at the time of a patient's admission. Thus, we do not hold that such a policy is never enforceable. But we are confident in this case that it would be wrong to allow the nursing home to discharge Mrs. Jobes. The evidence indicates that at this point it would be extremely difficult, perhaps impossible, to find another facility that would accept Mrs. Jobes as a patient. Therefore, to allow the nursing home to discharge Mrs. Jobes if her family does not consent to continued artificial feeding would essentially frustrate Mrs. Jobes' right of self-determination.

Some state statutes governing end-of-life decision-making include specific provisions for situations in which health care professionals or facilities object to the decisions of patients (or their surrogates). See, e.g., Uniform Health-Care Decisions Act, below. The option to inform patients in advance of particular policies and then seek transfer is the most common compromise. Many states also have enacted statutes allowing such conscience-based actions by health care professionals in particular contexts, such as medically assisted dying.

4. *What Information Must Be Provided?* If we take seriously the *Bouvia* suggestion that hospitals have an obligation to provide comfort to patients who choose to forgo treatment and thus die, do physicians have an obligation to inform patients of the various ways of dying that are available to them, and the consequences of choosing any one of them? See Margaret Battin, The Least Worst Death, 13 Hastings Ctr. Rep. 13–16 (April 1983).

5. *Concerns Regarding Overtreatment.* Growing concerns that patients are provided more treatment at the end of life than they really want gave rise to the promulgation of the Right to Know End of Life Options Act in California in 2009 and the stronger Palliative Care Information Act in New York in 2010. Both statutes are designed to assure that patients are given all of the

information they request (in California) or need (in New York) once they have been diagnosed as terminally ill. In California the Right to Know End of Life Options Act was opposed by Right to Life organizations and medical professional groups, which generally object to all statutory intrusion on the doctor-patient relationship. The stronger New York statute was opposed only by medical professional organizations, and it passed by large margins in both houses of the legislature. The New York statute is a simple statement of principle, more hortatory than enforceable:

Palliative Care Patient Information

New York Pub. Health Law § 2997–c

1. Definitions. . . .

(a) "Appropriate" means consistent with applicable legal, health and professional standards; the patient's clinical and other circumstances; and the patient's reasonably known wishes and beliefs.

* * *

(c) "Palliative care" means health care treatment, including interdisciplinary end-of-life care, and consultation with patients and family members, to prevent or relieve pain and suffering and to enhance the patient's quality of life, including hospice care . . . .

(d) "Terminal illness or condition" means an illness or condition which can reasonably be expected to cause death within six months, whether or not treatment is provided.

2. If a patient is diagnosed with a terminal illness or condition, the patient's attending health care practitioner shall offer to provide the patient with information and counseling regarding palliative care and end-of-life options appropriate to the patient, including but not limited to: the range of options appropriate to the patient; the prognosis, risks and benefits of the various options; and the patient's legal rights to comprehensive pain and symptom management at the end of life. The information and counseling may be provided orally or in writing. Where the patient lacks capacity to reasonably understand and make informed choices relating to palliative care, the attending health care practitioner shall provide information and counseling under this section to a person with authority to make health care decisions for the patient. The attending health care practitioner may arrange for information and counseling under this section to be provided by another professionally qualified individual.

Is this kind of a statute a good idea? Should other states follow the lead of California and New York? See, e.g., M.G.L.A. 111 § 227 (2012) (Massachusetts); O.R.S. § 413.273 (2015) (Oregon). Writing in The New England Journal of Medicine just three months after the New York statute became effective, two New York physicians argue that barriers to respectful conversations at the end of life are better addressed through professional standards of care than legislation.

Alan Astrow and Beth Propp, Perspective: The Palliative Care Information Act in Real Life (Topics: Public Health, NEJM, May 18, 2011). Are they right? Should the underlying question be one about patients' and families' experiences, physicians' experiences, or both? The original Right to Know End of Life Options bill introduced in California explicitly mentioned some alternatives about which patients should be informed in appropriate circumstances—for example, palliative sedation (described in the *Cruzan* case) and the voluntary stopping of eating and drinking (often referred to as "VSED") (discussed in Chapter 6). Political compromise required that mention of those specific alternatives be deleted from the bill. Would general language, like that in the New York statute, require that a patient be told of those end-of-life alternatives? Under what circumstances?

6.   *Ms. Bouvia's Decision.* After the California Supreme Court confirmed Ms. Bouvia's right to choose to die, she decided to accept the medical care necessary to treat her pain and to keep her alive. She appeared on television (on "60 Minutes") in 1998, where she expressed the hope that she would die soon. Why would someone seek judicial confirmation of a "right to die" and then not act upon it? Does it indicate that people waver on this issue? Does it suggest that knowing that one has the choice—when it becomes necessary— contributes to that person's well-being? Are those who seek a judicially confirmed "right to die" really seeking control over their destiny, not their death? Ironically, the existence of a right to die may be the reason that some people choose to live, just as the existence of medically assisted dying may be a reason that many people choose not to ingest the lethal dose of medication they have been legally prescribed. See Chapter 6.

7.   *Few State Statutes.* Is it surprising that the fundamental principle that competent adults can make all of their own health care decisions has made it into the statutes of only a very few states? For an exception to this general rule, see N.M. Stat. Ann. Section 24–7A–2. Especially after the concern shown for this issue in *Bouvia*, one might expect more legislatures to have confirmed this right. Have they failed to do so because the law is so clear that legislative confirmation is unnecessary, or because there is a real dispute about the substance of the principle?

NOTE: COUNTERVAILING STATE INTERESTS

The right to choose to forgo life-sustaining treatment is not absolute, even for competent adults. In Superintendent of Belchertown State School v. Saikewicz, 373 Mass. 728, 370 N.E.2d 417 (1977), the Massachusetts Supreme Judicial Court first identified the four "countervailing State interests" that could overcome a patient's choice: (1) preservation of life; (2) protection of the interests of innocent third parties; (3) prevention of suicide; and (4) maintenance of the ethical integrity of the medical profession.

Although *Saikewicz* involved an incompetent patient who was born with significant intellectual and developmental disabilities and had never been competent, the mantra of those four interests has also been applied in later cases involving competent patients—including *Bouvia*—but they have never been found to be sufficient to overcome the choice of a *competent* patient. You will see the state's interests discussed in almost every case in this chapter.

In *Saikewicz* the Massachusetts Supreme Judicial Court explored the significance of these four state interests and their limitations:

> It is clear that the most significant of the asserted State interests is that of the preservation of human life. Recognition of such an interest, however, does not necessarily resolve the problem where the affliction or disease clearly indicates that life will end soon, and inevitably be extinguished. The interest of the State in prolonging a life must be reconciled with the interest of an individual to reject the traumatic cost of that prolongation. There is a substantial distinction in the State's insistence that human life be saved where the affliction is curable, as opposed to the State interest where, as here, the issue is not whether but when, for how long, and at what cost to the individual that life may be briefly extended. Even if we assume that the State has an additional interest in seeing to it that individual decisions on the prolongation of life do not in any way tend to "cheapen" the value which is placed on the concept of living, we believe it is not inconsistent to recognize a right to decline medical treatment in a situation of incurable illness. The constitutional right to privacy, as we conceive it, is an expression of the sanctity of individual free choice and self-determination as fundamental constituents of life. The value of life as so perceived is lessened not by a decision to refuse treatment, but by the failure to allow a competent human being the right of a choice.
>
> A second interest of considerable magnitude, which the State may have some interest in asserting, is that of protecting third parties, particularly minor children, from the emotional and financial damage which may occur as a result of the decision of a competent adult to refuse life-saving or life-prolonging treatment. Thus, even when the State's interest in preserving an individual's life was not sufficient, by itself, to outweigh the individual's interest in the exercise of free choice, the possible impact on minor children would be a factor which might have a critical effect on the outcome of the balancing process.

* * *

> The last State interest requiring discussion[11] is that of the maintenance of the ethical integrity of the medical profession as well as allowing

---

[11] The interest in protecting against suicide seems to require little if any discussion. In the case of the competent adult's refusing medical treatment such an act does not necessarily constitute suicide since (1) in refusing treatment the patient may not have the specific intent to die, and (2) even if he did, to the extent that the cause of death was from natural causes, the patient did not set the death producing agent in motion with the intent of causing his own death.

hospitals the full opportunity to care for people under their control. The force and impact of this interest is lessened by the prevailing medical ethical standards. Prevailing medical ethical practice does not, without exception, demand that all efforts toward life prolongation be made in all circumstances. Rather, the prevailing ethical practice seems to be to recognize that the dying are more often in need of comfort than treatment. Recognition of the right to refuse necessary treatment in appropriate circumstances is consistent with existing medical mores; such a doctrine does not threaten either the integrity of the medical profession, the proper role of hospitals in caring for such patients or the State's interest in protecting the same. It is not necessary to deny a right of self-determination to a patient in order to recognize the interest of doctors, hospitals, and medical personnel in attendance on the patient. Also, if the doctrines of informed consent and right of privacy have as their foundations in the right to bodily integrity, and control of one's own fate, then those rights are superior to the institutional considerations. 370 N.E.2d at 425–427.

In fact, these four interests raise issues beyond those discussed in *Saikewicz*:

(1) *Preservation of life.* If the value of the preservation of life is the very question faced by the court in right-to-die cases, does it make sense to define it, *a priori,* as a value that is countervailing to the patient's desire to discontinue treatment?

The nature of the state's interest in the preservation of life was discussed in the *Cruzan* case, in which it was the only interest advanced by the state of Missouri. The Chief Justice said that

> a state may properly decline to make judgments about the 'quality' of life that a particular individual may enjoy, and simply assert an unqualified interest in the preservation of human life to be weighed against the constitutionally protected interests of the individual.

Not surprisingly, the dissenters viewed the state's interest in the preservation of life very differently. Justice Stevens objected to Missouri's policy of "equating [Cruzan's] life with the biological persistence of her bodily functions." He pointed out that,

> [l]ife, particularly human life, is not commonly thought of as a merely physiological condition or function. Its sanctity is often thought to derive from the impossibility of any such reduction. When people speak of life, they often mean to describe the experiences that comprise a person's history. . . .

Furthermore, the underlying State interest in this area lies in the prevention of irrational self-destruction. What we consider here is a competent, rational decision to refuse treatment when death is inevitable, and the treatment offers no hope of cure or preservation of life. There is no connection between the conduct here in issue and any State concern to prevent suicide.

Justice Brennan was especially offended by the notion that the generalized state interest in life could overcome the liberty interest to forgo life-sustaining treatment. One's rights, he argued, may not be sacrificed just to make society feel good:

> If Missouri were correct that its interests outweigh Nancy's interests in avoiding medical procedures as long as she is free of pain and physical discomfort, [ ] it is not apparent why a state could not choose to remove one of her kidneys without consent on the ground that society would be better off if the recipient of that kidney were saved from renal poisoning . . . , patches of her skin could also be removed to provide grafts for burn victims, and scrapings of bone marrow to provide grafts for someone with leukemia. . . . Indeed, why could the state not perform medical experiments on her body, experiments that might save countless lives, and would cause her no greater burden than she already bears by being fed through her gastrostomy tube? This would be too brave a new world for me and, I submit, for our constitution.

497 U.S. 261, 312–14 n. 13, 110 S. Ct. 2841, 2869–70 n. 13, 111 L. Ed. 2d 224. Chief Justice Rehnquist stated in *Glucksberg*, discussed in Chapter 6, that *Cruzan* had decided that states may choose to act to protect the sanctity of all life, independent of any inquiry into quality of life, and independent of the value of that life to the one living it.

(2) *Protection of innocent third parties.* Does the protection of the interests of innocent third parties have any meaning if courts are not willing to force people to stop pursuing their own interests and to serve some undefined communal goal? Is it merely a make-weight argument in a society as individualistic as ours? On the other hand, might children, for example, have a claim on the lives of their parents? Under what circumstances would such a claim be strongest?

(3) *Prevention of suicide.* Although *Glucksberg* confirmed that a state could make assisting suicide a crime, committing suicide is no longer a crime in any state. Is there still a consensus behind Justice Nolan's position, dissenting in Brophy v. New England Sinai Hosp., 398 Mass. 417, 497 N.E.2d 626, 640 (1986), that "suicide is direct self-destruction and is intrinsically evil. No set of circumstances can make it moral . . . ."?

(4) *Protecting the ethical integrity of the medical profession.* Finally, there is no longer any reason to believe that the ethics of the medical profession do not permit discontinuation of medical treatment to a competent patient who refuses it. See AMA Code of Medical Ethics Opinion 5.3, Withholding or Withdrawing Life-Prolonging Medical Treatment (2016). Even if there were, though, should the protection of the "ethical integrity of the medical profession" overcome an otherwise proper decision to forgo some form of treatment? If all other analyses point to allowing a patient to deny consent to some form of treatment, in what cases, if any, should the medical profession be able to require the treatment in the interest of its own self-defined integrity?

Are there special circumstances in which the interest of the patient ought not to be recognized or the interest of the state is especially important? Can the state require a criminal defendant to submit to medical treatment to make him competent to stand trial? See Sell v. United States, 539 U.S. 166, 123 S. Ct. 2174, 156 L. Ed. 2d 197 (2003), holding that forced medication may violate Due Process rights. Does the national interest allow the military to require its soldiers to undergo life-saving (or other) medical care so that they can be returned to the front? Can a prisoner refuse kidney dialysis that is necessary to save his life unless the prison administration moves him from a medium to minimum security prison? See Commissioner of Correction v. Myers, 379 Mass. 255, 399 N.E.2d 452 (1979) (interest in "orderly prison administration" outweighs any privacy right of the prisoner to refuse dialysis unless he were moved to another site); People ex rel. Illinois Dep't of Corrections v. Millard, 335 Ill. App. 3d 1066, 270 Ill. Dec. 407, 782 N.E.2d 966 (2003) (force feeding of prisoner allowed). For a general discussion of several courts' approaches to balancing a patient's right to refuse treatment with these countervailing state interests, see Alan Meisel, Kathy L. Cerminara and Thaddeus Mason Pope, The Right to Die, 3rd ed. (2004, and 2017 Supplement).

APPLICATION OF THE PRESIDENT AND DIRECTORS OF GEORGETOWN COLLEGE

United States Court of Appeals, District of Columbia Circuit, 1964.
331 F.2d 1000.

J. SKELLY WRIGHT, CIRCUIT JUDGE.

Mrs. Jones was brought to the hospital by her husband for emergency care, having lost two thirds of her body's blood supply from a ruptured ulcer. She had no personal physician, and relied solely on the hospital staff. She was a total hospital responsibility. It appeared that the patient, age 25, mother of a seven-month-old child, and her husband were both Jehovah's Witnesses, the teachings of which sect, according to their interpretation, prohibited the injection of blood into the body. When death without blood became imminent, the hospital sought the advice of counsel, who applied to the District Court in the name of the hospital for permission to administer blood. Judge Tamm of the District Court denied the application, and counsel immediately applied to me, as a member of the Court of Appeals, for an appropriate writ.

* * *

Mr. Jones, the husband of the patient . . . [s]aid, that if the court ordered the transfusion, the responsibility was not his.

* * *

I tried to communicate with her, advising her again as to what the doctors had said. The only audible reply I could hear was "Against my will." It was obvious that the woman was not in a mental condition to make a decision. I was reluctant [t]o press her because of the seriousness of her condition and because I felt that to suggest repeatedly the imminence of death without blood might place a strain on her religious convictions. I asked her whether she would oppose the blood transfusion if the court allowed it. She indicated, as best I could make out, that it would not then be her responsibility.

* * *

I thereupon signed the order allowing the hospital to administer such transfusions as the doctors should determine were necessary to save her life.

It has been firmly established that the courts can order compulsory medical treatment of children for any serious illness or injury, and that adults, sick or well, can be required to submit to compulsory treatment or prophylaxis, at least for contagious diseases, *e.g.,* Jacobson v. Massachusetts. [ ] And there are no religious exemptions from these orders . . . .

The right to practice religion freely does not include liberty to expose the community or the child to communicable disease or the latter to ill health or death. [ ]

Of course, there is here no sick child or contagious disease. However, the sick child cases may provide persuasive analogies because she was as little able competently to decide for herself as any child would be. Under the circumstances, it may well be the duty of a court of general jurisdiction, such as the United States District Court for the District of Columbia, to assume the responsibility of guardianship for her, as for a child, at least to the extent of authorizing treatment to save her life. And if, as shown above, a parent has no power to forbid the saving of his child's life, *a fortiori* the husband of the patient here had no right to order the doctors to treat his wife in a way so that she would die. . . .

[Another] set of considerations involved the position of the doctors and the hospital. Mrs. Jones was their responsibility to treat. The hospital doctors had the choice of administering the proper treatment or letting Mrs. Jones die in the hospital bed, thus exposing themselves, and the hospital, to the risk of civil and criminal liability in either case. . . .

[N]either the principle that life and liberty are inalienable rights, nor the principle of liberty of religion, provides an easy answer to the question whether the state can prevent martyrdom. Moreover, Mrs. Jones had no wish to be a martyr. And her religion merely prevented her consent to a transfusion. If the law undertook the responsibility of authorizing the transfusion without her consent, no problem would be raised with respect to her religious practice. Thus, the effect of the order was to preserve for Mrs. Jones the life she wanted without sacrifice of her religious beliefs.

The final, and compelling, reason for granting the emergency writ was that a life hung in the balance. There was no time for research and reflection. Death could have mooted the cause in a matter of minutes, if action were not taken to preserve the *status quo*. To refuse to act, only to find later that the law required action, was a risk I was unwilling to accept. I determined to act on the side of life.

NOTES AND QUESTIONS

1.   *State Action.* Unlike *Cruzan, Bouvia,* and *Wons,* below, *Georgetown College* did not involve a state facility. However, for purposes of federal constitutional analysis, state action can be found where a private person or entity takes control over a traditionally state function. The court in *Matter of Welfare of Colyer* explains:

> The existence of 'state action' for constitutional purposes depends on 'whether there is a sufficiently close nexus between the State and the challenged action of the regulated entity so that the action of the latter may be fairly treated as that of the State itself.' [ ]. Here, the presence of the state is manifested by its capability of imposing criminal sanctions on the hospital and its staff [ ], by its licensing of physicians [ ], by the required involvement of the judiciary in the guardianship appointment process [ ], and by the State's parens patriae responsibility to supervise the affairs of incompetents. [ ]. Taken together, these factors show a sufficient nexus between the state and the prohibitions against withholding or discontinuance of life sustaining treatment to call into play the constitutional right of privacy.

660 P.2d 738, 742 (1983). See also TL v. Cook Children's Med Ctr, cited in Section VI (discussing existence of state action despite the private nature of the hospital that sought to terminate life-sustaining treatment unilaterally).

2.   *Justification for Judicial Review.* The right to choose to forgo life-sustaining treatment is usually based upon the premise that a person rationally may decide that death is preferable to the pain, expense, and inconvenience of life. Given that the process of weighing the value of life and death is necessarily based in personal history, religious and moral values, and individual sensitivity to a number of different factors, and given that it finds its philosophical basis in the principle of autonomy, is there any justification for independent judicial evaluation of whether the balancing was properly, or even rationally, performed by the patient?

3.   *Religious Beliefs.* Most difficult cases have arisen over decisions based upon the dictates of religious principles. For example, Christian Scientists generally accept the healing power of prayer to the exclusion of medical assistance, so most Christian Scientists refuse most traditional medical care. Christian Scientists may be less likely than Jehovah's Witnesses to find themselves in litigation over their refusal of medical treatment, however, because they are less likely to be at the hospital seeking medical care.

For an account of the Christian Science position, see Mary Baker Eddy, Manual of the Mother Church 17–19 (1935) (first edition published in 1895). Given the ability of medicine at the time, it was likely that in its early days the Christian Science faith saved more lives than it cost by discouraging its adherents from seeing doctors. Research suggests that today Christian Scientists do not live quite as long as others. See W. Simpson, Comparative Longevity in a College Cohort of Christian Scientists, 262 JAMA 1657 (1989).

Jehovah's Witnesses, on the other hand, accept most medical care, but they do not accept blood transfusions, which they perceive to be a violation of the biblical prohibition on the ingestion of blood. The Jehovah's Witness belief that the ingestion of blood is prohibited finds its source in a number of Biblical passages. See Leviticus 17:10 ("As for any man . . . who eats any sort of blood, I shall certainly set my face against the soul that is eating the blood . . . "); Leviticus 17:14 ("You must not eat the blood of any sort of flesh . . . "); Acts 15:10; Genesis 9:4. These passages cause Witnesses to believe that receiving blood products will render them unable to obtain resurrection and eternal life.

Other sects also believe that God will provide any cure that is appropriate for each sick person, and their actions pose the same legal and ethical problems as do those of the Christian Scientists and the Jehovah's Witnesses. In England, for example, the "Peculiar People" presented the British courts with this issue long before the first case arose on this side of the Atlantic. See R v. Senior, All ER 511 (1895–9).

Should the courts ever attempt to decide whether an interpretation of the Bible is correct, or even just reasonable? Is it relevant that others consider the religious ban on the ingestion of blood or the rejection of all medical treatment to be irrational? Because courts were less able to empathize with patients who had unusual religious beliefs than with others, for many years courts were less willing to entertain the right to forgo life-sustaining treatment on religious grounds than on other grounds. Should a court treat a Christian Scientist or Jehovah's Witness who chooses for religious reasons to forgo necessary care any differently than it treats Elizabeth Bouvia?

4. *Judicial Rationalization?* In *Georgetown College*, Judge Wright concluded that Jehovah's Witnesses were not required by their religious code to forgo blood transfusions; they were merely required to refuse consent to those transfusions. Thus, a weak denial was taken as a plea for medical intervention against the patient's stated, but misleading, request to be left without adequate care. Would the decision, based on this reasoning, vindicate the principle of autonomy? Is it appropriate for Judge Wright, in his role as a federal judge, to determine that the Jehovah's Witness faith requires only that an adherent deny consent to a transfusion, not that she avoid actually having one? Is it relevant that Jehovah's Witness religious authorities reject this position?

The other reasons for the court's decision in *Georgetown College* seem equally fragile. The hospital could hardly claim that the risk of civil or criminal liability would require the transfusion after the hospital went to court to determine its legal responsibility. This claim is better understood as one asking for resolution of the liability issues should the hospital follow or override refusals of transfusions in future cases. Finally, the presumption that anyone so ill as to need a blood transfusion to save her life is likely to be incompetent is simply unsupported by fact. The need for a blood transfusion is hardly sufficient to justify the literal infantalization of a patient like Mrs. Jones. These bases for the decision demonstrate why this opinion by one of the great federal judges of the last century has come to be seen as one of the most painful examples of judicial rationalization.

5.   *Most Courts Recognize the Right to Refuse Treatment.* Most of the cases that have considered this issue have concluded that competent adult Jehovah's Witnesses may choose to forgo medical treatment, whatever the results of those decisions may be, because the patient bears the consequences of choosing to forgo life-sustaining treatment. See, e.g., Norwood Hosp. v. Munoz, 409 Mass. 116, 564 N.E.2d 1017 (1991); In re Brooks' Estate, 32 Ill. 2d 361, 205 N.E.2d 435 (1965); In re Osborne, 294 A.2d 372 (D.C. App.1972). Compare Fosmire v. Nicoleau, 75 N.Y.2d 218, 551 N.Y.S.2d 876, 551 N.E.2d 77 (1990) (Jehovah's Witness mother permitted to forgo a blood transfusion during child birth even though her life was thus put at risk) with Raleigh Fitkin-Paul Morgan Mem. Hosp v. Anderson, 42 N.J. 421, 201 A.2d 537 (1964), cert. denied, 377 U.S. 985, 84 S. Ct. 1894, 12 L. Ed. 2d 1032 (1964) (pregnant Jehovah's Witness not permitted to refuse a necessary transfusion).

Courts may be more disposed to recognize religious beliefs that prohibit forms of medical treatment when the courts are familiar with the beliefs (as in the case of Jehovah's Witnesses) and where there are a large number of believers who share the limitation. In In the Matter of J.M., 416 N.J. Super. 222, 3 A.3d 651 (Super. Ct. 2010) the Chancery Division of the New Jersey Superior Court was faced with an arguably incompetent patient who decided, inconsistently with the medical advice she received, that she did not need dialysis because "Jesus would save her." The psychiatrists split on her competence, but the court decided that her views were not based on her religion because there was no tenet of her religion that explicitly rejected dialysis and her evangelical pastor had attempted to talk her into accepting the dialysis. Her dependence on her view that Jesus would save her was thus not a religious view upon which a competent person could base a decision; rather, it was part of her delusional thought process. The court thus found her incompetent, transferred decision-making authority to a special medical guardian, and ordered that she undergo dialysis. Was that the proper analysis of that case?

Even when others (such as children) are indirectly affected, the courts now tend to recognize the competent adult's right to forgo treatment:

PUBLIC HEALTH TRUST OF DADE COUNTY V. WONS

Supreme Court of Florida, 1989.
541 So.2d 96.

KOGAN, JUSTICE.

* * *

The Court of Appeal has certified the following question as one of great public importance:

WHETHER A COMPETENT ADULT HAS A LAWFUL RIGHT TO REFUSE A BLOOD TRANSFUSION WITHOUT WHICH SHE MAY WELL DIE.

* * *

The issues presented by this difficult case challenge us to balance the right of an individual to practice her religion and protect her right of privacy against the state's interest in maintaining life and protecting innocent third parties.

Norma Wons entered . . . a medical facility operated by the Public Health Trust of Dade County, with a condition known as dysfunctional uterine bleeding. Doctors informed Mrs. Wons that she would require treatment in the form of a blood transfusion or she would, in all probability, die. Mrs. Wons, a practicing Jehovah's Witness and mother of two minor children, declined the treatment on ground that it violated her religious principles to receive blood from outside her own body. At the time she refused consent Mrs. Wons was conscious and able to reach an informed decision concerning her treatment.

The Health Trust petitioned the Circuit Court to force Mrs. Wons to undergo a blood transfusion. . . . [T]he court granted the petition, ordering the hospital doctors to administer the blood transfusion, which was done while Mrs. Wons was unconscious. The trial judge reasoned that minor children have a right to be reared by two loving parents, a right which overrides the mother's rights of free religious exercise and privacy. Upon regaining consciousness, Mrs. Wons appealed to the Third District which reversed the order. After holding that the case was not moot due to the recurring nature of Mrs. Wons condition . . . , the district court held that Mrs. Wons' constitutional rights of religion and privacy could not be overridden by the state's purported interests.

* * *

The Health Trust asserts that the children's right to be reared by two loving parents is sufficient to trigger the compelling state interest [in protection of innocent third parties]. While we agree that the nurturing and support by two parents is important in the development of any child, it is not sufficient to override fundamental constitutional rights. . . . As the district court noted in its highly articulate opinion below:

Surely nothing, in the last analysis, is more private or more sacred than one's religion or view of life, and here the courts, quite properly, have given great deference to the individual's right to make decisions vitally affecting his private life according to his own conscience. It is difficult to overstate this right because it is, without exaggeration, the very bedrock upon which this country was founded.

NOTES AND QUESTIONS

1. *Determining the Best Interests of the Children.* A concurring opinion in *Wons* depends in part upon the fact that the children would be cared for by relatives even if Mrs. Wons were to die. In that opinion the Chief Justice points out:

The medical profession may consider a blood transfusion a rather ordinary or routine procedure, but, given Mrs. Wons' religious beliefs, that procedure for her is extraordinary. . . . [W]e must not assume from her choice that Mrs. Wons was not considering the best interests of her children. She knows they will be well cared for by her family. As a parent, however, she also must consider the example she sets for her children, how to teach them to follow what she believes is God's law if she herself does not. The choice for her can not be an easy one, but it is hers to make. It is not for this court to judge the reasonableness or validity of her beliefs. Absent a truly compelling state interest to the contrary, the law must protect her right to make that choice.

2. *Parental Abandonment Rationale.* The dissent in *Wons* depended in part upon another portion of *Georgetown College,* above, where Judge Wright concluded that

[t]he state, as parens patriae, will not allow a parent to abandon a child, and so it should not allow this most ultimate of voluntary abandonments. The patient had a responsibility to the community to care for her infant. Thus the people had an interest in preserving the life of this mother.

Would this rationale support state intervention and an injunction to stop a mother (or a father) who had decided to take up hang-gliding, bronco-riding, working as a firefighter, or some other dangerous occupation? Why do you think it is applied in the case of a Jehovah's Witness, and not in any of these other cases?

3. *Refusing Treatment for Children.* The rule allowing patients to adhere to their religious faiths, even if that means that they choose to forgo life-sustaining treatment, is different for children. The courts have always held that children are not permitted to become martyrs to their parents' (or their own) religious beliefs. Where the refused treatment is not highly invasive, and where it is likely to return the child to full health—as in the case of a blood transfusion for a Jehovah's Witnesses child—courts universally order the

treatment. On the other hand, where the chance of success is lower—as in the case of certain kinds of chemotherapy for some childhood cancers—the courts are less likely to overrule the parents and the child. See Section V of this chapter, below.

IV.   HEALTH CARE DECISIONS FOR ADULTS WITHOUT DECISION-MAKING CAPACITY

A.   DECISIONAL CAPACITY

PROBLEM: DETERMINING THE DECISIONAL CAPACITY OF A DYING PATIENT

Theodore Flores is a 27-year-old who has been quadriplegic since a serious auto accident (in which he was driving while intoxicated) about a year ago. His spinal injury was so high and so substantial that he requires intermittent ventilator support, and he is fed through a gastrostomy tube that has been inserted through his abdominal wall and directly into his stomach. Although he has not worked since his accident, he had been employed as a pharmacist. He now communicates with others by winking his eyes or blowing through a straw connected to an alphabet board. The accident appears to have had no effect on his intellectual abilities, although he is now unable to concentrate for more than a few minutes, and he sometimes refuses to communicate to outsiders for days at a time.

Mr. Flores has now informed his physician that he wishes to have the ventilator disconnected and the feeding tube removed. His physician has informed him of the certainty that death will follow from either of these acts. Mr. Flores refuses to respond to such information, except to repeat his request. Mr. Flores has read most of the important recent medical journal articles about his condition, and his physician believes that Mr. Flores understands the risks, benefits, and alternatives more thoroughly than most similarly situated patients. Despite this, though, the physician is concerned about whether Mr. Flores has the capacity to decide to forgo ventilator support and nutrition and hydration. His concern grows out of several circumstances, including Mr. Flores's limited concentration span, his occasional unwillingness to communicate with anyone, and the following three factors:

(1)   A psychiatrist who has been seeing Mr. Flores regularly has informed his physician that Mr. Flores became seriously depressed about three weeks ago, when he began to realize that there would never be any improvement in his physical condition. As the psychiatrist pointed out, of course, most reasonable people would become depressed upon such a realization. The psychiatrist has recommended antidepressants for Mr. Flores, but he refuses to take them.

(2)   Since the accident, Mr. Flores has become a devoted follower of August Marsh, a religious leader who preaches that self-abnegation (and, particularly, self-abnegation leading to death) is the only way to gain

salvation. In particular, Rev. Marsh believes that the self-discipline of starvation (he usually recommends a period of a week) can bring substantial and, perhaps, eternal joy.

(3) Mr. Flores has always desperately wanted to father a child. His ability to contribute gametes to create a fertilized ovum is not precluded by his current condition. While the doctors have explained to Mr. Flores that his reproductive system is intact and unaffected by the accident, he insists that there is no way that he will be able to have children. No matter what the medical staff does, and despite Mr. Flores's generally sophisticated understanding of his medical condition, he does not understand how he will be able to become a father.

Does Mr. Flores possess sufficient decisional capacity to make the choice to forgo nutrition and hydration and ventilator support? What other information would you want to have before you make this determination? Remember, you are looking only for information relevant to a determination of decisional capacity, you are not looking for information that is relevant to the substantive decision. What process should be employed to make this capacity determination? Is a court order required to confirm Mr. Flores's capacity? Incapacity? What kind of evidence should be introduced in the hearing that would lead to an appropriate court order?

To determine whether a patient can choose to undergo (or forgo) medical care, someone must determine whether the patient has the capacity to make that choice. Because the theory behind decisional capacity serves the social principle of autonomy, capacity determinations should not be entirely medical; social, philosophical, and political factors should also be considered.

Until recently, most courts that addressed this issue referred to the "competency" of the patient. In other areas of law, such as those concerned with guardianships and conservatorships, "competency" was often employed as a term with all-or-nothing consequences. A person was either competent for all purposes and at all times, or incompetent for all purposes and at all times. When we evaluate the ability of a patient to make a health care decision, though, we are dealing with something far more subtle. A patient may have the capacity to make some simple health care decisions, and not to make other more complex decisions. Similarly, a patient may be able to make certain decisions at some times but not at others. To recognize these potential variabilities in competency, many courts and legislatures use the term "decisional capacity"—a term which focuses on the actual decision to be made—rather than "competency"—a term which focuses on the status of the patient. The change has been recognized in the law of guardianship and conservatorship in many states, too. In this text, the terms "decisional capacity" and "competency" are used interchangeably.

Courts have been reluctant to articulate a standard for decisional capacity. There are few reported opinions in which courts state and apply any formal principle. Courts have been much more likely to finesse the issue out of the law and back into medicine by inviting physicians, especially psychiatrists, to testify about the mental state and, thus, capacity of a patient. Ironically, medical textbooks point out that the standard for capacity is a legal, not a medical, one.

For many years, physicians were likely to find that a patient had decisional capacity to make a serious medical decision whenever that patient agreed with the physician. When the patient disagreed with the physician—especially if that the patient's choice would lead to the death of the patient—the physician, and subsequently the court, would be likely to find that the patient lacked capacity and then seek out some surrogate decision-maker more likely to agree with the physician. In reaction to this, and as a consequence of the frustration of attempting to develop any consistent and practical definition of competence, consumerist attorneys and physicians in the 1970s suggested that any patient who could indicate an affirmative or negative ought to be considered to have decisional capacity. Of course, this reactionary view is no more satisfactory than the previously prevailing view. Neither serves the purpose of protecting the individual personality of the patient and the authority of the patient to control his own life in a way that is consistent with his own values.

Despite the extremes described above, some scholars have attempted to categorize the possible tests for decisional capacity (or competency) that could be applied to patients of questionable capacity. Five different kinds competency tests were described in an early and influential article prepared by a psychiatrist, a lawyer, and a sociologist with extensive expertise in psychiatry. In Loren H. Roth, Alan Meisel, and Charles W. Lidz, Tests of Competency to Consent to Treatment, 134 Am. J. Psychiatry 279 (1977), the authors reviewed the legal and medical literature and concluded that there are five different tests applied to determine if a patient is competent:

(1) Does the patient evidence a choice? This test sets the lowest standard for competency. If the patient can communicate a choice, the patient is competent.

(2) Is the outcome of the patient's choice reasonable (or "right" or "responsible")?

(3) Whether or not the outcome is "reasonable," was the choice based on "rational" reasons?

(4) Does the patient have the ability to understand the information that must be communicated as a matter of informed consent—the risks, benefits, and alternatives of treatment?

(5) Does the patient actually understand that information?

Would you find any one of these tests for competency to be satisfactory? Which one? Might you apply different tests under different circumstances?

Roth, Meisel, and Lidz suggested that each test is influenced by the evaluator's analysis of whether the treatment would succeed—that is, whether the evaluator would consent for herself under the same circumstances. The authors concluded that where the benefit of treatment is likely to far outweigh the risk (i.e., the evaluator would choose to undergo it), there is likely to be a low standard for competency when the patient consents and a high standard for competency when the patient refuses. Analogously, where the risk greatly outweighs the benefit (again, of course, in the evaluator's mind), a low standard for competency will be applied if the patient refuses treatment, but a high standard will be applied when the patient consents.

In what might have been the most successful endeavor to unpack the theoretical basis of decision-making capacity, in 1980 the President's Commission for the Study of Ethical Problems in Medicine and Biomedical and Behavioral Research articulated the three elements of capacity:

(1) Possession of a set of values and goals,

(2) The ability to communicate and to understand information, and

(3) The ability to reason and to deliberate about one's choices.

President's Commission, 1 Making Health Care Decisions 57–60 (1980). Is this a complete list of the elements of capacity? Does the President's Commission approach serve the purposes that we require a patient to have capacity before we allow that patient to make health care decisions? As a practical matter, how easy is it to apply the President's Commission's test?

One case that squarely faced the question of whether a patient had the capacity to refuse life-sustaining treatment involved Robert Quackenbush, a 72-year-old recluse whose gangrenous leg would have to be amputated to avoid a certain, quick death. He was rambunctious, belligerent, and "a conscientious objector to medical therapy" who had shunned medical care for 40 years. In deciding that the patient was competent, the court depended on the testimony of two psychiatrists and the judge's own visit with Mr. Quackenbush.

> The testimony concerning Quackenbush's mental condition was elicited from two psychiatrists. The first, appearing for the hospital, was Dr. Michael Giuliano. Dr. Giuliano . . . saw Quackenbush once on January 6. The doctor's conclusions are that Quackenbush is suffering from an organic brain syndrome with psychotic elements. He asserts that the organic brain syndrome is acute—i.e., subject to change—and could be induced by the septicemia . . . . [Dr. Giuliano] concluded that

Quackenbush's mental condition was not sufficient to make an informed decision concerning the operation.

* * *

Dr. [Abraham] Lenzner [a specialist in geriatric psychiatry testifying at the request of the court] is of the opinion, based upon reasonable medical certainty, that Quackenbush has the mental capacity to make decisions, to understand the nature and extent of his physical condition, to understand the nature and extent of the operations, to understand the risks involved if he consents to the operation, and to understand the risks involved if he refuses the operation . . . .

I visited with Quackenbush for about ten minutes on January 12. During that period he did not hallucinate, his answers to my questions were responsive and he seemed reasonably alert. His conversation did wander occasionally but to no greater extent than would be expected of a 72-year-old man in his circumstances. He spoke somewhat philosophically about his circumstances and desires. He hopes for a miracle but realizes there is no great likelihood of its occurrence. He indicates a desire—plebian [sic], as he described it—to return to his trailer and live out his life. He is not experiencing any pain and indicates that if he does, he could change his mind about having the operation.

* * *

. . . My findings . . . are that Robert Quackenbush is competent and capable of exercising informed consent on whether or not to have the operation. I do not question the events and conditions described by Dr. Giuliano but find they were of a temporary, curative, fluctuating nature, and whatever their cause the patient's lucidity is sufficient for him to make an informed choice.

* * *

In re Quackenbush, 156 N.J. Super. 282, 383 A.2d 785, 788 (1978).

Did the court merely choose the more credible of the two psychiatrists, or did the court depend upon some intuitive conclusions that followed the judge's ten-minute visit with the patient? Which would be the more satisfying basis for the court's determination of competency?

To what extent do value judgments and prejudices enter into decisions regarding competency? Would the *Quackenbush* case have been a simpler one if Mr. Quackenbush were a retired lawyer leading a middle-class life rather than a belligerent hermit? One fascinating review of "right to die" cases suggests that gender may be an important factor, and that the legal system takes the expressed wishes of men more seriously than those of women. Steven Miles & Allison August, Courts, Gender, and "The Right to

Die," 18 L. Med. & Health Care 85 (1990). Stereotypes about the elderly and about dementia may also have an effect on evaluations of capacity by family members and doctors. For a discussion of evaluating capacity for individuals in these groups, see Jennifer Moye, Evaluating the Capacity of Older Adults: Psychological Models and Tools, 17 Nat'l Acad. Elder L. Attys 3 (2004).

How significant is it that terminally ill patients who seek a right to die are often depressed, or perceived to be so? Isn't it natural for someone dying of an incurable disease to be depressed? Two psychiatrists have suggested that physicians should attempt to treat depression in the seriously ill and those facing imminent death, but that psychiatrists also should recognize that some depression in terminally ill patients is not treatable. They also acknowledge that seriously ill patients do not lose their right to refuse medical treatment simply because they are experiencing depression. Mark Sullivan & Stuart Youngner, Depression, Competence, and the Right to Refuse Lifesaving Medical Treatment, 151 Am. J. Psych. 971, 977 (1994). See also Candice T. Player, Death with Dignity and Mental Disorder, 60 Ariz. L. Rev. 115 (2018). Do any of the suggested tests help you decide when depression renders a person incapable of making a health care decision?

B. ADVANCE DIRECTIVES

It is very difficult to serve the underlying goal of autonomy, if that goal is defined as personal choice, in patients who no longer have decisional capacity. One way to serve this principle is through the application of the doctrine of substituted judgment. Under this doctrine, a person, committee, institution, or other substitute decision-maker attempts to determine what the patient would do if the patient had decisional capacity. It may be possible to review the values of a formerly competent patient to determine whether that patient would choose to undergo or forgo proposed medical care. This can be done through a thoughtful analysis of the patient's values during life or through review of formal statements made by the patient when the patient had capacity. The most relevant considerations may be statements made by the patient about the proposed treatment itself. Indeed, such statements may provide the only *constitutionally* relevant information about an incompetent patient's wishes with regard to life-sustaining medical treatment after *Cruzan*.

Of course, there is no way to know with certainty what the now-incompetent patient would do under the precise circumstances at the time the decision must be made. Some have argued that the doctrine of substituted judgment is too speculative to be applied reliably and that there is simply no way to protect the autonomy of a patient without decisional capacity. Where there is no possible method for establishing what the autonomous patient would do, bioethicists (and increasingly, courts) move to another principle of bioethical decision-making, beneficence. In these circumstances, the alternative to serving autonomy is serving beneficence, and the alternative to the doctrine of substituted judgment is the doctrine of the "best interest" of the patient. As we shall see, the more difficult it becomes to decide what the patient would do if that patient had decisional capacity, the more likely it is that the court will apply the principle of beneficence rather than the principle of autonomy.

Although there are several kinds of statutes that allow competent individual patients to control some element of their health care when they lose decisional capacity, the most influential model statute within the U.S. is probably the Uniform Health-Care Decisions Act, which was adopted by the Uniform Law Commissioners in the early 1990s. Although it has not been fully adopted by most states, it forms the basis of several states' laws with regard to advance directives.

UNIFORM HEALTH-CARE DECISIONS ACT
Uniform Law Commissioners, 1993.

SECTION 1.   DEFINITIONS. In this [Act]:

(1)   "Advance health-care directive" means an individual instruction or a power of attorney for health care.

(2)   "Agent" means an individual designated in a power of attorney for health care to make a health-care decision for the individual granting the power.

(3)   "Capacity" means an individual's ability to understand the significant benefits, risks, and alternatives to proposed health care and to make and communicate a health-care decision.

* * *

(5)   "Health care" means any care, treatment, service, or procedure to maintain, diagnose, or otherwise affect an individual's physical or mental condition.

(6)   "Health-care decision" means a decision made by an individual or the individual's agent, guardian, or surrogate, regarding the individual's health care, including:

   (i)   selection and discharge of health-care providers and institutions;

   (ii)   approval or disapproval of diagnostic tests, surgical procedures, programs of medication, and orders not to resuscitate; and

   (iii) directions to provide, withhold, or withdraw artificial nutrition and hydration and all other forms of health care.

* * *

(9) "Individual instruction" means an individual's direction concerning a health-care decision for the individual.

* * *

(12) "Power of attorney for health care" means the designation of an agent to make health-care decisions for the individual granting the power.

* * *

(16) "Supervising health-care provider" means the primary physician or, if there is no primary physician or the primary physician is not reasonably available, the health-care provider who has undertaken primary responsibility for an individual's health care.

(17) "Surrogate" means an individual, other than a patient's agent or guardian, authorized under this [Act] to make a health-care decision for the patient.

SECTION 2. ADVANCE HEALTH-CARE DIRECTIVES.

(a) An adult or emancipated minor may give an individual instruction. The instruction may be oral or written. The instruction may be limited to take effect only if a specified condition arises.

(b) An adult or emancipated minor may execute a power of attorney for health care, which may authorize the agent to make any health-care decision the principal could have made while having capacity. The power must be in writing and signed by the principal. The power remains in effect notwithstanding the principal's later incapacity and may include individual instructions. . . .

(c) Unless otherwise specified in a power of attorney for health care, the authority of an agent becomes effective only upon a determination that the principal lacks capacity, and ceases to be effective upon a determination that the principal has recovered capacity.

(d) Unless otherwise specified in a written advance health-care directive, a determination that an individual lacks or has recovered capacity, or that another condition exists that affects an individual instruction or the authority of an agent, must be made by the primary physician.

(e) An agent shall make a health-care decision in accordance with the principal's individual instructions, if any, and other wishes to the extent known to the agent. Otherwise, the agent shall make the decision in accordance with the agent's determination of the principal's best interest. In determining the principal's best interest, the agent shall consider the principal's personal values to the extent known to the agent.

* * *

SECTION 4. OPTIONAL FORM. The [form included in this Act] may, but need not, be used to create an advance health-care directive.

* * *

SECTION 5. DECISIONS BY SURROGATE.

(a) A surrogate may make a health-care decision for a patient who is an adult or emancipated minor if the patient has been determined by the primary physician to lack capacity and no agent or guardian has been appointed or the agent or guardian is not reasonably available.

(b) An adult or emancipated minor may designate any individual to act as surrogate by personally informing the supervising health-care provider. In the absence of a designation, or if the designee is not reasonably available, any member of the following classes of the patient's family who is reasonably available, in descending order of priority, may act as surrogate:

    (1) the spouse, unless legally separated;

    (2) an adult child;

    (3) a parent; or

    (4) an adult brother or sister.

(c) If none of the individuals eligible to act as surrogate under subsection (b) is reasonably available, an adult who has exhibited special care and concern for the patient, who is familiar with the patient's personal values, and who is reasonably available may act as surrogate.

* * *

(e) If more than one member of a class assumes authority to act as surrogate, and they do not agree on a health-care decision and the supervising health-care provider is so informed, the supervising health-care provider shall comply with the decision of a majority of the members of that class who have communicated their views to the provider. If the class is evenly divided concerning the health-care decision and the supervising health-care provider is so informed, that class and all individuals having lower priority are disqualified from making the decision.

(f) A surrogate shall make a health-care decision in accordance with the patient's individual instructions, if any, and other wishes to the extent known to the surrogate. Otherwise, the surrogate shall make the decision in accordance with the surrogate's determination of the patient's best interest. In determining the patient's best interest, the surrogate shall consider the patient's personal values to the extent known to the surrogate.

* * *

SECTION 7.   OBLIGATIONS OF HEALTH-CARE PROVIDER.

* * *

(e)   A health-care provider may decline to comply with an individual instruction or health-care decision for reasons of conscience. A health-care institution may decline to comply with an individual instruction or health-care decision if the instruction or decision is contrary to a policy of the institution which is expressly based on reasons of conscience and if the policy was timely communicated to the patient or to a person then authorized to make health-care decisions for the patient.

(f)   A health-care provider or institution may decline to comply with an individual instruction or health-care decision that requires medically ineffective health care or health care contrary to generally accepted health-care standards applicable to the health-care provider or institution.

(g)   A health-care provider or institution that declines to comply with an individual instruction or health-care decision shall:

(1)   promptly so inform the patient, if possible, and any person then authorized to make health-care decisions for the patient;

(2)   provide continuing care to the patient until a transfer can be effected; and

(3)   unless the patient or person then authorized to make health-care decisions for the patient refuses assistance, immediately make all reasonable efforts to assist in the transfer of the patient to another health-care provider or institution that is willing to comply with the instruction or decision.

(h)   A health-care provider or institution may not require or prohibit the execution or revocation of an advance health-care directive as a condition for providing health care.

* * *

SECTION 11.   CAPACITY.

(a)   This [Act] does not affect the right of an individual to make health-care decisions while having capacity to do so.

(b)   An individual is presumed to have capacity to make a health-care decision, to give or revoke an advance health-care directive, and to designate or disqualify a surrogate.

* * *

SECTION 14.   JUDICIAL RELIEF. On petition of a patient, the patient's agent, guardian, or surrogate, a health-care provider or institution involved with the patient's care, or an individual described in

Section 5(b) or (c), the [appropriate] court may enjoin or direct a health-care decision or order other equitable relief.

NOTES AND QUESTIONS

1. *Should Decisions Regarding Life-Sustaining Treatment Be Treated Differently?* Should decisions to terminate life-sustaining medical treatment be treated differently from other kinds of health care decisions? The Uniform Health-Care Decisions Act does not distinguish between life-sustaining medical treatment and other forms of treatment. Should it make that distinction and treat decisions that may lead to the patient's death differently from decisions that will not?

2. *Role of Institutional Ethics Committees.* The Uniform Health-Care Decisions Act does not defer to institutional ethics committees under any circumstances. Should it have been written to recognize and defer to an "ethics review committee," as the more recent New York statute does, at least in the most serious of cases? See Family Health Care Decisions Act, NY Pub. Health sec. 2994–d. Might the fact that these committees now have formal statutory authority enhance the status of hospital and nursing home ethics committees? Does it make sense to look to the courts as an alternative to a determination of institutional ethics committees?

3. *Revising the Uniform Health-Care Decisions Act.* In 2021, the Uniform Law Commission convened a committee to amend or revise the Uniform Health-Care Decisions Act. Key issues for consideration include: whether the Act should include procedures for determining whether an individual lacks capacity to make their own decisions, and, if so, what those procedures should be; requirements for execution and revocation of advanced directives, especially when in electronic format; review of the default surrogate priority list; the relationship between advanced directives and a Physician Order for Life Sustaining Treatment (POLST), discussed below; psychiatric advanced directives; the authority of an agent or surrogate to apply for health benefits; the grounds for disqualification of an agent or default surrogate; whether additional qualifications should be required to make mental health decisions; the benefits and costs of statutory forms; and whether oral designation of surrogates should be honored. What revisions, if any, would you suggest and why?

NOTE: THE DEVELOPMENT OF ADVANCE DIRECTIVES AND SURROGATE CONSENT LAWS

The Uniform Health-Care Decisions Act, excerpted above, takes a comprehensive approach to the issue by combining several elements of existing state laws into a single legal document, as well as adding new provisions. Because the Uniform Act has not been fully adopted by most states, it remains important to understand the variations in state law among states that have not adopted the Act. This note provides an overview of the development of advance directive and surrogate consent laws, noting areas of variation among the states.

The Starting Point: Family Consent Policies and Laws. Over the past century, it became standard medical practice to seek consent to any medical procedure from close family members of an incompetent patient. There is no common law authority for this practice; it is an example of medical practice (and good common sense) being subtly absorbed by the law. The President's Commission suggests five reasons for this deference to family members:

(1)   The family is generally most concerned about the good of the patient.

(2)   The family will also usually be most knowledgeable about the patient's goals, preferences, and values.

(3)   The family deserves recognition as an important social unit that ought to be treated, within limits, as a responsible decision-maker in matters that intimately affect its members.

(4)   Especially in a society in which many other traditional forms of community have eroded, participation in a family is often an important dimension of personal fulfillment.

(5)   Since a protected sphere of privacy and autonomy is required for the flourishing of this interpersonal union, institutions and the state should be reluctant to intrude, particularly regarding matters that are personal and on which there is a wide range of opinion in society.

President's Commission, Deciding to Forego Life-Sustaining Treatment, 128 (1983). It is difficult to determine whether the resort to close relatives to give consent is merely a procedural device to discover what the patient, if competent, would choose, or whether it is based in an independent substantive doctrine. Although it seems essentially procedural—the family is most likely to know what the patient would choose—many courts are willing to accept most decisions of family members even when there is little support for the position that these family members are actually choosing what the patient would choose. Of course, the assumptions about family relationships are quite optimistic. Consulting with family members also neutralizes potential malpractice plaintiffs; this factor accounts for part of the longstanding popularity of this decision-making process among health care providers.

Over the past several decades some states have enacted formal family consent laws that authorize statutorily designated family members to make health care decisions for their relatives in circumscribed situations. These statutes often apply to a wide range of health care decisions (including, in most cases, decisions to forgo life-sustaining treatment), although sometimes they apply only when there has been a physician's certification of the patient's inability to make the health care decision. Sometimes they are limited to particular kinds of treatment (e.g., cardiopulmonary resuscitation) or exclude particular decisions (like discontinuing nutrition and hydration). In addition, family consent laws often provide immunity from liability for family members and physicians acting in good faith, as well as judicial authority to resolve disputes about the authority of the family members under the statutes.

The definition of "family member" and the position of each family member in the hierarchy vary from state to state. In some states those in a long-term spouse-like relationship with the patient are included in the list of family members who can make decisions for the incompetent patient; in some states they are not. Some lists include a residuary class of anyone who knows the values, interests, and wishes of the patient; some states list the physician as the residuary decision-maker; some provide for no residuary decision-maker. Some states give a general guardian top priority; some states place surrogates actually appointed by the patient (whether family members or not) ahead of the general guardian.

The Rise of Living Wills. Nearly two decades before the Uniform Health-Care Decisions Act was proposed, many people first became concerned about the potential abuses of powerful new forms of life-sustaining medical treatment. In the wake of the *Quinlan* case, several states had adopted statutes that formally recognized certain forms of written statements requesting that some kinds of medical care be discontinued. These statutes, generally referred to as "living will" statutes, "right to die" legislation, or "natural death" acts, provided a political outlet for the frustration that accompanied the empathy for Ms. Quinlan.

The statutes, which still provide the only statutory advance directive in some jurisdictions, vary among the states in several respects. In some states living wills may be executed by any person, at any time (and in some states they may be executed on behalf of minors), while in other states they require a waiting period, and may not be executed during a terminal illness. In most states they are of indefinite duration, although in some states they expire after a determined number of years.

Some statutes address only the terminally ill, others include those in "irreversible coma" or persistent vegetative state, and still others provide for different conditions to trigger the substantive provisions of the document. In many states, living wills are not effective while the patient is pregnant. See Chapter 3, Section VI. Some states require the formalities of a will for the living will to be recognized by statute, while other states require different formalities. Living will statutes that do not apply to those in persistent vegetative state, irreversible coma, or any other medical condition that may not be considered "terminal" are of no assistance to people in the position of Nancy Cruzan. Is there a reason to limit legislation to terminal conditions, or should such statutes be extended to other conditions where there is broad social consensus that patients should have the right to forgo life-sustaining treatment?

The statutes generally relieve physicians and other health care providers of any civil or criminal liability for withdrawal of treatment if they properly follow the requirements of the statute. A living will is always relevant as evidence of a patient's intent, although immunity does not apply unless the statute is followed.

Some of the statutes require any physician who cannot in good conscience carry out those provisions to transfer the patient to a physician who can. The statutes also provide that carrying out the provisions of a properly executed living will does not constitute homicide or suicide for any legal purpose. It is hard to know whether the absence of litigation over the terms of living wills means that these documents are working well or that they are not working at all. Because health care providers were not used to seeing these documents and were used to making these decisions on their own, many were reluctant to accept and follow the instructions in living wills. Some health care professionals are still reluctant to carry out advance directives, even when they are clearly legally authorized.

Many living will statutes specifically exclude "the performance of any procedure to provide nutrition or hydration" from the definition of death-prolonging or life-sustaining procedures, and thus do not extend any statutory protection to those who remove nutrition or hydration from a patient. For the most famous example, see Vernon's Ann. Mo. Stat. § 459.010(3). After the U.S. Supreme Court decision in *Cruzan,* are such exceptions legally meaningful? Are they constitutional? For an example of judicial avoidance of the unwelcome consequences of a nutrition and hydration exception to a living will statute, see McConnell v. Beverly Enterprises-Connecticut, 209 Conn. 692, 553 A.2d 596 (1989). See also In re Guardianship of Browning, 568 So. 2d 4 (Fla. 1990).

The Next Step: Durable Powers of Attorney for Health Care. Another means of identifying who should speak for the patient when the patient is incompetent is to allow competent individuals to designate a spokesperson or agent to act should the individual become incompetent. This designation may be accomplished through the individual's execution of a durable power of attorney, the most common form of advance directive available today.

Powers of attorney have been available over the past several centuries to allow for financial transactions to be consummated by agents of a principal. A power of attorney may be executed by any competent person. It provides that the agent designated shall have the right to act on behalf of the principal for purposes that are described and limited in the document itself. Thus, a principal may give an agent a power of attorney to enter into a particular contract, a particular kind of contract, or all contracts. The power may be limited by time, by geographic area, or in any other way. It may be granted to any person, who, upon appointment, becomes the agent and "attorney-in-fact" for the principal.

At common law, a power of attorney expired upon the "incapacity" of the principal. This was necessary to assure that the principal could maintain adequate authority over his agent. As long as a power of attorney expired upon the incapacity of the principal, the power of attorney had no value in making medical decisions. After all, a competent patient could decide for himself; there was no reason for him to delegate authority to an agent.

In the mid-1970s, it became clear that the value of the power of attorney could be increased if it could extend beyond the incapacity of the principal. For example, as an increasing number of very elderly people depended upon their children and others to handle their financial affairs, it became important that there be some device by which they could delegate their authority to these agents. For such principals, it was most important that the authority remain with their agents when they did become incapacitated. The Uniform Probate Code was therefore amended to provide for a durable power of attorney; that is, a power of attorney that would remain in effect (or even only become effective) upon the incapacity of the principal if the document clearly stated that direction.

There is no reported judicial opinion formally holding that the authority of a durable power of attorney executed under the Uniform Probate Code extends to health care decision-making. The President's Commission assumed, without any discussion, that it could be used for this purpose. See President's Commission, Deciding to Forego Life-Sustaining Treatment 145–149 (1983). The vast majority of states have now adopted statutes that formally authorize the execution of durable powers of attorney for health care decisions.

Health care providers, who were uncertain about how to deal with living wills that were often ambiguous and rarely written with knowledge of the patient's eventual diagnosis and prognosis, were more accepting of the durable power of attorney for health care, at least in theory. Providers need a sure decision by a clear decision-maker when they cannot get informed consent from a patient, and the durable powers allowed clearly identified decision-makers to make exactly the decision that confronted the patient at any given moment. Durable powers, then, protected patients, who could appoint trusted family members and friends to make decisions, and physicians, who needed someone they could rely upon to be present and actually make the decisions.

Some state statutes now allow an agent authorized by a durable power of attorney to make health care decisions for a principal even if the principal has capacity—as long as that is the explicit desire of the principal. Why should an agent make a decision for a principal *with* capacity? Doesn't that undermine the principle of autonomy? Such a provision gives health care providers a surrogate decision-maker to turn to in the case of a patient with capacity that is highly variable. Some argue that, in such cases, health care providers ought to be able to depend upon the consent of a patient-designated surrogate without doing a full competency analysis each time a health care decision is to be made. Of course, the decision of the surrogate can always be overruled by the patient herself if she has capacity.

The legal significance of a durable power of attorney for health care is defined by each state's durable power statute. Recall that in her concurring opinion in *Cruzan*, though, Justice O'Connor suggested that there may also be constitutional significance to a properly executed durable power of attorney.

The Development of the Uniform Health-Care Decisions Act. The Uniform Health-Care Decisions Act, excerpted above, takes a comprehensive approach to the issue by combining the living will (which is retitled the "individual instruction"), the durable power of attorney (now called the "power of attorney for health care"), a family consent law, and some provisions concerning organ donation together in one statute. Further, the statute integrates the current living will and durable power (and statement of desire to donate organs) into a single document. The Uniform Act provides a statutory form, but it also explicitly declares that the form is not a mandatory one, and that individuals may draft their own form that includes only some of the kinds of instructions permitted in the unified form. The new "individual instruction" can apply to virtually any health care decision, not just the end-of-life decisions to which living wills are typically applicable. Further, "health care decision" is defined very broadly.

The Uniform Health-Care Decisions Act also makes the execution of the unified document very easy. It has no witness requirement, and it does not require that the document be notarized. The drafters of the proposed act concluded that the formalities often associated with living wills and durable powers served to discourage their execution more than to deter fraud. Despite this approach of the Uniform Act, most adopting states have added some execution formalities. Do you think that executing these documents ought to require witnesses, an oath, a seal, a notary, or some other formalizing act?

The residual decision-making portion of the Uniform Act is very much like the family consent statutes that have now been adopted in some states, and this section of the Act applies only if there is no applicable individual instruction or appointed agent. While it provides for a common family hierarchy of decision-makers for decisionally incapacitated patients, it also provides that the family can be trumped by an "orally designated surrogate," who may be appointed by a patient informing her "supervising physician" that the surrogate is entitled to make health care decisions on her behalf. Thus, patients can effectively orally appoint decision-making agents who previously could only be appointed in a writing signed pursuant to a rigorous process.

Thus, under the Uniform Act, a health care decision will be made by the first available in this hierarchy:

(1) the patient, if the patient has decisional capacity,

(2) the patient, through an individual instruction,

(3) an agent appointed by the patient in a written power of attorney for health care, unless a court has given this authority explicitly to a guardian,

(4) a guardian appointed by the court,

(5) a surrogate appointed orally by the patient,

(6) a surrogate selected from the list of family members and others who can make health care decisions on behalf of the patient.

The Uniform Act explicitly provides that the decision-maker (whether an agent, guardian, or surrogate) should decide based on the principle of substituted judgment rather than the best interest principle. If it is impossible to apply the substituted judgment principle, the statute would allow the substitute decision-maker to apply the best interest principle.

The Uniform Act includes the normal raft of recordkeeping provisions, limitations on the reach of the criminal law, assurances regarding the insurance rights of those who execute the documents, and restrictions on the liability of those who act under the statute in good faith. A provision for $500 in liquidated damages in actions for breach of the Act may not encourage litigation when the statute is ignored, but the provision for attorney's fees in those cases might provide an incentive for lawyers to bring those cases. The Act applies only to adults.

Physicians' Orders Regarding End-of-Life Care. Health care providers have been inconsistent in their recognition and implementation of advance directives, even when there is no doubt about the authenticity or legality of the advance directive under local law. By custom, health care institutions and health care workers caring for patients rely on orders given by the health care professional responsible for the patient's treatment, not on documents signed outside of the health care system. Recognizing the strength of this custom, some patient advocates argue that the best way to protect a patient's interests at the end of life is to incorporate the patient's health care decisions into a physician's order.

One particularly effective way of achieving this protection is to incorporate the decisions into a Physician Order for Life Sustaining Treatment (POLST). A POLST (which goes by a variety of other names with other acronyms in other states—MOST, MOLST, SMOST, TPOP, or, in the Veteran's Administration hospitals, SAPO) may include information about a patient's decisions with regard to resuscitation orders, the extent of appropriate medical intervention, the use of antibiotics and other pharmaceuticals, the provision of nutrition and hydration, the desired place of treatment (home, hospital or nursing home), the identity of the authorized health care decision-maker, and other relevant issues likely to arise in each case.

Modeled on emergency medical services' DNR or DNAR ("do not resuscitate," or, more recently, "do not attempt resuscitation") orders, the POLST is a real medical order, signed by a health care provider—in some states it must be a physician, but in others it can be signed by other providers—with authority to issue that order. As a result, it is more likely to be implemented across health care settings: in the emergency room, at the rehabilitation hospital, in the nursing home, and in the field. It is designed to travel with the patient as the patient moves among these health care settings. An example of a POLST—this one called a "MOST" ("Medical Orders for Scope of Treatment") and used by providers in Colorado—is reprinted below. For discussion of the development of this MOST and a useful discussion of the utility of the document, see http://coloradoadvancedirectives.com/most-in-colorado/.

SEND ORIGINAL FORM WITH PERSON WHENEVER TRANSFERRED OR DISCHARGED

Colorado Medical Orders
for Scope of Treatment (MOST)

- **FIRST** follow these orders, **THEN** contact Physician, Advanced Practice Nurse (APN), or Physician Assistant (PA) for further orders if indicated.
- These Medical Orders are based on the person's medical condition & wishes.
- If Section A or B is not completed, full treatment for that section is implied.
- May only be completed by, or on behalf of, a person 18 years of age or older.
- Everyone shall be treated with dignity and respect.

Legal Last Name	
Legal First Name/Middle Name	
Date of Birth	Sex
Hair Color	Eye Color	Race/Ethnicity

In preparing these orders, please inquire whether patient has executed a living will or other advance directive.
If yes and available, review for consistency with these orders and update as needed. (See additional instructions on page 2.)

A
Check one box only

CARDIOPULMONARY RESUSCITATION (CPR)   ***Person has no pulse and is not breathing.***

☐ **Yes CPR:** Attempt Resuscitation   ☐ **No CPR:** Do Not Attempt Resuscitation

NOTE: Selecting 'Yes CPR' requires choosing "Full Treatment" in Section B.
When not in cardiopulmonary arrest, follow orders in Section B.

B
Check one box only

MEDICAL INTERVENTIONS   ***Person has pulse and/or is breathing.***

☐ **Full Treatment**—primary goal to prolong life by all medically effective means:
In addition to treatment described in Selective Treatment and Comfort-focused Treatment, use intubation, advanced airway interventions, mechanical ventilation, and cardioversion as indicated. Transfer to hospital if indicated. Includes intensive care.

☐ **Selective Treatment**—goal to treat medical conditions while avoiding burdensome measures:
In addition to treatment described in Comfort-focused Treatment below, use IV antibiotics and IV fluids as indicated. **Do not intubate.** May use noninvasive positive airway pressure. Transfer to hospital if indicated. **Avoid intensive care.**

☐ **Comfort-focused Treatment**—primary goal to maximize comfort:
Relieve pain and suffering with medication by any route as needed; use oxygen, suctioning, and manual treatment of airway obstruction. Do not use treatments listed in Full and Selective Treatment unless consistent with comfort goal. **Do not transfer to hospital for life-sustaining treatment. Transfer only if comfort needs cannot be met in current location.**

Additional Orders: _____

C
Check one box only

ARTIFICIALLY ADMINISTERED NUTRITION   *Always offer food & water by mouth if feasible.*

Any surrogate legal decision maker (Medical Durable Power of Attorney [MDPOA], Proxy-by-Statute, guardian, or other) must follow directions in the patient's living will, if any. Not completing this section *does not* imply any one of the choices—further discussion is required. **NOTE:** *Special rules for Proxy-by-Statute apply; see reverse side ("Completing the MOST form")* for details.

☐ Artificial nutrition by tube long term/permanent if indicated.
☐ Artificial nutrition by tube short term/temporary only. (May state term & goal in "Additional Orders")
☐ No artificial nutrition by tube.
Additional Orders: _____

D

DISCUSSED WITH (check all that apply):
☐ Patient
☐ Agent under Medical Durable Power of Attorney

☐ Proxy-by-Statute (per C.R.S. 15-18.5-103(6))
☐ Legal guardian
☐ Other: _____

SIGNATURES OF PROVIDER AND PATIENT, AGENT, GUARDIAN, OR PROXY-BY-STATUTE AND DATE (*MANDATORY*)

Significant thought has been given to these instructions. Preferences have been discussed and expressed to a healthcare professional. This document reflects those treatment preferences, which may also be documented in a Medical Durable Power OA, CPR Directive, living will, or other advance directive (attached if available). To the extent that previously completed advance directives do not conflict with these *Medical Orders for Scope of Treatment,* they shall remain in full force and effect.

If signed by surrogate legal decision maker, preferences expressed must reflect patient's wishes as best understood by surrogate.

Patient/Legal Decision Maker Signature (Mandatory)	Name (Print)	Relationship/ Decision maker status (Write "self" if patient)	Date Signed (Mandatory; Revokes all previous MOST forms)
Physician / APN / PA Signature (Mandatory)	Print Physician / APN / PA Name, Address, and Phone Number		Date Signed (Mandatory)
Colorado License #:			

HIPAA PERMITS DISCLOSURE OF THIS INFORMATION TO OTHER HEALTHCARE PROFESSIONALS AS NECESSARY

Authority for this form and process is granted by C.R.S. 15-18.7: Directives Concerning Medical Orders for Scope of Treatment, enacted 2010.

SEND ORIGINAL FORM WITH PERSON WHENEVER TRANSFERRED OR DISCHARGED			
ADDITIONAL INFORMATION: *Please provide contact information below, in case follow up or more information needed.*			
Patient Legal Last Name	*Patient Legal First Name*	*Patient Middle Name (if any)*	*Patient Date of Birth*
Primary Contact Person for the Patient	*Relationship and/or MDPOA, Proxy, Guardian*	*Phone Number/email/Other contact information*	
Healthcare Professional Preparing Form	*Preparer Title*	*Phone Number/Email*	*Date Prepared*
Patient Primary Diagnosis	*Hospice Program (if applicable) /Address*	*Hospice Phone Number*	

DIRECTIONS FOR HEALTH CARE PROFESSIONALS

For more information, please refer to the "Getting the MOST Out of the Medical Orders for Scope of Treatment: Guidelines for Healthcare Professionals," www.ColoradoMOST.com

Completing the MOST form:

- MOST form master may be downloaded from www.ColoradoMOST.com and photocopied onto **Astrobrights® "Vulcan Green"** or **"Terra Green"** 60lb paper. This special paper is strongly encouraged but not required. Visit www.ColoradoMOST.com for a link to paper suppliers.
- The form must be signed by a physician, advanced practice nurse, or physician assistant to be valid as medical orders. Physician assistants must include physician name and contact information. In the absence of a provider signature, however, the patient selections should be considered as valid, documented patient preferences for treatment.
- Verbal orders are acceptable with follow-up signature by physician, advanced practice nurse, or physician assistant in accordance with facility policy, but not to exceed 30 days.
- **Completion of the MOST form is _not_ mandatory.** "A healthcare facility shall not require a person to have executed a MOST form as a condition of being admitted to, or receiving medical treatment from, the healthcare facility" per C.R.S. 15-18.7-108.
- Patient preferences and medical indications shall guide the healthcare professional in completing the MOST form.
- Patients with capacity should participate in the discussion and sign these orders; a healthcare agent, Proxy-by-Statute, or guardian may complete these orders on behalf of an incapacitated patient, *making selections according to patient preferences, if known.*
- "Proxy-by-Statute" is a decision maker selected through a proxy process, per C.R.S. 15-18.5-103(6). Such a decision maker may not decline artificial nutrition or hydration (ANH) for an incapacitated patient without an attending physician and a second physician trained in neurology certifying that "the provision of ANH is merely prolonging the act of dying and is unlikely to result in the restoration of the patient to independent neurological functioning."
- Photocopy, fax, and electronic images of signed MOST forms are legal and valid.

Following the Medical Orders:

- Per C.R.S. 15-18.7-104: Emergency medical personnel, a healthcare provider, or healthcare facility _shall_ comply with an adult's properly executed MOST form that has been executed in this state or another state and is apparent and immediately available. The fact that the signing physician, advanced practice nurse, or physician assistant does not have admitting privileges in the facility where the adult is receiving care does not remove the duty to comply with these orders. Providers who comply with the orders are immune from civil and criminal prosecution in connection with any outcome of complying with the orders.
- If a healthcare provider considers these orders *medically* inappropriate, she or he should discuss concerns with the patient or surrogate legal decision maker and revise orders only after obtaining the patient or surrogate consent.
- If Section A or B is not completed, full treatment is implied for that section.
- **Comfort care is never optional.** Among other comfort measures, oral fluids and nutrition must be offered if tolerated.
- When "Comfort-focused Treatment" is checked in Section B, hospice or palliative care referral is strongly recommended.
- If a healthcare provider or facility cannot comply with these orders due to policy or ethical/religious objections, the provider or facility must arrange to transfer the patient to another provider or facility and provide appropriate care until transfer.

Reviewing the Medical Orders:

- These medical orders should be reviewed
  - ○ regularly by the person's attending physician or facility staff with the patient and/or patient's legal decision maker;
  - ○ on admission to or discharge from any facility or on transfer between care settings or levels;
  - ○ at any substantial change in the person's health status or treatment preferences; and
  - ○ when legal decision maker or contact information changes.
- If substantive changes are made, please complete a new form and void the replaced one.
- To void the form, draw a line across Sections A through C and write "VOID" in large letters. Sign and date.

REVIEW OF THIS COLORADO MOST FORM			
Review Date	**Reviewer**	**Location of Review**	**Review Outcome**
			☐ No Change ☐ New Form Completed
			☐ No Change ☐ New Form Completed
			☐ No Change ☐ New Form Completed
			☐ No Change ☐ New Form Completed

HIPAA PERMITS DISCLOSURE OF THIS INFORMATION TO OTHER HEALTHCARE PROFESSIONALS AS NECESSARY

Colorado Advance Directives Consortium, www.ColoradoMOST.com   2015

POLST forms are generally entered in the patient's medical record after the provider has discussed all of the relevant issues with the patient, the patient's family, the agent or surrogate authorized to make health care decisions, and others. The POLST form may include a summary of the values and goals of the patient that form the basis of the order, and the nature of the discussions that gave rise to the order. The patient or the patient's decision-maker is often asked to countersign the order, so it is clear that it represents the agreed view of the patient and the provider. Sometimes the patient's (or the patient's representative's) signature is required; sometimes it is not. In some jurisdictions there is no place for the countersignature at all. In order to assure that these forms are not lost in the patient's chart, they are often printed on distinctively colored (usually bright pink or bright green) paper. Such orders can be recognized by state law (as they are in a many states), or by institutional or community policy.

The process of entering a POLST requires discussion among all of the relevant parties, and it provides a way of integrating a patient's advance directive and the physician's order. It does not override a legally authorized advance directive, but it may make it easier for succeeding health care providers to be aware of advance directives made earlier by the patient; and it eliminates a provider's concern over the obligation to carry out an advance directive that appears to be inconsistent with an order of a physician. The POLST is usually filled out by those familiar with the current medical needs of the patient when the need for end-of-life care is imminent, and those factors may also account for the very high level of implementation of these documents.

Some people are concerned that the development of the POLST paradigm marks a return to the days of paternalistic medicine when it was the physician, not the patient, who decided the treatment to be applied in each case. Here, the provider's signature on the POLST effectively supplants the patient's signature on an advance directive, although the effect is ameliorated by the fact that the entry of a POLST requires a discussion with the patient or the substitute decision-maker, and, often, that person's signature as well. Does the POLST require a patient to give up some level of autonomy that the patient might have had when making decisions through other forms of advance directives? Is the fact that the POLST really will be honored enough to justify giving up that autonomy? For a good account of the state of POLSTs in the U.S. (and Canada and Australia, where similar concepts have developed), see Thaddeus Mason Pope and Melinda Hexum, Legal Briefing: POLST: Physicians Order for Life-Sustaining Treatment, 36 J. Clinical Ethics 353 (2012). See also Keith Sonderling, POLST: A Cure for the Common Advance Directive—It's Just What the Doctor Ordered, 33 Nova L. Rev. 451 (2009) and Susan Hickman et al., Hope for the Future: Achieving the Original Intent of Advance Directives, 35 Hastings Ctr. Rep. (6) Supp. 26–30 (2005).

Religious Objections to Some Provisions of Advance Directives. Some religious denominations have principled objections to facilitating certain decisions that might be made in advance directives. These objections are reflected in a range of state statutes, often because those statutes are products of political compromise. For example, as this note suggests, many states limit the applicability of advance directives when the patient is pregnant or when the patient wishes to reject some forms of nutrition or hydration. These limitations are consistent with the tenets of particular religious groups (although there are secular arguments to support them as well). Advance directive laws without these explicit limitations often include other "conscience" provisions, which serve to relieve health care institutions and individual providers from taking actions that violate their own values. See section 7(e) of the Uniform Health-Care Decisions Act, above.

What should happen when a patient with particular values makes a decision that is protected by law in that jurisdiction but is inconsistent with the religious values of a health care institution? What if the legally authorized decision-maker (through a durable power of attorney) or the patient (through an individual instruction) requests that all artificial nutrition and hydration be terminated if the patient is in a persistent vegetative state, but the patient finds himself in a religious hospital that refuses to honor that decision? What if a patient seeking palliative sedation finds himself at a religious hospital that, as a matter of religious and ethical principle, will not provide that form of treatment? What if the family of a religious person seeks the continuation of nutrition and hydration as long as the patient remains in a persistent vegetative state, treatment arguably required by the patient's faith, in an institution that considers it to be a medically useless and futile form of treatment?

Several advocacy groups now provide sample language that individuals can include in their advance directives to request that, if the patient must be hospitalized, the hospitalization take place—for example—in a Catholic hospital bound by the Ethical and Religious Directives (ERDs) of the Catholic Church, or alternatively, that the patient *not* be hospitalized in a Catholic hospital bound by the ERDs. Should providers be bound by that kind of request, on either side? Under the California Right to Know End of Life Options Act and the New York Palliative Care Information Act, must hospitalized patients be told of legal alternatives that might not be available at the hospital because they are not permitted by religious doctrine? Is it a violation of the First Amendment to require that those institutions provide care (or information about care) to which they have strong moral objection?

The Patient Self-Determination Act and Advance Directives. The federal Patient Self-Determination Act applies to hospitals, skilled nursing facilities, home health agencies, hospice programs, and HMOs that receive Medicaid or Medicare funding. It requires each of those covered by the Act to provide each patient with written information concerning:

> (i)   an individual's rights under State law (whether statutory or as recognized by the courts of the State) to make decisions concerning . . . medical care, including the right to accept or refuse medical or surgical treatment and the right to formulate advance directives . . . and

> (ii) the written policies of the provider or organization respecting the implementation of such rights.

42 U.S.C.A. § 1395cc(a)(1)(f)(1)(A). In addition, those covered must document in each patient's record whether that patient has signed an advance directive, assure that the state law is followed in the institution, and provide for education of both the staff and the public concerning living wills and durable powers of attorney.

A few states have also taken action to increase the utility of advance directives. For example, some states have central registries of advance directives, and a handful of states provide for drivers' licenses to show if a patient has an advance directive. At least one statute requires managed health care providers to discuss advance directives with their patients/enrollees. There has also been some attention given to how to make advance directives of various stripes enforced across state and provincial borders.

NOTES AND QUESTIONS

1. *The Values History and the "Five Wishes" Form.* Because it is almost impossible to predict exactly what kinds of medical care one might need and what kinds of medical conditions one will suffer, some courts and scholars have suggested that the most helpful kind of advance directive would be one that appoints a substitute decision-maker (through a durable power of attorney) and then describes generally the principal's interests and values that are likely to be significant in subsequent decision-making over medical treatment.

One particularly good device for encouraging such discussion (and for recording the results) is the "Values History Form," which asks prospective patients about their general values; their medical values; their relationships with family members, friends, and health care providers; their wishes in particular cases; and a host of other issues likely to become relevant if they become incompetent and health care decisions must be made on their behalf. There is no doubt that the existence of such a document would be of great value to a substitute decision-maker and to any court called upon to confirm that substitute's decision even though it doesn't carry the immunity provision or requirement of compliance that a living will does. See Joan McIver Gibson, Reflecting on Values, 51 Ohio St. L.J. 451 (1990).

In addition, over the past several years the "Five Wishes" document, a simple and easy-to-use advance directive that is arguably legally effective under the laws of most states, has become popular. The Five Wishes document, developed by the advocacy group Aging with Dignity, tells your family and doctors: (1) which person you want to make health care decisions for you when you can't make them; (2) the kind of medical treatment you want or don't want; (3) how comfortable you want to be; (4) how you want people to treat you; and (5) what you want your loved ones to know.

2. *Prior Preferences or Current Interests?* There has been considerable research on the efficacy of advanced directives generally and their value in particular circumstances—for example, when a patient is experiencing, or is likely to experience, dementia. Several research and advocacy groups now offer particular advance directive language that might be useful to those dementia

patients, as long as that language is also consistent with other state statutory requirements. See Paula Span, The New Old Age: One Day Your Mind May Fade. But You Can Plan Ahead, NY Times, January 23, 2018. While the law generally requires that the patient's prior preferences and values be used to guide decision-making, some have argued that the patient's current interests should be used instead. See, e.g., Rebecca Dresser, Precommitment: A Misguided Strategy for Securing Death with Dignity, 81 Tex. L. Rev. 1823 (2003) (describing problems with precommitment).

3.   *Supported Decision-Making.* Scholars have noted that the law of health care decision-making in the U.S. reflects a model of decision-making that is personal, private, and independent. Is this how most people make decisions? Or do they turn to trusted friends, family members, and others for advice and support? Supported decision-making, developed to allow people with disabilities to retain decision-making capacity as an alternative to surrogate decision-making or guardianship, reflects the latter approach. People using supported decision-making rely on trusted friends, family members, professionals, and others, to make their own decisions. An increasing number of states have adopted laws related to supported decision-making for people with disabilities. Peter Blanck & Jonathan G. Martinis, "The Right to Make Choices": The National Resource Center for Supported Decision-Making, 3 Inclusion 24 (2015). See also Megan S. Wright, Dementia, Autonomy, and Supported Healthcare Decisionmaking, 79 Md. L. Rev. 257 (2020); Megan S. Wright, End of Life and Autonomy: The Case for Relational Nudges in End-of-Life Decision-Making Law and Policy, 77 Md. L. Rev. 1062 (2018).

4.   *Outmoded Distinctions as to Treatment.* Distinctions between "active" and "passive" conduct and between "withholding" and "withdrawing" treatment are anachronistic distinctions which have not found a safe harbor in the law, just as they have been increasingly recognized as meaningless in ethics. An early New Jersey case, *In re Conroy*, 98 N.J. 321, 486 A.2d 1209 (N.J. 1985), considered each of these distinctions and summarized the ethical and legal literature and the reasons for rejecting the distinctions. As to the distinction between active and passive conduct, the *Conroy* court announced:

> We emphasize that in making decisions whether to administer life-sustaining treatment to patients such as Claire Conroy, the primary focus should be the patient's desires and experience of pain and enjoyment—not the type of treatment involved. Thus, we reject the distinction that some have made between actively hastening death by terminating treatment and passively allowing a person to die of a disease as one of limited use in a legal analysis of such a decision-making situation.

> Characterizing conduct as active or passive is often an elusive notion, even outside the context of medical decision-making . . . . The distinction is particularly nebulous, however, in the context of decisions whether to withhold or withdraw life-sustaining treatment. In a case like that of Claire Conroy, for example, would a physician who discontinued

nasogastric feeding be actively causing her death by removing her primary source of nutrients; or would he merely be omitting to continue the artificial form of treatment, thus passively allowing her medical condition, which includes her inability to swallow, to take its natural course? [ ] The ambiguity inherent in this distinction is further heightened when one performs an act within an over-all plan of non-intervention, such as when a doctor writes an order not to resuscitate a patient. . . .

For a similar reason, we also reject any distinction between withholding and withdrawing life-sustaining treatment. Some commentators have suggested that discontinuing life-sustaining treatment once it has been commenced is morally more problematic than merely failing to begin the treatment. Discontinuing life-sustaining treatment, to some, is an "active" taking of life, as opposed to the more "passive" act of omitting the treatment in the first instance.

This distinction is more psychologically compelling than logically sound. As mentioned above, the line between active and passive conduct in the context of medical decisions is far too nebulous to constitute a principled basis for decision-making. Whether necessary treatment is withheld at the outset or withdrawn later on, the consequence—the patient's death— is the same. Moreover, from a policy standpoint, it might well be unwise to forbid persons from discontinuing a treatment under circumstances in which the treatment could permissibly be withheld. Such a rule could discourage families and doctors from even attempting certain types of care and could thereby force them into hasty and premature decisions to allow a patient to die. [ ]

486 A.2d at 1233–1234.

As to the distinction between "ordinary" and "extraordinary" treatment, *Conroy* pointed out:

We also find unpersuasive the distinction relied upon by some courts, commentators, and theologians between "ordinary" treatment, which they would always require, and "extraordinary" treatment, which they deem optional. . . . The terms "ordinary" and "extraordinary" have assumed too many conflicting meanings to remain useful. To draw a line on this basis for determining whether treatment should be given leads to a semantical milieu that does not advance the analysis.

The distinction between ordinary and extraordinary treatment is frequently phrased as one between common and unusual, or simple and complex, treatment [ ]; "extraordinary" treatment also has been equated with elaborate, artificial, heroic, aggressive, expensive, or highly involved or invasive forms of medical intervention [ ]. Depending on the definitions applied, a particular treatment for a given patient may be considered both ordinary and extraordinary. [ ] Further, since the common/unusual and simple/complex distinctions among medical treatments "exist on continuums with no precise dividing line," [ ] and the continuum is constantly shifting due to progress in medical care, disagreement will often exist about whether a particular treatment is ordinary or extraordinary. In addition, the competent patient generally could refuse even ordinary treatment; therefore, an incompetent patient theoretically should also be able to make such a choice when the surrogate decision-making is effectuating the patient's subjective intent. In such cases, the ordinary/extraordinary distinction is irrelevant except insofar as the particular patient would have made the distinction.

The ordinary/extraordinary distinction has also been discussed in terms of the benefits and burdens of treatment for the patient. If the benefits of the treatment outweigh the burdens it imposes on the patient, it is characterized as ordinary and therefore ethically required; if not, it is characterized as extraordinary and therefore optional. [ ] This formulation is extremely fact-sensitive and would lead to different classifications of the same treatment in different situations.

. . . Moreover, while the analysis may be useful in weighing the implications of the specific treatment for the patient, essentially it merely restates the question: whether the burdens of a treatment so clearly outweigh its benefits to the patient that continued treatment would be inhumane.

468 A.2d at 1234–1235. See also Brophy v. New England Sinai Hosp., Inc., 398 Mass. 417, 497 N.E.2d 626 (1986) ("while we believe that the distinction between extraordinary and ordinary care is a factor to be considered, the use of such a distinction as the sole, or major, factor of decision tends . . . to create a distinction without meaning.")

DOCTORS HOSPITAL OF AUGUSTA V. ALICEA
Supreme Court of Georgia, 2016.
788 S.E.2d 392.

NAHMIAS, JUSTICE.

In March 2012, Jacqueline Alicea's 91-year-old grandmother, Bucilla Stephenson, died at the end of a two-week stay at Doctors Hospital of Augusta, LLC . . . . In May 2013, Alicea, acting as the administratix of her grandmother's estate, sued the Hospital and Dr. Phillip Catalano . . . . Alicea alleged among other things that they intubated her grandmother and put her on a mechanical ventilator, which prolonged her life when she was in a terminal condition and caused her unnecessary pain and suffering, contrary to her advance directive for health care and the specific directions of Alicea, her designated health care agent. The Defendants filed a motion for summary judgment, arguing among other things that the Georgia Advance Directive for Health Care Act ("Advance Directive Act" or "Act") [ ] provided them immunity from liability.

* * *

On November 12, 2009, Stephenson, who was then 89 years old, executed an advance directive for health care ("Advance Directive"), designating as her health care agent Alicea, the granddaughter with whom she lived. The Advance Directive specified that Alicea was "authorized to make all health-care decisions for me, including decisions to provide, withhold, or withdraw artificial nutrition and hydration, and all other forms of health care to keep me alive." The Advance Directive also said:

> My agent shall make health-care decisions for me in accordance with this power of attorney for health care, any instructions I give in this form, and my other wishes to the extent known to my agent. . . .

Stephenson repeatedly told her family members that "she was ready to go when the good Lord called her," and said "when it's my time, it's my time, don't prolong it." She told Alicea specifically that "[s]he did not want . . . to rely on a machine to have to live," including a ventilator to breathe for her. In 2007, Alicea's 80-year-old father had died at the Hospital after entering it with pneumonia and without an advance directive or other document concerning end-of-life decisions. Because Alicea's mother had Alzheimer's disease, Alicea ultimately had to make the decision to take her father off a ventilator. Stephenson did not want Alicea to have to make that kind of decision about her. In the paragraph of the Advance Directive addressing "end-of-life decisions," Stephenson initialed the option that said:

> Choice NOT to Prolong Life.

> I do not want my life to be prolonged if (1) I have an incurable[ ] and irreversible condition that will result in my death within a relatively short time, (2) I become unconscious and, to a reasonable degree of medical certainty, I will not regain consciousness, or (3) the likely risks and burdens of treatment would outweigh the expected benefits.

[ ] Two years passed. Then, around February 28, 2012, Stephenson developed a persistent cough. On Saturday, March 3, she woke up lethargic; she was minimally responsive and had urinated on herself. After Alicea and her husband assisted Stephenson in sitting up, she became more aware, and they helped her to the bathroom to clean up. As Alicea was bathing her, Stephenson lost control of her bowels, and her eyes rolled toward the back of her head. Alicea and her husband feared that Stephenson was having a stroke and drove her to the Hospital's emergency room; Alicea brought the Advance Directive with her.

Blood tests and a chest x-ray showed that Stephenson was suffering from pneumonia, sepsis, and acute renal failure, and she was admitted to the Hospital. Alicea gave the Hospital the Advance Directive, which was placed in Stephenson's medical record, but not in the front behind the admission tab as required by hospital policy to ensure its ready availability to all doctors and hospital staff. Alicea also gave the Hospital her contact information, including her home, work, and cell phone numbers and her husband's cell phone number, so that she could be reached whenever she was away from the Hospital. Alicea has presented evidence showing that from the time Stephenson arrived at the Hospital, she was unable or chose not to make significant health care decisions for herself, triggering Alicea's authority to make those decisions pursuant to the Advance Directive.

Around 9:00 a.m. the next day, Dr. Catalano, a surgeon with staff privileges at the Hospital, called Alicea to tell her that he was taking care of Stephenson and that she was being moved to the intensive care unit ("ICU"). Dr. Catalano said that he planned to perform a computed tomography ("CT") scan to better assess her condition. Alicea did not object to the CT scan, but she told Dr. Catalano about Stephenson's Advance Directive and specifically instructed that "by no means was CPR [cardio-pulmonary resuscitation] to ever be administered" and that "no heroic measures were to be used" to prolong Stephenson's life.

[Another physician at the hospital] asked Alicea about ventilation specifically, and Alicea directed him to call her before intubating Stephenson and putting her on a ventilator. [That physician] wrote twice in his progress note that Stephenson was "no CPR" and that Alicea had to be called "before patient is intubated."

On Monday morning, March 5, Dr. Catalano called Alicea and requested her verbal consent for a "surgical" thoracentesis to drain more fluid from Stephenson's lung cavity. . . . Dr. Catalano had not read the Advance Directive or the progress notes in Stephenson's medical chart; he did not tell Alicea, and she did not know, that this procedure would require intubation and the use of a ventilator. Had Alicea known that intubation was required, she would not have consented to the surgery. During the surgery, Dr. Catalano found that much of Stephenson's right lung was necrotic (dead tissue), and he removed two-thirds of the lung. Stephenson was extubated in the recovery room, and Alicea was not told that she had been intubated and put on a ventilator.

Two days later, on March 7, Stephenson was experiencing respiratory distress in the early morning hours . . . . Around 4:00 a.m., the nursing staff called Dr. Catalano at home. Dr. Catalano decided to have Stephenson intubated and put on a ventilator to prevent her from going into respiratory or cardiac arrest. A nurse asked Dr. Catalano if he wanted to call Alicea before ordering the life-prolonging intubation, but he rebuffed her, saying, "I'm not going to call her at six o'clock in the morning and scare the hell out of her. I'll wait till, you know, she wakes up and then I'm going to call her and tell her what happened." Dr. Catalano then spoke to the on-duty doctor and directed him to intubate Stephenson, telling the doctor, "I don't want her to die."[3] The on-duty doctor arrived at Stephenson's bedside . . . ,

---

[3]   In his deposition, Dr. Catalano explained his thought process at the time as follows:

performed the intubation, and hooked Stephenson up to a ventilator. No effort was made to contact Alicea before or after Stephenson was intubated.

When Alicea's husband stopped by the Hospital that morning around 8:00 a.m. to check on Stephenson, he was surprised to see her on a ventilator. He called Alicea, who was shocked by the news, and told her that the nursing staff could not find the Advance Directive. . . . It took the nursing staff 15 to 20 minutes of searching to locate the Hospital's copy of the Advance Directive, and one nurse remarked to Alicea's husband, "Boy, somebody has really messed up. I found it."

When Alicea got to the Hospital, she . . . held up the Advance Directive and told the doctor that her grandmother had expressed her wishes, which were contrary to what had happened, and that Alicea had specifically said to call her before putting Stephenson on a ventilator. . . .

Alicea asked [the director of the ICU] about possible next steps. He told her that she could decide to have Stephenson taken off the ventilator and extubated, which would cause her grandmother to suffocate and die (as had happened with Alicea's father), and that the only other option was to perform another surgery to clean out Stephenson's lung cavity more. Had Stephenson been allowed to die that morning, Alicea "would have understood that it was her time and God took her." Having been deprived of the opportunity to let nature take its course, Alicea consented to the surgical procedure and others recommended by Dr. Catalano and the Hospital staff over the next week, including the placement of a feeding tube, a bronchoscopy to remove pus from Stephenson's airway, and a tracheostomy to provide an alternate airway and to remove secretions.

On March 14, Alicea was informed that Stephenson's kidneys were shutting down and that she needed dialysis, and Alicea gathered the family at the Hospital to discuss the situation. . . . Alicea then authorized the removal of the ventilator and the provision of comfort measures only, and three days later, on March 17, 2012, Stephenson died.

[ ] On May 14, 2013, Alicea, acting as the administratrix of Stephenson's estate, filed a complaint against the Hospital and Dr. Catalano, raising claims of breach of agreement, professional and ordinary negligence, medical battery, intentional infliction of emotional distress, and breach of fiduciary duty. The Hospital's alleged liability was based on respondeat superior for the actions of its agents and employees. The

---

I'm thinking, well, I mean what's the worst. If the family does not want her on the respirator, well, we can just pull the tube out. And, you know, we've wasted an hour or two of her staying in the hospital ICU but, on the other hand, if we try to make calls she'd be dead. I mean I can't call the family. . . . I said, well, let's just do what's right for the patient. My God, we can always undo it. But . . . if the patient dies, you know, that's my ultimate loss. There's no way I can get her back. So when this happened I really didn't go into any of the code/no code/do not intubate/resuscitate. Save the patient's life first and then we'll do whatever it takes to make the family and that patient whatever, but we can't undo death. So that's what I was thinking.

complaint alleged that Dr. Catalano and other medical personnel associated with the Hospital had subjected the terminally ill Stephenson to unnecessary medical procedures, in particular her intubation and placement on a ventilator on . . . in violation of her Advance Directive and the directions of Alicea as her designated health care agent.

In support of her claims, Alicea relied on an expert on gerontology, geriatrics, and palliative care. The expert concluded that when Stephenson arrived at the Hospital on March 4, she already "had an incurable and irreversible condition that was likely to result in her death within a relatively short period of time thereafter," and that "her condition was such that the likely risk and burdens of any invasive procedures and treatment outweighed any expected benefits." Consequently, the expert opined that the Defendants were required under the standard of care to refrain from taking steps to prolong her life in accordance with her Advance Directive as well as the instructions of Alicea, her designated health care agent. According to Alicea's expert, Dr. Catalano breached the standard of care by, among other things, failing to review Stephenson's Advance Directive and the progress notes in her medical chart to determine if Alicea had given any directions for Stephenson's care and by failing to obtain basic consent from Alicea before the March 7 intubation. The expert further opined that the nurses employed by the Hospital had violated the standard of care by failing to contact Alicea before the March 7 intubation and failing to call Dr. Catalano's attention to Stephenson's Advance Directive and the notation in the progress notes regarding intubation.

[T]he Defendants filed a motion for summary judgment, contending among other things that they were immune from liability based on the March 7 surgical procedure under [the provisions of the Georgia Advance Directive Act].

* * *

[ ] The Defendants argue that the Court of Appeals misconstrued the Advance Directive Act as requiring health care providers to act in "good faith reliance" on the designated health care agent's directions and decisions in order to qualify for the immunity from civil liability [ ]. Our rejection of this argument depends in part on our understanding of the Act's overall purpose and operation, so we will [first] outline those features before turning to a detailed examination of [the statute].

(a) In 2007, the General Assembly enacted the statute . . . . As the uncodified first section of the 2007 statute explained:

> The General Assembly has long recognized the right of the individual to control all aspects of his or her personal care and medical treatment, including the right to insist upon medical treatment, decline medical treatment, or direct that medical treatment be withdrawn.

[ ] Thus,

> the clear expression of an individual's decisions regarding health care, whether made by the individual or an agent appointed by the individual, is of critical importance not only to citizens but also to the health care and legal communities, third parties, and families.

. . . Thus, a clear objective of the Act is to ensure that in making decisions about a patient's health care, it is the will of the patient or her designated agent, and not the will of the health care provider, that controls.

* * *

The Act . . . sets forth several rules for how decisions are to be made in caring for a patient with an advance directive. If the patient's attending physician determines in good faith that the patient is able to understand the general nature of the health care procedure being consented to or refused, the patient's own decision about that procedure prevails over contrary instructions by a health care agent. . . . However,

> [w]henever a health care provider believes a declarant is unable to understand the general nature of the health care procedure which the provider deems necessary, the health care provider shall consult with any available health care agent known to the health care provider who then has power to act for the declarant under an advance directive for health care.

* * *

The health care decision now at issue in this case is the decision that needed to be made on March 7, 2012, about whether Stephenson should be intubated and put on a ventilator to prolong her life. Stephenson was unable to, and clearly did not, make that decision for herself, so Alicea had the authority to make that decision for Stephenson under her Advance Directive, which Alicea had given to the Hospital's staff and had discussed with them and with Dr. Catalano. With respect to the duties of "[e]ach health care provider and each other person with whom a health care agent interacts under an advance directive for health care" in this situation, the Act says the following [ ]:

> A health care decision made by a health care agent in accordance with the terms of an advance directive for health care shall be complied with by every health care provider to whom the decision is communicated, subject to the health care provider's right to administer treatment for the [patient's] comfort or alleviation of pain; provided, however, that if the health care provider is unwilling to comply with the health care agent's decision, the health care provider shall promptly inform the health care agent who shall then be responsible for arranging for the [patient's] transfer to another health care provider. A health care provider who is unwilling to comply with the health care agent's

decision shall provide reasonably necessary consultation and care in connection with the pending transfer.

Thus, a health care provider in this situation generally must comply with the health care agent's decision, with two exceptions. The first, not pertinent here, is as to pain treatment. The second recognizes that the provider may be "unwilling" to comply with the agent's decision, on medical, moral, or other grounds. But the unwilling provider is not entitled to then make the health care decision for the patient himself, or to just walk away. The Act requires such a provider to "promptly inform" the agent of his unwillingness to comply and also to "provide reasonably necessary consultation and care" in connection with the transfer of the patient to another care-giver as arranged by the agent—presumably a transfer to a provider (who may be in the same facility) who will comply with the agent's decision.

The Advance Directive Act then includes a series of immunity provisions [ ]. The Defendants seek to rely [first, on] a general release of liability for

[e]ach health care provider, health care facility, and any other person who acts in good faith reliance on any direction or decision by the health care agent . . . to the same extent as though such person had interacted directly with the declarant as a fully competent person.

The Act then says:

Without limiting the generality of the foregoing, the following specific provisions shall also govern, protect, and validate the acts of the health care agent and each such health care provider, health care facility, and any other person acting in good faith reliance on such direction or decision: . . . .

After the colon come five specific immunity provisions.[ ]

The first three of these statutory immunity provisions, it is important to recognize, mirror the statutory duties imposed on health care providers by [the statute]. Corresponding to the first clause [ ] which requires that "[a] health care decision made by a health care agent in accordance with the terms of an advance directive for health care shall be complied with by every health care provider to whom the decision is communicated," [the statute] grants providers immunity from civil or criminal liability or professional discipline "solely for complying with any direction or decision by the health care agent, even if death or injury to the declarant ensues." And corresponding to the proviso [ ] for health care providers who are "unwilling to comply with the health care agent's decision," [the statute] give such providers similarly broad immunity—so long as they promptly inform the agent of the "refusal or failure" to comply with the agent's direction or decision and assist with the patient's continued care to the extent of continuing to provide reasonably necessary consultation and care in connection with a pending transfer of the patient, acting substantially in accord with reasonable medical standards, and cooperating in any transfer of the patient as authorized by [the statute itself].

* * *

[O]nly health care providers who act in good faith reliance on the agent's directions are entitled to immunity [ ], the introductory clause speaks of "each such health care provider . . . acting in good faith reliance on such direction or decision"—the "such" referring back to the providers discussed in the preceding general-release sentence, that is, "[e]ach health care provider who acts in good faith reliance on any direction or decision by the health care agent."

* * *

The [statute] plainly authorizes a health care provider to make no effort to comply with an agent's direction—to refuse or fail entirely to comply—so long as the provider promptly informs the agent of that choice and takes the other steps of care and cooperation that the Act requires. . . . What is critical, in our view, is that a provider claiming to have acted in "good faith reliance" on the agent's direction or decision can show that he acted in dependence on that direction or decision, not without reference to the agent's wishes. . . .

Recall that a primary purpose of the Advance Directive Act is to ensure that in making decisions about a patient's health care, it is the will of the patient or her designated agent, rather than the will of the health care provider, that controls. OCGA § 31–32–8 (1) enforces this purpose by declaring that a health care provider who

> believes a declarant is unable to understand the general nature of the health care procedure which the provider deems necessary . . . shall consult with any available health care agent known to the health care provider who then has power to act for the declarant under an advance directive for health care.

If a provider is aware of what the agent has decided, and then proceeds as the statute mandates [ ]—either by complying with that decision or by taking the steps required when he is unwilling to comply with the decision—then he may look to the immunity provisions [ ] for protection. But a provider cannot claim this immunity when his action was not based in good faith on the agent's direction, just because the decision he made for the patient happens to be one that arguably complied or failed to comply with what the agent would have decided. . . .

. . . The health care decision in question is the decision to intubate Stephenson and put her on a ventilator as a life-prolonging measure around 4:00 a.m. on the morning of March 7, 2012. Although there is evidence to the contrary, there is ample evidence that in ordering that procedure, Dr. Catalano was not acting in good faith reliance—in honest dependence—on any decision Alicea had made as Stephenson's health care agent, either to comply with it or to refuse or fail to comply with it and then promptly inform Alicea of his unwillingness. Instead, the evidence would support a finding that Dr. Catalano made the health care decision himself, in the exercise of his own medical and personal judgment. . . .

[T]he Advance Directive Act is all about letting patients and their health care agents, rather than the health care provider, control such decisions. . . .

Because there is at least a disputed issue of fact as to whether Dr. Catalano acted with good faith reliance on any decision made by his patient's health care agent, Dr. Catalano cannot on motion for summary judgment claim the immunity that [is available] to providers who honestly depend on such a decision to either comply with it or promptly inform the patient that they are unwilling to comply with it. Likewise, the Hospital points to no evidence that its staff acted based on a decision by Alicea with respect to the March 7 intubation; when Dr. Catalano made the decision himself, the staff simply proceeded based on his directive.

NOTES AND QUESTIONS

1.   *Subsequent Settlement.* In this case, the matter was settled just before the trial was scheduled to begin in 2017. One report suggests the plaintiff originally sought $200,000 to compensate for additional medical expenses, as well as punitive damages and attorneys' fees. In fact, the case settled for about $1,000,000. While normally the settlement figure itself would be kept confidential by agreement of the parties, in this case the plaintiffs insisted that the amount be made public so that it would serve as a warning to other physicians and institutions in Georgia who might be tempted to ignore agents and advance directives. See S. Hodson, Hospital Settles Lawsuit Failing to Honor Patient's Wishes on Extending Life, Augusta Chronicle, May 25, 2017. See also Paula Span, Filing Suit for "Wrongful Life," New York Times (Jan. 26. 2021) (noting other cases that have recognized wrongful prolongation of life claims).

2.   *Calculating Damages.* Should damages that are accrued because physicians ignore agents be assessed in the same way as damages that are accrued because physicians ignore patients themselves? Should damages that accrue from the failure of a hospital to follow a POLST be treated just like damages that accrue because physicians ignore agents and patients? The Georgia statute authorizing agents to make decisions for incompetent patients in this case is based on the Uniform Health-Care Decisions Act, excerpted earlier in this chapter. Would you expect the same rule on damages to apply in other states that have adopted this form of that Uniform Act? Should the

Uniform Act be amended to explicitly provide that damages are available under these circumstances?

3. *Hospital Liability?* The Georgia Supreme Court suggests that the hospital is liable in this case because it is vicariously liable for the acts of the physicians. Is that always true? Were the physicians here agents of the hospital? Apparent or ostensible agents of the hospital? Might the hospital be directly liable for not enforcing its policy on the placement of advance directives at the front of the patient chart? Might the hospital be liable for not having an adequate policy recognizing the decision-making authority of agents and surrogate?

4. *Failure of Informed Consent?* When a physician or an institution ignores the request of a patient, agent, or surrogate, does that automatically create an informed consent action on behalf of the patient? Are informed consent damages different from other damages (like ordinary tort damages or medical malpractice damages) that might be awarded to a patient under these circumstances?

5. *Immunity Provision.* This case focused on the meaning of the immunity provision in the statute. Is that provision necessary? Why do you think it was included in the statute? If there is no liability in any case, why would a provider have to worry about immunity?

C. DECISION-MAKING IN THE ABSENCE OF ADVANCE DIRECTIVES OR STATUTORILY DESIGNATED SURROGATES

IN RE EICHNER

New York Court of Appeals, 1981.
420 N.E.2d 64.

WACHTLER, JUDGE.

For over 66 years Brother Joseph Fox was a member of the Society of Mary, a Catholic religious order which, among other things, operates Chaminade High School in Mineola. . . .

While [an] operation was being performed . . . he suffered cardiac arrest, with resulting loss of oxygen to the brain and substantial brain damage. He lost the ability to breathe spontaneously and was placed on a respirator which maintained him in a vegetative state. The attending physicians informed Father Philip Eichner, who was the president of Chaminade and the director of the society at the school, that there was no reasonable chance of recovery and that Brother Fox would die in that state.

After retaining two neurosurgeons who confirmed the diagnosis, Father Eichner requested the hospital to remove the respirator. The hospital, however, refused to do so without court authorization. Father Eichner then applied . . . to be appointed committee of the person and property of Brother Fox, with authority to direct removal of the respirator. The application was supported by the patient's 10 nieces and nephews, his only surviving relatives. The court appointed a guardian ad litem and directed that notice be served on various parties, including the District Attorney.

At the hearing the District Attorney opposed the application and called medical experts to show that there might be some improvement in the patient's condition. All the experts agreed, however, that there was no reasonable likelihood that Brother Fox would ever emerge from the vegetative coma or recover his cognitive powers.

There was also evidence, submitted by the petitioner, that before the operation rendered him incompetent the patient had made it known that under these circumstances he would want a respirator removed. Brother Fox had first expressed this view in 1976 when the Chaminade community discussed the moral implications of the celebrated *Karen Ann Quinlan* case, in which the parents of a 19-year-old New Jersey girl who was in a vegetative coma requested the hospital to remove the respirator [ ]. These were formal discussions prompted by Chaminade's mission to teach and promulgate Catholic moral principles. At that time it was noted that the Pope had stated that Catholic principles permitted the termination of extraordinary life support systems when there is no reasonable hope for the patient's recovery and that church officials in New Jersey had concluded that use of the respirator in the *Quinlan* case constituted an extraordinary measure under the circumstances. Brother Fox expressed agreement with those views and stated that he would not want any of this "extraordinary business" done for him under those circumstances. Several years later, and only a couple of months before his final hospitalization, Brother Fox again stated that he would not want his life prolonged by such measures if his condition were hopeless.

[A trial court had held that Father Eichner could refuse ongoing life-sustaining treatment as a surrogate on behalf of Brother Fox because there was evidence that Brother Fox, when competent, had expressed his opposition to such treatment for himself if he was ever in a persistent vegetative state, and the trial court found this evidence to be "unchallenged at every turn and unimpeachable in its sincerity." The District Attorney appealed, questioning—among other things—the standard of proof required of a surrogate refusing life-sustaining treatment on behalf of an incompetent patient on the grounds that the patient had previously expressed a wish not to receive such treatment. The Court ruled that Father Eichner was required to present "clear and convincing" evidence of Brother Fox's wishes, and then the Court considered the evidence that had been presented at trial.]

* * *

In this case the proof was compelling. There was no suggestion that the witnesses who testified for the petitioner had any motive other than to see that Brother Fox' stated wishes were respected. The finding that he carefully reflected on the subject, expressed his views and concluded not to have his life prolonged by medical means if there were no hope of recovery is supported by his religious beliefs and is not inconsistent with his life of unselfish religious devotion. These were obviously solemn pronouncements and not casual remarks made at some social gathering, nor can it be said that he was too young to realize or feel the consequences of his statements [ ]. That this was a persistent commitment is evidenced by the fact that he reiterated the decision but two months before his final hospitalization. There was, of course, no need to speculate as to whether he would want this particular medical procedure to be discontinued under these circumstances. What occurred to him was identical to what happened in the *Karen Ann Quinlan* case, which had originally prompted his decision. In sum, the evidence clearly and convincingly shows that Brother Fox did not want to be maintained in a vegetative coma by use of a respirator.

* * *

NOTES AND QUESTIONS

1.   *Substituted Judgment.* The Illinois Supreme Court has described the principle of substituted judgment clearly and simply:

> Under substituted judgment, a surrogate decisionmaker attempts to establish, with as much accuracy as possible, what decision the patient would make if he were competent to do so. Employing this theory, the surrogate first tries to determine if the patient had expressed explicit intent regarding this type of medical treatment prior to becoming incompetent. [ ] Where no clear intent exists, the patient's personal value system must guide the surrogate. . . .

In re Estate of Longeway, 133 Ill. 2d 33, 139 Ill. Dec. 780, 549 N.E.2d 292, 299 (1989).

2.   *Evidence of Patient Wishes.* In applying the principle of substituted judgment, most courts look wherever they can to determine the patient's wishes. In Brophy v. New England Sinai Hosp., Inc., 398 Mass. 417, 497 N.E.2d 626 (1986), the Massachusetts Supreme Court based its conclusion that food and hydration could be withheld from a comatose adult on the substituted judgment analysis done by the lower court.

> [After a full hearing] the judge found on the basis of ample evidence which no one disputes, that Brophy's judgment would be to decline the provision of food and water and to terminate his life. In reaching that conclusion, the judge considered various factors including the following: (1) Brophy's expressed preferences; (2) his religious convictions and their relation to refusal of treatment; (3) the impact on his family; (4) the probability of

adverse side effects; and (5) the prognosis, both with and without treatment. The judge also considered present and future incompetency as an element which Brophy would consider in his decision-making process. The judge relied on several statements made by Brophy prior to the onset of his illness. Although he never had discussed specifically whether a G-tube or feeding tube should be withdrawn in the event that he was diagnosed as being in a persistent vegetative state following his surgery, the judge inferred that, if presently competent, Brophy would choose to forgo artificial nutrition and hydration by means of a G-tube. The judge found that Brophy would not likely view his own religion as a barrier to that choice.

3. *Other Factors That Have Been Considered.* Other factors that have been considered include the patient's diagnosis, life history, ability to knowingly participate in treatment, potential quality of life, and, more generally, the patient's values and attitude toward health care. See, e.g., Mack v. Mack, 329 Md. 188, 618 A.2d 744 (1993) (focusing on the "moral views, life goals, and values" of the patient, and her "attitudes toward sickness, medical procedures, suffering and death."); DeGrella v. Elston, 858 S.W.2d 698 (Ky. 1993). For a thorough and well-annotated list of relevant factors that have been considered by the courts see Alan Meisel, Kathy L. Cerminara, and Thaddeus Mason Pope, The Right to Die, 3rd ed. (2004, and 2017 Supplement).

4. *Decision-Making for Adults Who Have Never Been Competent.* Where the patient has never been competent—where the patient has been severely intellectually and developmentally disabled from birth, for example—the courts still make an attempt to determine what the patient's choice would be when deciding what treatment would be legally appropriate. Of course, it is exceptionally difficult to imagine what a person who has never been competent would want to do if that person were suddenly competent. Are you concerned that our society might be more willing to err on the side of removing life-sustaining treatment when the patient has a severe disability? Perhaps the courts should review all decisions to remove life-sustaining medical treatment from patients in this class, even if the courts need not be involved in similar cases with previously competent patients. See Protection and Advocacy Sys., Inc. v. Presbyterian Healthcare Servs., 128 N.M. 73, 989 P.2d 890 (1999). See also Kathy Cerminara, Critical Essay: Musings on the Need to Convince Some People With Disabilities that End-of-Life Decision Making Advocates Are Not Out to Get Them, 37 Loy. U. Chi. L.J. 343 (2006); Adrienne Asch, Recognizing Death While Affirming Life: Can End of Life Reform Uphold a Disabled Person's Interest in Continued Life?, Hastings Ctr. Spec. Rep. on Improving End of Life Care (Jennings et al., eds. 2005).

5. *Distinguishing the Two Standards.* How difficult is it for a surrogate decision-maker to distinguish what the patient would really want from what that decision-maker would want if she were in the position of that patient? Is it really possible to clearly distinguish the subjective "substituted judgment" standard from an objective standard that asks what a reasonable person would do under the circumstances? Courts do struggle to distinguish the "substituted

judgment" and "best interest" principles, and even those that appear to adopt the "best interest" approach may qualify it by requiring that the best interest of the patient be defined in terms of the wishes, values, and desires of the patient. Some state courts appear to adopt the best interest test, but they actually take the "substituted judgment" approach. See Conservatorship of Drabick, 200 Cal. App. 3d 185, 245 Cal. Rptr. 840 (1988); In re Gordy, 658 A.2d 613 (Del. Ch. 1994).

NOTE: THE SPECIAL STATUS OF NUTRITION AND HYDRATION

The issue of withdrawing nutrition and hydration has become an especially contentious one. Recall, for example, that the living will statutes of some states treat procedures to provide nutrition and hydration differently from other forms of life-sustaining medical care. Medical sources generally recognize the irrelevancy of distinguishing between nutrition and hydration and other forms of medical treatment. Generally, courts also have concluded that the termination of nutrition and hydration is no different from the termination of other forms of mechanical support. For example, *Conroy* suggested:

> Some commentators, . . . have made yet [another] distinction, between the termination of artificial feedings and the termination of other forms of life-sustaining medical treatment. . . . According to the Appellate Division:
>
> > If, as here, the patient is not comatose and does not face imminent and inevitable death, nourishment accomplishes the substantial benefit of sustaining life until the illness takes its natural course. Under such circumstances nourishment always will be an essential element of ordinary care which physicians are ethically obligated to provide. [ ]

> Certainly, feeding has an emotional significance. As infants we could breathe without assistance, but we were dependent on others for our lifeline of nourishment. Even more, feeding is an expression of nurturing and caring, certainly for infants and children, and in many cases for adults as well.

> Once one enters the realm of complex, high-technology medical care, it is hard to shed the "emotional symbolism" of food. . . . Analytically, artificial feeding by means of a nasogastric tube or intravenous infusion can be seen as equivalent to artificial breathing by means of a respirator. Both prolong life through mechanical means when the body is no longer able to perform a vital bodily function on its own.

> Furthermore, while nasogastric feeding and other medical procedures to ensure nutrition and hydration are usually well tolerated, they are not free from risks or burdens; they have complications that are sometimes serious and distressing to the patient.

> Finally, dehydration may well not be distressing or painful to a dying patient. For patients who are unable to sense hunger and thirst,

withholding of feeding devices such as nasogastric tubes may not result in more pain than the termination of any other medical treatment. . . . Thus, it cannot be assumed that it will always be beneficial for an incompetent patient to receive artificial feeding or harmful for him not to receive it. . . .

Under the analysis articulated above, withdrawal or withholding of artificial feeding, like any other medical treatment, would be permissible [under some circumstances]. A competent patient has the right to decline any medical treatment, including artificial feeding, and should retain that right when and if he becomes incompetent. In addition, in the case of an incompetent patient who has given little or no trustworthy indication of an intent to decline treatment . . . the pain and invasiveness of an artificial feeding device, and the pain of withdrawing that device, should be treated just like the results of administering or withholding any other medical treatment.

98 N.J. 321, 486 A.2d 1209, 1235–1237.

In McConnell v. Beverly Enterprises-Connecticut, 209 Conn. 692, 553 A.2d 596 (1989), the court authorized the withdrawal of feeding by a gastrostomy tube despite a statute that appeared to say that under such circumstances "nutrition and hydration must be provided." The court reasoned that the nutrition and hydration that was implicated in the statute was that provided by "a spoon or a straw," and that feeding by gastrostomy tube was no different than any other mechanical or electronic medical intervention.

In 1990, at least, a majority of the Supreme Court (the four dissenters and concurring Justice O'Connor in *Cruzan*) viewed nutrition and hydration as another form of medical care. As Justice O'Connor pointed out, "artificial feeding cannot readily be distinguished from other forms of medical treatment. Whether or not the techniques used to pass food and water into the patient's alimentary tract are termed 'medical treatment,' it is clear they all involve some degree of intrusion and restraint." She concluded that "the liberty guaranteed by the due process clause must protect, if it protects anything, an individual's deeply personal decision to reject medical treatment, including the artificial delivery of food and water."

In his dissent, Justice Brennan reached the same conclusion, vividly describing the medical processes involved:

The artificial delivery of nutrition and hydration is undoubtedly medical treatment. The technique to which Nancy Cruzan is subject—artificial feeding through a gastrostomy tube—involves a tube implanted surgically into her stomach through incisions in her abdominal wall. It may obstruct the intestinal tract, erode and pierce the stomach wall, or cause leakage of the stomach's contents into the abdominal cavity. [ ] The tube can cause pneumonia from reflux of the stomach's contents into the lung. [ ] Typically, and in this case, commercially prepared formulas are used, rather than fresh food. [ ] The type of formula and method of administration must be experimented with to avoid gastrointestinal problems. [ ] The patient must be monitored daily by medical personnel as to weight, fluid intake and fluid output; blood tests must be done weekly.

Artificial delivery of food and water is regarded as medical treatment by the medical profession and the federal government. . . . The federal government permits the cost of the medical devices and formulas used in enteral feeding to be reimbursed under Medicare. [ ] The formulas are regulated by the Federal Drug Administration as "medical foods," [ ] and the feeding tubes are regulated as medical devices [ ].

497 U.S. at 306–308, 110 S. Ct. at 2866–67.

To many, nutrition and hydration remain symbols of the bonds between human beings; the care we provide to the ones we love includes, at the very least, food and water. For others, the provision of artificial nutrition and hydration has religious and moral significance. Consider the following documents defining the Catholic position on this issue.

UNITED STATES CONFERENCE OF CATHOLIC BISHOPS, ETHICAL AND RELIGIOUS DIRECTIVES FOR CATHOLIC HEALTH CARE SERVICES

(6th ed. 2018).

* * *

The Church's teaching authority has addressed the moral issues concerning medically assisted nutrition and hydration. We are guided on this issue by Catholic teaching against euthanasia, which is "an action or an omission which of itself or by intention causes death, in order that all suffering may in this way be eliminated."[ ] While medically assisted nutrition and hydration are not morally obligatory in certain cases, these forms of basic care should in principle be provided to all patients who need them, including patients diagnosed as being in a "persistent vegetative state" (PVS), because even the most severely debilitated and helpless patient retains the full dignity of a human person and must receive ordinary and proportionate care.

* * *

[DIRECTIVE 58] In principle, there is an obligation to provide patients with food and water, including medically assisted nutrition and hydration for those who cannot take food orally. This obligation extends to patients in chronic and presumably irreversible conditions (e.g., the "persistent vegetative state") who can reasonably be expected to live indefinitely if given such care. [ ] Medically assisted nutrition and hydration become morally optional when they cannot reasonably be expected to prolong life or when they would be "excessively burdensome for the patient or [would] cause significant physical discomfort, for example resulting from complications in the use of the means employed."[ ] For instance, as a patient draws close to inevitable death from an underlying progressive and fatal condition, certain measures to provide nutrition and hydration may become excessively burdensome and therefore not obligatory in light of their very limited ability to prolong life or provide comfort.

* * *

CONGREGATION FOR THE DOCTRINE OF THE FAITH, RESPONSES TO CERTAIN QUESTIONS OF THE UNITED STATES CONFERENCE OF CATHOLIC BISHOPS CONCERNING ARTIFICIAL NUTRITION AND HYDRATION
(2007)

First question: Is the administration of food and water (whether by natural or artificial means) to a patient in a "vegetative state" morally obligatory except when they cannot be assimilated by the patient's body or cannot be administered to the patient without causing significant physical discomfort?

Response: Yes. The administration of food and water even by artificial means is, in principle, an ordinary and proportionate means of preserving life. It is therefore obligatory to the extent to which, and for as long as, it is shown to accomplish its proper finality, which is the hydration and nourishment of the patient. In this way suffering and death by starvation and dehydration are prevented.

Second question: When nutrition and hydration are being supplied by artificial means to a patient in a "permanent vegetative state", may they be discontinued when competent physicians judge with moral certainty that the patient will never recover consciousness?

Response: No. A patient in a "permanent vegetative state" is a person with fundamental human dignity and must, therefore, receive ordinary and proportionate care which includes, in principle, the administration of water and food even by artificial means.

For interpretations of the content and application of this statement within the Roman Catholic tradition, see John Hardt & Kevin O'Rourke, Nutrition and Hydration: The [Congregation for the Doctrine of Faith], In Perspective, 88 Health Progress 1 (2007), arguing, inter alia, that Catholic Canon Law requires that the statement be interpreted strictly as it restricts rights; that the form of the statement has less force than other vehicles for establishing directives; that the use of the term "in principle" indicates that there are exceptions. In particular, in contrast to earlier statements on nutrition and hydration, this statement is specifically confined to circumstances of confirmed diagnosis of PVS and addresses only nutrition and hydration and not other medical interventions. See also Daniel Sulmasy, Preserving Life? The Vatican and PVS, 134 Commonweal (No. 21) 16 (Dec. 7, 2007).

PROBLEM: NOT QUITE PERSISTENT VEGETATIVE STATE

When unhelmeted 57-year-old Tad Gonzales ran his motorcycle into a bridge support post on an interstate highway, he was revived at the scene and rushed to Big Central Hospital. There he underwent several hours of emergency surgery designed to preserve his life and repair the very substantial head injuries he sustained. A few days after the surgery, he began to regain consciousness, and a week later he was able to be removed from the ventilator which was supporting his breathing, although he still needs ventilator assistance from time to time. During his first two weeks in the hospital, Tad showed little improvement. He was not in a coma, but he was only occasionally responsive, and he demonstrated no awareness of Ralph, his partner of twenty-two years, or his two brothers and his parents, even though Tad had always shared a close relationship with all of them.

After two weeks, Big City Hospital transferred Tad to Commercial Affiliated Nursing Home, where he began receiving physical and occupational therapy. After a year of care there, he has shown little improvement. He still does not seem to recognize Ralph, who visits daily, or anyone else. He is able to sit up and his eyes sometimes seem to be following images on a screen. He grunts when he is hungry or uncomfortable. He cannot eat or drink, and he is fed through a feeding tube in his abdomen. He cannot control his bowels or bladder. He has regularly suffered from urinary tract infections and asthma during his nursing home stay, although his need for the ventilator is becoming rarer. On a few, increasingly rare, occasions, his heart stopped beating, but he was resuscitated immediately. Doctors believe that there is little chance of substantial improvement in his condition (although, in the words of one doctor, "Who knows? Anything can happen.") Tad's doctors estimate that his life expectancy could be another twenty years or more.

Tad's doctor and Ralph have developed a good working relationship, and the doctor has called upon Ralph to approve any change in Tad's treatment regimen even though Tad never signed any kind of advance directive. Ralph has now informed the doctor that he believes that Tad would want the use of the feeding tube (and the occasional use of the ventilator) discontinued, and that he would not want to be resuscitated (i.e., if his heart were to stop beating, he would not want health care providers to try to start it again). When the doctor challenged him on this, Ralph respectfully ordered the doctor to terminate feeding, and to note that Tad was not to be mechanically ventilated or resuscitated. Ralph explained that he had spoken with Tad often about "this kind of thing," and that Tad had said that the indignity of being fed, or wearing diapers, or being bed-bound, was not worth the value of life to him. In particular, Ralph remembers Tad describing one of their friends, another motorcyclist who became a quadriplegic and suffered some intellectual impairment after an accident, as "better off dead." Tad and Ralph promised each other that neither would ever let that happen to the other.

Tad's parents and one of his brothers agree that there is no doubt that Tad would want all treatment terminated under these circumstances. Tad's other brother disagrees. As he has articulated it, "How can we know what he would want? He could never have imagined himself in this situation, and we shouldn't read anything into what he said about other people. Maybe he thought those other people could feel more pain than he can."

Tad's doctor has approached you, the hospital counsel, to ask you what she should do. Should she discontinue feeding? The use of the ventilator? Should the hospital pursue an action in court? What position should the hospital take if Ralph, or Tad's dissenting brother, pursues an action in court? How should that action be resolved?

NOTE: THE ROLE OF THE COURT IN RESOLVING DISPUTES ABOUT THE PROVISION OF LIFE-SUSTAINING MEDICAL CARE

The drafters of the Uniform Health-Care Decisions Act make it clear in their comments that one purpose of the statute is to assure that these intimate health care decisions remain within the realm of the patient, the patient's family and close friends, and the health care providers, and that others not be permitted to disrupt that process. The court would very rarely have a role in any decision-making under the Unform Act, and outsiders (including outside organizations) who do not think a patient is adequately protected have no standing to seek judicial intervention. See Protection and Advocacy Sys., Inc. v. Presbyterian Healthcare Servs., 128 N.M. 73, 989 P.2d 890 (1999).

Should courts have a role in resolving disputes about the provision of life-sustaining medical care? On the one hand, there is a fear that the absence of judicial oversight will lead to arbitrary, biased, or discriminatory decisions that end the lives of already vulnerable people. On the other hand, any attempt to bring every case that could result in the termination of life-sustaining treatment to the courts would result in an intolerable caseload and delay the deaths of many patients who desperately seek that relief. In addition, there is little reason to believe that courts have any wisdom that will make them better (or less susceptible to bias) than a patient's family at making these decisions.

Should judicial review of all decisions to terminate life-sustaining treatment be required? Should such judicial review be required in some cases, for example, in cases involving allegations of bias or discrimination? In cases in which there is no written advance directive? In which there is no agreement among family members? In which the patients disagree with the health care providers? In which the health care providers disagree among themselves? In which the decision-maker is self-interested? In which there is an ambiguity in the previous statements of the patient? Is there any way to adequately categorize those cases in which judicial review ought to be required? If judicial review is not required, should some other form of review—by an ethics committee or an independent board, for example—be required in its stead?

When resort to a court is required, what procedure ought to be employed by the court? Should the court's involvement be limited to the appointment of a decision-maker, should the court make the decision itself, or should the court review every decision (or some decisions) made by an appointed decision-maker? Should the action be a special statutory action, an injunction action, or a guardianship? Does the form of the action really make much difference? Should the court always appoint a guardian ad litem for the patient? What would the role of such a guardian be? Every state has a nursing home ombudsman. Should the ombudsman be notified whenever discontinuation of treatment of a nursing home patient is requested? Should the ombudsman participate in every such case?

In In re Guardianship of Hamlin, 102 Wash. 2d 810, 689 P.2d 1372 (1984), the Washington Supreme Court announced two entirely separate processes— one to be followed where there is "total agreement among the patient's family, treating physicians and prognosis committee as to the course of medical treatment," and one to be followed where there is "an incompetent with no known family, who has never made his wishes known." No judicial process or formal guardianship is required in the first case where "the incompetent patient is in . . . a persistent vegetative state with no reasonable chance of recovery and . . . the patient's life is being maintained by life support systems." In the second situation a guardian must be appointed by the court, but that guardian need not obtain judicial confirmation of any particular decision "if the treating physicians and prognosis committee are unanimous that life-sustaining efforts should be withheld or withdrawn and the guardian concurs." Of course, most cases fall between the two extremes discussed in *Hamlin*.

Who should have standing to commence a judicial action seeking to review a decision to terminate life-sustaining medical treatment? All relatives? Health care providers? Health care institutions? Patient advocacy groups? Right-to-Life or Right-to-Die or disability advocacy groups? Who should be able to join such litigation as a party? Is the Uniform Health-Care Decisions Act approach to this issue—that there will be no judicial hearing unless a relative or health care provider is willing to file an action—the best way of dealing with this issue? See Protection and Advocacy Sys., Inc. v. Presbyterian Healthcare Servs., 128 N.M. 73, 989 P.2d 890 (1999) (no standing in advocacy group to challenge unanimous decision of family members and health care providers).

If there is to be a judicial process, should it be an adversary process? While Chief Justice Rehnquist appeared to think that the adversary process would be helpful in these cases (see *Cruzan*, above), that part of the opinion gave Justice Stevens pause:

> The Court recognizes that "the state has been involved as an adversary from the beginning" in this case only because Nancy Cruzan "was a patient at a state hospital when this litigation commenced." . . . It seems to me, however, that the Court draws precisely the wrong conclusion from this insight. The Court apparently believes that the absence of the state from the litigation would have created a problem, because agreement

among the family and the independent guardian *ad litem* as to Nancy Cruzan's best interests might have prevented her treatment from becoming the focus of a "truly adversarial" proceeding. [ ] It may reasonably be debated whether some judicial process should be required before life-sustaining treatment is discontinued; this issue has divided the state courts. [ ] . . . I tend, however, to agree . . . that the intervention of the state in these proceedings as an *adversary* is not so much a cure as it is part of the disease.

Cruzan v. Director, Missouri Dept. of Health, 497 U.S. 261, 341, n.13, 110 S. Ct. 2841, 2884, n.13, 111 L. Ed. 2d 224 (1990) (Stevens, J., dissenting).

A decade earlier, the Florida Supreme Court had expressed the same reservations:

Because the issue with its ramifications is fraught with complexity and encompasses the interests of the law, both civil and criminal, medical ethics and social morality, it is not one which is well suited for a solution in an adversary judicial proceeding.

Satz v. Perlmutter, 379 So. 2d 359, 360 (Fla. 1980). See also *Protection and Advocacy System*, cited at the beginning of this note. Are the "advantages" of an adversary proceeding truly advantageous in these agonizing cases? See Nancy Dubler, Conflict and Consensus at the End of Life, in Hastings Ctr. Spec. Rep. on Improving End of Life Care (Jennings et al., eds. 2005); Alan Meisel, The Role of Litigation in End of Life Care: A Reappraisal, in Hastings Ctr. Spec. Rep. on Improving End of Life Care (Jennings et al., eds. 2005).

V. HEALTH CARE DECISION-MAKING FOR CHILDREN

PROBLEM: CHOOSING TO FORGO CANCER TREATMENT

Michael Armstrong was a healthy, socially well-adjusted, intelligent 14-year-old when he discovered that he had testicular cancer that had also produced a mass in his liver. The prognosis was not terribly optimistic: with intensive intervention he would have about a 50% chance of surviving for two years, and a 25% chance of a "cure" (i.e., of surviving five years). "Intensive intervention" would include several months of painful and debilitating chemotherapy, surgery to remove part of his liver and his testes, and a substantial amount of radiation therapy. Without treatment, Michael would probably die within six months.

Michael has informed his oncologist that he has decided to forgo the proposed surgery, even though he recognizes that this will result in his death. He explained that he reached this conclusion after reading everything accessible to him about his cancer and the proposed treatment, after praying with his religious advisor and mentor, and after discussing the issue with his family, friends, and two patients (names provided by the oncologist with the consent of those patients) who had previously undergone the same surgery and chemotherapy. He also admits that the teasing he has taken over the primary location of his disease, and his consequent extreme embarrassment, were not irrelevant factors in his decision.

Michael's parents, who are divorced and have joint legal and physical custody, are split on this issue. His father believes that it is Michael's decision to make, and he supports Michael's decision. His mother insists that the oncologist and the hospital provide the proposed treatment. In addition, a social worker in the local child protective services office (which was informed about the case by one of Michael's teachers) has informed all of the parties that, in her opinion, Michael would be a neglected child if the treatment were not administered.

Michael's parents have now filed cross petitions seeking to vindicate their positions. Michael has filed a petition as well. The oncologist and the hospital have filed a declaratory judgment action seeking an order of the court before taking any action, and the child protective services agency has filed a neglect petition against the father. All of these actions have been consolidated and all assigned to the same trial court judge to resolve. How should the judge proceed? What kind of hearing, if any, should the judge hold? How should she rule?

NEWMARK V. WILLIAMS
Supreme Court of Delaware, 1991.
588 A.2d 1108.

MOORE, JUSTICE.

Colin Newmark, a three year old child, faced death from a deadly aggressive and advanced form of pediatric cancer known as Burkitt's Lymphoma. We were presented with a clash of interests between medical science, Colin's tragic plight, the unquestioned sincerity of his parents' religious beliefs as Christian Scientists, and the legal right of the State to protect dependent children from perceived neglect when medical treatment is withheld on religious grounds. The Delaware Division of Child Protective Services ("DCPS") petitioned the Family Court for temporary custody of Colin to authorize the Alfred I. DuPont Institute ("DuPont Institute"), a nationally recognized children's hospital, to treat Colin's condition with chemotherapy. His parents, Morris and Kara Newmark, are well educated and economically prosperous. As members of the First Church of Christ, Scientist ("Christian Science") they rejected medical treatment proposed for Colin, preferring instead a course of spiritual aid and prayer. The parents rely upon provisions of Delaware law, which exempt those who treat their children's illnesses "solely by spiritual means" from the abuse and neglect statutes. Thus, they opposed the State's petition. [ ] The Newmarks also claimed that removing Colin from their custody would violate their First Amendment right, guaranteed under the United States Constitution, to freely exercise their religion.

* * *

[The Court explained how the Newmarks discovered that Colin suffered from Burkitt's Lymphoma, an extremely fast growing and dangerous form of pediatric cancer, when they took him to the doctor "out of concern for their potential criminal liability . . . ."]

We have concluded that Colin was not an abused or neglected child under Delaware law. Parents enjoy a well established legal right to make important decisions for their children. Although this right is not absolute, the State has the burden of proving by clear and convincing evidence that intervening in the parent-child relationship is necessary to ensure the safety or health of the child, or to protect the public at large. DCPS did not meet this heavy burden. This is especially true where the purpose of the custody petition was to administer, over the objections of Colin's parents, an extremely risky, toxic and dangerously life threatening medical treatment offering less than a 40% chance for "success".

* * *

Dr. Meek [an attending physician and board-certified pediatric hematologist-oncologist] opined that chemotherapy offered a 40% chance of "curing" Colin's illness. She concluded that he would die within six to eight months without treatment. The Newmarks . . . advised Dr. Meek that they would place him under the care of a Christian Science practitioner and reject all medical treatment for their son. Accordingly, they refused to authorize the chemotherapy. There was no doubt that the Newmarks sincerely believed, as part of their religious beliefs, that the tenets of their faith provided an effective treatment.

II.

We start with an overview of the relevant Delaware statutory provisions. Delaware law defines a neglected child as:

[A] child whose physical, mental or emotional health and well-being is threatened or impaired because of inadequate care and protection by the child's custodian, who has the ability and financial means to provide for the care but does not or will not provide adequate care; or a child who has been abused or neglected . . .

[The statute] further defines abuse and neglect as:

Physical injury by other than accidental means, injury resulting in a mental or emotional condition which is a result of abuse or neglect, negligent treatment, sexual abuse, maltreatment, mistreatment, nontreatment, exploitation or abandonment, of a child under the age of 18. (Emphasis added).

Sections of the Delaware Code, however, contain spiritual treatment exemptions which directly affect Christian Scientists. Specifically, the exemptions state:

> No child who in good faith is under treatment solely by spiritual means through prayer in accordance with the tenets and practices of a recognized church or religious denomination by a duly accredited practitioner thereof shall for that reason alone be considered a neglected child for purposes of this chapter.

[ ] These exceptions reflect the intention of the Delaware General Assembly to provide a "safe harbor" for parents, like the Newmarks, to pursue their own religious beliefs. [We recognize] that the spiritual treatment exemptions reflect, in part, "the policy of this State with respect to the quality of life" a desperately ill child might have in the caring and loving atmosphere of his or her family, versus the sterile hospital environment demanded by physicians seeking to prescribe excruciating, and life threatening, treatments of doubtful efficacy.

. . . [W]e recognize the possibility that the spiritual treatment exemptions may violate the ban against the establishment of an official State religion guaranteed under both the Federal and Delaware Constitutions. Clearly, in both reality and practical effect, the language providing an exemption only to those individuals practicing "in accordance" with the "practices of a recognized church or religious denomination by a duly accredited practitioner thereof" is intended for the principal benefit of Christian Scientists. Our concern is that it possibly forces us to impermissibly determine the validity of an individual's own religious beliefs.

Neither party challenged the constitutionality of the spiritual treatment exemptions in either the Family Court or on appeal. Thus, except to recognize that the issue is far more complicated than was originally presented to us, we must leave such questions for another day.

III.

Addressing the facts of this case, we turn to the novel legal question whether, under any circumstances, Colin was a neglected child when his parents refused to accede to medical demands that he receive a radical form of chemotherapy having only a forty percent chance of success. [The court then explained that it would apply a balancing test to answer this question.]

* * *

A.

Any balancing test must begin with the parental interest. The primacy of the familial unit is a bedrock principle of law. [ ] We have repeatedly emphasized that the parental right is sacred which can be invaded for only the most compelling reasons.

* * *

Courts have also recognized that the essential element of preserving the integrity of the family is maintaining the autonomy of the parent-child relationship. [ ] In Prince v. Commonwealth of Massachusetts, [ ] the United States Supreme Court announced:

> It is cardinal with us that the custody, care and nurture of the child reside first in the parents, whose primary function and freedom include preparation for obligations the state can neither supply nor hinder. [ ]

Parental autonomy to care for children free from government interference therefore satisfies a child's need for continuity and thus ensures his or her psychological and physical well-being. [ ]

Parental authority to make fundamental decisions for minor children is also a recognized common law principle. A doctor commits the tort of battery if he or she performs an operation under normal circumstances without the informed consent of the patient. [ ] Tort law also assumes that a child does not have the capacity to consent to an operation in most situations. [ ] Thus, the common law recognizes that the only party capable of authorizing medical treatment for a minor in "normal" circumstances is usually his parent or guardian. [ ]

Courts, therefore, give great deference to parental decisions involving minor children. In many circumstances the State simply is not an adequate surrogate for the judgment of a loving, nurturing parent. [ ] As one commentator aptly recognized, the "law does not have the capacity to supervise the delicately complex interpersonal bonds between parent and child." [ ]

B.

We also recognize that parental autonomy over minor children is not an absolute right. Clearly, the State can intervene in the parent-child relationship where the health and safety of the child and the public at large are in jeopardy. [ ] Accordingly, the State, under the doctrine of parens patriae, has a special duty to protect its youngest and most helpless citizens.

The parens patriae doctrine is a derivation of the common law giving the State the right to act on behalf of minor children in certain property and marital disputes. [ ] More recently, courts have accepted the doctrine of parens patriae to justify State intervention in cases of parental religious objections to medical treatment of minor children's life threatening conditions. [ ] The Supreme Court of the United States [pointed out] that parental autonomy, under the guise of the parents' religious freedom, was not unlimited. [ ] Rather, the Court held:

> Parents may be free to become martyrs themselves. But it does not follow they are free, in identical circumstances, to make martyrs of their children before they have reached the age of full and legal discretion when they can make that choice for themselves. [ ]

The basic principle underlying the parens patriae doctrine is the State's interest in preserving human life. [ ] Yet this interest and the parens patriae doctrine are not unlimited. In its recent Cruzan opinion, the Supreme Court of the United States announced that the state's interest in preserving life must "be weighed against the constitutionally protected interests of the individual." [ ]

The individual interests at stake here include both the Newmarks' right to decide what is best for Colin and Colin's own right to life. We have already considered the Newmarks' stake in this case and its relationship to the parens patriae doctrine. The resolution of the issues here, however, is incomplete without a discussion of Colin's interests.

C.

All children indisputably have the right to enjoy a full and healthy life. Colin, a three year old boy, unfortunately lacked the ability to reach a detached, informed decision regarding his own medical care. This Court must therefore substitute its own objective judgment to determine what is in Colin's "best interests." [ ]

There are two basic inquiries when a dispute involves chemotherapy treatment over parents' religious objections. The court must first consider the effectiveness of the treatment and determine the child's chances of survival with and without medical care. [ ] The court must then consider the nature of the treatments and their effect on the child. [ ]

The "best interests" analysis is hardly unique or novel. Federal and State courts have unhesitatingly authorized medical treatment over a parent's religious objection when the treatment is relatively innocuous in comparison to the dangers of withholding medical care. [ ] Accordingly, courts are reluctant to authorize medical care over parental objection when the child is not suffering a life threatening or potential life threatening illness. [ ]

The linchpin in all cases discussing the "best interests of a child," when a parent refuses to authorize medical care, is an evaluation of the risk of the procedure compared to its potential success. This analysis is consistent with the principle that State intervention in the parent-child relationship is only justifiable under compelling conditions. [ ] The State's interest in forcing a minor to undergo medical care diminishes as the risks of treatment increase and its benefits decrease.

* * *

Applying the foregoing considerations to the "best interests standard" here, the State's petition must be denied. The egregious facts of this case indicate that Colin's proposed medical treatment was highly invasive, painful, involved terrible temporary and potentially permanent side effects, posed an unacceptably low chance of success, and a high risk that the treatment itself would cause his death. The State's authority to intervene in this case, therefore, cannot outweigh the Newmarks' parental prerogative and Colin's inherent right to enjoy at least a modicum of human dignity in the short time that was left to him.

IV.

Dr. Meek originally diagnosed Colin's condition as Burkitt's Lymphoma. She testified that the cancer was "a very bad tumor" in an advanced disseminated state and not localized to only one section of the body. She accordingly recommended that the hospital begin an "extremely intensive" chemotherapy program scheduled to extend for at least six months.

[The court then explained how intensive such a chemotherapy program would be, and how such a treatment program would itself threaten Colin's life.]

Dr. Meek prescribed "maximum" doses of at least six different types of cancer-fighting drugs during Colin's chemotherapy. This proposed "maximum" treatment represented the most aggressive form of cancer therapy short of a bone marrow transplant. The side effects would include hair loss, reduced immunological function creating a high risk of infection in the patient, and certain neurological problems. The drugs also are toxic to bone marrow.

The record demonstrates that this form of chemotherapy also would adversely affect other parts of Colin's body.

* * *

The physicians planned to administer the chemotherapy in cycles, each of which would bring Colin near death. Then they would wait until Colin's body recovered sufficiently before introducing more drugs.

* * *

Dr. Meek also wanted the State to place Colin in a foster home after the initial phases of hospital treatment. Children require intensive home monitoring during chemotherapy. For example, Dr. Meek testified that a usually low grade fever for a healthy child could indicate the presence of a potentially deadly infection in a child cancer patient. She believed that the Newmarks, although well educated and financially responsible, were incapable of providing this intensive care because of their firm religious objections to medical treatment.

Dr. Meek ultimately admitted that there was a real possibility that the chemotherapy could kill Colin. In fact, assuming the treatment did not itself prove fatal, she offered Colin at "best" a 40% chance that he would "survive."[12] Dr. Meek additionally could not accurately predict whether, if Colin completed the therapy, he would subsequently suffer additional tumors.

A.

No American court, even in the most egregious case, has ever authorized the State to remove a child from the loving, nurturing care of his parents and subject him, over parental objection, to an invasive regimen of treatment which offered, as Dr. Meek defined the term, only a forty percent chance of "survival."

* * *

B.

The aggressive form of chemotherapy that Dr. Meek prescribed for Colin was more likely to fail than succeed. The proposed treatment was also highly invasive and could have independently caused Colin's death. Dr. Meek also wanted to take Colin away from his parents and family during the treatment phase and place the boy in a foster home. This certainly would have caused Colin severe emotional difficulties given his medical condition, tender age, and the unquestioned close bond between Colin and his family.

In sum, Colin's best interests were served by permitting the Newmarks to retain custody of their child. Parents must have the right at some point to reject medical treatment for their child. Under all of the circumstances here, this clearly is such a case. The State's important and legitimate role in safeguarding the interests of minor children diminishes in the face of this egregious record.

Parents undertake an awesome responsibility in raising and caring for their children. No doubt a parent's decision to withhold medical care is both deeply personal and soul wrenching. It need not be made worse by the invasions which both the State and medical profession sought on this record. Colin's ultimate fate therefore rested with his parents and their faith.

---

[12] Dr. Meek testified that there was no available medical data to conclude that Colin could survive to adulthood. Rather, she stated that the term "survival", as applied to victims of leukemia or lymphoma, refers only to the probability that the patient will live two years after chemotherapy without a recurrence of cancer.

NOTES AND QUESTIONS

1.   *Majority Approach.* Most judicial encounters with parents' rights to refuse medical treatment for their children follow the pattern established in *Newmark.* The court first announces the legal presumption that parents can make important decisions, including health care decisions, for their children. This is a common law right in every state, and it is protected by the U.S. Constitution as well. See Troxel v. Granville, 528 U.S. 1151, 120 S. Ct. 1225, 145 L. Ed. 2d 1068 (2000), Parham v. J.R., 442 U.S. 584, 99 S. Ct. 2493, 61 L. Ed. 2d 101 (1979); Santosky v. Kramer, 455 U.S. 745, 102 S. Ct. 1388, 71 L. Ed. 2d 599 (1982). However, this presumption can be overcome if the conditions of the state's child protective services statute are met.

Most state child abuse and neglect statutes are similar to the Delaware statute discussed in *Newmark*, and they provide—with greater or lesser specificity—that the parents' failure to provide adequate health care for their children constitutes neglect. If parents neglect their children, the state may commence a legal process (as the state did in *Newmark*) to obtain legal custody of the child to assure that the child is no longer medically neglected. The state need not take physical custody of the child, and the child can (as a technical legal matter) remain in the physical custody of the parents even while receiving the medical care to which the parents object.

2.   *What Standard Did the Court Apply?* What is the legal test that the court applied to determine that Colin Newmark should not be ordered to undergo the proposed chemotherapy? Did the court determine that it was in Colin's best interests not to undergo this treatment, or did the court apply a balancing test and determine that the parents' interest in maintaining custody of their child (and in making health care decisions for their child) outweighed the state's interest in providing Colin with potentially life-saving treatment? The choice between these two alternative forms of analysis may make no difference in this particular case, but can you think of a case in which the choice between these two legal positions would be dispositive?

3.   *Age of the Child.* As a general matter, courts reviewing young children's decisions to forgo life-sustaining treatment apply the "best interests" test (with a substantial bow to the articulated desires of the parents, who are deemed to have the best interests of their children in mind). While the application of the "substituted judgment" standard seems appropriate as the children approach majority, a few states attempt to apply the "substituted judgment" theory to very young children, too. Compare, e.g., Care and Protection of Beth, 412 Mass. 188, 587 N.E.2d 1377 (1992) and Custody of a Minor, 385 Mass. 697, 434 N.E.2d 601 (1982) (applying the substituted judgment test) with In re K.I., B.I. and D.M., 735 A.2d 448 (D.C. 1999) (applying the best interests test). See In re Christopher I., 106 Cal. App. 4th 533, 131 Cal. Rptr. 2d 122 (2003) (applying the best interests test to a child in the custody of the state).

Does the "substituted judgment" standard make any sense when it is applied to infants or very young children? Although it may be absurd to search for the values, interests, desires, and expectancies of a newborn or a three-month old, or even a four-year old, might not a seven- or eight-year-old child's parents be able to talk about all of those attributes? Of course, it is hard to imagine anyone other than the parents (or their legal substitutes) being able to evaluate such factors. Does that mean that a court's decision to apply a substituted judgment standard to small children amounts to a *de facto* decision to defer to the parent's wishes?

In Matter of AB, 196 Misc. 2d 940, 768 N.Y.S.2d 256 (Sup. Ct. 2003), a New York trial court confirmed the right of a mother of a child in persistent vegetative state to discontinue life-sustaining medical treatment—but only because the best interests test had been met:

> Thus, this Court holds that it is [the mother's] right, as a parent and natural guardian of AB, to exercise her responsibility and prerogative to make this decision to withhold extraordinary life-prolonging measures, with the assistance of treating physicians. CD's parental choice, made in the best interest of her child, to allow her daughter to pass away peacefully and with dignity are to be honored. This decision respects the values of family privacy without compromising a patient's rights or overstepping the State's legitimate interests.

> Having sought judicial intervention, CD has proven by clear and convincing evidence that it is in the best interest of her child to remove the mechanical ventilator. As CD sought intervention, this Court has employed the best interest standard, weighing whether the burdens of prolonged life outweigh any physical pleasure, emotional enjoyment, or intellectual satisfaction that the child may still be able to derive from life.

768 N.Y.S.2d at 271–272. In the *Matter of AB* case, AB's mother had sought a court order only because the hospital refused to permit the removal of treatment unless a court order supported that decision. Is the mother free to make health care decisions for her child—but only if the court agrees that the decisions are in the best interest of the child? Who is the real decision-maker in this case?

In a subsequent case, ethics consultants, a guardian ad litem appointed for the child by the court, the institution's ethics committee, and the parents all agreed that the termination of treatment was proper. Because one board eligible doctor and one nurse at the institution thought that the treatment would not constitute an "extraordinary burden" on the patient, though, the trial judge determined that the consensus determination of all other interested parties did not rise to the level of clear and convincing evidence that removal of the treatment was in the best interest of the teenager. See In re D.H., 15 Misc. 3d 565, 834 N.Y.S.2d 623 (Sup. 2007).

4. *Determining the Best Interest of a Child.* How is a court to determine what is in the best interest of a child? Does the court look to its own values, to values derived from community sources, to values held by the child's family, or elsewhere? For example, in the case of a neurologically devastated child with

no cognitive facilities and no hope of regaining any—a child who is entirely insensate and cannot feel pleasure or pain—is the terminal removal of a feeding tube or a ventilator *always* in the child's best interest, *never* in the child's best interest, or *sometimes* in the child's best interest? If you chose "sometimes," what other factors are relevant?

In *Matter of AB*, the court depended on standards provided by the New York Health Care Decisions for Mentally Retarded Persons Act and AMA guidelines. The AMA best interests guidelines, which are designed to address decision-making for newborns, are available at American Medical Association, Code of Medical Ethics, Opinion 2.2.4, Treatment Decisions for Seriously Ill Newborns (2016). In In re Christopher I., 106 Cal. App. 4th 533, 131 Cal. Rptr. 2d 122 (Cal. App. 2003), the court announced an exhaustive list of factors, drawn from local and national state court guidelines:

> We conclude that a court making the decision of whether to withhold or withdraw life-sustaining medical treatment from a dependent child should consider the following factors: (1) the child's present levels of physical, sensory, emotional and cognitive functioning; (2) the quality of life, life expectancy and prognosis for recovery with and without treatment, including the futility of continued treatment; (3) the various treatment options, and the risks, side effects, and benefits of each; (4) the nature and degree of physical pain or suffering resulting from the medical condition; (5) whether the medical treatment being provided is causing or may cause pain, suffering, or serious complications; (6) the pain or suffering to the child if the medical treatment is withdrawn; (7) whether any particular treatment would be proportionate or disproportionate in terms of the benefits to be gained by the child versus the burdens caused to the child; (8) the likelihood that pain or suffering resulting from withholding or withdrawal of treatment could be avoided or minimized; (9) the degree of humiliation, dependence and loss of dignity resulting from the condition and treatment; (10) the opinions of the family, the reasons behind those opinions, and the reasons why the family either has no opinion or cannot agree on a course of treatment; (11) the motivations of the family in advocating a particular course of treatment; and (12) the child's preference, if it can be ascertained, for treatment.

131 Cal. Rptr. at 551, 552. How helpful are these lists of factors? What factors do you believe a court should employ in determining if the removal of life-sustaining medical care is in the best interest of a child?

5.   *Mature Minor Doctrine.* Parental rights to make health care decisions for their children are terminated when the child reaches majority. Those rights may be diminished or terminated earlier if the child is a "mature minor," a condition governed by statute (in some states) or the common law (in other states) or both. At the very least, a child is a mature minor when he can "present clear and convincing evidence that he [is] mature enough to exercise an adult's judgment and [understand] the consequences of his decision." *Newmark*, at 1116 n. 9. See In re E.G., 133 Ill. 2d 98, 139 Ill. Dec. 810, 549

N.E.2d 322 (1989), where a Jehovah's Witness child "just months shy of her eighteenth birthday" was found to be sufficiently mature to choose to forgo a blood transfusion in a case with a rather slim chance of long-term survival. See also Belcher v. Charleston Area Med. Ctr., 188 W. Va. 105, 422 S.E.2d 827 (1992) (17-year-old may have been mature enough to demand a DNR order; the issue was for the jury in a subsequent informed consent action). For a discussion of the propriety of statutory recognition of the decision-making authority of teenagers, see Molly J. Walker Wilson, "Legal and Psychological Considerations in Adolescents' End-Of-Life Choices" 109 Nw. U. L. Rev. Online 203 (2015); Rhonda Hartman, Coming of Age: Devising Legislation for Adolescent Medical Decision-Making, 28 Am. J.L. & Med. 409 (2002).

Courts may give some weight to the statements of older children even if those courts are reluctant to declare those children "emancipated" or "mature minors." Although children under 16 typically are not found sufficiently mature to independently choose to forgo life-sustaining treatment, many courts have given considerable weight to the statements made by children considerably younger than that. See, e.g., In re Guardianship of Crum, 61 Ohio Misc. 2d 596, 580 N.E.2d 876 (Prob.Ct., Franklin Co. 1991). See also the discussion of children's assent to participation in research in Chapter 7.

At least one state explicitly authorizes children who understand the relevant health care factors to make their own decisions with regard to end-of-life care, even if they have not been emancipated and are not authorized to make other health care decisions. See N.M. Stat. Ann. § 24–7A–6.1(C), providing:

> [I]f an unemancipated minor has capacity sufficient to understand the nature of that unemancipated minor's medical condition, the risks and benefits of treatment and the contemplated decision to withhold or withdraw life-sustaining treatment, that unemancipated minor shall have the authority to withhold or withdraw life-sustaining treatment.

Should a child who cannot consent to an appendectomy under state law have the authority to consent to removal of life-sustaining treatment? Are children who are most likely to meet the statutory standard—those teenagers who have spent years suffering from a degenerative disease, for example—more likely than others to be able to recognize and apply their own life values to these important decisions? For a particularly interesting account of the roles of parents and children in consenting to body-reshaping medical care, see Alicia Ouellette, Shaping Parental Authority over Children's Bodies, 85 Ind. L.J. 955 (2010).

6. *Overriding Parental Religious Objections.* As the *Newmark* court explained, courts have "unhesitatingly authorized medical treatment over a parent's religious objection when the treatment is relatively innocuous in comparison with the dangers of withholding medical care." Of course, what constitutes an innocuous treatment for most of us might constitute a very serious intrusion for others. Most people are not terribly concerned by the prospect of a blood transfusion; for others, it may eliminate a possibility of

eternal salvation. Despite this, courts have not flinched when they have been presented with requests for treatment that is "medically necessary"—i.e., when the child will surely die if the treatment is not provided, and just as surely will live if the treatment is provided. Courts routinely and consistently order blood transfusions for children of objecting Jehovah's Witness parents, at least when they meet this requirement of medical necessity. In this sense, the courts treat Jehovah's Witness children very differently from Jehovah's Witness adults. On the other hand, courts generally do not order treatment for children who do not face life-threatening conditions. The hardest cases, like the *Newmark* case, are those cases in which the proposed treatment is highly invasive, but the child faces a life-threatening condition. As the chance of temporary remission and long-term cure increases, the courts are more likely to order the treatment. Is this consistent with evaluation under the factors listed in note 4, above?

7.   *When Parents Seek Treatment for Children.* While courts regularly face parental decisions that their children not receive treatment, they are sometimes confronted with the reverse situation. In In re K.I., B.I. and D.M., 735 A.2d 448 (D.C. 1999) the District of Columbia Court of Appeals approved a DNR order for a neglected, comatose two-year-old child who was born "neurologically devastated." The child was taken from his intoxicated mother after he had been left alone for days without his necessary heart and lung medication. The mother opposed the hospital's request that the child be given DNR status. Applying the best interests (rather than the substituted judgment) standard, the court recognized that the mother might be criminally liable for homicide if the child were to die, and it thus disregarded her request that there be aggressive attempts at resuscitation. See also Care and Protection of Sharlene, 445 Mass. 756, 840 N.E.2d 915 (2006).

Can you imagine other circumstances in which a parent might inappropriately demand treatment for a child? Should these cases be treated any differently from those in which the parent denies consent for (rather than demands) the recommended care? See also *In re Christopher I*, above, in which the father, whose abuse had resulted in the child's serious neurological injury, argued that the child's life should be maintained, and J.N. v. Superior Court, 156 Cal. App. 4th 523, 67 Cal. Rptr. 3d 384 (Cal. App. 2007), where the court ultimately permitted the withdrawal of a breathing tube but not the entry of a "do not attempt resuscitation" order in an abuse and neglect case involving an arguably criminally "shaken baby." A family court also allowed for the de-escalation of treatment of a child placed in the custody of the state due to abuse and neglect in Hunt v. Div. of Family Servs., 146 A.3d 1051 (Del. 2015). See also Division of Family Services v. Truselo, 846 A.2d 256 (Del. Fam. Ct. 2000).

8.   *"Spiritual Healing."* Like Delaware, most states have promulgated statutes that provide that "spiritual healing" *per se* cannot constitute child abuse or neglect for purposes of the state's criminal or child protective services statutes. In fact, for some time the existence of such protections was required as a condition of receiving some forms of federal funding. These spiritual healing statutes were the consequence of lobbying by the Christian Science

church, and they were written using terms (e.g., "accredited practitioner") that have well defined meanings within that church. Are such statutes constitutional? Might they violate the Establishment Clause of the First Amendment because they give a preference to Christian Scientists and protect them in ways that they do not protect others? Is it relevant that the statute only protects members of "recognized" religions, and that the state is called upon to determine which religions it will "recognize"? Might they violate the Establishment Clause because they disfavor children within certain religious communities?

Most commentators who have considered the issue have determined that these statutes are not constitutional (at least as they are applied to civil abuse and neglect proceedings) because they violate either the Establishment Clause or the Equal Protection Clause of the Fourteenth Amendment. For good accounts of the best arguments on this issue, see Shaakirrah R. Sanders, Religious Healing Exemptions and the Jurisprudential Gap Between Substantive Due Process and Free Exercise Rights, 8 U.C. Irvine L. Rev. 633 (2018); Ann MacLean Massie, The Religion Clauses and Parental Health Care Decision-making for Children: Suggestions for a New Approach, 21 Hastings Const. L. Q. 725 (1994). See also State v. Miskimens, 22 Ohio Misc. 2d 43, 490 N.E.2d 931 (Com. Pl. 1984); Walker v. Superior Court, 222 Cal. Rptr. 87 (Cal. App. 1986), aff'd 47 Cal. 3d 112, 253 Cal. Rptr. 1, 763 P.2d 852 (1988).

9.   *Decision-Making for Newborns.* The law generally views decision-making for newborns as a separate inquiry from decision-making for other children. Questions about the propriety of discontinuing treatment for newborns and infants often revolve around whether denying treatment would constitute discrimination based on the newborn's medical condition. During the 1980s, the case of Baby Doe focused attention on denial of care to seriously ill newborns based on disability. In 1983, Baby Doe was born with Down syndrome and a tracheoesophageal fistula that would require repair to allow him to consume nutrition orally. The parents decided not to authorize the necessary surgery because of the Down syndrome, and the baby was given phenobarbital and morphine until he died, six days after birth. A lawsuit by the hospital in the local children's court was rendered moot by Baby Doe's death.

Several political groups, including right-to-life groups and disability advocacy groups, were outraged by the circumstances of Baby Doe's death. The Secretary of HHS issued emergency regulations pursuant to Section 504 of the federal Rehabilitation Act, discussed in Chapter 2, to assure that no hospital would deny treatment to seriously ill newborns based on disability. After a furious political battle, a compromise was reached in the form of the Child Abuse Amendments of 1984, 42 U.S.C.A. § 5102, that conditions each state's receipt of some federal funding on the maintenance of procedures for dealing with reports of the medical neglect of newborns. Regulations issued under the Child Abuse Amendments interpret "medical neglect" to include "the withholding of medically indicated treatment from a disabled infant with a life threatening condition." 45 C.F.R. § 1340.15(b)(1). Further, the regulations provide that the " 'withholding of medically indicated treatment' means the failure to respond to the infant's life threatening conditions by providing treatment (including appropriate nutrition, hydration, and medication) which in the treating physicians' . . . reasonable medical judgment will be most likely to be effective in ameliorating or correcting all such conditions." 45 C.F.R. § 1340.15(b)(2). There is no obligation to provide care to an infant who is "chronically and irreversibly comatose," when the treatment would "merely prolong the dying process" of the infant and not ameliorate or correct the underlying medical problem (and thus be futile), or when the treatment would be "virtually futile" and "inhumane." *Id.*

PROBLEM: CONJOINED TWINS

After a woman was diagnosed as carrying conjoined twins, she left her home village of Xaghra on the Mediterranean island of Gozo to seek medical care at a hospital in the United Kingdom well known for its care of such birth anomalies. At birth, it was clear that one baby's heart and lungs were providing for the circulation in both babies. That baby, Jodie, was generally healthy, while her conjoined sibling, Mary, suffered from severe brain damage as well as other substantial disabilities. The doctors concluded that Jodie could lead a fairly normal life if she were separated from her sister and provided a series of reconstructive surgeries. Of course, the separation would result in Mary's immediate death. Alternatively, the failure to separate the twins would result in both of their deaths within the next several months.

The devout Catholic parents denied consent for the surgery, saying "[w]e cannot begin to accept or contemplate that one of our children should die to enable the other one to survive." Others also pointed out that the decision to allow Mary to die during the separation surgery could be met with horror in Xaghra and make it very difficult for the family to return home. The physicians, however, were distressed that both children would be allowed to die when one could be saved, and they sought an order from the court to authorize the separation surgery.

Would it be homicide to do the surgery with the certain knowledge that it would result in Mary's death? Is it medical neglect to deny consent to surgery that will save one child but result in the death of the second? Should the decision be left to the parents, or should the state intervene and exercise its *parens patriae* authority? What would you do if you were (1) a parent of these children? (2) the primary care physician for these children? (3) the judge before whom the case was presented?

The court was well briefed in this case. In addition to counsel retained by the parents and the public health service, a guardian ad litem was appointed for each child. What position should each of those guardians ad litem have taken? For description of this case (and how it came out), see Jacob Appel, Recent Developments in Health Law, 28 J.L. Med. & Ethics 312 (2000); Alexander MacLeod, Medical, Religious Values Clash Over Conjoined Twins, Christian Science Monitor, September 29, 2000, p. 7.

VI.   "FUTILE" TREATMENT

BETANCOURT V. TRINITAS HOSPITAL

Superior Court of New Jersey, Appellate Division, 2010.
1 A.3d 823.

PER CURIAM.

* * *

. . . On January 22, 2008, Rueben [Betancourt] underwent surgery at defendant to remove a malignant tumor from his thymus gland. . . . [T]he surgery went well, but while Rueben was recovering in the post-operative intensive care unit, the ventilation tube that was supplying him with oxygen somehow became dislodged. As a result, his brain was deprived of oxygen, and he developed anoxic encephalopathy, a condition that left him in a persistent vegetative state.

Rueben was subsequently discharged from defendant and admitted to other facilities that attempted rehabilitative treatments. He was readmitted to defendant on July 3, 2008, however, with a diagnosis of renal failure. Further attempts at placement in another facility proved fruitless, and he remained at defendant until his death on May 29, 2009.

At the time of his death, Rueben had not executed an advanced directive under the New Jersey Advanced Directives for Health Care Act. [ ] He had neither designated a health care representative nor memorialized "specific wishes regarding the provision, withholding or withdrawal of any form of health care, including life-sustaining treatment." [ ]

Witnesses for both parties to the dispute presented disparate views of Rueben's condition, the impact of treatment and prognosis. At the hearing, Rueben's attending physician, Dr. Arthur E. Millman, indicated that Rueben was a seventy-three-year-old man who was suffering from multi-system organ failure; his kidneys had failed, his lungs had failed, he was intermittently septic, he had hypertensive heart disease and congestive heart disease, and his skin was breaking down. He had "truly horrific decubitus ulcers" that had progressed to the bone, developing into osteomyelitis. Rueben was on a ventilator and received renal dialysis three times per week; he was fed through a tube into his stomach, given antibiotics and was turned frequently in his bed.

Millman stated that Rueben's most overwhelming problem was his permanent anoxic encephalopathy. He described Rueben's neurological state as "non-cognitive" with no higher mental functioning. He did believe, however, that Rueben was responsive to pain because he had personally witnessed Rueben's reactions to it. There had been no change in Rueben's neurological condition since he was admitted in July 2008, and Millman believed that the likelihood of his return to cognizant function was "virtually zero."

Dr. Bernard Schanzer, Chief of Neurology at defendant, corroborated most of Millman's views concerning Rueben's neurological condition. He explained that the cortical part of Rueben's brain had been irreversibly damaged. As a result, Rueben was in a permanent vegetative state, unable to speak or respond to verbal cues, and although Rueben's eyes were open and he appeared awake, he was not alert or aware of his environment. Schanzer disagreed with Millman, however, concerning Rueben's ability to experience pain. He believed that Rueben did not feel pain, and Rueben's responses to stimuli were due to basic reflexes of the brain stem and spinal cord. He opined that there was no chance that Rueben would ever regain a cognitive state.

Dr. Maria Silva Khazaei, a nephrologist, concluded that Rueben was suffering from end-stage renal disease, and there was no likelihood of improvement. She opined that it was contrary to accepted standards of medical care to continue dialysis treatments because they only prolonged Rueben's dying process.

Not surprisingly, plaintiff's consulting nephrologist had a different opinion. Dr. Carl Goldstein, a nephrologist retained by plaintiff, stated that Rueben's current plan of dialysis "comports in every way with the prevailing standards of care." He explained that the dialysis had been effective in removing excess fluid and waste products from Rueben's body. Rueben was tolerating the treatment well, and it was not harmful or dangerous to him.

Dr. William J. McHugh, Medical Director at defendant, was a member of the hospital's prognosis committee. The committee had been consulted concerning the efficacy of continuing Rueben's treatment; as a result, McHugh reviewed many, but not all, of the relevant medical records. He concluded that Rueben had "no outlook" because no affirmative treatment would improve his condition. As opposed to Millman, who believed that Rueben would probably die within a matter of months regardless of continued treatment, McHugh stated that Rueben's death "may take some time." In fact, he opined that if treatment were continued at the present level, Rueben "could go on for quite a while." On cross-examination, McHugh admitted that Rueben's present medical treatment was harmful only in the sense that the doctors were continuing to treat a hopeless situation.

Other members of the hospital's prognosis committee weighed in as well. Dr. Paul Veiana, president of the defendant's medical staff, examined Rueben the day before the hearing while Rueben was "wheeling" down to dialysis. Based on his review, he concluded that the doctors were not treating Rueben—they were just treating a body. He stated that the everyday drawing of blood and injections violated Rueben's body, and as a Christian, he believed that a body should not be so desecrated.

On several occasions, the hospital administration sought agreement from Rueben's family to place a DNR order and cease dialysis treatment, but they staunchly refused. It also made "exhaustive efforts" to transfer Rueben to another facility, but no other facility was willing to accept him. Ultimately, defendant acted unilaterally, placing the DNR order in Rueben's chart as well as surgically removing a dialysis port from Rueben's body.

At the hearing [on a motion to appoint plaintiff, Rueben's daughter Jacqueline Betancourt, as Rueben's guardian and for a temporary restraining order enjoining the hospital from withholding treatment], plaintiff provided information about Rueben. Before his illness, Rueben lived with his wife and his two adult sons. Plaintiff resided next door and saw her father every day. The family had always been very close, and Rueben was "dedicated" to his wife and children.

Plaintiff described Rueben's history of medical treatment at the hospital, asserting that it was the hospital's fault that he suffered a brain injury. She visited her father in the hospital almost daily and saw him make movements and gestures that led her to believe that he was awake and alert. She did not, however, believe that he was suffering. The family determined that they did not want a DNR order placed in Rueben's chart and did not want the dialysis treatment to be stopped. Rather, they wanted to make the decision as to whether Rueben was "ready to go." Plaintiff explained: "[M]y father is a fighter. He will not give up."

Robin, Rueben's thirty-six-year-old son, offered that his father was his "only . . . real friend" and that he loved him very much. He recalled, anecdotally, that he and his father had discussed the Terri Schiavo case when it was in the news, and his father had said that it was the right of Schiavo's parents—not her doctors—to decide what to do. Robin stated that his father reacted to him during hospital visits, and that those reactions were not simply reflexes. He described how his father had different facial expressions depending on what was happening around him and how his father's pulse would slow down when family members spoke to him or played music. He said the family did not trust the doctors to make the decision as to when to terminate his father's life.

Maria, Rueben's wife of thirty-seven years, was convinced that her husband reacted positively when she spoke to him or touched him. She believed that he would want "to continue living until God wished."

Nonetheless, the trial judge acknowledged that the temporary restraining order procedure should rarely be used to direct affirmative relief, but he found that the matter presented an "extreme situation" in which he needed to move quickly in order to maintain the status quo. The judge ordered defendant to re-establish the level of treatment that had been provided to Rueben prior to the discontinuation of dialysis and also to remove a DNR order that had been placed in his chart. He then ordered a hearing, which was held approximately two weeks later.

Following the hearing, Judge Malone issued a written opinion in which he concluded that decisions concerning the proper course of treatment for Rueben could not be made by the hospital; rather, such decisions should be made by a surrogate who could take Rueben's personal value systems into account when determining what medical treatment was appropriate. He granted plaintiff's application, appointed plaintiff as her father's guardian and permanently restrained the hospital from discontinuing treatment to Rueben. This was memorialized in a March 20, 2009 order. This appeal followed.

On May 29, 2009, Rueben died. Plaintiff filed a motion to dismiss the appeal as moot, and we reserved decision on the motion pending consideration of the merits of the appeal. We now grant the motion and dismiss the appeal.

[The portion of the opinion discussing mootness is omitted. The court determined that the sparse record and the unique circumstances of the case—the plaintiff thought that Reuben's death was caused by the defendant's negligence, and they thought that the hospital's conduct was generated by the fact that they owed so much for the care provided by defendant—made it appropriate to dismiss on mootness grounds even though the case presented a matter of substantial public importance that would likely to arise again and could easily evade judicial review. The Supreme Court of New Jersey denied review.]

TEX. HEALTH & SAFETY CODE § 166.046; § 166.052; § 166.053 [THE "TEXAS FUTILITY STATUTE," A PART OF THE TEXAS ADVANCE DIRECTIVES ACT]

* * *

§ 166.046—Procedure if Not Effectuating a Directive or Treatment Decision

(a) If an attending physician refuses to honor a patient's advance directive or a health care or treatment decision made by or on behalf of a patient, the physician's refusal shall be reviewed by an ethics or medical committee. The attending physician may not be a member of that committee. The patient shall be given life-sustaining treatment during the review.

(b) The patient or the person responsible for the health care decisions of the individual who has made the decision regarding the directive or treatment decision:

(1) may be given a written description of the ethics or medical committee review process and any other policies and procedures related to this section adopted by the health care facility;

(2) shall be informed of the committee review process not less than 48 hours before the meeting called to discuss the patient's directive, unless the time period is waived by mutual agreement;

(3) at the time of being so informed, shall be provided:

(A) a copy of the appropriate statement [describing this process]; and

(B) a copy of the registry list of health care providers and referral groups that have volunteered their readiness to consider accepting transfer or to assist in locating a provider willing to accept transfer . . .; and

(4) is entitled to:

(A) attend the meeting;

(B) receive a written explanation of the decision reached during the review process;

(C) receive a copy of the portion of the patient's medical record related to the treatment received by the patient in the facility for the lesser of:

(i) the period of the patient's current admission to the facility; or

(ii) the preceding 30 calendar days; and

(D) receive a copy of all of the patient's reasonably available diagnostic results and reports related to the medical record provided under Paragraph (C).

(c) The written explanation . . . must be included in the patient's medical record.

(d) If the attending physician, the patient, or the person responsible for the health care decisions of the individual does not agree with the decision reached during the review process under Subsection (b), the physician shall make a reasonable effort to transfer the patient to a physician who is willing to comply with the directive. If the patient is a patient in a health care facility, the facility's personnel shall assist the physician in arranging the patient's transfer to:

(1) another physician;

(2) an alternative care setting within that facility; or

(3) another facility.

(e)   If the patient or the person responsible for the health care decisions of the patient is requesting life-sustaining treatment that the attending physician has decided and the ethics or medical committee has affirmed is medically inappropriate treatment, the patient shall be given available life-sustaining treatment pending transfer under Subsection (d). This subsection does not authorize withholding or withdrawing pain management medication, medical procedures necessary to provide comfort, or any other health care provided to alleviate a patient's pain. The patient is responsible for any costs incurred in transferring the patient to another facility. The attending physician, any other physician responsible for the care of the patient, and the health care facility are not obligated to provide life-sustaining treatment after the 10th day after both the written decision and the patient's medical record required under Subsection (b) are provided to the patient or the person responsible for the health care decisions of the patient unless ordered to do so under Subsection (g), except that artificially administered nutrition and hydration must be provided unless, based on reasonable medical judgment, providing artificially administered nutrition and hydration would:

(1)   hasten the patient's death;

(2)   be medically contraindicated such that the provision of the treatment seriously exacerbates life-threatening medical problems not outweighed by the benefits of the provision of the treatment;

(3)   result in substantial irremediable physical pain not outweighed by the benefit of the provision of the treatment;

(4)   be medically ineffective in prolonging life; or

(5)   be contrary to the patient's or surrogate's clearly documented desire not to receive artificially administered nutrition or hydration.

(e–1)   If during a previous admission to a facility a patient's attending physician and the review process . . . have determined that life-sustaining treatment is inappropriate, and the patient is readmitted to the same facility within six months from the date of the decision reached during the review process conducted upon the previous admission, Subsections (b) through (e) need not be followed if the patient's attending physician and a consulting physician who is a member of the ethics or medical committee of the facility document on the patient's readmission that the patient's condition either has not improved or has deteriorated since the review process was conducted.

(f)   Life-sustaining treatment under this section may not be entered in the patient's medical record as medically unnecessary treatment until the time period provided under Subsection (e) has expired.

(g)   At the request of the patient or the person responsible for the health care decisions of the patient, the appropriate district or county court shall extend the time period provided under Subsection (e) only if the court finds, by a preponderance of the evidence, that there is a reasonable expectation that a physician or health care facility that will honor the patient's directive will be found if the time extension is granted.

(h) ... This section does not apply to hospice services provided by a home and community support services agency ....

* * *

§ 166.052—Statements Explaining Patient's Right to Transfer

(a) In cases in which the attending physician refuses to honor an advance directive or health care or treatment decision requesting the provision of life-sustaining treatment, the statement required ... shall be in substantially the following form:

> **When There Is A Disagreement About Medical Treatment: The Physician Recommends Against Certain Life-Sustaining Treatment That You Wish To Continue**
>
> You have been given this information because you have requested life-sustaining treatment* for yourself as the patient or on behalf of the patient, as applicable, which the attending physician believes is not medically appropriate. This information is being provided to help you understand state law, your rights, and the resources available to you in such circumstances. It outlines the process for resolving disagreements about treatment among patients, families, and physicians. It is based upon [this section].
>
> When an attending physician refuses to comply with an advance directive or other request for life-sustaining treatment because of the physician's judgment that the treatment would be medically inappropriate, the case will be reviewed by an ethics or medical committee. Life-sustaining treatment will be provided through the review.
>
> You will receive notification of this review at least 48 hours before a meeting of the committee related to your case. You are entitled to attend the meeting. With your agreement, the meeting may be held sooner than 48 hours, if possible.
>
> You are entitled to receive a written explanation of the decision reached during the review process.

---

* "Life-sustaining treatment" means treatment that, based on reasonable medical judgment, sustains the life of a patient and without which the patient will die. The term includes both life-sustaining medications and artificial life support, such as mechanical breathing machines, kidney dialysis treatment, and artificially administered nutrition and hydration. The term does not include the administration of pain management medication or the performance of a medical procedure considered to be necessary to provide comfort care, or any other medical care provided to alleviate a patient's pain.

If after this review process both the attending physician and the ethics or medical committee conclude that life-sustaining treatment is medically inappropriate and yet you continue to request such treatment, then the following procedure will occur:

1.   The physician, with the help of the health care facility, will assist you in trying to find a physician and facility willing to provide the requested treatment.

2.   You are being given a list of health care providers, licensed physicians, health care facilities, and referral groups that have volunteered their readiness to consider accepting transfer, or to assist in locating a provider willing to accept transfer. . . . You may wish to contact providers, facilities, or referral groups on the list or others of your choice to get help in arranging a transfer.

3.   The patient will continue to be given life-sustaining treatment until the patient can be transferred to a willing provider for up to 10 days from the time you were given both the committee's written decision that life-sustaining treatment is not appropriate and the patient's medical record. The patient will continue to be given after the 10-day period treatment to enhance pain management and reduce suffering, including artificially administered nutrition and hydration, unless, based on reasonable medical judgment, providing artificially administered nutrition and hydration would hasten the patient's death, be medically contraindicated such that the provision of the treatment seriously exacerbates life-threatening medical problems not outweighed by the benefit of the provision of the treatment, result in substantial irremediable physical pain not outweighed by the benefit of the provision of the treatment, be medically ineffective in prolonging life, or be contrary to the patient's or surrogate's clearly documented desires.

4.   If a transfer can be arranged, the patient will be responsible for the costs of the transfer.

5.   If a provider cannot be found willing to give the requested treatment within 10 days, life-sustaining treatment may be withdrawn unless a court of law has granted an extension.

6.   You may ask the appropriate district or county court to extend the 10-day period if the court finds that there is a reasonable expectation that you may find a physician or health care facility willing to provide life-sustaining treatment if the extension is granted. Patient medical records will be provided to the patient or surrogate in accordance with [the] Texas Health and Safety Code.

* * *

(c) An attending physician or health care facility may, if it chooses, include any additional information concerning the physician's or facility's policy, perspective, experience, or review procedure.

§ 166.053—Registry to Assist Transfers

(a) The [Texas Department of State Health Services] shall maintain a registry listing the identity of and contact information for health care providers and referral groups . . . that have voluntarily notified the department they may consider accepting or may assist in locating a provider willing to accept transfer of a patient under [this statute].

(b) The listing of a provider or referral group in the registry described in this section does not obligate the provider or group to accept transfer of or provide services to any particular patient.

* * *

FROEDTERT HOSPITAL—MEDICAL COLLEGE OF WISCONSIN, FUTILE MEDICAL CARE POLICY
(revised, 2020 version).

* * *

A. If, in the judgment of the Attending Physician and a staff physician consultant, life-sustaining medical treatment would be futile, the Attending Physician may write an order withholding or withdrawing the treatment after notifying the patient or patient's Surrogate(s). Appropriate palliative care measures should be instituted.

B. A life-sustaining medical intervention should be considered "futile" if it cannot be expected to restore or maintain vital organ function or to achieve the expressed goals of the patient when decisional.

C. Life-sustaining medical treatment includes cardiopulmonary resuscitation ("CPR"), mechanical ventilation, artificial nutrition and hydration, blood products, renal dialysis, vasopressors, or any other treatment that prolongs dying.

D. Consultation with Palliative Medicine, Social Services, and Chaplaincy, as appropriate, is strongly encouraged. If there are remaining questions, the physician should consult the Ethics Committee. If the patient (or Surrogate) disagrees, the Attending Physician should consider whether transfer to another attending physician, or another health care facility willing to accept the patient, is feasible. If transfer to a physician or facility willing to accept the patient is not feasible, further life-sustaining medical treatment may be withdrawn.

* * *

F.   The Attending Physician must contact the Office of the Chief Medical Officer ("CMO") verbally when this policy is invoked and document in the patient's legal health record ("LHR") a progress note . . . .

NOTES AND QUESTIONS

1.   *How Would You Resolve the Case?* The *Betancourt* case was never resolved on its merits. Should the court have decided it on the merits even though it was moot? How should it have been decided? You have been presented with the factual portion of the opinion; how would you write the substantive legal analysis? How would this issue have been resolved if the hospital had a policy like the Froedtert policy?

2.   *Who Decides?* Who has the final authority to make that decision under the Texas statute and the hospital policy? Who would you allow to be the final arbiter of whether a proposed treatment is futile? For a discussion of the advantages and disadvantages of different decision-making bodies in resolving futility disputes, see Janet L. Dolgin, Medical Disputes and Conflicting Values: Is There a "Right to Die" Later?, 2020 B.Y.U. L. Rev. 95 (2020).

3.   *The Limits of Autonomy?* As noted in Section II, the law of health care decision-making has relied heavily on the principle of autonomy. The right of a patient to make his or her own informed decisions to accept or refuse medical care, the substituted judgment standard, and access to medically assisted death (covered in Chapter 6) all reflect this principle. Medical futility disputes, however, raise questions about the limits of patient autonomy. How would you explain this apparent departure? Is it a conflict between autonomy and beneficence? Or between autonomy and other values? Is it similar to the specific provisions for situations in which health care professionals or facilities object to the health care decisions of patients (or their surrogates) covered earlier in this chapter? Why or why not?

4.   *Defining "Futility."* When is requested care truly "futile"? Under the hospital policy, how is futility defined? Who has the authority to make that decision? Under the Texas statute, how is futility defined? Which definition makes the most sense, given the purposes of futility laws and policies? How would you define futility? See Janet L. Dolgin, Remaking the 'Right to Die': Give Me Liberty but Do Not Give Me Death, 73 S.M.U. L. Rev. 47 (2020); Meir Katz, When Is Medical Care Futile? The Institutional Competence of the Medical Profession Regarding the Provision of Life-Sustaining Medical Care, 90 Neb. L. Rev.1 (2011).

5.   *Medical vs. Ethical Futility.* The term futility is subject to a variety of interpretations. Treatment is *scientifically futile* (or *medically futile*) when it cannot achieve the medical result that is expected by the patient (or by the family) making the request. As a general matter, scientifically futile treatment need not be offered or provided to a patient. A seriously ill cancer patient need not be provided with laetrile, a useless drug that has been popularized by those who would prey upon desperate patients and their families, even if that

treatment is requested. A child with a viral illness need not be prescribed an antibiotic, even if the child's parents request one, because, as a matter of science, the antibiotic will not be effective in treating that illness. Doctors need not do a CAT scan on a patient with a cold, even if that is what the patient wants, because there is no reason to believe that there will be any connection between what can be discovered on the scan and the appropriate treatment of the cold. As a general matter, health care providers, who are trained in the science of medicine, are entitled to determine which treatments are scientifically futile.

A harder question arises when a patient requests treatment that is not scientifically futile, but that is, in the opinion of the health care provider, *ethically futile*. Treatment is ethically futile if it will not serve the underlying interests of the patient. For example, some providers believe that it is ethically futile to keep a patient's body aerated and nourished when that patient is in a persistent vegetative state. These health care providers believe that it is beyond the scope of medicine to sustain mere corporeal existence. Some health care providers believe that it would be ethically futile to engage in CPR under circumstances in which the most that can be accomplished through that intervention would be to prolong the patient's life by a few hours. Families may disagree with physicians over what constitutes ethically futile treatment. Is there any reason to adopt the provider's perspective, rather than the family's, as the ethically "correct" one?

The Council on Ethical and Judicial Affairs of the AMA has determined that:

> Physicians are not required to offer or provide interventions that, in their best medical judgment, cannot reasonably be expected to yield the intended clinical benefit or achieve agreed-on goals for care. Respecting patient autonomy does not mean that patients should receive specific interventions simply because they (or their surrogates) request them.

Code of Medical Ethics Opinion 5.5, *Medically Ineffective Interventions* (2016).

Is the Council's position a convincing one, or is it merely a device to transfer the authority to make ethically charged decisions from patients to physicians? See Patrick Moore, An End-of-Life Quandary in Need of a Statutory Response: When Patients Demand Life-Sustaining Treatment That Physicians Are Unwilling to Provide, 48 B.C. L. Rev. 433 (2007).

6.   *Risk of Bias and Discrimination.* How can we ensure that unexamined assumptions and value judgments do not find their way into medical decisions to withhold or withdraw care deemed futile? A 2019 report by the National Council on Disability found that such decisions are compromised by negative biases and lack of knowledge about the quality of life of people with disabilities, and they often lack objectivity and procedural safeguards. Nat'l Council on Disability, Medical Futility and Disability Bias (2019).

The death of Michael Hickson in 2020 focused attention on these concerns. Mr. Hickson, a 46-year-old Black man and father of five, sought treatment for serious COVID-19 disease. The hospital refused to provide life-saving care over the objection of his wife, Melissa Hickson. News coverage included reports of a recording of a conversation between Melissa Hickson and a physician at the hospital who informed her that the hospital would not provide medical treatment because it deemed Mr. Hickson to have a poor quality of life due to his pre-existing quadriplegia and head injury. Joseph Shapiro, One Man's COVID-19 Death Raises the Worst Fears of Many People with Disabilities, Nat'l Public Radio, July 31, 2020. In June 2021, Melissa Hickson filed a complaint raising claims under federal laws prohibiting discrimination based on disability and under state laws. Complaint, Hickson v. St. David's Healthcare Partnership, No. 1:21-cv-514 (W.D. Tex., Austin Div. 2021).

Federal disability antidiscrimination laws apply to medical futility decisions, but they have received little attention in this context to date. See also Chapter 2 for discussion of federal laws that prohibit discrimination based on disability.

7. *Procedural Requirements.* Under the Texas statute, what process must be followed before a patient seeking futile care is discharged from the hospital? You might prepare a flow chart including all of the steps required by the statute. Do these requirements seem reasonable? Is ten days enough to find an alternative placement for one who is about to have futile treatment removed?

8. *Challenges to Texas Statute.* The Texas statute has given rise to litigation. In one celebrated case, the mother of Sun Hudson, an infant, sought judicial intervention to avoid the discontinuation of treatment that physicians believed to be futile. In another case involving an adult, Spiro Nikolouzos had been transferred once after ventilator support was found to be futile. When the receiving institution reached the same conclusion and began to prepare for his discharge, his family sought more time to arrange for yet a second transfer. Both cases were decided on procedural grounds largely unrelated to the substantive provisions of the statute, but each ultimately recognized the authority of the health care provider to make futility determinations. See Hudson v. Texas Children's Hosp., 177 S.W.3d 232 (Tex. Ct. App. 2005); Nikolouzos v. St. Luke's Episcopal Hosp., 162 S.W.3d 678 (Tex. Ct. App. 2005).

Several other litigated cases have resulted in a great deal of press coverage. In the recent case of Tinslee Lewis, an infant patient and her mother sought injunctive relief from a hospital's unilateral discontinuation of life-sustaining treatment, alleging that the statutory committee review process did not provide adequate procedural due process protections. In July 2020, the court of appeals directed the trial court to render an order granting the temporary injunction pending a decision on the merits. T.L. v. Cook Children's Med. Ctr., 607 S.W.3d 9 (2020), cert. denied 141 S. Ct. 1069 (2021).

9. *Uniform Health-Care Decisions Act Model.* Some states with statutes modeled on the Uniform Health-Care Decisions Act, reprinted above, provide

that physicians need not give futile care. Recall that section 7(f) of that Uniform Act provides that "[a] health-care provider or institution may decline to comply with an individual instruction or health-care decision that requires medically ineffective health care or health care contrary to generally accepted health-care standards applicable to the health-care provider or institution." See also Md. Code Ann., Health-Gen. section 5–611 and Va. Code Ann. section 54.1–2990. The Texas statute differs from these earlier statutes by providing an authoritative and conclusive endpoint for disputes. See Thaddeus Mason Pope, Procedural Due Process and Intramural Hospital Dispute Resolution Mechanisms: The Texas Advance Directives Act, 10 St. Louis U.J. of Health L. & Pol'y 93 (2016) (closely analyzing Texas statute and suggesting amendments to better comport with fundamental notions of procedural due process).

10. *Notice Requirements.* In 2012 Froedtert Hospital was cited by a CMS inspector for not telling a patient about the existence of the hospital's futility policy. In that case a dying patient had been hospitalized 21 times in the past year and was in such failing health that the physicians concluded that any further life-sustaining treatment would be futile under the hospital policy. The family objected and continued to insist that all available treatment, including near daily blood transfusions, be continued. The patient was given DNR status without the agreement of the family. The CMS inspector did not find that the Froedtert futility policy violated any law, but, rather, that the hospital violated the Patient Self Determination Act, described earlier in this chapter, by not telling the patient and his family of the existence of the futility policy. Ultimately the patient's family requested the transfer of the patient to another facility, and Froedtert cooperated in arranging for that transfer.

11. *Mediation?* Robert Truog, a thoughtful physician and bioethicist, writes that medical futility issues are based in "power, trust, hope, money, and suffering." Robert Truog, Medical Futility, 25 Ga. St. U.L. Rev. 985, 986 (2009). Is he right? If he is right, is this fact helpful in figuring out how to resolve disputes over medical futility? Mediation is often employed to help resolve these disputes in health care institutions. How helpful is mediation in dealing with conflict based in power, trust, hope, money, and suffering?

CHAPTER 6

MEDICALLY ASSISTED DYING

■ ■ ■

I. INTRODUCTION

There has been a significant debate in the U.S. and elsewhere over the past quarter of a century over the propriety of medically assisted death, that is, medical care designed to help a patient die how and when the patient wants to die. As the last chapter established, every state allows a physician to assist a competent adult patient who wishes to refuse or discontinue life-sustaining medical care, at least under most circumstances. But may a health care provider also provide affirmative intervention to a patient who wishes to hasten death? If so, under what circumstances?

The language we use to discuss these questions has become especially divisive over the past few years, and we must be careful how we use words that have become laden with political and philosophical meaning. "Euthanasia," for example, generally refers to an affirmative act that directly and immediately causes the death of a patient for the benefit of that patient. "Involuntary euthanasia" is sometimes used to describe euthanasia against the will of the patient, although sometimes it means euthanasia without the formal and expressed consent of the patient. "Voluntary euthanasia" is sometimes used to refer to euthanasia upon request of the patient, although most careful thinkers avoid using that term altogether. "Active euthanasia" sometimes refers to some affirmative act of euthanasia, while "passive euthanasia" generally refers, technically incorrectly, to withholding or withdrawing life-sustaining treatment.

"Suicide" can refer to any act taken by a person to intentionally end his own life (including the act of a patient taking a prescribed lethal dose of medication), but sometimes it refers only to an act taken by someone acting irrationally, or under the influence of extraordinary emotional distress or mental disease (and not, for example, to the act of a terminally ill, competent patient who knowingly takes a prescribed lethal dose to end his suffering). Both "euthanasia" and "suicide" have highly negative connotations, and the people who use those terms generally oppose the propriety of the act. The words "death" and "killing" in this context also have strong negative connotations, and those terms are used to describe medically assisted dying primarily by those who oppose its legality.

On the other hand, "death with dignity," "patient choice," or "end-of-life option," which often refer to these same acts, have positive connotations for many. That is why advocates of the practice use similar language on state initiatives. Similarly, "Right to Die" is generally a popular concept, and that term is used almost exclusively by those who approve of the use of this practice. For years, however, the phrase was used to describe the withdrawal of life-sustaining treatment. Using the same phrase in this context implicitly argues that there is no difference between the two.

Opponents of medically assisted dying often use the terms "assisted suicide" or "physician assisted suicide," and they have raised a range of concerns about the practice.

As with life and death decision-making addressed in Chapter 5, advocates for disability rights have brought attention to the impact of individual and structural biases that may distort decision-making by and for people with disabilities and other marginalized populations in this context. What safeguards are needed to address these concerns? How does medically assisted death intersect with issues of equity and justice in health care, more generally?

As you read the materials that follow, make sure that you identify with specificity exactly what practice is being considered. While there are a few countries that permit euthanasia under certain limited circumstances, that has never been a legally available option in the U.S., and it is rarely part of the legal debate. In this country, the debate has centered on the propriety (and legality) of a medically prescribed lethal dose of medication designed to be ingested by the competent, fully informed patient who originally requested a prescription for it from a physician. It is that practice that forms the core of the discussion in this chapter. While, increasingly, physicians, judges and scholars are calling this assistance "Aid in Dying," the authors have chosen to call it "Medically Assisted Dying" because this term is not used in advocacy either for or against these actions.

II.   THE CONSTITUTIONAL FRAMEWORK

WASHINGTON V. GLUCKSBERG

Supreme Court of the United States, 1997.
521 U.S. 702.

REHNQUIST, C. J., delivered the opinion of the Court, in which O'CONNOR, SCALIA, KENNEDY, and THOMAS, JJ., joined. O'CONNOR, J., filed a concurring opinion, in which GINSBURG and BREYER, JJ., joined in part. STEVENS, J., SOUTER, J., GINSBURG, J., and BREYER, J., filed opinions concurring in the judgment.

CHIEF JUSTICE REHNQUIST delivered the opinion of the Court.

The question presented in this case is whether Washington's prohibition against "causing" or "aiding" a suicide offends the Fourteenth Amendment to the United States Constitution. We hold that it does not.

* * *

The plaintiffs assert [ ] "the existence of a liberty interest protected by the Fourteenth Amendment which extends to a personal choice by a mentally competent, terminally ill adult to commit physician-assisted suicide." [ ] Relying primarily on Planned Parenthood v. Casey, [ ] and Cruzan v. Director, Missouri Dept. of Health, [ ] the District Court agreed, [ ] and concluded that Washington's assisted-suicide ban is unconstitutional because it "places an undue burden on the exercise of [that] constitutionally protected liberty interest." [ ] The District Court also decided that the Washington statute violated the Equal Protection Clause's requirement that " 'all persons similarly situated . . . be treated alike.' "[ ]

A panel of the Court of Appeals for the Ninth Circuit reversed, emphasizing that "in the two hundred and five years of our existence no constitutional right to aid in killing oneself has ever been asserted and upheld by a court of final jurisdiction." [ ] The Ninth Circuit reheard the case en banc, reversed the panel's decision, and affirmed the District Court. [ ] Like the District Court, the en banc Court of Appeals emphasized our Casey and Cruzan decisions. [ ] The court also discussed what it described as "historical" and "current societal attitudes" toward suicide and assisted suicide, [ ] and concluded that "the Constitution encompasses a due process liberty interest in controlling the time and manner of one's death—that there is, in short, a constitutionally-recognized 'right to die.' "[ ] After "weighing and then balancing" this interest against Washington's various interests, the court held that the State's assisted-suicide ban was unconstitutional "as applied to terminally ill competent [ ] adults who wish to hasten their deaths with medication prescribed by their physicians." [ ] We granted certiorari [ ] and now reverse.

I

We begin, as we do in all due-process cases, by examining our Nation's history, legal traditions, and practices. [ ] In almost every State—indeed, in almost every western democracy—it is a crime to assist a suicide. The States' assisted-suicide bans are not innovations. Rather, they are longstanding expressions of the States' commitment to the protection and preservation of all human life. [ ] Indeed, opposition to and condemnation of suicide—and, therefore, of assisting suicide—are consistent and enduring themes of our philosophical, legal, and cultural heritages. [ ]

More specifically, for over 700 years, the Anglo-American common-law tradition has punished or otherwise disapproved of both suicide and assisting suicide. . . . [The Chief Justice then reviews the common law of England, the American colonies, and U.S. states, from the 13th century to the present, with regard to suicide.]

* * *

Attitudes toward suicide itself have changed since [the 13th Century prohibitions on suicide] . . . but our laws have consistently condemned, and continue to prohibit, assisting suicide. Despite changes in medical technology and notwithstanding an increased emphasis on the importance of end-of-life decisionmaking, we have not retreated from this prohibition. Against this backdrop of history, tradition, and practice, we now turn to respondents' constitutional claim.

II

The Due Process Clause guarantees more than fair process, and the "liberty" it protects includes more than the absence of physical restraint. [ ] The Clause also provides heightened protection against government interference with certain fundamental rights and liberty interests. [ ] In a long line of cases, we have held that, in addition to the specific freedoms protected by the Bill of Rights, the "liberty" specially protected by the Due Process Clause includes the rights to marry, [ ]; to have children, [ ]; to direct the education and upbringing of one's children, [ ]; to marital privacy, [ ]; to use contraception, [ ]; to bodily integrity, [ ] and to abortion, [ ]. We have also assumed, and strongly suggested, that the Due Process Clause protects the traditional right to refuse unwanted lifesaving medical treatment. [ ]

But we "have always been reluctant to expand the concept of substantive due process because guideposts for responsible decisionmaking in this unchartered area are scarce and open-ended." [ ] By extending constitutional protection to an asserted right or liberty interest, we, to a great extent, place the matter outside the arena of public debate and legislative action. We must therefore "exercise the utmost care whenever we are asked to break new ground in this field" [ ] lest the liberty protected by the Due Process Clause be subtly transformed into the policy preferences of the members of this Court [ ].

Our established method of substantive-due-process analysis has two primary features: First, we have regularly observed that the Due Process Clause specially protects those fundamental rights and liberties which are, objectively, "deeply rooted in this Nation's history and tradition" [ ] and "implicit in the concept of ordered liberty," such that "neither liberty nor justice would exist if they were sacrificed" [ ]. Second, we have required in substantive-due-process cases a "careful description" of the asserted fundamental liberty interest. [ ] Cruzan, supra, at 277–278. Our Nation's history, legal traditions, and practices thus provide the crucial "guideposts for responsible decisionmaking" [ ] that direct and restrain our exposition of the Due Process Clause. As we stated recently in Flores, the Fourteenth Amendment "forbids the government to infringe . . . 'fundamental' liberty interests at all, no matter what process is provided, unless the infringement is narrowly tailored to serve a compelling state interest." [ ]

* * *

Turning to the claim at issue here, the Court of Appeals stated that "properly analyzed, the first issue to be resolved is whether there is a liberty interest in determining the time and manner of one's death" [ ] or, in other words, "is there a right to die?" [ ]. Similarly, respondents assert a "liberty to choose how to die" and a right to "control of one's final days," [ ] and describe the asserted liberty as "the right to choose a humane, dignified death" [ ] and "the liberty to shape death" [ ]. As noted above, we have a tradition of carefully formulating the interest at stake in substantive-due-process cases. For example, although Cruzan is often described as a "right to die" case [ ] we were, in fact, more precise: we assumed that the Constitution granted competent persons a "constitutionally protected right to refuse lifesaving hydration and nutrition." [ ] The Washington statute at issue in this case prohibits "aiding another person to attempt suicide," [ ] and, thus, the question before us is whether the "liberty" specially protected by the Due Process Clause includes a right to commit suicide which itself includes a right to assistance in doing so.

. . . With this "careful description" of respondents' claim in mind, we turn to Casey and Cruzan.

[The Chief Justice next discusses the Cruzan case, where, he says,] "we assumed that the United States Constitution would grant a competent person a constitutionally protected right to refuse lifesaving hydration and nutrition."

* * *

The right assumed in Cruzan, however, was not simply deduced from abstract concepts of personal autonomy. Given the common-law rule that forced medication was a battery, and the long legal tradition protecting the decision to refuse unwanted medical treatment, our assumption was entirely consistent with this Nation's history and constitutional traditions. The decision to commit suicide with the assistance of another may be just as personal and profound as the decision to refuse unwanted medical treatment, but it has never enjoyed similar legal protection. Indeed, the two acts are widely and reasonably regarded as quite distinct. [ ] In Cruzan itself, we recognized that most States outlawed assisted suicide—and even more do today—and we certainly gave no intimation that the right to refuse unwanted medical treatment could be somehow transmuted into a right to assistance in committing suicide. [ ]

Respondents also rely on Casey. There, the Court's opinion concluded that "the essential holding of Roe v. Wade should be retained and once again reaffirmed." [ ] We held, first, that a woman has a right, before her fetus is viable, to an abortion "without undue interference from the State"; second, that States may restrict post-viability abortions, so long as exceptions are made to protect a woman's life and health; and third, that the State has legitimate interests throughout a pregnancy in protecting the health of the woman and the life of the unborn child. [ ] In reaching this conclusion, the opinion discussed in some detail this Court's substantive-due-process tradition of interpreting the Due Process Clause to protect certain fundamental rights and "personal decisions relating to marriage, procreation, contraception, family relationships, child rearing, and education," and noted that many of those rights and liberties "involve the most intimate and personal choices a person may make in a lifetime." [ ]

* * *

That many of the rights and liberties protected by the Due Process Clause sound in personal autonomy does not warrant the sweeping conclusion that any and all important, intimate, and personal decisions are so protected, [ ] and Casey did not suggest otherwise.

The history of the law's treatment of assisted suicide in this country has been and continues to be one of the rejection of nearly all efforts to permit it. That being the case, our decisions lead us to conclude that the asserted "right" to assistance in committing suicide is not a fundamental liberty interest protected by the Due Process Clause. The Constitution also requires, however, that Washington's assisted-suicide ban be rationally related to legitimate government interests. [ ] This requirement is unquestionably met here. As the court below recognized, [ ] Washington's assisted-suicide ban implicates a number of state interests. [ ]

First, Washington has an "unqualified interest in the preservation of human life."

* * *

Relatedly, all admit that suicide is a serious public-health problem, especially among persons in otherwise vulnerable groups. [ ] The State has an interest in preventing suicide, and in studying, identifying, and treating its causes. [ ]

* * *

The State also has an interest in protecting the integrity and ethics of the medical profession. . . . [T]he American Medical Association, like many other medical and physicians' groups, has concluded that "physician-assisted suicide is fundamentally incompatible with the physician's role as healer." [ ] And physician-assisted suicide could, it is argued, undermine the trust that is essential to the doctor-patient relationship by blurring the time-honored line between healing and harming. [ ]

Next, the State has an interest in protecting vulnerable groups—including the poor, the elderly, and disabled persons—from abuse, neglect, and mistakes. . . . [One respected state task force] warned that "legalizing physician-assisted suicide would pose profound risks to many individuals who are ill and vulnerable. . . . The risk of harm is greatest for the many individuals in our society whose autonomy and well-being are already compromised by poverty, lack of access to good medical care, advanced age, or membership in a stigmatized social group." [ ] If physician-assisted suicide were permitted, many might resort to it to spare their families the substantial financial burden of end-of-life health-care costs.

. . . The State's assisted-suicide ban reflects and reinforces its policy that the lives of terminally ill, disabled, and elderly people must be no less valued than the lives of the young and healthy, and that a seriously disabled person's suicidal impulses should be interpreted and treated the same way as anyone else's. [ ]

Finally, the State may fear that permitting assisted suicide will start it down the path to voluntary and perhaps even involuntary euthanasia. . . . [Justice Rehnquist then discussed how this fear could arise out of the practice in the Netherlands.]

We need not weigh exactingly the relative strengths of these various interests. They are unquestionably important and legitimate, and Washington's ban on assisted suicide is at least reasonably related to their promotion and protection. We therefore hold that [the Washington ban on assisting suicide] does not violate the Fourteenth Amendment, either on its face or "as applied to competent, terminally ill adults who wish to hasten their deaths by obtaining medication prescribed by their doctors."[24] [ ]

* * *

Throughout the Nation, Americans are engaged in an earnest and profound debate about the morality, legality, and practicality of physician-assisted suicide. Our holding permits this debate to continue, as it should in a democratic society. The decision of the en banc Court of Appeals is

---

[24] Justice Stevens states that "the Court does conceive of respondents' claim as a facial challenge—addressing not the application of the statute to a particular set of plaintiffs before it, but the constitutionality of the statute's categorical prohibition. . . . "[ ] We emphasize that we today reject the Court of Appeals' specific holding that the statute is unconstitutional "as applied" to a particular class. [ ] Justice Stevens agrees with this holding, [ ] but would not "foreclose the possibility that an individual plaintiff seeking to hasten her death, or a doctor whose assistance was sought, could prevail in a more particularized challenge," ibid. Our opinion does not absolutely foreclose such a claim. However, given our holding that the Due Process Clause of the Fourteenth Amendment does not provide heightened protection to the asserted liberty interest in ending one's life with a physician's assistance, such a claim would have to be quite different from the ones advanced by respondents here.

reversed, and the case is remanded for further proceedings consistent with this opinion.

It is so ordered.

JUSTICE O'CONNOR, concurring [in both this case, Glucksberg, and Vacco, the case excerpted after this one in this book].*

Death will be different for each of us. For many, the last days will be spent in physical pain and perhaps the despair that accompanies physical deterioration and a loss of control of basic bodily and mental functions. Some will seek medication to alleviate that pain and other symptoms.

The Court frames the issue in this case as whether the Due Process Clause of the Constitution protects a "right to commit suicide which itself includes a right to assistance in doing so," [ ] and concludes that our Nation's history, legal traditions, and practices do not support the existence of such a right. I join the Court's opinions because I agree that there is no generalized right to "commit suicide." But respondents urge us to address the narrower question whether a mentally competent person who is experiencing great suffering has a constitutionally cognizable interest in controlling the circumstances of his or her imminent death. I see no need to reach that question in the context of the facial challenges to the New York and Washington laws at issue here. [ ] The parties and amici agree that in these States a patient who is suffering from a terminal illness and who is experiencing great pain has no legal barriers to obtaining medication, from qualified physicians, to alleviate that suffering, even to the point of causing unconsciousness and hastening death. [ ] In this light, even assuming that we would recognize such an interest, I agree that the State's interests in protecting those who are not truly competent or facing imminent death, or those whose decisions to hasten death would not truly be voluntary, are sufficiently weighty to justify a prohibition against physician-assisted suicide. [ ]

Every one of us at some point may be affected by our own or a family member's terminal illness. There is no reason to think the democratic process will not strike the proper balance between the interests of terminally ill, mentally competent individuals who would seek to end their suffering and the State's interests in protecting those who might seek to end life mistakenly or under pressure. As the Court recognizes, States are presently undertaking extensive and serious evaluation of physician-assisted suicide and other related issues. [ ] In such circumstances, "the . . . challenging task of crafting appropriate procedures for safeguarding . . . liberty interests is entrusted to the 'laboratory' of the States . . . in the first instance." [ ]

---

* Justice Ginsburg concurs in the Court's judgments substantially for the reasons stated in this opinion. Justice Breyer joins this opinion except insofar as it joins the opinions of the Court.

In sum, there is no need to address the question whether suffering patients have a constitutionally cognizable interest in obtaining relief from the suffering that they may experience in the last days of their lives. There is no dispute that dying patients in Washington and New York can obtain palliative care, even when doing so would hasten their deaths. The difficulty in defining terminal illness and the risk that a dying patient's request for assistance in ending his or her life might not be truly voluntary justifies the prohibitions on assisted suicide we uphold here.

JUSTICE STEVENS, concurring in the judgments [in both Glucksberg and Vacco].

The Court ends its opinion with the important observation that our holding today is fully consistent with a continuation of the vigorous debate about the "morality, legality, and practicality of physician-assisted suicide" in a democratic society. [ ] I write separately to make it clear that there is also room for further debate about the limits that the Constitution places on the power of the States to punish the practice.

I

The morality, legality, and practicality of capital punishment have been the subject of debate for many years. In 1976, this Court upheld the constitutionality of the practice in cases coming to us from Georgia, Florida, and Texas. In those cases we concluded that a State does have the power to place a lesser value on some lives than on others; there is no absolute requirement that a State treat all human life as having an equal right to preservation. Because the state legislatures had sufficiently narrowed the category of lives that the State could terminate, and had enacted special procedures to ensure that the defendant belonged in that limited category, we concluded that the statutes were not unconstitutional on their face. In later cases coming to us from each of those States, however, we found that some applications of the statutes were unconstitutional.

Today, the Court decides that Washington's statute prohibiting assisted suicide is not invalid "on its face," that is to say, in all or most cases in which it might be applied. That holding, however, does not foreclose the possibility that some applications of the statute might well be invalid.

* * *

History and tradition provide ample support for refusing to recognize an open-ended constitutional right to commit suicide. Much more than the State's paternalistic interest in protecting the individual from the irrevocable consequences of an ill-advised decision motivated by temporary concerns is at stake. There is truth in John Donne's observation that "No man is an island." The State has an interest in preserving and fostering the benefits that every human being may provide to the community—a community that thrives on the exchange of ideas, expressions of affection, shared memories and humorous incidents as well as on the material contributions that its members create and support. The value to others of a person's life is far too precious to allow the individual to claim a constitutional entitlement to complete autonomy in making a decision to end that life. Thus, I fully agree with the Court that the "liberty" protected by the Due Process Clause does not include a categorical "right to commit suicide which itself includes a right to assistance in doing so." [ ]

But just as our conclusion that capital punishment is not always unconstitutional did not preclude later decisions holding that it is sometimes impermissibly cruel, so is it equally clear that a decision upholding a general statutory prohibition of assisted suicide does not mean that every possible application of the statute would be valid. A State, like Washington, that has authorized the death penalty and thereby has concluded that the sanctity of human life does not require that it always be preserved, must acknowledge that there are situations in which an interest in hastening death is legitimate. Indeed, not only is that interest sometimes legitimate, I am also convinced that there are times when it is entitled to constitutional protection.

II

In Cruzan [ ] the Court assumed that the interest in liberty protected by the Fourteenth Amendment encompassed the right of a terminally ill patient to direct the withdrawal of life-sustaining treatment. As the Court correctly observes today, that assumption "was not simply deduced from abstract concepts of personal autonomy." [ ] Instead, it was supported by the common-law tradition protecting the individual's general right to refuse unwanted medical treatment. [ ] We have recognized, however, that this common-law right to refuse treatment is neither absolute nor always sufficiently weighty to overcome valid countervailing state interests. . . .

Cruzan, however, was not the normal case. Given the irreversible nature of her illness and the progressive character of her suffering, Nancy Cruzan's interest in refusing medical care was incidental to her more basic interest in controlling the manner and timing of her death. In finding that her best interests would be served by cutting off the nourishment that kept her alive, the trial court did more than simply vindicate Cruzan's interest in refusing medical treatment; the court, in essence, authorized affirmative conduct that would hasten her death. When this Court reviewed the case and upheld Missouri's requirement that there be clear and convincing evidence establishing Nancy Cruzan's intent to have life-sustaining nourishment withdrawn, it made two important assumptions: (1) that there was a "liberty interest" in refusing unwanted treatment protected by the Due Process Clause; and (2) that this liberty interest did not "end the inquiry" because it might be outweighed by relevant state interests. [ ] I agree with both of those assumptions, but I insist that the source of Nancy Cruzan's right to refuse treatment was not just a common-law rule. Rather, this right is an aspect of a far broader and more basic concept of freedom that is even older than the common law. This freedom embraces, not merely a person's right to refuse a particular kind of unwanted treatment, but also her interest in dignity, and in determining the character of the memories that will survive long after her death. In recognizing that the State's interests did not outweigh Nancy Cruzan's liberty interest in refusing medical treatment, Cruzan rested not simply on the common-law right to refuse medical treatment, but—at least implicitly—on the even more fundamental right to make this "deeply personal decision," [ ].

* * *

While I agree with the Court that Cruzan does not decide the issue presented by these cases, Cruzan did give recognition, not just to vague, unbridled notions of autonomy, but to the more specific interest in making decisions about how to confront an imminent death. Although there is no absolute right to physician-assisted suicide, Cruzan makes it clear that some individuals who no longer have the option of deciding whether to live or to die because they are already on the threshold of death have a constitutionally protected interest that may outweigh the State's interest in preserving life at all costs. The liberty interest at stake in a case like this differs from, and is stronger than, both the common-law right to refuse medical treatment and the unbridled interest in deciding whether to live or die. It is an interest in deciding how, rather than whether, a critical threshold shall be crossed.

III

The state interests supporting a general rule banning the practice of physician-assisted suicide do not have the same force in all cases. First and foremost of these interests is the " 'unqualified interest in the preservation of human life' "[ ].

... Although as a general matter the State's interest in the contributions each person may make to society outweighs the person's interest in ending her life, this interest does not have the same force for a terminally ill patient faced not with the choice of whether to live, only of how to die. . . .

Similarly, the State's legitimate interests in preventing suicide, protecting the vulnerable from coercion and abuse, and preventing euthanasia are less significant in this context. I agree that the State has a compelling interest in preventing persons from committing suicide because of depression, or coercion by third parties. But the State's legitimate interest in preventing abuse does not apply to an individual who is not victimized by abuse, who is not suffering from depression, and who makes a rational and voluntary decision to seek assistance in dying.

* * *

The final major interest asserted by the State is its interest in preserving the traditional integrity of the medical profession. The fear is that a rule permitting physicians to assist in suicide is inconsistent with the perception that they serve their patients solely as healers. But for some patients, it would be a physician's refusal to dispense medication to ease their suffering and make their death tolerable and dignified that would be inconsistent with the healing role . . . .

. . . I do not . . . foreclose the possibility that an individual plaintiff seeking to hasten her death, or a doctor whose assistance was sought, could prevail in a more particularized challenge. Future cases will determine whether such a challenge may succeed.

IV

* * *

There may be little distinction between the intent of a terminally-ill patient who decides to remove her life-support and one who seeks the assistance of a doctor in ending her life; in both situations, the patient is seeking to hasten a certain, impending death. The doctor's intent might also be the same in prescribing lethal medication as it is in terminating life support. . . .

Thus, although the differences the majority notes in causation and intent between terminating life-support and assisting in suicide support the Court's rejection of the respondents' facial challenge, these distinctions may be inapplicable to particular terminally ill patients and their doctors. Our holding today in Vacco v. Quill that the Equal Protection Clause is not violated by New York's classification, just like our holding in Washington v. Glucksberg that the Washington statute is not invalid on its face, does not foreclose the possibility that some applications of the New York statute may impose an intolerable intrusion on the patient's freedom.

There remains room for vigorous debate about the outcome of particular cases that are not necessarily resolved by the opinions announced today. How such cases may be decided will depend on their specific facts. In my judgment, however, it is clear that the so-called "unqualified interest in the preservation of human life," [ ] is not itself sufficient to outweigh the interest in liberty that may justify the only possible means of preserving a dying patient's dignity and alleviating her intolerable suffering.

JUSTICE SOUTER, concurring in the judgment.

* * *

When the physicians claim that the Washington law deprives them of a right falling within the scope of liberty that the Fourteenth Amendment guarantees against denial without due process of law, they are not claiming some sort of procedural defect in the process through which the statute has been enacted or is administered. Their claim, rather, is that the State has no substantively adequate justification for barring the assistance sought by the patient and sought to be offered by the physician. Thus, we are dealing with a claim to one of those rights sometimes described as rights of substantive due process and sometimes as unenumerated rights, in view of the breadth and indeterminacy of the "due process" serving as the claim's textual basis. The doctors accordingly arouse the skepticism of those who find the Due Process Clause an unduly vague or oxymoronic warrant for judicial review of substantive state law, just as they also invoke two centuries of American constitutional practice in recognizing unenumerated, substantive limits on governmental action. . . .

* * *

[Justice Souter explains that he is adopting Justice Harlan's approach to the Constitutional evaluation and protection of unenumerated rights under the Due Process Clause, as articulated in his dissent in Poe v. Ullman.] My understanding of unenumerated rights in the wake of the Poe dissent and subsequent cases avoids the absolutist failing of many older cases without embracing the opposite pole of equating reasonableness with past practice described at a very specific level. [ ] That understanding begins with a concept of "ordered liberty," [ ] comprising a continuum of rights to be free from "arbitrary impositions and purposeless restraints" [ ].

* * *

This approach calls for a court to assess the relative "weights" or dignities of the contending interests, and to this extent the judicial method is familiar to the common law. Common law method is subject, however, to two important constraints in the hands of a court engaged in substantive due process review. First, such a court is bound to confine the values that it recognizes to those truly deserving constitutional stature, either to those expressed in constitutional text, or those exemplified by "the traditions from which [the Nation] developed," or revealed by contrast with "the traditions from which it broke." [ ]

The second constraint, again, simply reflects the fact that constitutional review, not judicial lawmaking, is a court's business here. The weighing or valuing of contending interests in this sphere is only the first step, forming the basis for determining whether the statute in question falls inside or outside the zone of what is reasonable in the way it resolves the conflict between the interests of state and individual.

* * *

The State has put forward several interests to justify the Washington law as applied to physicians treating terminally ill patients, even those competent to make responsible choices: protecting life generally [ ], discouraging suicide even if knowing and voluntary [ ], and protecting terminally ill patients from involuntary suicide and euthanasia, both voluntary and nonvoluntary [ ].

It is not necessary to discuss the exact strengths of the first two claims of justification in the present circumstances, for the third is dispositive for me. . . . [Justice Souter explains why the Washington State Legislature, on the basis of information now available, could have reasonably decided that a statute forbidding assisting suicide might protect terminally ill patients.]

* * *

The Court should accordingly stay its hand to allow reasonable legislative consideration. While I do not decide for all time that respondents' claim should not be recognized, I acknowledge the legislative institutional competence as the better one to deal with that claim at this time.

JUSTICE BREYER, concurring in the judgments [in both Glucksberg and Vacco].

I believe that Justice O'Connor's views, which I share, have greater legal significance than the Court's opinion suggests. I join her separate opinion, except insofar as it joins the majority. . . .

I agree with the Court in Vacco v. Quill [ ] that the articulated state interests justify the distinction drawn between physician assisted suicide and withdrawal of life-support. I also agree with the Court that the critical question in both of the cases before us is whether "the 'liberty' specially protected by the Due Process Clause includes a right" of the sort that the respondents assert. [ ] I do not agree, however, with the Court's formulation of that claimed "liberty" interest. The Court describes it as a "right to commit suicide with another's assistance." [ ] But I would not reject the respondents' claim without considering a different formulation, for which our legal tradition may provide greater support. That formulation would use words roughly like a "right to die with dignity." But irrespective of the exact words used, at its core would lie personal control over the manner of death, professional medical assistance, and the avoidance of unnecessary and severe physical suffering—combined.

* * *

I do not believe, however, that this Court need or now should decide whether or a not . . . [a right to die with dignity] is "fundamental." That is because, in my view, the avoidance of severe physical pain (connected with death) would have to comprise an essential part of any successful claim and because . . . the laws before us do not force a dying person to undergo that kind of pain. [ ] Rather, the laws of New York and of Washington do not prohibit doctors from providing patients with drugs sufficient to control pain despite the risk that those drugs themselves will kill. [ ] And under these circumstances the laws of New York and Washington would overcome any remaining significant interests and would be justified, regardless.

* * *

Were the legal circumstances different—for example, were state law to prevent the provision of palliative care, including the administration of drugs as needed to avoid pain at the end of life—then the law's impact upon serious and otherwise unavoidable physical pain (and accompanying death) would be more directly at issue. And as JUSTICE O'CONNOR suggests, the Court might have to revisit its conclusions in these cases.

* * *

VACCO V. QUILL
Supreme Court of the United States, 1997.
521 U.S. 793.

CHIEF JUSTICE REHNQUIST delivered the opinion of the Court.

In New York, as in most States, it is a crime to aid another to commit or attempt suicide, but patients may refuse even lifesaving medical treatment. The question presented by this case is whether New York's prohibition on assisting suicide therefore violates the Equal Protection Clause of the Fourteenth Amendment. We hold that it does not.

. . . Respondents, and three gravely ill patients who have since died, sued the State's Attorney General in the United States District Court. They urged that because New York permits a competent person to refuse life-sustaining medical treatment, and because the refusal of such treatment is "essentially the same thing" as physician-assisted suicide, New York's assisted-suicide ban violates the Equal Protection Clause. [ ]

The District Court disagreed . . . .

The Court of Appeals for the Second Circuit reversed. [ ] The court determined that, despite the assisted-suicide ban's apparent general applicability, "New York law does not treat equally all competent persons who are in the final stages of fatal illness and wish to hasten their deaths," because "those in the final stages of terminal illness who are on life-support systems are allowed to hasten their deaths by directing the removal of such systems; but those who are similarly situated, except for the previous attachment of life-sustaining equipment, are not allowed to hasten death by self-administering prescribed drugs." [ ] The Court of Appeals then examined whether this supposed unequal treatment was rationally related to any legitimate state interests, and concluded that "to the extent that [New York's statutes] prohibit a physician from prescribing medications to be self-administered by a mentally competent, terminally-ill person in the final stages of his terminal illness, they are not rationally related to any legitimate state interest." [ ] We granted certiorari [ ] and now reverse.

The Equal Protection Clause commands that no State shall "deny to any person within its jurisdiction the equal protection of the laws." This provision creates no substantive rights. [ ] Instead, it embodies a general rule that States must treat like cases alike but may treat unlike cases accordingly. [ ] If a legislative classification or distinction "neither burdens a fundamental right nor targets a suspect class, we will uphold [it] so long as it bears a rational relation to some legitimate end." [ ]

New York's statutes outlawing assisting suicide affect and address matters of profound significance to all New Yorkers alike. They neither infringe fundamental rights nor involve suspect classifications. [ ] These laws are therefore entitled to a "strong presumption of validity." [ ]

On their faces, neither New York's ban on assisting suicide nor its statutes permitting patients to refuse medical treatment treat anyone differently than anyone else or draw any distinctions between persons. Everyone, regardless of physical condition, is entitled, if competent, to refuse unwanted lifesaving medical treatment; no one is permitted to assist a suicide. Generally speaking, laws that apply evenhandedly to all "unquestionably comply" with the Equal Protection Clause. [ ]

The Court of Appeals, however, concluded that some terminally ill people—those who are on life-support systems—are treated differently than those who are not, in that the former may "hasten death" by ending treatment, but the latter may not "hasten death" through physician-assisted suicide. [ ] This conclusion depends on the submission that ending or refusing lifesaving medical treatment "is nothing more nor less than assisted suicide." [ ] Unlike the Court of Appeals, we think the distinction between assisting suicide and withdrawing life-sustaining treatment, a distinction widely recognized and endorsed in the medical profession and in our legal traditions, is both important and logical; it is certainly rational. [ ]

The distinction comports with fundamental legal principles of causation and intent. First, when a patient refuses life-sustaining medical treatment, he dies from an underlying fatal disease or pathology; but if a patient ingests lethal medication prescribed by a physician, he is killed by that medication. [ ]

Furthermore, a physician who withdraws, or honors a patient's refusal to begin, life-sustaining medical treatment purposefully intends, or may so intend, only to respect his patient's wishes and "to cease doing useless and futile or degrading things to the patient when [the patient] no longer stands to benefit from them." [ ] The same is true when a doctor provides aggressive palliative care; in some cases, painkilling drugs may hasten a patient's death, but the physician's purpose and intent is, or may be, only to ease his patient's pain. A doctor who assists a suicide, however, "must, necessarily and indubitably, intend primarily that the patient be made dead." [ ] Similarly, a patient who commits suicide with a doctor's aid necessarily has the specific intent to end his or her own life, while a patient who refuses or discontinues treatment might not. [ ]

The law has long used actors' intent or purpose to distinguish between two acts that may have the same result. [ ] Put differently, the law distinguishes actions taken "because of" a given end from actions taken "in spite of" their unintended but foreseen consequences. [ ]

Given these general principles, it is not surprising that many courts, including New York courts, have carefully distinguished refusing life-sustaining treatment from suicide. . . .

Similarly, the overwhelming majority of state legislatures have drawn a clear line between assisting suicide and withdrawing or permitting the refusal of unwanted lifesaving medical treatment by prohibiting the former and permitting the latter. [ ] And "nearly all states expressly disapprove of suicide and assisted suicide either in statutes dealing with durable powers of attorney in health-care situations, or in 'living will' statutes." [ ] Thus, even as the States move to protect and promote patients' dignity at the end of life, they remain opposed to physician-assisted suicide.

* * *

This Court has also recognized, at least implicitly, the distinction between letting a patient die and making that patient die. In Cruzan [ ] we concluded that "the principle that a competent person has a constitutionally protected liberty interest in refusing unwanted medical treatment may be inferred from our prior decisions," and we assumed the existence of such a right for purposes of that case [ ]. But our assumption of a right to refuse treatment was grounded not, as the Court of Appeals supposed, on the proposition that patients have a general and abstract "right to hasten death," [ ] but on well established, traditional rights to bodily integrity and freedom from unwanted touching [ ]. In fact, we observed that "the majority of States in this country have laws imposing criminal penalties on one who assists another to commit suicide." [ ]

Cruzan therefore provides no support for the notion that refusing life-sustaining medical treatment is "nothing more nor less than suicide."

For all these reasons, we disagree with respondents' claim that the distinction between refusing lifesaving medical treatment and assisted suicide is "arbitrary" and "irrational.[11] Granted, in some cases, the line between the two may not be clear, but certainty is not required, even were it possible. Logic and contemporary practice support New York's judgment that the two acts are different, and New York may therefore, consistent with the Constitution, treat them differently. By permitting everyone to refuse unwanted medical treatment while prohibiting anyone from assisting a suicide, New York law follows a longstanding and rational distinction.

New York's reasons for recognizing and acting on this distinction—including prohibiting intentional killing and preserving life; preventing suicide; maintaining physicians' role as their patients' healers; protecting vulnerable people from indifference, prejudice, and psychological and financial pressure to end their lives; and avoiding a possible slide towards euthanasia—are discussed in greater detail in our opinion in Glucksberg, ante. These valid and important public interests easily satisfy the constitutional requirement that a legislative classification bear a rational relation to some legitimate end.

The judgment of the Court of Appeals is reversed.

* * *

JUSTICE SOUTER, concurring in the judgment.

Even though I do not conclude that assisted suicide is a fundamental right entitled to recognition at this time, I accord the claims raised by the patients and physicians in this case and Washington v. Glucksberg a high degree of importance, requiring a commensurate justification. [ ] The reasons that lead me to conclude in Glucksberg that the prohibition on assisted suicide is not arbitrary under the due process standard also support the distinction between assistance to suicide, which is banned, and practices such as termination of artificial life support and death-hastening pain medication, which are permitted. I accordingly concur in the judgment of the Court.

---

[11] Respondents also argue that the State irrationally distinguishes between physician-assisted suicide and "terminal sedation," a process respondents characterize as "inducing barbiturate coma and then starving the person to death." [ ] Petitioners insist, however, that " 'although proponents of physician-assisted suicide and euthanasia contend that terminal sedation is covert physician-assisted suicide or euthanasia, the concept of sedating pharmacotherapy is based on informed consent and the principle of double effect.' "[ ] Just as a State may prohibit assisting suicide while permitting patients to refuse unwanted lifesaving treatment, it may permit palliative care related to that refusal, which may have the foreseen but unintended "double effect" of hastening the patient's death. [ ]

* * *

NOTES AND QUESTIONS

1. *Defining the Interest at Stake.* In *Glucksberg*, proponents of the right to medically assisted death asserted a constitutionally protected liberty interest that extended to a personal choice by a competent, terminally ill adult to choose the time and manner of their death with the aid of a physician. Writing for the Court, Justice Rehnquist reframed the asserted interest as a more generalized right to assisted suicide. In her concurrence, Justice O'Connor agreed that there is no generalized right to commit suicide but stated that there was no need to address "the narrower question of whether a mentally competent person who is experiencing great suffering has a constitutionally cognizable interest in controlling the circumstances of his or her imminent death." What is the legal and ethical significance of these different framings of the interest at issue?

2. *Discerning "Legal Tradition."* In *Glucksberg*, the Court held that the asserted right was not a fundamental liberty interest protected by the Due Process Clause because it was not a right deeply rooted in U.S. history, legal traditions, and practices. To what extent have states (like Oregon) changed their laws with respect to medically assisted death such that the "legal tradition" may have changed since *Glucksberg*? What kind of change would be required to conclude that the liberty interest has become fundamental? In other cases, the Court has been willing to recognize constitutionally protected interests in the absence of a tradition of legal protection. For a discussion of such cases before and after *Glucksberg*, see Kenji Yoshino, A New Birth of Freedom? Obergefell v. Hodges, 129 Harv. L. Rev. 147 (2015); Erwin Chemerinsky, Washington v. Glucksberg Was Tragically Wrong, 106 Mich. L. Rev. 1501 (2008).

3. *Contentious Debate.* These cases generated many highly emotional responses. Although the Supreme Court's unanimous decision brought a semblance of propriety back to the discussion of these issues, supporters and opponents of medically assisted dying continue to attack the arguments of their opponents—and, as in the case of the abortion debate—they continue to attack their opponents, too. Some of the commentary on the Ninth Circuit opinions was especially personal. Judge Reinhardt (who wrote the primary decision finding the Washington law to be unconstitutional) was roundly criticized for his ACLU connections, which, some said, made it impossible for him to fairly decide the case. On the other hand, Judge Noonan (who would have upheld the statute for the first panel) had been criticized for his Right to Life connections and his Catholic faith which, others argued, made it impossible for him to be impartial.

Should judges recuse themselves from cases involving these difficult and controversial bioethics issues if they have deeply held personal beliefs about the underlying practice—here, medically assisted dying? Does it make a difference if they were members (or officers, or high-ranking employees) of organizations which have taken explicit positions on the underlying issues? On the particular case in litigation? Should they recuse themselves if the issue is one on which the religion to which they subscribe has taken a formal position? Should Catholic judges recuse themselves from abortion and medically assisted dying cases? Should judges who belong to the United Church of Christ (which has been strongly pro-choice for decades) recuse themselves from abortion cases? Should the member of a congregation whose rabbi helped organize a march to support access to contraception through the ACA recuse herself from all contraception rights cases? Is their obligation any different from the obligation of a judge who is a dedicated ACLU (or American Family Association or Republican Party) member and who confronts a case upon which the ACLU (or the American Family Association, or the Republican Party) has taken a firm position?

4.   *Determining Legislative Intent of Old Laws.* Judge Calabresi concurred in the Second Circuit decision in *Vacco* but on entirely different grounds. Depending on the theory of statutory construction that he had explained 15 years earlier in his text, A Common Law for the Age of Statutes (1982), he concluded that the history of the New York manslaughter statute suggested that there was no reason to believe that its framers ever intended it to apply to cases of competent terminally ill patients seeking aid in dying from physicians. Still, as he pointed out, "neither *Cruzan,* nor *Casey,* nor the language of our Constitution, nor our constitutional tradition clearly makes these laws invalid."

So, what should the court do with a "highly suspect" but "not clearly invalid" statute that may no longer serve the purposes for which it was originally promulgated? The answer, according to Judge Calabresi, is the "constitutional remand." Under this approach, a court would not determine the validity of the state law in the absence of clear and current statements by the people or elected officials of the state, or, if the court made a determination, it would leave open the possibility of reconsideration. Thus, Judge Calabresi finds the New York statute unconstitutional, but he "takes no position" on whether verbatim identical statutes would be constitutional "were New York to reenact them while articulating the reasons for the distinctions it makes. . . . " Is this a reasonable way to deal with ancient statutes effectively criminalizing medically assisted dying? Is this argument still available to those challenging state statutes that forbid assisting suicide? For a history of criminal laws related to medically assisted dying, see Alan Meisel, A History of the Law of Assisted Dying in the United States, 73(1) SMU L. Rev. 119 (2020).

5.   *Intersection with Criminal Law.* Medically assisted dying may constitute murder, manslaughter, some other form of homicide, or no crime at all, depending on the language of the state statute and the nature of the physician's act. While most states criminalize assisting suicide, it is not always easy to determine what conduct is prohibited by those statutes. Consider one representative statute:

Every person who deliberately aids, or advises, or encourages another to commit suicide, is guilty of a felony. Cal. Pen. Code § 401.

Would this statute apply to a physician who clamps a feeding tube? To a physician who withholds antibiotics? To a physician who prescribes morphine to a patient in persistent pain, provides enough morphine to constitute a lethal dose? To a physician who prescribes that same morphine and tells the patient what would constitute a lethal dose? To those who publish instructions on how to hasten death for the use of those who are terminally ill or in excruciating pain? To those who make generally available information about how to commit suicide at home? See, e.g., State v. Melchert-Dinkel, 844 N.W.2d 13 (Minn. 2014) (rejecting application of statute prohibiting "encouraging" or "advising"—but not "assisting"—another in committing suicide on free speech grounds); Final Exit Network v. State, 290 Ga. 508, 722 S.E.2d 722 (2012) (rejecting application of statute prohibiting offers of assistance of another in committing suicide on free speech grounds); McCollum v. CBS, Inc., 202 Cal. App. 3d 989, 249 Cal. Rptr. 187 (1988) (rejecting on free speech grounds application of the statute to those who play rock music with lyrics that suggest that suicide is acceptable).

6. *Disability Critiques.* As discussed in Chapter 5, the disability rights movement has become a key player in controversies over medically assisted death and other health care decisions with life and death consequences. A 2019 report by the National Council on Disability documented a host of concerns with the legalization of medically assisted death, asserting that "some people's lives, particularly those of people with disabilities, will be ended without their fully informed and free consent, through mistakes, abuse, insufficient knowledge, and the unjust lack of better options." See Nat'l Council on Disability, Medical Futility and Disability Bias (2019) at p. 14–15. The report describes a lack of adequate safeguards and oversight, as well as the impact of individual and structural biases that distort decision-making by and for people with disabilities in this context.

The report also notes concerns that people of color are particularly at risk of being harmed by these laws. See also Mary Crossley, Ending-Life Decisions: Some Disability Perspectives, 33 Ga. St. Univ. L. Rev. 893 (2017) (drawing parallels with concerns about racial justice and policing that have been expressed by the African-American community). Of course, we should not assume that all people with disabilities, or people who are members of any other group, hold the same views on a given issue, including medically assisted death.

7. *Gender Considerations.* Might women, specifically, be put at risk in a society that permits medically assisted dying? That argument is made by Susan Wolf, who regularly has argued that women's requests should be better respected by the health care system and that requests to remove life-sustaining treatment should be heeded.

As I have argued, there is a strong right to be free of unwanted bodily invasion. Indeed, for women, a long history of being harmed specifically through unwanted bodily invasion such as rape presents particularly compelling reasons for honoring a woman's refusal of invasion and effort to maintain bodily intactness. When it comes to the question of whether women's suicides should be aided, however, or whether women should be actively killed, there is no right to command physician assistance, the dangers of permitting assistance are immense, and the history of women's subordination cuts the other way. Women have historically been seen as fit objects for bodily invasion, self-sacrifice, and death at the hands of others. The task before us is to challenge all three.

Certainly some women, including some feminists, will see this problem differently. That may be especially true of women who feel in control of their lives, are less subject to subordination by age or race or wealth, and seek yet another option to add to their many. I am not arguing that women should lose control of their lives and selves. Instead, I am arguing that when women request to be put to death or ask help in taking their own lives, they become part of a broader social dynamic of which we have properly learned to be extremely wary. These are fatal practices. We can no longer ignore questions of gender or insights of feminist argument.

Susan Wolf, Gender, Feminism and Death: Physician Assisted Suicide and Euthanasia, in S. Wolf, Feminism and Bioethics: Beyond Reproduction 308 (1996). For an account of Professor Wolf's thoughtful approach to death and dying, see Daniel Bergner, Death in the Family, N.Y. Times Magazine (December 2, 2007) (reporting that 72% of the 75 individuals assisted by euthanasia advocate Kevorkian were women).

Susan Wolf also wrote a moving account of her own father's death, including her own response when he inquired whether it could be "accelerated" beyond withdrawing treatment. Her answer to her father was no, and she explains why in her essay; but she doesn't romanticize his dying and the great difficulties he and his family faced in getting the end-of-life care he needed. Confronting Physician-Assisted Suicide and Euthanasia: My Father's Death, 38 Hastings Ctr. Rept. 23 (2008).

8. *Medical Organizations.* Organized medical groups generally oppose any medical participation in euthanasia or assisted death. As to euthanasia, the AMA Council on Ethical and Judicial Affairs provides:

Code of Medical Ethics Opinion 5.8, Euthanasia (2013)

Euthanasia is the administration of a lethal agent by another person to a patient for the purpose of relieving the patient's intolerable and incurable suffering.

It is understandable, though tragic, that some patients in extreme duress—such as those suffering from a terminal, painful, debilitating illness—may come to decide that death is preferable to life.

However, permitting physicians to engage in euthanasia would ultimately cause more harm than good.

Euthanasia is fundamentally incompatible with the physician's role as healer, would be difficult or impossible to control, and would pose serious societal risks. Euthanasia could readily be extended to incompetent patients and other vulnerable populations.

The involvement of physicians in euthanasia heightens the significance of its ethical prohibition. The physician who performs euthanasia assumes unique responsibility for the act of ending the patient's life.

Instead of engaging in euthanasia, physicians must aggressively respond to the needs of patients at the end of life. Physicians:

(a)   Should not abandon a patient once it is determined that cure is impossible.

(b)   Must respect patient autonomy.

(c)   Must provide good communication and emotional support.

(d)   Must provide appropriate comfort care and adequate pain control.

As to Physician Assisted Suicide, the Council has adopted this position:

Code of Medical Ethics Opinion 5.7, Physician Assisted Suicide (2016)

Physician-assisted suicide occurs when a physician facilitates a patient's death by providing the necessary means and/or information to enable the patient to perform the life-ending act (eg, the physician provides sleeping pills and information about the lethal dose, while aware that the patient may commit suicide).

It is understandable, though tragic, that some patients in extreme duress—such as those suffering from a terminal, painful, debilitating illness—may come to decide that death is preferable to life. However, allowing physicians to participate in assisted suicide would cause more harm than good.

Physician-assisted suicide is fundamentally incompatible with the physician's role as healer, would be difficult or impossible to control, and would pose serious societal risks.

Instead of engaging in assisted suicide, physicians must aggressively respond to the needs of patients at the end of life. Physicians:

(a)   Should not abandon a patient once it is determined that cure is impossible.

(b)   Must respect patient autonomy.

(c)   Must provide good communication and emotional support.

(d)   Must provide appropriate comfort care and adequate pain control.

Does the AMA oppose both of these kinds of medically assisted dying because they are equally morally reprehensible, or because they are too morally complicated? See Christine Cassel and Diane Meier, Morals and Moralism in the Debate Over Euthanasia and Assisted Suicide, 323 NEJM 750, 751 (1990) (opining that a strong prohibition against medically assisted death may limit ethical consideration of complex individual cases).

9.   *Palliative Sedation.* In a footnote in *Vacco*, Justice Rehnquist raises the issue of the legal status of "terminal sedation," which is sometimes described as "palliative sedation" because it is employed to provide palliation both in patients who are quickly approaching death and in those who are suffering but not yet otherwise terminal. Patients whose suffering cannot be ameliorated in any other way can be sedated with enough medication so that they are put into a medication-induced coma. This sedation may be accompanied by the withdrawal of other forms of treatment, including artificial nutrition and hydration. This withdrawal may (and, with the withdrawal of nutrition and hydration, will) lead to the patient's death. Justice Rehnquist suggests that terminal sedation, which has not given rise to the ethical objections that have come with medically assisted dying, is distinguishable from an affirmative act intended to bring about death, applying the principle of "double effect." Do you agree?

Under *Vacco*, whether palliative sedation is treated like the withdrawal of life-sustaining treatment or like an affirmative act causing death makes all the legal difference. As an ethical and legal matter, should terminal sedation be treated like the withdrawal of treatment or like medically assisted dying? For differing views on this question, see Thomas D. Riisfeldt, Weakening the Ethical Distinction between Euthanasia, Palliative Opioid Use and Palliative Sedation, 45 J. Med. Ethics 125 (2019) and responses. See also Paulo Rodrigues et al., Palliative Sedation for Existential Suffering: A Systematic Review of Argument-Based Ethics Literature, 55(6) J. Pain & Symptom Mgmt. 1577 (2018).

In 2008, over a decade after *Glucksberg* was decided, the AMA adopted a policy providing that it is ethically appropriate for physicians to offer palliative sedation, which the AMA called "sedation to unconsciousness in end-of-life care" at least to some terminally ill patients. The relevant opinion was amended in 2016:

Code of Medical Ethics Opinion 5.6, Sedation to Unconsciousness in End-of-Life Care (2016)

The duty to relieve pain and suffering is central to the physician's role as healer and is an obligation physicians have to their patients. When a terminally ill patient experiences severe pain or other distressing clinical symptoms that do not respond to aggressive, symptom-specific palliation it can be appropriate to offer sedation to unconsciousness as an intervention of last resort.

Sedation to unconsciousness must never be used to intentionally cause a patient's death.

When considering whether to offer palliative sedation to unconsciousness, physicians should:

(a) Restrict palliative sedation to unconsciousness to patients in the final stages of terminal illness.

(b) Consult with a multi-disciplinary team (if available), including an expert in the field of palliative care, to ensure that symptom-specific treatments have been sufficiently employed and that palliative sedation to unconsciousness is now the most appropriate course of treatment.

(c) Document the rationale for all symptom management interventions in the medical record.

(d) Obtain the informed consent of the patient (or the authorized surrogate when the patient lacks decision-making capacity).

(e) Discuss with the patient (or surrogate) the plan of care relative to:

(1) Degree and length of sedation

(2) Specific expectations for continuing, withdrawing, or withholding future life-sustaining treatments

(f) Monitor care once palliative sedation to unconsciousness is initiated.

The 2016 revision of this Ethical Opinion made a couple of notable changes to the earlier version of the opinion. First, the following paragraph, part of the 2008 version of the Ethical Opinion, was omitted from the more recent version:

Palliative sedation is not an appropriate response to suffering that is primarily existential, defined as the experience of agony and distress that may arise from such issues as death anxiety, isolation and loss of control. Existential suffering is better addressed by other interventions. For example, palliative sedation is not the way to address suffering created by social isolation and loneliness; such suffering should be addressed by providing the patient with needed social support.

Does the omission suggest that sedation to unconsciousness is appropriate for existential suffering under some circumstances? Is it? Can all suffering (except physical pain) be treated through the provision of social support? Should palliative sedation be reserved for those with physical, not psychic, pain? See A. de Graeff and M. Dean, Palliative Sedation Therapy in the Last Weeks of Life: A Literature Review and Recommendations for Standards, 10 J. Palliative Med. 67 (2007); Allysa L. Ciancio et al., The Use of Palliative Sedation to Treat Existential Suffering: A Scoping Review on Practices, Ethical Considerations, and Guidelines, 35 J. Palliative Care 13 (2020). Some studies suggest that palliative sedation does not actually shorten the lives of cancer patients. M. Maltoni et al., Palliative Sedation in End-of-Life Care and Survival: A Systematic Review, 30 J. Clin. Oncology 1378 (2012); So-Jung Park, et al., Association between Continuous Deep Sedation and Survival Time in Terminally Ill Cancer Patients, 29 Supportive Care in Cancer 525 (2020).

Second, in 2016 the following portion of the Ethical Opinion was removed:

> When symptoms cannot be diminished through all other means of palliation, including symptom-specific treatments, it is the ethical obligation of a physician to offer palliative sedation to unconsciousness as an option for the relief of intractable symptoms.

Why was this sentence removed? Is it no longer an ethical obligation to offer palliative sedation to a patient? Should there be such an obligation if the physician believes that it is unethical to provide that form of care?

10. *Voluntary Stopping of Eating and Drinking.* Voluntary stopping of eating and drinking (VSED) has attracted increasing attention as a widely accessible alternative to medically assisted dying. As Chapter 5 established, every state allows a competent adult patient to refuse or discontinue life-sustaining medical care, and a physician to assist in implementing that decision, at least under most circumstances. Further, courts generally have concluded that the termination of artificial nutrition and hydration procedures is no different from the termination of other forms of mechanical support. But what if the patient is not receiving *artificial* nutrition and hydration or other life-sustaining care? Many assert a competent adult patient may opt for VSED as a method of hastening the patient's own death, and a health care provider may provide support to the patient, under such circumstances. For a discussion of clinical, ethical and legal issues surrounding VSED, see Timothy E. Quill et al, Voluntarily Stopping Eating and Drinking: A Compassionate, Widely Available Option for Hastening Death (2021).

III. STATE LEGISLATION AND LITIGATION TO SUPPORT MEDICALLY ASSISTED DYING

The debate over the proper role of physicians in assisting their patients in death has been carried on through legislative and citizen initiative processes as well as through litigation. "Death with Dignity" initiatives were narrowly defeated in California in 1991 and in Washington in 1992. However, Oregon's "Death with Dignity" initiative was approved by voters in the November 1994 election, and it thus became part of the statutory law of Oregon.

DEATH WITH DIGNITY ACT
Or. Rev. Stat. §§ 127.800–.897.

127.800. Definitions.

The following words and phrases, whenever used in ORS 127.800 to 127.897, have the following meanings:

(1) "Adult" means an individual who is 18 years of age or older.

(2) "Attending physician" means the physician who has primary responsibility for the care of the patient and treatment of the patient's terminal disease.

(3) "Capable" means that in the opinion of a court or in the opinion of the patient's attending physician or consulting physician, psychiatrist or psychologist, a patient has the ability to make and communicate health care decisions to health care providers, including communication through persons familiar with the patient's manner of communicating if those persons are available.

(4) "Consulting physician" means a physician who is qualified by specialty or experience to make a professional diagnosis and prognosis regarding the patient's disease.

(5) "Counseling" means one or more consultations as necessary between a state licensed psychiatrist or psychologist and a patient for the purpose of determining that the patient is capable and not suffering from a psychiatric or psychological disorder or depression causing impaired judgment.

(6) "Health care provider" means a person licensed, certified or otherwise authorized or permitted by the law of this state to administer health care or dispense medication in the ordinary course of business or practice of a profession, and includes a health care facility.

(7) "Informed decision" means a decision by a qualified patient, to request and obtain a prescription to end his or her life in a humane and dignified manner, that is based on an appreciation of the relevant facts and after being fully informed by the attending physician of:

(a) His or her medical diagnosis;

(b) His or her prognosis;

(c) The potential risks associated with taking the medication to be prescribed;

(d) The probable result of taking the medication to be prescribed; and

(e) The feasible alternatives, including, but not limited to, comfort care, hospice care and pain control.

(8) "Medically confirmed" means the medical opinion of the attending physician has been confirmed by a consulting physician who has examined the patient and the patient's relevant medical records.

(9) "Patient" means a person who is under the care of a physician.

(10) "Physician" means a doctor licensed to practice medicine * * *

(11) "Qualified patient" means a capable adult who is a resident of Oregon and has satisfied the requirements of ORS 127.800 to 127.897 in

order to obtain a prescription for medication to end his or her life in a humane and dignified manner.

(12) "Terminal disease" means an incurable and irreversible disease that has been medically confirmed and will, within reasonable medical judgment, produce death within six months.

127.805.Who may initiate a written request for medication.

(1) An adult who is capable, is a resident of Oregon, and has been determined by the attending physician and consulting physician to be suffering from a terminal disease, and who has voluntarily expressed his or her wish to die, may make a written request for medication for the purpose of ending his or her life in a humane and dignified manner in accordance with ORS 127.800 to 127.897.

(2) No person shall qualify under the provisions of ORS 127.800 to 127.897 solely because of age or disability.

127.810.Form of the written request.

(1) A valid request for medication under ORS 127.800 to 127.897 shall be in substantially the form described in ORS 127.897, signed and dated by the patient and witnessed by at least two individuals who, in the presence of the patient, attest that to the best of their knowledge and belief the patient is capable, acting voluntarily, and is not being coerced to sign the request.

(2) One of the witnesses shall be a person who is not:

(a) A relative of the patient by blood, marriage or adoption;

(b) A person who at the time the request is signed would be entitled to any portion of the estate of the qualified patient upon death under any will or by operation of law; or

(c) An owner, operator or employee of a health care facility where the qualified patient is receiving medical treatment or is a resident.

(3) The patient's attending physician at the time the request is signed shall not be a witness.

(4) If the patient is a patient in a long term care facility at the time the written request is made, one of the witnesses shall be an individual designated by the facility and having the qualifications specified by the Department of Human Services by rule.

127.815.Attending physician responsibilities.

(1) The attending physician shall:

(a) Make the initial determination of whether a patient has a terminal disease, is capable, and has made the request voluntarily;

(b) Request that the patient demonstrate Oregon residency pursuant to ORS 127.860;

(c) To ensure that the patient is making an informed decision, inform the patient of:

(A) His or her medical diagnosis;

(B) His or her prognosis;

(C) The potential risks associated with taking the medication to be prescribed;

(D) The probable result of taking the medication to be prescribed; and

(E) The feasible alternatives, including, but not limited to, comfort care, hospice care and pain control;

(d) Refer the patient to a consulting physician for medical confirmation of the diagnosis, and for a determination that the patient is capable and acting voluntarily;

(e) Refer the patient for counseling if appropriate pursuant to ORS 127.825;

(f) Recommend that the patient notify next of kin;

(g) Counsel the patient about the importance of having another person present when the patient takes the medication prescribed pursuant to ORS 127.800 to 127.897 and of not taking the medication in a public place;

(h) Inform the patient that he or she has an opportunity to rescind the request at any time and in any manner, and offer the patient an opportunity to rescind at the time the patient makes the patient's second oral request pursuant to ORS 127.840;

(i) Verify, immediately prior to writing the prescription for medication under ORS 127.800 to 127.897, that the patient is making an informed decision;

(j) Fulfill the medical record documentation requirements of ORS 127.855;

(k) Ensure that all appropriate steps are carried out in accordance with ORS 127.800 to 127.897 prior to writing a prescription for medication to enable a qualified patient to end his or her life in a humane and dignified manner; and

(A) Dispense medications directly . . . or [(B) through a pharmacist].

(2) Notwithstanding any other provision of law, the attending physician may sign the patient's report of death.

127.820. Consulting physician confirmation.

Before a patient is qualified under ORS 127.800 to 127.897, a consulting physician shall examine the patient and his or her relevant medical records and confirm, in writing, the attending physician's diagnosis that the patient is suffering from a terminal disease, and verify that the patient is capable, is acting voluntarily and has made an informed decision.

127.825. Counseling referral.

If in the opinion of the attending physician or the consulting physician a patient may be suffering from a psychiatric or psychological disorder or depression causing impaired judgment, either physician shall refer the patient for counseling. No medication to end a patient's life in a humane and dignified manner shall be prescribed until the person performing the counseling determines that the patient is not suffering from a psychiatric or psychological disorder or depression causing impaired judgment.

127.830. Informed decision.

No person shall receive a prescription for medication to end his or her life in a humane and dignified manner unless he or she has made an informed decision as defined in ORS 127.800 (7). Immediately prior to writing a prescription for medication under ORS 127.800 to 127.897, the attending physician shall verify that the patient is making an informed decision.

127.835. Family notification.

The attending physician shall recommend that the patient notify the next of kin of his or her request for medication pursuant to ORS 127.800 to 127.897. A patient who declines or is unable to notify next of kin shall not have his or her request denied for that reason.

127.840. Written and oral requests.

(1) In order to receive a prescription for medication to end his or her life in a humane and dignified manner, a qualified patient shall have made an oral request and a written request, and reiterate the oral request to his or her attending physician no less than 15 days after making the initial oral request.

(2) Notwithstanding subsection (1) of this section, if the qualified patient's attending physician has medically confirmed that the qualified patient will, within reasonable medical judgment, die within 15 days after making the initial oral request under this section, the qualified patient may reiterate the oral request to his or her attending physician at any time after making the initial oral request.

(3) At the time the qualified patient makes his or her second oral request, the attending physician shall offer the patient an opportunity to rescind the request.

127.845. Right to rescind request.

A patient may rescind his or her request at any time and in any manner without regard to his or her mental state. No prescription for medication under ORS 127.800 to 127.897 may be written without the attending physician offering the qualified patient an opportunity to rescind the request.

127.850. Waiting periods.

(1) No less than 15 days shall elapse between the patient's initial oral request and the writing of a prescription under ORS 127.800 to 127.897. No less than 48 hours shall elapse between the patient's written request and the writing of a prescription under ORS 127.800 to 127.897.

(2) Notwithstanding subsection (1) of this section, if the qualified patient's attending physician has medically confirmed that the qualified patient will, within reasonable medical judgment, die before the expiration of at least one of the waiting periods described in subsection (1) of this section, the prescription for medication under ORS 127.800 to 127.897 may be written at any time following the later of the qualified patient's written request or second oral request under ORS 127.840.

* * *

127.860. Residency requirement.

Only requests made by Oregon residents under ORS 127.800 to 127.897 shall be granted. Factors demonstrating Oregon residency include but are not limited to [being licensed to drive, registering to vote, owning property, and paying taxes in Oregon.]

* * *

127.880. Construction of Act.

Nothing in ORS 127.800 to 127.897 shall be construed to authorize a physician or any other person to end a patient's life by lethal injection, mercy killing or active euthanasia. Actions taken in accordance with ORS 127.800 to 127.897 shall not, for any purpose, constitute suicide, assisted suicide, mercy killing or homicide, under the law.

* * *

127.897. Form of the request.

A request for a medication as authorized by ORS 127.800 to 127.897 shall be in substantially the following form:

REQUEST FOR MEDICATION TO END MY LIFE
IN A HUMANE AND DIGNIFIED MANNER

I, _____, am an adult of sound mind.

I am suffering from _____, which my attending physician has determined is a terminal disease and which has been medically confirmed by a consulting physician.

I have been fully informed of my diagnosis, prognosis, the nature of medication to be prescribed and potential associated risks, the expected result, and the feasible alternatives, including comfort care, hospice care and pain control.

I request that my attending physician prescribe medication that will end my life in a humane and dignified manner.

INITIAL ONE:

_____ I have informed my family of my decision and taken their opinions into consideration.

_____ I have decided not to inform my family of my decision.

_____ I have no family to inform of my decision.

I understand that I have the right to rescind this request at any time.

I understand the full import of this request and I expect to die when I take the medication to be prescribed. I further understand that although most deaths occur within three hours, my death may take longer and my physician has counseled me about this possibility.

I make this request voluntarily and without reservation, and I accept full moral responsibility for my actions.

[SIGNATURE LINE; WITNESS
LINES]

* * *

NOTE: OREGON DEATH WITH DIGNITY
ACT, 2020 ANNUAL REPORT

Under Oregon's Death with Dignity Act (DWDA), the Oregon Public Health Division is required to collect information on compliance and to issue an annual report. All of the annual reports are available at the Oregon Department of Health website. The key findings from the 2020 report, released February 2021, are summarized below.

- Since the law was passed in 1997, a total of 2,895 people have had DWDA prescriptions written and 1,905 people have died from ingesting medications prescribed under the DWDA. Both the number of prescriptions and the number of deaths have risen nearly every year since 1998.

- In 2020, 370 people received DWDA prescriptions, a 25% increase over the prior year. Of the 370 people, 223 (60%) died as a result of ingesting the medication. Sixty-seven (67) people (18%) did not take the medications and later died of other causes.

- An additional 22 people died in 2020 as a result of ingesting medication prescribed in 2019.

- Ingestion status is unknown for 80 people who were prescribed DWDA medications in 2020. Thirty-six of these patients died, but ingestion status is unknown. For the remaining 44 patients, both death and ingestion status are unknown.

- Of the 245 DWDA deaths during 2020, most (81%) were aged 65 years or older; the median age was 74 years. As in previous years, most were white (97%), well-educated (42% had a least a baccalaureate degree), and had cancer (66%). Three people were referred for formal psychiatric or psychological evaluation.

- Most (92%) patients died at home; and most (95%) were enrolled in hospice care either at the time the DWDA prescription was written or at the time of death. Excluding unknown cases, all (100%) had some form of health care insurance, although the number of patients who had private insurance (26%) continued to decline from previous years, and the number of patients who had only Medicare or Medicaid insurance (74%) continued to increase from previous years.

- As in previous years, the three most frequent concerns (as reported by physicians or family members of patients) were: decreasing ability to participate in activities that made life enjoyable (94%), loss of autonomy (93%), and loss of dignity (72%).

- Prescribing physicians were present at the time of death for 29 patients (12%), and 55 patients (22%) had other health care providers present.

- Data on time from ingestion to death is available for 138 DWDA deaths (56%) in 2020. Among those patients, time from ingestion until death ranged from six minutes to eight hours, with a median time of 50 minutes. Time from ingestion to death is influenced by the specific medications prescribed under the DWDA, which have changed over time.

- A total of 142 physicians wrote the 370 prescriptions in 2020 (range of 1–31 prescriptions per physician).

- During 2020, no referrals were made to the Oregon Medical Board for failure to comply with DWDA requirements.

NOTES AND QUESTIONS

1.   *Other Provisions*. The Oregon Death with Dignity Act provides that no contract or statute can affect a person's request for medically assisted death, and that no insurance policy can be conditioned upon, or affected by, a patient's decision to choose (or reject) medically assisted death. The measure includes a section providing immunity for those who follow the requirements of the statute, and imposing liability on those who violate it.

2.   *Minor Amendments*. The Oregon statute has been amended, but the changes are not substantial. The legislature added the definition of "capable," added some new language designed to encourage patients to discuss the matter with their families, and provided some factors to be considered in determining residency. The amendments also made clear the broad extent of the institutional conscience exception to the statute, which permits health care institutions to limit physicians from engaging in assisted death on their premises or in their organizations. It also changed the written consent form to assure that patients recognize that death will probably, but not always, take place about three hours after taking the medication.

3.   *Legal Challenges*. The Oregon initiative was immediately challenged for a wide range of reasons. Some argued that it discriminated against the disabled, for example, by coercing them into choosing medically assisted dying, and others argued it discriminated against the disabled because those with physical disabilities that made it impossible to take oral medications would not be able to use the statute. The U.S. District Court in Oregon issued a preliminary injunction against enforcing the initiative shortly after its passage and issued a permanent injunction several months later. Ultimately, the Ninth Circuit reversed the District Court, finding that those challenging the Oregon initiative had no standing to raise the issue in federal court. Lee v. Oregon, 107 F.3d 1382 (9th Cir.1997).

4.   *Federalism Concerns*. The federal government has not ignored medically assisted dying, either. Even before the Oregon statute became effective, Congress passed the Assisted Suicide Prevention Restriction Act of 1997, which outlawed the use of federal money to aid medically assisted dying, directly or indirectly.

Shortly after the Oregon Death with Dignity Act became effective, some suggested that any physician who prescribed a lethal drug under that statute would be prescribing that drug without a "legitimate medical purpose," and thus would be acting inconsistently with the federal Controlled Substances Act (CSA). A physician's violation of the Act could lead to both the loss of prescribing authority and criminal indictment. In 1998, after the matter had been pending for some time, the U.S. States Department of Justice published a report concluding that use of controlled substances under the Oregon statute would satisfy the "legitimate medical purpose" requirement of the federal Act.

Members of the House and Senate then introduced the Lethal Drug Abuse Prevention Act of 1998, which would have expanded the authority of the Drug

Enforcement Agency to investigate the lethal use of controlled substances, which the Act provided could not be used with the intent of causing death. Supporters of medically assisted dying joined many of their staunchest opponents and mainstream medical organizations (including the AMA) in opposing the bill because, they said, it would be likely to chill physicians from providing adequate pain relief at the end of life.

Although the bill failed, it was resurrected in slightly milder form in 2000 as the Pain Relief Promotion Act (PRPA), which included a well-publicized section announcing that the provision of medication with the intent to manage pain (and not the intent to cause death) was protected. The 2000 version of the bill also provided for the education of health care professionals on issues related to pain management, and it was supported by the AMA (but opposed by the American Bar Association, the American Cancer Society, and most groups advocating for improved pain management). Although the stated purpose of PRPA was to promote adequate pain relief practices, its effect would be (and, some say, its real purpose was) to render it impossible for physicians in Oregon to carry out the provisions of the Death with Dignity Act. The PRPA died when Congress adjourned in late 2000.

In 2001, the Attorney General, John Ashcroft, issued an interpretive rule that reversed the 1998 Department of Justice position on the issue of the application of the CSA and implemented an enforcement action. Within a day of the announcement of this change in the federal position, Oregon sought relief from the Attorney General's decision in the federal court. A private action seeking an injunction against the Ashcroft position was filed shortly thereafter on behalf of an Oregon oncologist. The District Court immediately restrained the U.S. from enforcing the new interpretation of the CSA, and in 2006 the Supreme Court determined that the Attorney General's interpretive rule was beyond the scope of his authority and was improperly promulgated. Gonzales v. Oregon, 546 U.S. 243, 126 S. Ct. 904, 163 L. Ed. 2d 748 (2006). The Court left open the possibility that the Attorney General could properly promulgate a substantive rule that would bar the operation of the Oregon Death with Dignity Act, and that Congress could amend the Controlled Substances Act to achieve this end. Neither has happened to date.

5.   *Adequate Oversight?* Does the Annual Report address whether the requirements of Section 127.815, or any other section of the DWDA, actually were followed in practice? How would the Department determine the answer to that question? Those who are skeptical about whether the Annual Report accurately presents the impact of the Act point out that the report can describe only those cases that are reported to the state. They argue that cases that do not comply with the specific requirements of the statute will not be reported. It is in this potential body of unreported cases that evidence for concerns over the effectiveness of the statute's boundaries would be found. See, e.g., Nat'l Council on Disability, Medical Futility and Disability Bias (2019); Margaret K. Dore, "Death with Dignity"; A Recipe for Elder Abuse and Homicide (Albeit Not by Name), 11(2) Marquette Elder's Advisor 387 (Spring 2010) (noting

possibility that someone other than the patient could administer the prescribed medications).

6. *Reported Reasons.* As the Annual Report published in 2020 indicates, according to patient physicians and families, most Oregonians who have sought medically assisted dying have done so because of decreasing ability to participate in activities that made life enjoyable (94%) and loss of autonomy (93%), as well as perceived loss of dignity (72%), not because of physical pain. An excerpt from Oregon's Second Annual Report, published in 2000, explains this:

> Responses from both physician and family interviews indicate that patient's decisions to request PAS were motivated by multiple interrelated concerns. Physical suffering was discussed by several families as a cause of loss of autonomy, inability to participate in activities that made life enjoyable, or a "non-existent" quality of life. For example, "She would have stuck it out through the pain if she thought she'd get better . . . [but she believed that] when quality of life has no meaning, it's no use hanging around." For another participant, a feeling of being trapped because of ALS contributed to concern about loss of autonomy. Family members frequently commented on loss of control of bodily functions when discussing loss of autonomy. Those reporting patient concern about being a burden on friends and family also reported concern about loss of autonomy and control of bodily functions. Reasons for requesting a prescription were sometimes so interrelated they were difficult to categorize. According to one family member being asked to distinguish reasons for the patient's decision, "It was everything; it was nothing; [he was suffering terribly]."

> Difficulty categorizing and differences in interpreting the nature of the concerns made physician and family member responses hard to compare quantitatively. Nonetheless, family interviews corroborate physician reports from both years that patients are greatly concerned about issues of autonomy and control. In addition, responses of both physicians and family consistently pointed to patient concerns about quality of life and the wish to have a means of controlling the end of life should it become unbearable. As one family member said, "She always thought that if something was terminal, she would [want to] control the end . . . It was not the dying that she dreaded, it was getting to that death."

Might physicians and family members, or both, be mistaken about what motivated a relative's decisions? In any case, could the expressed concerns have been relieved by better medical and nursing care, or by patient-directed supports and services that allowed for participation in enjoyable activities to the extent possible? Should we be concerned that decisions may be influenced by a belief that "dignity" is not possible for people who rely on caregivers or other supports to live their lives? Or are these decisions that, even if influenced by social and political factors, should be left to individuals?

7. *Requirement of Contemporaneous Competence.* The Oregon statute requires that the requesting adult be "capable" at the time of the request. The requirement of contemporaneous competence excludes individuals who are no longer competent from exercising the option for medically assisted death through an advance directive. Some have argued that advance directives opting for medically assisted death, where legally permitted, should be permitted. Do you agree? For a discussion of the application of advance directives in this context, focusing on individuals with dementia, see Paul T. Menzel & Bonnie Steinbock, Advance Directives, Dementia, and Physician-Assisted Death, 41 J.L. Med. & Ethics 484 (2013). See also Megan S. Wright, Equality of Autonomy?: Physician Aid in Dying and Supported Decision-Making, 63 Ariz. L. Rev. 157 (2021) (examining whether terminally ill people with cognitive impairments such as dementia should be able to access medically assisted death through supported decision-making).

8. *Legislation in Other States.* Advocates for medically assisted dying in Washington made that state the second to use the initiative process to provide for medically assisted dying with the passage of the Washington Death with Dignity Act, again through the initiative process and a statewide vote, in 2008. Former Washington Governor Booth Gardner, who had Parkinson's disease, became the leading advocate for an Oregon-like statute. For the story of Governor Gardner's "last campaign" from the perspective of someone who opposes medically assisted dying, see Daniel Bergner, Death in the Family, N.Y. Times Magazine (December 2, 2007). The Washington Death with Dignity statute, R.C.W. section 70.245, became effective in 2009.

In 2013 Vermont became the first state to adopt a statute permitting aid in dying through the regular legislative process (rather than through an initiative) in the Patient Choice and Control at the End of Life Act, Vt. Stat. Ann. tit. 18, §§ 5281–93. California followed with the End of Life Option Act in 2015, Cal. Health & Safety Code § 443, and the District of Columbia City Council promulgated a Death with Dignity Act in 2016, D.C. Law 21–182. Although there were threats of a Congressional veto of that legislation—a remedy available to Congress only with regard to the District, and not with regard to states—attempts to do so failed. Congress could still effectively repeal the D.C. statute through separate legislation, but it has not done so.

Voters passed the Colorado End of Life Options Act in 2016, Colo. Rev. Stat. §§ 25–48–101–23, and it became effective at the end of that year. The Hawai'i Legislature passed an Oregon-like statute in 2018. Our Care, Our Choice Act, Haw. Rev. Stat. § 327L–1. In 2019, New Jersey enacted the Medical Aid in Dying for the Terminally Ill Act, N.J. State Ann. §§ 26:16–1 et seq., and Maine enacted the Maine Death with Dignity Act, Me. Stat. tit. 22, § 2140. Finally, New Mexico passed the Elizabeth Whitefield End of Life Options Act, H.B. 46, in 2021.

In all, 11 jurisdictions in the U.S. now formally permit medically assisted dying, at least under some circumstances—Oregon, Washington, and Colorado by initiative; Vermont, California, Hawai'i, the District of Columbia, New

Jersey, Maine and New Mexico by ordinary legislative action; and, as we shall see in the Note below, Montana by judicial interpretation of current statutory law. For a discussion of variation among these laws in terms of eligibility requirements, procedural conditions, and other mandates see Thaddeus Mason Pope, Medical Aid in Dying: Key Variations Among U.S. State Laws, 14(1) J. Health & Life Sci. L. 25–59 (2020).

Legislatures in several other states are currently considering legislation that would permit medically assisted dying. Of course, not all efforts to pass such legislation will be effective. Many states have rejected such legislation, and some have rejected multiple proposals. A complete listing of proposals in the U.S. is maintained by the Patients Rights Council at their website.

9.   *Adequate Pain Treatment.* Is there some common ground available between those, on the one hand, who believe that permitting medically assisted dying is necessary for patients to be properly treated at the end of life, and those, on the other hand, who believe that medically assisted dying must be outlawed for patients to be properly treated? Both groups agree that pain is often inadequately treated at the end of life, in part because physicians fear legal action for homicide (if pain relief results in the death of the patient) or for distribution of drugs (if the condition of a patient requires a larger dose of narcotic medication than is standard).

In some states advocates on both sides of the medically assisted dying issue have joined together to support intractable pain relief statutes, which are designed to protect health care providers who deliver adequate pain relief from adverse licensing and criminal actions. These statutes generally provide that a health care provider will not be liable in a state disciplinary proceeding or a criminal action for the aggressive prescription of pain medication as long as the use of that medication is in accord with accepted guidelines for pain management. Several states have promulgated intractable pain relief acts, and several more are considering them. For a model pain relief act, see 24 J.L., Med. Ethics 317 (1996). See also Ann Alpers, Criminal Act or Palliative Care? Prosecutions Involving the Care of the Dying, 26 J.L. Med. Ethics 308 (1998) (identifying factors that create a risk of prosecution); Kelly K. Dineen & James M. DuBois, Between a Rock and a Hard Place: Can Physicians Prescribe Opioids to Treat Pain Adequately While Avoiding Legal Sanction?, 42 Am. J. L. & Med. 7 (2016) (proposing a model for characterizing misprescribers).

For a series of articles on the relationship between pain relief and medically assisted dying, with the conclusion that we ought to create a system that both provides excellent palliative care in every case and allows for medically assisted dying in the rare cases in which it is necessary, see Timothy Quill and Margaret Battin, eds., Physician-Assisted Dying: The Right to Excellent End-of-Life Care and Patient Choice (2004). See also Kathleen Foley, The Case Against Assisted Suicide: For the Right to End of Life Care (2002) (written by a leading palliative care physician). For an account of the meaning of euthanasia and suicide in our society and others, and a description of how these arguments are likely to be discussed in the future, see Margaret Battin, Ending Life: Ethics and the Way We Die (2005).

10. *Palliative Care and Cannabis.* Federal and state policy may conflict on one kind of palliative care—the use of marijuana and cannabis products. Some cancer patients, patients with glaucoma, AIDS patients, patients with multiple sclerosis, those with migraine headaches, and others find that they can obtain relief from some of the symptoms of their condition—or from some of the side effects of the treatments for their condition—through the use of marijuana. In particular, some cancer patients find that marijuana helps them overcome the nausea that follows the use of many chemotherapeutic agents.

As of May 2021, 36 states and four territories allow for the medical use of cannabis products (under highly varying state standards). A complete listing of medical marijuana laws is maintained by the National Conference of State Legislatures at their website. However, the manufacture and distribution of cannabis is still a felony under the Federal Controlled Substances Act (CSA). In 2001, the Supreme Court determined that there was no medical necessity defense available to those who manufactured or distributed marijuana to patients in violation of the federal law. U.S. v. Oakland Cannabis Buyers' Cooperative, 532 U.S. 483, 121 S. Ct. 1711, 149 L. Ed. 2d 722 (2001). Four years later the Supreme Court confirmed that the CSA marijuana prohibition fell within Congress's commerce power. Gonzales v. Raich, 545 U.S. 1, 125 S. Ct. 2195, 162 L. Ed. 2d 1 (2005).

Wavering federal approaches toward enforcement of the federal law have added complexity to the increasing conflict between state reforms and federal laws. See Robert A. Mikos, The Evolving Federal Response to State Marijuana Reforms, 26 Widener L. Rev. 1 (2020). Public opinion is also shifting, increasing pressure on the federal government to act. According to a 2021 poll conducted by the Pew Research Center, an overwhelming majority of U.S. adults (91%) say either that marijuana should be legal for medical *and* recreational use (60%) or that it should be legal for medical use only (31%). Three major bills were introduced in Congress in 2019, each taking a different approach to cannabis decriminalization, legalization, and reform. For a discussion of those bills through a health equity lens, see Nicole Huberfeld, Health Equity, Federalism, and Cannabis Policy, 101 B.U. L. Rev. 897 (2021). See also Rebecca Haffajee and Amanda Mauri, Cannabis Liberalization in the U.S.: The Policy Landscape, Health Aff. Health Policy Brief (July 1, 2021).

11. *International Views.* The public interest in this issue is not limited to the U.S. Medically assisted dying has been tolerated in the Netherlands, as a legal matter, since 1969, and it was formally legalized by the Parliament in 2001 in cases of intractable suffering (not just pain), where the patient has been informed of all of the alternative treatments available, has consulted two physicians, and has followed other requirements established by the national medical association. Belgium joined the Netherlands in legalizing medically assisted dying through the legislative process, and the practice is now officially tolerated under some circumstances in Switzerland and Spain. The

Constitutional Court of Colombia has approved legislation that provides for medically assisted dying, and there has been an active debate in France, Venezuela and Australia over the propriety of permitting medically assisted dying.

Australia's Northern Territory's Parliament passed The Rights of the Terminally Ill Act (1995), which permitted what some have called "voluntary euthanasia" under some circumstances; however, the national Parliament effectively overturned that territorial statute. The national Parliament has no such control over legislation coming from the states (rather than territories) in Australia, and voluntary assisted death is permitted in the states of Victoria and Western Australia. The states of Tasmania and South Australia also have passed bills which, upon their effective dates, will authorize voluntary assisted death, also in limited circumstances. Voluntary assisted death is not lawful in other Australian States and Territories, though this may change as Australian jurisdictions consider legal reform in this area.

Canada has allowed for medically assisted dying since the Parliament permitted it, encouraged (or, perhaps, compelled) by a 2015 decision of the Supreme Court of Canada. The governing statute has many of the requirements of the Oregon and Washington statutes, although it is generally more liberal in permitting the practice, and it allows a patient to choose euthanasia (where the final dose is actually delivered by a health care provider) rather than taking the medication himself, if that is his desire. The vast majority of Canadian patients given the option to choose medical delivery of the lethal dose (rather than self-administration) have done so. The Canadian statute was encouraged by the 2015 decision in Carter v. Canada (AG), 2015 SCC 5, which found that not allowing a competent adult access to relief from suffering at the end of life by outlawing medically assisted dying violated the Canadian Charter of Rights and Freedoms.

The European Court of Human Rights considered a challenge to the laws of the United Kingdom that outlawed medically assisted dying in Pretty v. United Kingdom, 35 Eur. H. R. Rep. 1 (2002). Applying an analysis similar, in many respects, to that in *Glucksberg* and *Vacco*, the Court found that the United Kingdom had the authority to enforce its law. An English court again confirmed the legal prohibition against medically assisted dying in 2012. For a brief description of the current state of the law in other countries, see Alan Meisel, Kathy Cerminara, and Thaddeus Mason Pope, The Right to Die (3d ed. 2005, Supplement 2020).

PROBLEM: DRAFTING LEGISLATION

You are the nonpolitical legislative counsel to a state legislature. Currently that state has a statute prohibiting assisting suicide which, it is assumed, would outlaw any conduct intended to result in the death of another person. You have been asked by several members of the legislature to draft bills relating to medically assisted dying. One member has asked you to draft a bill that would outlaw all medically assisted dying, under all circumstances.

A second member wants you to draft a bill that would put as few limitations on medically assisted dying as possible; this libertarian member believes that the decision should be left to individual doctors and patients. Another member has asked you to draft a statute that would protect health care providers from potential liability for participating in medically assisted dying, and that would give providers an option to avoid participating in medically assisted dying if they chose not to do so. Yet another member has asked you to draft a statute that would prohibit managed care organizations from directly or indirectly giving any incentives to their members to choose medically assisted dying. Finally, one long-time incumbent has asked you to draft a consensus statute— one with enough political support across the spectrum that it has a reasonable chance of passing. Presumably, this would be a statute that would permit some narrow form of medically assisted dying under carefully controlled and narrowly defined circumstances.

How would you go about drafting these statutes? How would they be different? What facts (about the legal landscape in this state, about the politics of the current officeholders, about the religious backgrounds of those within the state, about other issues) would you want to have before you started drafting these statutes?

NOTE: ESTABLISHING A STATE RIGHT TO MEDICALLY ASSISTED DYING THROUGH LITIGATION

As it becomes clear that each state is going to determine the legal status of medically assisted dying within its borders, advocates of legal medically assisted dying have looked to state judiciaries as well as state legislatures to establish such a right. While the issue of the legalization of medically assisted dying has been litigated in a host of different forms over the last few decades, the highest courts of only five states—Florida, Alaska, Montana, New Mexico, and New York—have had a chance to directly address the issue in the last few years. State constitutional claims were raised in all five states, and they were rejected in all five cases. One of these courts, in Montana, also did a statutory analysis and found that there was no violation of any statute when a provider wrote a lethal prescription under circumstances similar to those permitted by statute in Oregon. Thus, the Supreme Court of Montana found that current Montana law did not outlaw medically assisted dying, even if the Montana Constitution did not independently protect a person who was seeking that form of treatment.

The other state courts to consider the issue—the Florida, Alaska, and New Mexico Supreme Courts and the New York Court of Appeals—found no state constitutional right to medically assisted dying, and, unlike the court in Montana, also found that there was no statutory right to medically assisted dying within their states because any exercise of medically assisted dying would constitute the crime of aiding or abetting (or assisting) suicide under state statute.

The Montana case, Baxter v. Montana, 354 Mont. 234, 224 P.3d 1211 (2009), rejected the argument (which had been accepted at the trial court) that outlawing medically assisted dying would violate the protection of individual dignity guaranteed by Article II, section 4 of the Montana Constitution, which provides that "[t]he dignity of the human being is inviolable." While at least one judge would have found that the individual dignity clause was sufficient to protect medically assisted dying, the majority decided the case on statutory grounds. The court determined that under Montana law consent is a defense, even to homicide crimes, as long as the defendant's conduct is not "against public policy." Here, the court determined, medically assisted dying was not contrary to public policy; indeed, the Montana Legislature had never spoken about it in a way that approved or disapproved of it. Thus, while the Montana Legislature could outlaw medically assisted dying, it had not done so as a matter of statutory law, and no state constitutional analysis was necessary.

At the next plenary Montana legislative session following *Baxter*, in 2011, the Legislature was squarely confronted both with bills that would outlaw medically assisted dying, and with a bill that would formally legalize it and impose the same kinds of restrictions that are imposed in Washington and Oregon. The bills were carefully followed by the press, and there was a good deal of advocacy by partisans on both sides of the issue. Ultimately, the Legislature rejected all of the bills on both sides of the issue and left the law as it stood the day that *Baxter* was decided.

In 2013 the Montana Legislature again faced bills to both institutionalize (on one side) and prohibit (on the other side) aid in dying. Other legislative proposals have been introduced since, without success. It seems like the default position will always be the legislative winner in Montana; whatever the law is today will remain the law indefinitely because there is inadequate legislative support either to make medically assisted death legal (if it is not) or illegal (if it is now legal).

In fact, given the difficulty of pushing legislation through a state legislature and across a Governor's desk, the state "default" position—the law in the absence of any legislation dealing directly with medically assisted dying—will determine whether medically assisted dying is permitted. This "default" position, of course, must be determined by the courts. The Montana Supreme Court described the "default" position as one in which there was no statutory provision outlawing medically assisted dying—so medically assisted dying is permitted in Montana. Other state judiciaries to consider the issue found the "default" position to be that medically assisted dying is outlawed by state statutes criminalizing assisting suicide. Thus, medically assisted dying is not permitted in Florida, Alaska, New Mexico or New York. Because neither side on this issue has the political strength to change the current statutory law on medically assisted dying—whatever that law may be—the status of the state law is really determined by the state courts in determining the "default" position in that state.

Those opposed to medically assisted dying also sought to use the administrative process to reinstate the ban in Montana. An attempt to have the Montana Board of Medical Examiners declare that they would impose professional sanctions against any Montana physician who provided medically assisted dying in accord with the *Baxter* opinion was met with this response in 2012:

Physician Aid in Dying

The Montana Board of Medical Examiners has been asked if it will discipline physicians for participating in such aid-in-dying. This statement reflects the Board's position on this controversial question.

The Board recognizes that its mission is to protect the citizens of Montana against the unprofessional, improper, unauthorized and unqualified practice of medicine by ensuring that its licensees are competent professionals. [ ] In all matters of medical practice, including end-of-life matters, physicians are held to professional standards. If the Board receives a complaint related to physician aid-in-dying, it will evaluate the complaint on its individual merits and will consider, as it would any other medical procedure or intervention, whether the physician engaged in unprofessional conduct as defined by the laws and rules pertinent to the Board.

Mont. Bd. Of Med. Examiners, Position Statement No. 20 (2012). This statement was applauded by those who support aid-in-dying and who believe that it should be treated like any other medical procedure. Montanans Against Assisted Suicide (MAAS) and for Living with Dignity, an advocacy group opposed to aid-in-dying, asked the Board to withdraw the statement, which, those advocates said, was issued without proper notice, without statutory authority, and in violation of the principle of separation of powers. They argued that the Position Statement put doctors and the public at risk. MAAS then sought a declaratory ruling that the Position Statement was invalid, but the action was dismissed after the Board rescinded the statement. See Montanans Against Assisted Suicide v. Bd. Of Med. Examiners, 379 Mont. 11, 347 P.3d 1244 (2015).

Litigation to establish the kind of right recognized in *Baxter* has not been successful outside of Montana. In Krischer v. McIver, 697 So. 2d 97 (Fla. 1997), a terminally ill AIDS patient and his physician sought an injunction against the prosecution of the physician for assisting in his patient's suicide. The Florida Supreme Court rejected a claim that the privacy provision of the Florida Constitution included the right to have a physician assist in one's death. The court announced that a properly drawn statute authorizing physician-assisted suicide would be constitutionally permissible, but that principles of separation of powers left the decision about whether it should be made legal to the legislature. The Chief Justice filed a vigorous dissent, arguing that, ". . . the right of privacy attaches with unusual force at the death bed. . . . What possible interest does society have in saving life when there is nothing of life to save but a final convulsion of agony? The state has no business in this arena." 697 So. 2d at 111. See also Sampson v. State, 31 P.3d 88 (Alaska 2001).

The New Mexico Supreme Court faced an argument that its assisting suicide statute was a violation of state constitutional law in Morris v. Brandenburg, 376 P.3d 836, 2016-NMSC-027 (2016). In that case, a patient and her physicians brought a declaratory judgment action against state officials alleging that the statute criminalizing assisting suicide was unconstitutional. Judgment was entered for the plaintiffs after a multi-day trial, and the Court of Appeals reversed. The New Mexico Supreme Court determined that medically assisted dying would constitute "assisting suicide," a crime in New Mexico, and thus it was illegal unless that criminal statute itself violated a federal or state constitutional provision. The court rejected due process arguments based on both the U.S. and state constitutions by deferring to the U.S. Supreme Court's analysis in *Glucksberg*, above. The court separately rejected the application of the New Mexico constitution's provision which assures that the state will protect such "natural, inherent and inalienable" rights as those of "seeking and obtaining safety and happiness." The court concluded that any right to medically assisted dying would have to come from its legislature.

The New York Court of Appeals faced the same issue in Myers v. Schneiderman, 85 N.E.3d 57, 30 N.Y.3d 1 (N.Y. 2017). Like the New Mexico court, the New York Court of Appeals deferred to the *Glucksberg* analysis of the due process and equal protection issues and, in addition, found that there was a rational basis for applying the law prohibiting assisted suicide. Like the New Mexico court, the New York Court of Appeals determined that the application of the criminal assisted suicide statute to those offering medically assisted dying is a matter for the New York Legislature, at least given the current state of the record. Although that application of the criminal law is not required by state constitutional law, it is not prohibited by state constitutional law either.

PROBLEM: MEDICALLY ASSISTED DYING, A LEGAL ISSUE OR A MEDICAL ISSUE?

Our country's conflict over medically assisted dying has been fought in the legal arena. Is there a right to engage in medically assisted dying? Do homicide and suicide statutes prohibit the practice of medically assisted dying? Those with more libertarian views tend to support finding such a right; those with strongly developed religious views or a more expansive view of criminal law are less likely to find such a right. The battles over legislation and most of the reported litigation reflect the rights-oriented legal battle over this issue.

Instead, should we be asking whether medically assisted dying is within the medical standard of care? If we were to take this approach, we would treat medically assisted dying as one of many end-of-life care possibilities, and we would let providers determine if it is appropriate on a case-by-case basis. Instead of arguing about the issue in the courts, we would leave it to health care professionals, who could offer to their patients whatever came within the standard of care.

Do you think the issue is best analyzed as one of legal rights and obligations, ultimately to be decided by courts, or as one of good medicine, ultimately to be determined by the physician and her patient in each case?

For an account of an early effort by physicians to establish a standard of care in this area, see T.E. Quill, C.K. Cassel and D.E. Meier, Care of the Hopelessly Ill: Proposed Clinical Criteria for Physician-Assisted Suicide, 327 NEJM 1380 (1992). For a more recent legal suggestion that physicians should be developing standards of practice in this area, written by an advocate of medically assisted death, see Kathryn Tucker, Aid in Dying: Guidance for an Emerging End-of-Life Practice, 142 Chest 218 (2012).

CHAPTER 7

REGULATION OF RESEARCH
INVOLVING HUMAN SUBJECTS

■ ■ ■

I. INTRODUCTION

Medical research has contributed to relieving great human suffering. It also has been the cause of injustice and misery to individuals and groups who have been treated as grist for the research mill or as conscripts in the war against disease.

The ethical framework for the relationship between researcher and subject differs from that for physician and patient. While the goal of improving the well-being of the individual patient is held as the essential goal in the physician-patient relationship, that is simply not the case in medical research. Instead, research has as a primary goal the production of new knowledge for the good of society and typically requires risk to the individual, although formal ethical principles for medical research require respect for the autonomy and well-being of the individual subject as well. This balance rests on the fulcrum of the prohibition against research without the voluntary consent of the subject, but distrusts that consent alone is up to the task of protecting individuals while benefiting the common good.

Problems are especially likely to arise when subjects are also patients, and their physicians are also researchers. In this context, the goals of research and the goals of treatment may become incompatible in ways that are not readily apparent. Research studies, for example, adopt strict protocols to assure validity of results that may differ from the more flexible approach taken in a treatment context. In addition, claims of trust as a basis for the relationship of patient to physician rely on the primacy of the physician's duty to act in the patient's best interest, an underlying assumption that is inapt for the researcher-subject relationship. As a result, modeling the consent process for medical research upon the informed consent process for treatment creates an ill-fitting foundation for judging adequacy of consent. Similarly, applying the legal framework for rights and duties in medical treatment to the context of medical research can produce unsatisfying results.

The regulatory framework for research in the U.S. reflects that same tension between the gains in knowledge and health promised by research and the value of protecting the individual subject from harm. Research regulation constantly confronts competing claims about the nature of the research enterprise: Is medical research such a privileged endeavor for the common good that restraints should be minimal; or is competition for accomplishment, reputation, and new products so strong that robust oversight is required?

II. STANDARDS GOVERNING PROTECTION OF HUMAN SUBJECTS

A. THE NUREMBERG CODE

The Nuremberg Code is the foundational document for the legal and moral principles that govern research with human subjects. This Code was born of tragedy on an enormous scale. The shocking horrors that could be perpetrated in the name of medical experimentation became apparent to the world through the Nazi war crimes trials. The Medical Trial detailed experiments in which "volunteers," usually concentration camp inmates, were exposed to extremely low atmospheric pressure until they died; were forced to drink seawater or breathe mustard gas; were intentionally infected with malaria and typhus; were placed in ice water until they froze; underwent macabre surgeries, often without anesthesia; or were subjected to other tortures. Perpetrators justified their actions as legitimate medical research. Most of the twenty physicians accused of participating in or directing this experimentation were convicted, and some were hanged. A few were acquitted and one, the infamous Dr. Mengele, escaped.

In one of the tribunal's most significant actions, the court sitting in Nuremberg promulgated principles for legal and ethical medical research. Perhaps surprisingly, prosecutors at Nuremberg had struggled to find an authoritative formal legal canon governing medical research in either the U.S. or Germany, and Nuremberg code emerged from that experience. See Sandra Johnson, Nazi Experiments, the Nuremberg Code, and the United States, in Nazi Law: From Nuremberg to Nuremberg (John Michalczyk (ed.), 2018).

NUREMBERG CODE: PERMISSIBLE MEDICAL EXPERIMENTS

The great weight of the evidence before us is to the effect that certain types of medical experiments on human beings, when kept within reasonably well-defined bounds, conform to the ethics of the medical profession generally. The protagonists of the practice of human experimentation justify their views on the basis that such experiments yield results for the good of society that are unprocurable by other methods or means of study. All agree, however, that certain basic principles must be observed in order to satisfy moral, ethical, and legal concepts:

1. The voluntary consent of the human subject is absolutely essential.

This means that the person involved should have legal capacity to give consent; should be so situated as to be able to exercise free power of choice, without the intervention of any element of force, fraud, deceit, duress, over-reaching, or other ulterior form of constraint or coercion; and should have sufficient knowledge and comprehension of the elements of the subject matter involved as to enable him to make an understanding and enlightened decision. This latter element requires that before the acceptance of an affirmative decision by the experimental subject there should be made known to him the nature, duration, and purpose of the experiment; the method and means by which it is to be conducted; all inconveniences and hazards reasonably to be expected; and the effects upon his health or person which may possibly come from his participation in the experiment.

The duty and responsibility for ascertaining the quality of the consent rests upon each individual who initiates, directs, or engages in the experiment. It is a personal duty and responsibility which may not be delegated to another with impunity.

2. The experiment should be such as to yield fruitful results for the good of society, unprocurable by other methods or means of study, and not random and unnecessary in nature.

3. The experiment should be so designed and based on the results of animal experimentation and a knowledge of the natural history of the disease or other problem under study that the anticipated results will justify the performance of the experiment.

4. The experiment should be so conducted as to avoid all unnecessary physical and mental suffering and injury.

5. No experiment should be conducted where there is an *a priori* reason to believe that death or disabling injury will occur; except, perhaps, in those experiments where the experimental physicians also serve as subjects.

6. The degree of risk to be taken should never exceed that determined by the humanitarian importance of the problem to be solved by the experiment.

7. Proper preparations should be made and adequate facilities provided to protect the experimental subject against even remote possibilities of injury, disability, or death.

8. The experiment should be conducted only by scientifically qualified persons. The highest degree of skill and care should be required

through all stages of the experiment by those who conduct or engage in the experiment.

9.   During the course of the experiment the human subject should be at liberty to bring the experiment to an end if he has reached the physical or mental state where continuation of the experiment seems to him to be impossible.

10. During the course of the experiment the scientist in charge must be prepared to terminate the experiment at any stage, if he has probable cause to believe, in the exercise of the good faith, superior skill, and careful judgment required of him, that a continuation of the experiment is likely to result in injury, disability, or death to the experimental subject.

B.   LEGAL STANDARDS IN THE U.S.

The Nuremberg Code did not have an immediate impact in the U.S., perhaps because the Nazi doctors were viewed as monsters working far outside the tradition of medical ethics in this country. See George Annas, Mengele's Birthmark: The Nuremberg Code in United States Courts, 7 J. Contemp. Health L. & Pol'y 17 (1991). In fact, the U.S. government was conducting research at the same time that would violate the principles of the Nuremberg Code. In just one example, the government tested the human response to radiation by injecting patients (many of whom were inaccurately described as terminally ill) with plutonium or uranium to observe its physical effect. These subjects were never informed of the nature of the experiments. See Advisory Committee on Human Radiation Experiments, The Human Radiation Experiments (1996); Susan Smith, Mustard Gas and American Race-Based Human Experimentation in World War II, 36 J. L. Med. & Ethics 517 (2008).

It wasn't until the publication of a landmark article detailing abuses of human subjects in the U.S. that it was recognized that medical research abuse was possible, if not common, in this country. Henry K. Beecher, Ethics and Clinical Research, 274 NEJM 1354 (1966). Just a few years later, in 1972, the most infamous violation of research ethics in twentieth-century America came to light. In the Tuskegee Syphilis Study, begun in 1932, the U.S. Public Health Service (USPHS) studied the natural course of untreated syphilis in 399 impoverished African American men. Even when penicillin, the first effective treatment for syphilis, became available, the USPHS physicians refused to offer that treatment to most of their subjects; and many of these men were expressly and regularly discouraged from seeking treatment. For nearly 40 years, the USPHS condemned these men and their wives, partners, and families to the ravages of untreated syphilis. See James H. Jones, Bad Blood: The Tuskegee Syphilis Experiment (1981); Ruquaiija Yearby, Exploitation in Medical Research:

The Enduring Legacy of the Tuskegee Syphilis Study, 67 Case W. Res. L. Rev. 1171 (2017).

The Tuskegee experiment is iconic, but it is only one episode in a long history of exploitation of African Americans in medical research. See Harriet A. Washington, Medical Apartheid: The Dark History of Medical Experimentation on Black Americans from Colonial Times to the Present (2007). Much experimentation with African-American subjects was motivated more by a voyeuristic fascination rather than a scientific research design, and it routinely relied on the disadvantages produced first by slavery and then by poverty and segregation. See also Camille A. Nelson, American Husbandry: Legal Norms Impacting the Production of (Re)productivity, 19 Yale J. L. & Fem. 1 (2007), detailing the work of Dr. Marion Sims, the "father of modern gynecology," who developed his surgical techniques through repeated unanesthetized vaginal surgeries on slave women; Section IV.E, below for a description of the case of Henrietta Lacks. Racialized medicine in the U.S. permeated these experiments and influenced pre-War Nazi health laws, which in turn laid the foundation for the Nazi experiments. See Johnson, above.

These stories do not stand alone as examples of exploitation in medical research in the U.S. Other publicized cases include a study at the Jewish Chronic Disease Hospital, in which live cancer cells were injected into patients without their knowledge, and the Willowbrook State Hospital hepatitis study, in which parents seeking hospital admission for their children were required to allow researchers to infect their child with the disease. See David Rothman, Strangers at the Bedside: A History of How Law and Bioethics Transformed Medical Decisionmaking (1991). These and other episodes, and the patterns of vulnerability and advantage-taking they reveal, influenced the regulatory framework ultimately adopted in the U.S. to govern research. For a comprehensive treatment of the history of legal standards for research in the U.S., see Carl H. Coleman et al., The Ethics and Regulation of Research with Human Subjects (2d ed. 2015).

1. Federal Regulation of Research

Congress enacted the National Research Act in 1974, shortly after Congressional hearings on the Tuskegee study. The Act established the National Commission for Protection of Human Subjects of Biomedical and Behavioral Research to identify "basic ethical principles" that should guide the conduct of research on human subjects; to develop procedures to assure that the research would be consistent with those principles; and to recommend guidelines for human subject research funded by the precursor to HHS. The Commission issued several reports between 1975 and 1978, the most significant of which is the Belmont Report, Ethical Principles and Guidelines for the Protection of Human Subjects of Research (1979). This document established an analytical framework that guided subsequent federal regulations; continues to influence research regulation today; and became the primary approach for decision making in bioethics far beyond questions relating to research.

The main federal regulations governing research with human subjects, first adopted jointly by a number of federal agencies in 1991, traditionally have been called the "Common Rule." While the Common Rule forms the core of federal regulation of medical research, other federal statutes, such as HIPAA, apply to a variety of issues that can arise in the research enterprise.

2. State Law

State law also plays an important role in controls on medical research: the Common Rule does not preempt state law that provides additional protections for individuals participating as subjects. A good number of individual states have enacted legislation to govern the conduct of research within their boundaries. Some states, for example, have enacted statutes requiring that all medical research performed in that state be subject to the standards of the federal regulations even if the research is not federally funded. See, e.g., Md. Code, Health-General § 13–2001. A good number of states have enacted legislation setting limits on the use of some types of tissue in research. See, e.g., Mo. Const. Art. 3 § 38d, regarding stem cell research. Most have statutes governing the release of state-held data or tissue for research. See, e.g., 410 Ill. Comp. Stat. § 525/9; N.D. Admin. Code § 33–06–16–05. Some states also regulate research performed on persons with cognitive disabilities. See, e.g., Ala. Code § 22–56–4.

The most influential impact of state law on medical research, however, comes in litigation under state common law. Plaintiffs claiming violation of standards for medical research most often rely on negligence, informed consent, breach of contract, or other state law claims. See Sections IV.C, D, and E, below.

III. THE COMMON RULE

The 1991 Common Rule saw no substantial revision for decades. After nearly six years of notice and comment, however, seventeen federal agencies jointly promulgated revised regulations in 2017. Significant revisions included changes in the required elements of consent, expanded categories of exempt research, and requirements for research with biospecimens and health information, among others. Jerry Menikoff et al., The Common Rule Updated, 376 NEJM 613 (2017). The revised Common Rule went into effect on January 21, 2019.

The Food and Drug Administration (FDA), in the most significant exception among federal agencies, abides by regulations on research that continue to depart significantly from the Common Rule in a few respects.

21 CFR § 50.1. All studies funded by or submitted to the FDA for its approval of drugs, devices, or other substances must meet the FDA's standards for research. Where a study is subject to both the FDA and another federal agency (e.g., pharmaceutical research funded by the National Institutes of Health (NIH)) and the regulations of the two agencies differ, the stricter of the two standards (in terms of protection of subjects) applies.

The FDA is working toward harmonizing its requirements with the Common Rule, as required by the 21st Century Cures Act, P.L. 114–255, § 3023 (2016). For example, the FDA issued guidance permitting Institutional Review Boards (IRBs) to waive or alter informed consent requirements for certain minimal risk clinical investigations. See FDA, Institutional Review Board Waiver or Alteration of Informed Consent for Clinical Investigations Involving No More Than Minimal Risk to Human Subjects; Guidance for Sponsors, Investigators, and Institutional Review Boards, 82 Fed. Reg. 34535–01 (July 25, 2017).

What follows is an edited version of the revised Common Rule. It includes the regulations needed to resolve the Problems presented in the rest of this chapter. (The numbers indicated by the underline are unique to each agency, so for example, the HHS regulations are found at 45 C.F.R. § 46.101 et seq. The regulations presented below include revisions promulgated in 2017. Where the 2017 revisions depart substantially and substantively from the earlier regulations, however, this difference is noted in the text of Notes and Problems in the rest of this chapter.) In addition to these regulations, federal agencies that regulate research issue informal guidance, policy statements, enforcement letters, and other materials that provide the agency's interpretation of the requirements of the regulations. These documents are accessible on the agencies' websites. See, e.g., www. hhs.gov/ohrp.

§ ___.101   To what does this policy apply?

(a) Except as detailed in § ___.104, this policy applies to all research involving human subjects conducted, supported, or otherwise subject to regulation by any Federal department or agency . . . . It also includes research conducted, supported, or otherwise subject to regulation by the Federal Government outside the United States. . . .

* * *

(e) Compliance with this policy requires compliance with pertinent federal laws or regulations that provide additional protections for human subjects.

(f) This policy does not affect any state or local laws or regulations (including tribal law . . .) that may otherwise be applicable and that provide additional protections for human subjects.

(g) This policy does not affect any foreign laws or regulations that may otherwise be applicable and that provide additional protections to human subjects of research.

(h) When research covered by this policy takes place in foreign countries, procedures normally followed in the foreign countries to protect human subjects may differ from those set forth in this policy. In these circumstances, if a department or agency head determines that the procedures prescribed by the institution afford protections that are at least equivalent to those provided in this policy, the department or agency head may approve the substitution of the foreign procedures in lieu of the procedural requirements provided in this policy. . . .

(i) Unless otherwise required by law, department or agency heads may waive the applicability of some or all of the provisions of this policy to specific research activities or classes of research activities otherwise covered by this policy, provided the alternative procedures to be followed are consistent with the principles of the Belmont Report. . . .

* * *

§ ___.102 Definitions for purposes of this policy.

* * *

(e) (1) Human subject means a living individual about whom an investigator . . . conducting research: (i) Obtains information or biospecimens through intervention or interaction with the individual, and uses, studies, or analyzes the information or biospecimens; or (ii) Obtains, uses, studies, analyzes, or generates identifiable private information or identifiable biospecimens.

(2) Intervention includes both physical procedures by which information or biospecimens are gathered (e.g., venipuncture) and manipulations of the subject or the subject's environment that are performed for research purposes.

(3) Interaction includes communication or interpersonal contact between investigator and subject.

(4) Private information includes information about behavior that occurs in a context in which an individual can reasonably expect that no observation or recording is taking place, and information that has been provided for specific purposes by an individual and that the individual can reasonably expect will not be made public (e.g., a medical record).

(5) Identifiable private information is private information for which the identity of the subject is or may readily be ascertained by the investigator or associated with the information.

(6) An identifiable biospecimen is a biospecimen for which the identity of the subject is or may readily be ascertained by the investigator or associated with the biospecimen.

* * *

(j) Minimal risk means that the probability and magnitude of harm or discomfort anticipated in the research are not greater in and of themselves than those ordinarily encountered in daily life or during the performance of routine physical or psychological examinations or tests.

* * *

(l) Research means a systematic investigation, including research development, testing, and evaluation, designed to develop or contribute to generalizable knowledge. . . [T]he following activities are deemed not to be research:

* * *

(2) Public health surveillance activities, including the collection and testing of information or biospecimens, conducted, supported, requested, ordered, required, or authorized by a public health authority. Such activities are limited to those necessary to allow a public health authority to identify, monitor, assess, or investigate potential public health signals, onsets of disease outbreaks, or conditions of public health importance (including trends, signals, risk factors, patterns in diseases, or increases in injuries from using consumer products). Such activities include those associated with providing timely situational awareness and priority setting during the course of an event or crisis that threatens public health (including natural or man-made disasters).

* * *

§ ___.104 Exempt research.

(d) [T]he following categories of human subjects research are exempt from this policy:

* * *

(2) Research that only includes interactions involving educational tests . . . , survey procedures, interview procedures, or observation of public behavior . . . [under certain conditions relating to identifiability of the subject, risk of liability, financial or reputational harm to the subject, and protection of confidentiality and privacy. This category of exemption is further limited in the case of child subjects].

* * *

(4)   Secondary research for which consent is not required: Secondary research uses of identifiable private information or identifiable biospecimens, if at least one of the following criteria is met:

(i)   The identifiable private information or identifiable biospecimens are publicly available;

(ii) Information, which may include information about biospecimens, is recorded by the investigator in such a manner that the identity of the human subjects cannot readily be ascertained directly or through identifiers linked to the subjects, the investigator does not contact the subjects, and the investigator will not re-identify subjects;

(iii) The research involves only information collection and analysis involving the investigator's use of identifiable health information when that use is regulated under [HIPAA], for the purposes of "health care operations" or "research" . . . or for "public health activities and purposes" [as defined in HIPAA]; or

(iv) The research is conducted by, or on behalf of, a Federal department or agency using government-generated or government-collected information obtained for nonresearch activities, if [the data meets certain federal statutory standards for collection and security of identifiable private data.]

* * *

(7)   Storage or maintenance for secondary research for which broad consent is required: Storage or maintenance of identifiable private information or identifiable biospecimens for potential secondary research use if an IRB conducts a limited IRB review [an expedited review by the Chair or a designated member of the IRB] and makes the determinations required by § ___.111(a)(8).

(8)   Secondary research for which broad consent is required: Research involving the use of identifiable private information or identifiable biospecimens for secondary research use, if the following criteria are met:

(i)   Broad consent for the storage, maintenance, and secondary research use of the identifiable private information or identifiable biospecimens was obtained in accordance with § ___.116(a)(1) through (4), (a)(6), and (d);

(ii) Documentation of informed consent or waiver of documentation of consent was obtained . . . ;

(iii) An IRB conducts a limited IRB review [an expedited review by the Chair or a designated member of the IRB] and makes the determination required by § ___.111(a)(7) and makes the determination that the research to be conducted is within the scope of the broad consent . . . ; and

(iv) The investigator does not include returning individual research results to subjects as part of the study plan. This provision does not prevent an investigator from abiding by any legal requirements to return individual research results.

* * *

§ ___.107   IRB membership.

(a) Each IRB shall have at least five members, with varying backgrounds to promote complete and adequate review of research activities commonly conducted by the institution. The IRB shall be sufficiently qualified through the experience and expertise of its members (professional competence), and the diversity of its members, including race, gender, and cultural backgrounds and sensitivity to such issues as community attitudes, to promote respect for its advice and counsel in safeguarding the rights and welfare of human subjects. The IRB shall be able to ascertain the acceptability of proposed research in terms of institutional commitments (including policies and resources) and regulations, applicable law, and standards of professional conduct and practice. The IRB shall therefore include persons knowledgeable in these areas. If an IRB regularly reviews research that involves a category of subjects that is vulnerable to coercion or undue influence, such as children, prisoners, individuals with impaired decision-making capacity, or economically or educationally disadvantaged persons, consideration shall be given to the inclusion of one or more individuals who are knowledgeable about and experienced in working with these categories of subjects.

(b) Each IRB shall include at least one member whose primary concerns are in scientific areas and at least one member whose primary concerns are in nonscientific areas.

(c) Each IRB shall include at least one member who is not otherwise affiliated with the institution and who is not part of the immediate family of a person who is affiliated with the institution.

(d) No IRB may have a member participate in the IRB's initial or continuing review of any project in which the member has a conflicting interest, except to provide information requested by the IRB.

(e) An IRB may, in its discretion, invite individuals with competence in special areas to assist in the review of issues that require expertise beyond or in addition to that available on the IRB. These individuals may not vote with the IRB.

* * *

§ ___.109   IRB review of research.

(a) An IRB shall review and have authority to approve, require modifications in (to secure approval), or disapprove all research activities covered by this policy. . . .

* * *

(e) An IRB shall conduct continuing review of research . . . at intervals appropriate to the degree of risk, not less than once per year, except [for studies eligible for expedited or limited review or studies that involve only data analysis or "accessing follow-up clinical data from procedures that subjects would undergo as part of clinical care"].

* * *

(g) An IRB shall have authority to observe or have a third party observe the consent process and the research.

* * *

§ ___.111   Criteria for IRB approval of research.

(a) [T]he IRB shall determine that all of the following requirements are satisfied:

(1) Risks to subjects are minimized: (i) By using procedures that are consistent with sound research design and that do not unnecessarily expose subjects to risk, and (ii) Whenever appropriate, by using procedures already being performed on the subjects for diagnostic or treatment purposes.

(2) Risks to subjects are reasonable in relation to anticipated benefits, if any, to subjects, and the importance of the knowledge that may reasonably be expected to result. In evaluating risks and benefits, the IRB should consider only those risks and benefits that may result from the research (as distinguished from risks and benefits of therapies subjects would receive even if not participating in the research). . . .

(3) Selection of subjects is equitable. In making this assessment the IRB should take into account the purposes of the research and the setting in which the research will be conducted. The IRB should be particularly cognizant of the special problems of research that involves a category of subjects who are vulnerable to coercion or undue influence, such as children, prisoners, individuals with impaired decision-making capacity, or economically or educationally disadvantaged persons.

(4) Informed consent will be sought from each prospective subject, or the subject's legally authorized representative, in accordance with, and to the extent required by, § ___.116.

* * *

(6) When appropriate, the research plan makes adequate provision for monitoring the data collected to ensure the safety of subjects.

(7) When appropriate, there are adequate provisions to protect the privacy of subjects and to maintain the confidentiality of data.

(8) For purposes of conducting the limited IRB review required by § ___.104(d)(7)), the IRB need not make the determinations at paragraphs (a)(1) through (7) of this section, and shall make the following determinations: (i) Broad consent for storage, maintenance, and secondary research use of identifiable private information is obtained in accordance with the requirements of § ___.116(a)(1)–(4), (a)(6), and (d); (ii) [consent is documented]; and (iii) [if there is a change in the way information or biospecimens are stored, there are adequate provisions for confidentiality and privacy.]

* * *

(b) When some or all of the subjects are likely to be vulnerable to coercion or undue influence, such as children, prisoners, individuals with impaired decision-making capacity, or economically or educationally disadvantaged persons, additional safeguards have been included in the study to protect the rights and welfare of these subjects.

§ ___.112   Review by institution.

Research covered by this policy that has been approved by an IRB may be subject to further appropriate review and approval or disapproval by officials of the institution. However, those officials may not approve the research if it has not been approved by an IRB.

§ ___.113   Suspension or termination of IRB approval of research.

An IRB shall have authority to suspend or terminate approval of research that is not being conducted in accordance with the IRB's requirements or that has been associated with unexpected serious harm to subjects. . . .

* * *

§ ___.116   General requirements for informed consent.

(a) . . . Except as provided elsewhere in this policy:

(1) Before involving a human subject in research covered by this policy, an investigator shall obtain the legally effective informed consent of the subject or the subject's legally authorized representative.

(2) An investigator shall seek informed consent only under circumstances that provide the prospective subject . . . sufficient opportunity to discuss and consider whether or not to participate and that minimize the possibility of coercion or undue influence.

(3) The information that is given to the subject . . . shall be in language understandable to the subject . . . .

(4) The prospective subject . . . must be provided with the information that a reasonable person would want to have in order to make an informed decision about whether to participate, and an opportunity to discuss that information.

(5) Except for broad consent obtained in accordance with paragraph (d) of this section:

(i) Informed consent must begin with a concise and focused presentation of the key information that is most likely to assist a prospective subject . . . in understanding the reasons why one might or might not want to participate in the research. . . .

(ii) Informed consent as a whole must present information in sufficient detail relating to the research, and must be organized and presented in a way that does not merely provide lists of isolated facts, but rather facilitates the prospective subject's . . . understanding of the reasons why one might or might not want to participate.

(6) No informed consent may include any exculpatory language through which the subject . . . is made to waive or appear to waive any of the subject's legal rights, or releases or appears to release the investigator, the sponsor, the institution, or its agents from liability for negligence.

(b) Basic elements of informed consent. Except as provided in paragraph (d), (e), or (f) of this section, in seeking informed consent the following information shall be provided to each subject . . . :

(1) A statement that the study involves research, an explanation of the purposes of the research and the expected duration of the subject's participation, a description of the procedures to be followed, and identification of any procedures that are experimental;

(2) A description of any reasonably foreseeable risks or discomforts to the subject;

(3) A description of any benefits to the subject or to others that may reasonably be expected from the research;

(4) A disclosure of appropriate alternative procedures or courses of treatment, if any, that might be advantageous to the subject;

(5) A statement describing the extent, if any, to which confidentiality of records identifying the subject will be maintained;

(6) For research involving more than minimal risk, an explanation as to whether any compensation and an explanation as to whether any medical treatments are available if injury occurs and, if so, what they consist of, or where further information may be obtained; . . .

* * *

(8)   A statement that participation is voluntary, refusal to participate will involve no penalty or loss of benefits to which the subject is otherwise entitled, and the subject may discontinue participation at any time without penalty or loss of benefits to which the subject is otherwise entitled; and

(9)   One of the following statements about any research that involves the collection of identifiable private information or identifiable biospecimens:

(i)   A statement that identifiers might be removed from the identifiable private information or identifiable biospecimens and that, after such removal, the information or biospecimens could be used for future research studies or distributed to another investigator for future research studies without additional informed consent from the subject . . . , if this might be a possibility; or

(ii)   A statement that the subject's information or biospecimens collected as part of the research, even if identifiers are removed, will not be used or distributed for future research studies.

(c)   Additional elements of informed consent. Except as provided in paragraph (d) . . . or (f) of this section, one or more of the following elements of information, when appropriate, shall also be provided to each subject
. . . .

(1)   A statement that the particular treatment or procedure may involve risks to the subject (or to the embryo or fetus, if the subject is or may become pregnant) that are currently unforeseeable;

(2)   Anticipated circumstances under which the subject's participation may be terminated by the investigator without regard to the subject's . . . consent;

* * *

(4)   The consequences of a subject's decision to withdraw from the research and procedures for orderly termination of participation by the subject;

(5)   A statement that significant new findings developed during the course of the research that may relate to the subject's willingness to continue participation will be provided to the subject;

(6)   The approximate number of subjects involved in the study;

(7)   A statement that the subject's biospecimens (even if identifiers are removed) may be used for commercial profit and whether the subject will or will not share in this commercial profit;

(8) A statement regarding whether clinically relevant research results, including individual research results, will be disclosed to subjects, and if so, under what conditions; and

(9) For research involving biospecimens, whether the research will (if known) or might include whole genome sequencing.

(d) Elements of broad consent for the storage, maintenance, and secondary research use of identifiable private information or identifiable biospecimens (collected for either research studies other than the proposed research or nonresearch purposes) is permitted as an alternative to the informed consent requirements in paragraphs (b) and (c) of this section. If the subject . . . is asked to provide broad consent, the following shall be provided to each subject . . . :

(1) The information required in paragraphs (b)(2), (b)(3), (b)(5), and (b)(8) and, when appropriate, (c)(7) and (9) of this section;

(2) A general description of the types of research that may be conducted . . . . This description must include sufficient information such that a reasonable person would expect that the broad consent would permit the types of research conducted;

(3) A description of the identifiable private information or identifiable biospecimens that might be used in research, whether sharing of [this material] might occur, and the types of institutions or researchers that might conduct research with [this material];

(4) A description of the period of time that [this material] may be stored and maintained (which period of time could be indefinite), and a description of the period of time that [this material] may be used for research purposes (which period of time could be indefinite);

(5) Unless the subject . . . will be provided details about specific research studies, a statement that they will not be informed of the details of any specific research studies that might be conducted using [this material], including the purposes of the research, and that they might have chosen not to consent to some of those specific research studies; [and]

(6) Unless it is known that clinically relevant research results, including individual research results, will be disclosed to the subject in all circumstances, a statement that such results may not be disclosed to the subject. . . .

* * *

(f) (1) [A]n IRB may waive the requirement to obtain informed consent for research under paragraphs (a) through (c) of this section, provided the IRB satisfies the requirements of paragraph (f)(3) of this section. If an individual was asked to provide broad consent . . . in accordance with the requirements at paragraph (d) of this section, and refused to consent, an IRB cannot waive consent [for those activities].

(2) Alteration. An IRB may approve a consent procedure that omits some, or alters some or all, of the elements of informed consent set forth in paragraphs (b) and (c) of this section provided the IRB satisfies the requirements of paragraph (f)(3) of this section. An IRB may not omit or alter any of the requirements described in paragraph (a) of this section. If a broad consent procedure is used, an IRB may not omit or alter any of the elements required under paragraph (d) of this section.

(3) Requirements for waiver and alteration. In order for an IRB to waive or alter consent as described in this subsection, the IRB must find and document that:

(i) The research involves no more than minimal risk to the subjects;

(ii) The research could not practicably be carried out without the requested waiver or alteration;

(iii) If the research involves using identifiable private information or identifiable biospecimens, the research could not practicably be carried out without using such information or biospecimens in an identifiable format;

(iv) The waiver or alteration will not adversely affect the rights and welfare of the subjects; and

(v) Whenever appropriate, the subjects . . . will be provided with additional pertinent information after participation.

* * *

(i) Preemption. The informed consent requirements in this policy are not intended to preempt any applicable Federal, state, or local laws (including tribal laws . . .) that require additional information to be disclosed in order for informed consent to be legally effective.

* * *

IV.   APPLYING THE COMMON RULE

A.   INSTITUTIONAL REVIEW BOARDS

The Common Rule delegates the frontline enforcement of the federal regulations to institutional review boards (IRBs), which most commonly are committees of the research institution, appointed and managed by the institution itself. What is the rationale for delegating oversight to the research entity itself?

Criticisms of the efficiency and effectiveness of IRB review contributed significantly to momentum for revision of the Common Rule. For a debate over the role of IRBs, see Symposium, Censorship and Institutional Review Boards, 101 Nw. U. L. Rev. 399 (2007); see also *Grimes* in Section IV.D. The revised Common Rule generally left the structure of IRBs unchanged, but for one exception. For studies conducted at more than one institution after January 20, 2020, the revised regulations require approval by a single IRB rather than the IRBs of each institution involved in the multisite study (called "cooperative research" in the regulations). § ___.114. In addition, the revision expands the range of studies that are exempt from the Rule, and therefore from full IRB review. See § ___.104(d)(4), (7) and (8), above.

The Office for Human Research Protections (OHRP) enforces the HHS regulations through investigation of complaints and episodic audits of the human subject protection programs (including the IRBs) of organizations conducting research subject to the regulations. The FDA uses similar processes to enforce its regulations. The biggest stick that OHRP carries is the authority to suspend or terminate all HHS funding of research. It suspended funding for twenty universities during 1999–2000. This extreme remedy is still used, but only rarely. The Office of Inspector General of HHS also enforces the federal regulations on research through its prosecution of fraud and false claims. Even with this oversight, the reviews performed by an IRB are the most significant tool in assuring compliance with the federal regulations.

PROBLEM: FORMING AN IRB

Assume that you have been charged with forming the first IRB for a community hospital that has begun to apply for small federal research grants and contracts with pharmaceutical firms to conduct clinical trials. Consider §§ ___.107, .109, .111, .112, and .113, above, in forming your recommendations to these questions:

1.   What is the ideal size for your IRB to incorporate the competencies and perspectives described in the regulations? Should hospital counsel be a member? Should nonscientific members of the IRB defer to the scientific members on the questions of risk assessment, research design, or continuing review of the results being produced by the research protocol? Can they do otherwise? What sort of competency should the nonscientific member have?

2.   Assume that your hospital serves a large proportion of economically disadvantaged patients and that research subjects are likely to be recruited from among that population. Does that influence the composition of your IRB?

3.   Assume that your new IRB has already ruffled some feathers with its review process. In particular, it has asked for more information, several times, on research proposed by a prominent physician on the hospital's medical staff. This researcher has approached the hospital's CEO for assistance. What can the CEO do?

4.   Instead of establishing its own IRB, the hospital could contract with and pay a free-standing independent IRB for review of protocols. See Janet Lis and Melinda Murray, The Ins and Outs of Independent IRBs, 2 J. Health & Life Sci. L. 73 (2008). See also Colleen Zern, Is the Customer Always Right? HHS Proposed Regulation Allows IRBs to Place Customer Service Ahead of the Welfare of Research Participants, 32 St. Louis U. Pub. L. Rev. 411 (2013), criticizing independent IRBs. What are the advantages and disadvantages of this approach for the hospital? For researchers? For patients who will be participating in research?

B. DEFINING "RESEARCH"

The Common Rule applies only to activities that meet its definition of research. Furthermore, the Common Rule does not apply to research that is not conducted by or funded by a federal agency; and the FDA regulations only apply to research that is funded by the FDA or submitted to the FDA for action regarding drugs or devices requiring FDA approval. An organization that conducts research subject to the Common Rule must file a Federalwide Assurance (FWA) every five years. In the FWA, the organization represents that it complies with applicable federal regulations on research covered by the Rule and maps essential components of that compliance. Most academic institutions filing an FWA voluntarily commit to compliance for all their research studies, whether federally funded or not.

PROBLEM: IS IT RESEARCH?

Use the Common Rule §§ ___.101(a), .102(e) and (l), and .104, above, to resolve whether each scenario constitutes research subject to the regulations.

1.   An oncologist treating a patient with colon cancer for whom customary chemotherapy drugs have proven ineffective offers his patient the option of trying another drug. The drug is currently used for other cancers and is being tested for colon cancer, but the research is not yet complete. Although the oncologist is not participating in any of the clinical trials of this drug, preliminary unpublished reports and informal discussions with researchers testing the drug show promise. Once a drug is approved for one purpose, physicians may prescribe it for other purposes, a practice known as off-label prescribing. If the doctor gives his patient this drug, is it research as defined in the Common Rule? See Nancy Kass et al., The Research-Treatment Distinction: A Problematic Approach for Determining Which Activities Should Have Ethical Oversight, 43 Hastings Ctr. Rep. S4 (2013). See also Section IV.C, below.

2.   It can be difficult to intubate seriously overweight adult patients, especially in an emergency. Assume that an experienced nurse anesthetist suggested that physicians use a pediatric tube for a particular patient, with great success. Because of their experience with this approach, the hospital decides to adopt this as a regular practice if a first attempt at intubation fails.

The nurse who developed this approach has decided to present the hospital's experience with this practice at a regional anesthesiology conference and has published an article in a professional journal. Is this research? Scenario based on Atul Gawande, "When Doctors Make Mistakes" in Complications: A Surgeon's Notes on an Imperfect Science (2002).

3.   A large multispecialty physician practice group is considering adding medical scribes to their offices. These individuals would be present as the physician works with the patient and would input findings, prescriptions, and directions into the patient's electronic medical record. This should allow physicians to focus on the patient rather than the keyboard. Because of the size of the financial investment required, the practice wants to make sure that adding scribes would be efficient and would improve accuracy and thoroughness of medical records in their particular practice setting. They have hired a small number of scribes and have established a number of variables, requiring examination of patient records and observing physician-patient interaction, which will be analyzed to judge whether further investment will add value. The group will use the results only within their own practice. They have no interest in sharing their findings. Who are the subjects of this study? Is this research subject to the Common Rule?

4.   A large university medical center shared redacted medical records of all patients treated over a multi-year period with Google as part of a project to develop a predictive health model designed to help hospitals reduce their rates of readmission. Is this research under the Common Rule? See Dinerstein v. Google, LLC, 484 F. Supp. 3d 561 (N.D. Ill. 2020). For analyses of the failure of law to account adequately for the interests of individuals whose personal data is used for research and development, see Michael Froomkin, Big Data: Destroyer of Informed Consent, 21 Yale J. L. & Tech. 27 (2019); Tara Sklar & Mabel Crescioni, Research Participants' Rights to Data Protection in the Era of Open Science, 69 DePaul L. Rev. 699 (2020); David M. Parker et. al., Privacy and Informed Consent for Research in the Age of Big Data, 123 Penn St. L. Rev. 703 (2019). See also Joshua Rolnick et al., Ethical Oversight of Quality Improvement and the Research-QI Boundary: A New Common Rule Changes Little, IRB: Ethics and Human Research 39/3 (May-June 2017).

5.   Starting in 2018, the Seattle Flu Study (SFS) routinely enrolled research subjects and collected and stored nasal swabs to improve the detection of and response to local flu outbreaks. When Seattle experienced its first COVID case in early 2020, SFS used recently collected nasal swabs to test for COVID-19 and notified those research participants whose COVID-19 tests were positive. Is this research under the Common Rule?

C. RISK AND CONSENT

The Common Rule sets standards for consent to research in §§ ___.111 and 116, above. Standardization in consent forms, to improve efficiency and effectiveness, was considered during the revision of the regulations, but that didn't occur. Instead, the regulations add a new requirement that consent forms be posted on a public website once enrollment in the study has closed. § ___.116(h). The regulations also significantly altered the consent requirements for "secondary use" of "identifiable private information" and "identifiable biospecimens." See Section IV.F, below.

The issue of risk is intertwined with consent in several ways. First, the federal regulations do not permit research, even with consenting adults, if the risks to the subject are not reasonable. See § ___.111(a)(2), above. Assuming a competent adult is acting voluntarily and without coercion, what, if anything, justifies the restrictive federal regulations concerning risk? See Richard A Epstein, Defanging IRBs: Replacing Coercion with Information, 101 Nw. U. L. Rev. 735 (2007), arguing that IRBs should assure only that information to the subject about the risks is accurate and understandable, leaving to the individual the decision to enroll or not. What criteria would you use to decide whether risk is excessive, either for yourself personally or as a matter of applying the regulation? Are these two different questions?

The consent process must communicate the nature and degree of risk (and benefit) presented in the study in an accurate and understandable manner. See *Looney*, below. In some circumstances, however, natural optimism on the part of patients, and even researchers, influences how the information is heard. The term "therapeutic misconception" was coined over thirty years ago to describe the observation that patients tend to believe that clinical trials testing only the safety of a new drug (called Phase I trials) will benefit them directly by providing treatment for their condition despite clear statements to the contrary. See Paul Appelbaum et al., The Therapeutic Misconception: Informed Consent to Psychiatric Research, 5 Int'l J.L. & Psych. 319 (1982); Lynn Jansen, Two Concepts of Therapeutic Optimism, 37 J. Med. Ethics 563 (2011). One IRB chair observed that, during an emergency, IRB members also can fall prey to therapeutic misconception. See Walter Dehority, Therapeutic Misconception, Misestimation and COVID-19, 11 Narrative Inq. in Bioeth. 61 (2021).

The calculus of risk can change over the life of a study. IRBs must perform ongoing risk assessments for certain approved studies (§ ___.109(e), above, with exceptions adopted in the 2017 revision) and must suspend research that is "associated with unexpected serious harm to subjects" (§ ___.113, above). This monitoring should assure that a study is halted when unanticipated levels of risk or unexpectedly early and strong positive results occur. How does the latter present a risk of harm?

Some research studies include medical tests that will produce results that are "clinically relevant" to individual subjects. In the context of medical treatment, a patient expects that her physician would inform her of significant test results. If this common expectation is carried to the research context, individuals may assume, to their detriment, that if the tests reveal a serious threat to their health, they would be informed. The revised Common Rule addresses this situation as an element of informed consent but does not require disclosure of results. See § ___.116(c)(8) and (d)(6), above. See Susan Wolf et al., Returning a Research Participant's Genomic Results to Relatives: Analysis and Recommendations, 43 J. L. Med. & Ethics 440 (2015).

Research on vaccines or therapeutics for COVID-19 demonstrates the difficulty of keeping research subjects informed about risks and alternatives given how quickly new information emerged about the virus, its variants, the resulting disease, and the effectiveness of masks and vaccines. For a discussion of how best to interpret and apply the Common Rule in such circumstances, see Holly Fernandez Lynch et. al., Regulatory Flexibility for Covid-19 Research, 7 J.L. & Biosciences (Jan.–June, 2020). For an ethical analysis of "challenge trials"—human experiments in which subjects are deliberately exposed to a disease—see Daniel P. Sulmasy, Are SARS-CoV-2 Human Challenge Trials Ethical?, 181 JAMA Intern Med. 1031 (2021).

PROBLEM: HOW INFORMED? HOW VOLUNTARY?

Consider the following scenarios under § ___.116, above.

1. *Taking Advantage of Medicaid Beneficiaries?* A teaching hospital is likely to enroll a large proportion of Medicaid beneficiaries and uninsured individuals in a particular study. The medication being tested has been approved for one use, but not for the one being tested in this study. Sales of the medication overall currently exceed $300 million annually even though Medicaid does not pay for the drug for the use being studied. Does the fact that Medicaid does not cover the medication make the enrollment of Medicaid beneficiaries coercive? If the protocol promises subjects hospital care that some otherwise would not receive due to inability to pay, is that coercive? See generally Ruqaiijah Yearby, Involuntary Consent: Conditioning Access to Health Care on Participation in Clinical Trials, 44 J. L. Med. & Ethics 445 (2016).

2. *When Is Payment to a Subject Too Much?* To recruit adequate numbers of subjects, an investigator at the same hospital plans on paying subjects $50 per hour for the time the subject spends traveling to and from the office as well as the time spent in undergoing tests and evaluation. How does payment relate to consent? Is the $50 excessive? Adequate? Fair? Emily Largent and Holly Lynch, Paying Research Participants: Regulatory Uncertainty, Conceptual Confusion, and a Path Forward, 17 Yale J. Health Pol'y & Ethics 61 (2017).

3. *Access to Investigational Therapies.* A terminally ill cancer patient is told by her oncologist that "nothing more can be done" to treat the disease in her case. When she contacts a cancer center at a teaching hospital, she is

invited to participate in a Phase I trial to test the safety of a new drug. The investigators have some reason to believe that the drug holds promise because of its performance in animal studies, but the drug has not been tested in humans. They plan to randomly assign subjects to receive either the drug or a placebo, and neither the physicians nor the subjects will know which arm of the study they are participating in. What issues do you see in securing informed consent under § ___.116? What if the patient tells the researcher that her willingness to continue in the trial depends on whether she is in the placebo arm? On access to investigational drugs, see Abigail Alliance for Better Access to Developmental Drugs v. von Eschenbach, 495 F.3d 695 (D.C. Cir. 2007), cert. denied, 552 U.S. 1159 (2008), rejecting plaintiff's claim of a Constitutional right to receive drugs that are in Phase I clinical trials without enrolling in the trial. After *Abigail Alliance*, the FDA increased access to experimental drugs outside of clinical trials. The 21st Century Cures Act, P.L. 114–255 (2016) adopted a number of policies to accelerate drug approval and relax standards for data required in the approval process. See also Rebecca Dresser, The "Right to Try" Investigational Drugs: Science and Stories in the Access Debate, 93 Tex. L. Rev. 1631 (2015).

4.   *Disclosing Off-Label Availability*. An individual has read about some promising results in a clinical trial of a new treatment being tested for her serious medical condition and has inquired about enrolling in the study. The medication is not new, but it has not been approved for her particular medical condition. Does the Common Rule require that she be told that her own treating physician could prescribe the medication for her off-label? See, e.g., Jerry Menikoff and Edward Richards, What the Doctor Didn't Say: The Hidden Truth About Medical Research (2006).

LOONEY V. MOORE
Court of Appeals for the 11th Circuit, 2017.
861 F.3d 1303.

Through their parents, Plaintiffs DreShan Collins, Christian Lewis, and Jaylen Malone, brought claims against Defendants for harms allegedly visited on Plaintiffs when the latter were enrolled in a clinical study while being treated for health issues accompanying their premature births. . . .

Plaintiffs brought claims against the various Defendants for negligence, negligence per se, breach of fiduciary duty, products liability, and lack of informed consent. The district court granted summary judgment on all claims. Like the district court, we conclude that Plaintiffs have failed to establish that participation in the clinical study caused any injuries, which means that the negligence, negligence per se, breach of fiduciary duty, and products liability claims were properly dismissed.

The viability of the claim alleging lack of informed consent, however, is less clear. . . .

I. BACKGROUND

The University of Alabama at Birmingham was the lead study site for a national clinical research trial known as the Surfactant, Positive Pressure, and Oxygenation Randomized Trial ("SUPPORT"). Designed by Dr. [Waldemar] Carlo and approved by the IRB Defendants, the SUPPORT study was created to analyze the effects of differing oxygen saturation levels on premature infants. At the time of the study, it was nationally accepted (and neither party contests) that the recognized standard of care was to keep the oxygen saturation levels of low-birth-weight infants at between 85% and 95%. This standard notwithstanding, it was also known that a prolonged period of high oxygen saturation can result in oxygen toxicity which leads to an increased risk of "retinopathy of prematurity". On the other hand, a prolonged period of low oxygen saturation can result in neuro-developmental impairment and death. Given the difficulties of calibrating the optimal oxygen range, the SUPPORT study sought to determine whether, within the accepted standard of care, there was a more precise range of oxygen saturation that would better reduce the risk of exposing an infant to either too much or too little oxygen.

The SUPPORT study randomly divided eligible and enrolled premature infants into two groups. One group was to be kept at an oxygen saturation level between 85–89%, which is the low end of the standard-of-care range, while the other would be kept at an oxygen saturation level between 90–95%, which is the high end of that range. Further, to ensure double-blind data collection, the study would employ specialized oximeters . . . that would "mask" to an onlooker the true oxygen saturation levels of the infants. The oximeters would, however, signal an alarm whenever an infant's oxygen level strayed below 85% or above 95%.

Publishing the results of the study in the New England Journal of Medicine, the study authors concluded that infants in the high-oxygen group were more likely to be diagnosed with retinopathy while infants in the low-oxygen group were more likely to die. There was no statistically significant difference in the incidence of neuro-developmental impairments between the high and low groups.

To enroll in the study, Plaintiffs' guardians had to execute informed consent documents that were drafted and approved by Defendants. After the study's completion, however, the Department of Health and Human Services authored a letter questioning whether these informed consent documents had properly disclosed all of the risks associated with enrollment in the SUPPORT study.

. . . Plaintiffs allege that they suffered serious injuries as a result of their participation in the study. Specifically, Plaintiffs Lewis and Malone were assigned to the low-oxygen group, with prolonged periods of low oxygen saturation being associated with neuro-developmental impairment and death. Fortunately, neither infant died, but they did develop neurological issues. Plaintiff Collins was assigned to the high-oxygen group, with prolonged high-oxygen saturation being associated with retinopathy, which can lead to blindness. Plaintiff Collins did develop retinopathy, but fortunately he did not suffer permanent vision loss. Following discovery, Defendants moved for summary judgment asserting that, based on the undisputed material facts, Plaintiffs had failed to demonstrate that participation in the SUPPORT study had caused the injuries alleged in the Complaint. The district court agreed that Plaintiffs had failed to prove that their injuries were caused by participation in the SUPPORT study, as opposed to being a consequence of their premature births.

II. ANALYSIS

We agree that under applicable Alabama law and taking all inferences in the light most favorable to Plaintiffs, Plaintiffs have failed to show that enrollment in the SUPPORT study caused their injuries. What is not clear is whether the absence of an actual physical injury caused by the SUPPORT study dooms Plaintiffs' argument that Defendants are nonetheless liable because they failed to obtain Plaintiffs' informed consent to participate in that study. . . .

. . . As far as we can tell . . . Alabama law has yet to explicitly address the question whether proof of a medical injury is also required before a plaintiff can claim that his consent to a medical procedure was not informed. Specifically, if a plaintiff cannot prove that he suffered any injury as a result of a particular medical procedure, can he still potentially prevail if he shows that the doctor failed to obtain his informed consent to that procedure? In other words, is there a free-standing tort arising from a lack of informed consent, even if there is no injury resulting from the procedure at issue? Assuming that injury is required to sustain an informed consent claim arising out of medical treatment, does that rule still apply if the medical treatment is provided as part of a clinical study?

The answer to the above questions will depend on how an informed consent claim is characterized. If an informed consent claim is considered to be a type of medical malpractice claim [under the Alabama malpractice statute] . . . then it is clear that a plaintiff must show the existence of an actual injury resulting from the procedure before he can raise a viable informed consent claim. If an informed consent claim is not classified as a malpractice type of claim—or if it exits [in] the malpractice arena when the uninformed consent pertains to participation in a medical study administered as part of the medical treatment—then we must search for an answer in Alabama common law. And if, as Plaintiffs argue, an informed consent claim arising out of a medical study is neither a malpractice nor a negligence claim, then we must identify precisely what type of claim it is and determine what Alabama law would prescribe as its elements.

* * *

. . . Because substantial doubt exists here about the answer to a material state law question, we respectfully certify the following question [to the Alabama Supreme Court] for a determination of state law:

> Must a patient whose particular medical treatment is dictated by the parameters of a clinical study, and who has not received adequate warnings of the risks of that particular protocol, prove that an injury actually resulted from the medical treatment in order to succeed on a claim that his consent to the procedure was not informed?

NOTES AND QUESTIONS

1.   *The Question Remains Unanswered.* The Eleventh Circuit certified an intriguing question to the Alabama Supreme Court, but that court declined to answer it. In its subsequent opinion in the case, the Eleventh Circuit makes only a glancing reference to the distinction between medical treatment and medical experimentation, referring to plaintiffs' claim as "an informed consent claim arising out of participation in a medical study administered as part of the medical treatment" provided plaintiffs. There is no other discussion of the issue. The court concludes that, in its interpretation of Alabama law, informed consent claims are a form of negligence claim and require proof of physical injury. 2018 WL 1547260. See also the discussion of the litigation involving Jesse Gelsinger and John Moore in Section IV.E, below.

2.   *Private Right of Action.* Should the federal regulations be enforceable in litigation by research participants, such as those in *Looney,* or should the government hold exclusive enforcement authority? Almost all of the cases considering the question have held that the federal regulations do not create an implied private right of action. Compare *Grimes,* below. See Valerie Koch, A Private Right of Action for Informed Consent in Research, 45 Seton Hall L. Rev. 173 (2015); Richard S. Saver, Medical Research and Intangible Harm, 74 U. Cinn. L. Rev. 941 (2006). If a private right of action is desirable, should Congress or the state legislature enact legislation providing for statutory damages in the absence of physical harm? See Cal. Health & Safety Code § 24176, providing for fines for violation of state standards for consent for research but only for research not subject to the federal regulations. Actions brought for failure of consent under the FDA's regulations on research may be barred by federal preemption of state products liability law. See Day v. Howmedica Osteonics Corp., 2015 WL 13469348 (D. Colo. 2015); but see Mink v. Smith & Nephew, Inc., 860 F.3d 1319 (11th Cir. 2017), overturning dismissal of a claim of misrepresentation.

3.   *Assessing the Disclosure.* The written consent for SUPPORT stated that "[s]ometimes higher ranges [of oxygen] are used and sometimes lower ranges are used. All of them are acceptable ranges." Is this statement accurate?

Complete? The federal Office for Human Research Protections determined that consent in SUPPORT failed to meet regulatory standards as it did not clearly identify risks and how the study might impact treatment decisions. OHRP noted that, while the entire range of oxygenation was within the current standard of care, treating physicians, including those at some of the research sites, often avoid the extreme poles of the acceptable range to reduce risks. OHRP Determination Letter (June 6, 2013). See also Robert Morse and Robin Wilson, Realizing Informed Consent in Times of Controversy: Lessons from the SUPPORT Study, 44 J. L. Med. & Ethics 402 (2016). The OHRP letter states:

> Doctors are required . . . to do what they view as being best for their individual patients. Researchers do not have the same obligation: Our society relaxes that requirement because of the need to conduct research, the results of which are important to us all. As a modest but crucial trade-off . . . society requires that researchers tell subjects how participating in the study might alter the risks to which they are exposed.

OHRP, however, took no compliance action against research institutions participating in SUPPORT and stated that they would not take any action against studies with a similar design until OHRP developed and posted formal guidance. OHRP has posted Draft Guidance on Disclosing Reasonably Foreseeable Risks in Research Evaluating Standards of Care, 79 Fed. Reg. 63629 (Oct. 24, 2014).

4. *Research Design.* The SUPPORT study is an example of comparative effectiveness research, as the study tested the outcomes of different treatment approaches in common practice at the time. D. Magnus, The SUPPORT Controversy and the Debate over Research within the Standard of Care, 13 Am. J. Bioethics 1 (2013), part of a symposium issue on SUPPORT. The June 2013 OHRP determination letter stated that OHRP "does not and has never questioned whether the design of the SUPPORT study was ethical," but only that consent was inadequate.

5. *Research Injury Compensation.* Research participants may be injured in the course of research, just as patients may be injured in the course of treatment either through unavoidable adverse effects or through negligence. The Common Rule does not require that institutions compensate subjects injured as a result of their research participation. If, however, a research protocol provides for compensation in the event of an injury, the Common Rule requires that this be disclosed to potential subjects as part of the informed consent process. See § ___.116(b)(6), above. See also Leslie Henry, Moral Gridlock: Conceptual Barriers to No-Fault Compensation for Injured Research Subjects, 41 J. L. Med. & Ethics 411 (2013).

D. EQUITABLE SELECTION OF SUBJECTS AND VULNERABLE POPULATIONS

The Common Rule requires that IRBs assure that the selection of subjects for research be "equitable," and that special attention be paid when subjects are selected from "vulnerable populations," including but not limited to those listed in the regulation. § ___.111(a)(3), above. In addition, if research subjects are selected from populations vulnerable to undue influence or coercion, the IRB must assure that "additional safeguards" are put in place to protect these subjects. § ___.111(b), above. If an IRB regularly reviews research with vulnerable populations, it is to "give consideration" to the composition of its own membership. See Section IV.A, above.

Review these regulations before reading *Grimes*. Develop your own working definition of "equitable" and "vulnerability" to apply to the case. See Ana Iltis, Introduction: Vulnerability in Biomedical Research (Symposium), 37 J. L. Med. & Ethics 6 (2009); Ruqaiijah Yearby, Missing the "Target:" Preventing the Unjust Inclusion of Vulnerable Children for Medical Research Studies, 42 Am. J. Law & Med. 797 (2016).

GRIMES V. KENNEDY KRIEGER INSTITUTE, INC.
Court of Appeals of Maryland, 2001.
782 A.2d 807.

CATHELL, JUDGE.

* * *

[According to the record before the Court on the motion for summary judgment in] these present cases, [the Kennedy Krieger Institute,] a prestigious research institute associated with Johns Hopkins University . . . created a nontherapeutic research program[2] whereby it required certain classes of homes to have only partial lead paint abatement modifications performed, and in at least some instances, including at least one of the cases at bar, arranged for the landlords to receive public funding by way of grants or loans to aid in the modifications. The research institute then encouraged, and in at least one of the cases at bar, required, the landlords to rent the premises to families with young children. In the event young children already resided in one of the study houses, it was contemplated that a child would remain in the premises, and the child was encouraged to remain, in order for his or her blood to be periodically analyzed. In other words, the continuing presence of the children that were the subjects of the study was required in order for the study to be complete. Apparently, the children and their parents involved in the cases *sub judice*

---

[2] At least to the extent that commercial profit motives are not implicated, therapeutic research's purpose is to directly help or aid a patient who is suffering from a health condition the objectives of the research are designed to address—hopefully by the alleviation, or potential alleviation, of the health condition. Nontherapeutic research generally utilizes subjects who are not known to have the condition the objectives of the research are designed to address, and/or is not designed to directly benefit the subjects utilized in the research, but, rather, is designed to achieve beneficial results for the public at large (or, under some circumstances, for profit).

were from a lower economic strata and were, at least in one case, minorities.

The purpose of the research was to determine how effective varying degrees of lead paint abatement procedures were. Success was to be determined by periodically, over a two-year period of time, measuring the extent to which lead dust remained in, or returned to, the premises after the varying levels of abatement modifications, and, as most important to our decision, by measuring the extent to which the theretofore healthy children's blood became contaminated with lead, and comparing that contamination with levels of lead dust in the houses over the same periods of time. . . .

* * *

. . . There was no complete and clear explanation in the consent agreements signed by the parents of the children that the research to be conducted was designed, at least in significant part, to measure the success of the abatement procedures by measuring the extent to which the children's blood was being contaminated. It can be argued that the researchers intended that the children be the canaries in the mines but never clearly told the parents. . . .

. . . Institutional Review Boards (IRB) are oversight entities within the institutional family to which an entity conducting research belongs. In research experiments, an IRB can be required in some instances by either federal or state regulation, or sometimes by the conditions attached to governmental grants that are used to fund research projects. Generally, their primary functions are to assess the protocols of the project to determine whether the project itself is appropriate, whether the consent procedures are adequate, whether the methods to be employed meet proper standards, whether reporting requirements are sufficient, and the assessment of various other aspects of a research project. One of the most important objectives of such review is the review of the potential safety and the health hazard impact of a research project on the human subjects of the experiment, especially on vulnerable subjects such as children. Their function is *not* to help researchers seek funding for research projects.

. . . In a letter dated May 11, 1992, [the IRB sent a message to the researcher of this project stating, in part]:

* * *

2. *Federal guidelines are really quite specific regarding using children as controls in projects in which there is no potential benefit* [to those children]. To call a subject a normal control is to indicate that there is no real benefit to be received [by the child]. . . . So we think it would be much more acceptable to indicate that the "control group" is being studied to determine what exposure outside the home may play

in a total lead exposure; thereby, indicating that these control individuals are gaining some benefit, namely learning whether safe housing alone is sufficient to keep the blood-lead levels in acceptable bounds. We suggest that you modify ... the consent form[s] ... accordingly. [Emphasis added [by the court].]

[T]his statement shows two things: (1) that the IRB had a partial misperception of the difference between therapeutic and nontherapeutic research and the IRB's role in the process and (2) that the IRB was willing to aid researchers in getting around federal regulations designed to protect children used as subjects in nontherapeutic research. . . .

* * *

While the validity of the consent agreement and its nature as a contract, the existence or nonexistence of a special relationship, and whether the researchers performed their functions under that agreement pursuant to any special relationships are important issues in these cases that we will address, the very inappropriateness of the research itself cannot be overlooked. It is apparent that the protocols of research are even more important than the method of obtaining parental consent and the extent to which the parents were, or were not, informed. If the research methods, the protocols, are inappropriate then, especially when the IRB is willing to help researchers avoid compliance with applicable safety requirements for using children in nontherapeutic research, the consent of the parents, or of any consent surrogates, in our view, cannot make the research appropriate or the actions of the researchers and the Institutional Review Board proper.

. . . The experiment was simply a "for the greater good" project. The specific children's health was put at risk, in order to develop low-cost abatement measures that would help all children, the landlords, and the general public as well.

* * *

It is clear to this Court that the scientific and medical communities cannot be permitted to assume sole authority to determine ultimately what is right and appropriate in respect to research projects involving young children free of the limitations and consequences of the application of Maryland law. The Institutional Review Boards, IRBs, are, primarily, in-house organs. In our view, they are not designed, generally, to be sufficiently objective in the sense that they are as sufficiently concerned with the ethicality of the experiments they review as they are with the success of the experiments. . . .

The conflicts are inherent. This would be especially so when science and private industry collaborate in search of material gains. Moreover, the special relationship between research entities and human subjects used in the research will almost always impose duties.

. . . It may well be that in the end, the trial courts will determine that no damages have been incurred in the instant cases and thus the actions will fail for that reason. In that regard, we note that there are substantial factual differences [among the] cases. But the actions, themselves, are not defective on the ground that no legal duty can, according to the trial courts, possibly exist. . . .

I.   The Cases

. . . [Plaintiffs] allege that KKI [Kennedy Krieger Institute] discovered lead hazards in their respective homes and, having a duty to notify them, failed to warn in a timely manner or otherwise act to prevent the children's exposure to the known presence of lead. Additionally, plaintiffs alleged that they were not fully informed of the risks of the research.

[The trial court granted summary judgment on the basis that KKI had no legal duty to warn the subjects of their exposure to lead.]

The trial court was incorrect. Such research programs normally create special relationships and/or can be of a contractual nature that create duties. The breaches of such duties may ultimately result in viable negligence actions. Because, at the very least, there are viable and genuine disputes of material fact concerning whether a special relationship, or other relationships arising out of agreements, giving rise to duties existed between KKI and both sets of appellants, we hold that the Circuit Court erred in granting KKI's motions for summary judgment in both cases before this Court. . . .

II.   Facts & Procedural Background

In summary, KKI conducted a study of five test groups of twenty-five houses each. The first three groups consisted of houses known to have lead present. The amount of repair and maintenance conducted increased from Group 1 to Group 2 to Group 3. The fourth group consisted of houses, which had at one time lead present but had since allegedly received a complete abatement of lead dust. The fifth group consisted of modern houses, which had never had the presence of lead dust. The twenty-five homes in each of the first three testing levels were then to be compared to the two control groups: the twenty-five homes in Group 4 that had previously been abated and the 25 modern homes in Group 5. The research study was specifically designed to do less than full lead dust abatement in some of the categories of houses in order to study the potential effectiveness, if any, of lesser levels of repair and maintenance.

If the children were to leave the houses upon the first manifestation of lead dust, it would be difficult, if not impossible, to test, over time, the rate of the level of lead accumulation in the blood of the children attributable to the manifestation. . . . Thus, it would benefit the accuracy of the test, and thus KKI, the compensated researcher, if children remained in the houses over the period of the study even after the presence of lead dust in the houses became evident.

* * *

B.   Case No. 128 [Ms. Viola Hughes and Daughter Ericka Grimes]

* * *

Nowhere in the consent form was it clearly disclosed to the mother that the researchers contemplated that, as a result of the experiment, the child might accumulate lead in her blood, and that in order for the experiment to succeed it was necessary that the child remain in the house as the lead in the child's blood increased or decreased, so that it could be measured. The Consent Form states in relevant part:

"PURPOSE OF STUDY:

As you may know, lead poisoning in children is a problem in Baltimore City and other communities across the country. Lead in paint, house dust and outside soil are major sources of lead exposure for children. Children can also be exposed to lead in drinking water and other sources. We understand that your house is going to have special repairs done in order to reduce exposure to lead in paint and dust. On a random basis, homes will receive one of two levels of repair. We are interested in finding out how well the two levels of repair work. The repairs are not intended, or expected, to completely remove exposure to lead.

We are now doing a study to learn about how well different practices work for reducing exposure to lead in paint and dust. We are asking you and over one hundred other families to allow us to test for lead in and around your homes up to 8 to 9 times over the next two years provided that your house qualifies for the full two years of study. Final eligibility will be determined after the initial testing of your home. We are also doing free blood lead testing of children aged 6 months to 7 years, up to 8 to 9 times over the next two years. We would also like you to respond to a short questionnaire every 6 months. This study is intended to monitor the effects of the repairs and is not intended to replace the regular medical care your family obtains.

* * *

BENEFITS

To compensate you for your time answering questions and allowing us to sketch your home we will mail you a check in the amount of $5.00. In the future we would mail you a check in the amount of $15 each time the full questionnaire is completed. The dust, soil, water, and blood samples would be tested for lead at [KKI] at no charge to you. *We would provide you with specific blood-lead results. We would contact you to discuss a summary of house test results and steps that you could take to reduce any risks of exposure.* [Emphasis added [by the court].]"

Pursuant to the plans of the research study, KKI collected dust samples [from the Grimes' house] on March 9, 1993, August 23, 1993, March 9, 1994, September 19, 1994, April 18, 1995, and November 13, 1995.[22] The March 9, 1993 dust testing revealed what the researchers referred to as "hot spots" where the level of lead was "higher than might be found in a completely renovated [abated] house." This information about the "hot spots" was not furnished to Ms. Hughes until December 16, 1993, more than nine months after the samples had been collected and, as we discuss, *infra,* not until after Ericka Grimes's blood was found to contain elevated levels of lead.

KKI drew blood from Ericka Grimes for lead content analysis on April, 9, 1993, September 15, 1993, and March 25, 1994. Unlike the lead concentration analysis in dust testing, the results of the blood testing were typically available to KKI in a matter of days. KKI notified Ms. Hughes of the results of the blood tests by letters dated April 9, 1993, September 29, 1993, and March 28, 1994, respectively. The results of the April 9, 1993 test found Ericka Grimes blood to be [within normal range]. * * * However, on two subsequent retests, long after KKI had identified "hot spots," but before KKI informed Ms. Hughes of the "hot spots," Ericka Grimes's blood lead level registered [highly elevated] on March 25, 1994. Ms. Hughes and her daughter vacated the Monroe Street property in the Summer of 1994, and, therefore, no further blood samples were obtained by KKI after March 25, 1994.

* * *

III. Discussion

* * *

Because of the way the cases *sub judice* have arrived, as appeals from the granting of summary judgments, there is no complete record of the specific compensation of the researchers involved. . . . Neither is there in the record any development of what pressures, if any, were exerted in respect to the researchers obtaining the consents of the parents and conducting the experiment. Nor, for the same reason, is there a sufficient indication as to the extent to which the Institute has joined with

---

[22] For some unexplained reason, processing the dust samples typically took several months. KKI notified Ms. Hughes of the dust sample results via letters dated December 16, 1993, December 17, 1993, May 19, 1994, October 28, 1994, July 19, 1995, and January 18, 1996, respectively.

commercial interests, if it has, for the purposes of profit, that might potentially impact upon the researcher's motivations and potential conflicts of interest—motivations that generally are assumed, in the cases of prestigious entities such as John Hopkins University, to be for the public good rather than a search for profit.

We do note that the institution involved . . . is a highly respected entity, considered to be a leader in the development of treatments, and treatment itself, for children infected with lead poisoning. With reasonable assurance, we can note that its reputation alone might normally suggest that there was no realization or understanding on the Institute's part that the protocols of the experiment were questionable, except for the letter from the IRB requesting that the researchers mischaracterize the study.

* * *

C.   Negligence

* * *

The relationship that existed between KKI and both sets of appellants in the case at bar was that of medical researcher and research study subject. Though not expressly recognized in the Maryland Code or in our prior cases as a type of relationship which creates a duty of care, evidence in the record suggests that such a relationship involving a duty or duties would ordinarily exist, and certainly could exist, based on the facts and circumstances of each of these individual cases. . . .

IV.   The Special Relationships

A.   The Consent Agreement

* * *

By having appellants sign the Consent Form, both KKI and appellants expressly made representations, which, in our view, created a bilateral contract between the parties. At the very least, it suggests that appellants were agreeing with KKI to participate in the research study with the expectation that they would be compensated, albeit, more or less, minimally, be informed of all the information necessary for the subject to freely choose whether to participate, and continue to participate, and receive promptly any information that might bear on their willingness to continue to participate in the study. This includes full, detailed, prompt, and continuing warnings as to all the potential risks and hazards inherent in the research or that arise during the research. KKI, in return, was getting the children to move into the houses and/or to remain there over time, and was given the right to test the children's blood for lead. As consideration to KKI, it got access to the houses and to the blood of children that had been encouraged to live in a "risk" environment. In other words, KKI received a measuring tool—the children's blood. Considerations existed, mainly money, food coupons, trinkets, bilateral promises, blood to be tested in order to measure success. "Informed consent" of the type used here, which imposes obligation and confers consideration on both researcher and subject (in these cases, the parents of the subjects) may differ from the more one-sided "informed consent" normally used in actual medical practice. . . .

B.   The Sufficiency of the Consent Form

* * *

A reasonable parent would expect to be clearly informed that it was at least contemplated that her child would ingest lead dust particles, and that the degree to which lead dust contaminated the child's blood would be used as one of the ways in which the success of the experiment would be measured. The fact that if such information was furnished, it might be difficult to obtain human subjects for the research, does not affect the need to supply the information, or alter the ethics of failing to provide such information. A human subject is entitled to *all* material information. . . . The "informed" consent was not valid because full material information was not furnished to the subjects or their parents.

* * *

D.   The Federal Regulations

A duty may be prescribed by a statute, or a special relationship creating duties may arise from the requirement for compliance with statutory provisions. [F]ederal regulations have been enacted that impose standards of care that attach to federally funded or sponsored research projects that use human subjects. [ ] . . . [The court excerpts the Common Rule, 45 C.F.R. § 46.116, which is reproduced, above.]

* * *

. . . [T]he risks associated with exposing children to lead-based paint were not only foreseeable, but were well known by KKI . . . . Moreover, in the present cases, the consent forms did not directly inform the parents that it was possible, even contemplated, that some level of lead, a harmful substance depending upon accumulation, might contaminate the blood of the children.

Clearly, KKI, as a research institution, is required to obtain a human participant's fully informed consent, using sound ethical principles. It is clear from the wording of the applicable federal regulations that this requirement of informed consent continues during the duration of the research study and applies to new or changing risks. . . .

* * *

V. The Ethical Appropriateness of the Research

* * *

Researchers cannot ever be permitted to completely immunize themselves by reliance on consents, especially when the information furnished to the subject, or the party consenting, is incomplete in a material respect. A researcher's duty is not created by, or extinguished by, the consent of a research subject or by IRB approval. The duty to a vulnerable research subject is independent of consent, although the obtaining of consent is one of the duties a researcher must perform. All of this is especially so when the subjects of research are children. Such legal duties, and legal protections, might additionally be warranted because of the likely conflict of interest between the goal of the research experimenter and the health of the human subject, especially, but not exclusively, when such research is commercialized. There is always a potential substantial conflict of interest on the part of researchers as between them and the human subjects used in their research. If participants in the study withdraw from the research study prior to its completion, then the results of the study could be rendered meaningless. There is thus an inherent reason for not conveying information to subjects as it arises, that might cause the subjects to leave the research project. That conflict dictates a stronger reason for full and continuous disclosure.

* * *

. . . Practical inequalities exist between researchers, who have superior knowledge, and participants "who are often poorly placed to protect themselves from risk." [ ] "[G]iven the gap in knowledge between investigators and participants and the inherent conflict of interest faced by investigators, participants cannot and should not be solely responsible for their own protection." [ ]

VI. Parental Consent for Children to Be Subjects of
Potentially Hazardous Nontherapeutic Research

The issue of whether a parent can consent to the participation of her or his child in a nontherapeutic health-related study that is known to be potentially hazardous to the health of the child raises serious questions with profound moral and ethical implications. What right does a parent have to knowingly expose a child not in need of therapy to health risks or otherwise knowingly place a child in danger, even if it can be argued it is for the greater good? . . . We have long stressed that the "best interests of the child" is the overriding concern of this Court in matters relating to children. . . .

To think otherwise, to turn over human and legal ethical concerns solely to the scientific community, is to risk embarking on slippery slopes, that all too often in the past, here and elsewhere, have resulted in practices we, or any community, should be ever unwilling to accept.

* * *

[The court describes several judicial opinions involving the donation of organs or tissue from a minor child or incompetent person to a family member.]

What is of primary importance to be gleaned in [these transplantation] cases is not that the parents or guardians consented to the procedures, but that they first sought permission of the courts, and received that permission, before consenting to a nontherapeutic procedure in respect to some of their minor children, but that was therapeutic to other of their children.

In the case sub judice, no impartial judicial review or oversight was sought by the researchers or by the parents. [Emphasis in original.] . . .

* * *

VII. Conclusion

We hold that in Maryland a parent, appropriate relative, or other applicable surrogate, cannot consent to the participation of a child or other person under legal disability in nontherapeutic research or studies in which there is any risk of injury or damage to the health of the subject.

We hold that informed consent agreements in nontherapeutic research projects, under certain circumstances can constitute contracts; and that, under certain circumstances, such research agreements can, as a matter of law, constitute "special relationships" giving rise to duties, out of the breach of which negligence actions may arise. We also hold that, normally, such special relationships are created between researchers and the human subjects used by the researchers. Additionally, we hold that governmental regulations can create duties on the part of researchers towards human subjects out of which "special relationships" can arise. Likewise, such duties and relationships are consistent with the provisions of the Nuremberg Code.

The determination as to whether a "special relationship" actually exists is to be done on a case by case basis. [ ] The determination as to whether a special relationship exists, if properly pled, lies with the trier of fact. We hold that there was ample evidence in the cases at bar to support a fact finder's determination of the existence of duties arising out of contract, or out of a special relationship, or out of regulations and codes, or out of all of them, in each of the cases.

* * *

RAKER, JUDGE, concurring in result only:

* * *

I cannot join in the majority's sweeping factual determinations that the risks associated with exposing children to lead-based paint were foreseeable and well known to appellees and that appellees contemplated lead contamination in participants' blood; that the children's health was put at risk; that there was no complete and clear explanation in the consent agreements that the research to be conducted was designed to measure the success of the abatement procedures by measuring the extent to which the children's blood was being contaminated and that a certain level of lead accumulation was anticipated; that the parental consent was ineffective; that the consent form was insufficient because it lacked certain specific warnings; that the consent agreements did not provide that appellees would provide repairs in the event of lead dust contamination subsequent to the original abatement measures; that the Institutional Review Board involved in these cases abdicated its responsibility to protect the safety of the research subjects by misconstruing the difference between therapeutic and nontherapeutic research and aiding researchers in circumventing federal regulations; that Institutional Review Boards are not sufficiently objective to regulate the ethics of experimental research; that it is never in the best interest of any child to be placed in a nontherapeutic research study that might be hazardous to the child's health; that there was no therapeutic value in the research for the child subjects involved; that the research did not comply with applicable regulations; or that there was more than a minimal risk involved in this study. I do not here condone the conduct of appellee, and it may well be that the majority's conclusions are warranted by the facts of these cases, but the record before us is limited. Indeed, the majority recognizes that the record is "sparse." The critical point is that these are questions for the jury on remand and are not properly before this Court at this time.

. . . I cannot join the majority in holding that, in Maryland, a parent or guardian cannot consent to the participation of a minor child in a nontherapeutic research study in which there is *any* risk of injury or damage to the health of the child without prior judicial approval and oversight. Nor can I join in the majority's holding that the research conducted in these cases was *per se* inappropriate, unethical, and illegal. Such sweeping holdings are far beyond the question presented in these appeals, and their resolution by the Court, at this time, is inappropriate. I also do not join in what I perceive as the majority's wholesale adoption of the Nuremberg Code into Maryland state tort law. . . .

Accordingly, I join the majority only in the judgment to reverse the Circuit Courts' granting of summary judgments to appellees.

NOTES AND QUESTIONS

1. *Categorizing Research with Children.* HHS regulations on research with children divide that research into three categories based on the level of risk presented: (1) "research not involving greater than the minimal risk;" (2) "research involving greater than minimal risk but presenting the prospect of direct benefit to the individual subjects;" and (3) "research involving greater than minimal risk and no prospect of direct benefit to individual subjects, but likely to yield generalizable knowledge about the subject's disorder or condition." 45 C.F.R. §§ 46.401–46.409. "Minimal risk" is defined in § ___.102(j), above.

Where there is more than minimal risk but the research will also benefit the child, the risk must be justified "by the anticipated benefit to the subjects," and "[t]he relation of the anticipated benefit to the risk [must be] at least as favorable to the subjects as that presented by available alternative approaches."

Research providing no direct benefit to the individual child subject and involving more than minimal risk is not allowed unless: there is only a "minor" increase over minimal risk; the "experiences" that accompany the research are those that the child-subjects are likely to undergo anyway; and the research is aimed at a disorder or condition actually suffered by the subject. On this point, see Ana Iltis, Justice, Fairness, and Membership in a Class: Conceptual Confusions and Moral Puzzles in the Regulation of Human Subjects Research, 39 J. L. Med. & Ethics 488 (2011).

2. *Risk and Public Health Research.* Was the *Grimes* court correct in its conclusion that federal regulations on research with children prohibited the lead study? See Loretta Kopelman, Pediatric Research Regulations Under Legal Scrutiny: Grimes Narrows Their Interpretation, 30 J.L. Med. & Ethics 38 (2002), arguing that the key issue in *Grimes* is the interpretation of "risk" as used in the regulations and that *Grimes* was correct in its application to the particular protocol. But see Lainie F. Ross, In Defense of the Hopkins Lead Abatement Studies, 30 J.L. Med. & Ethics 50 (2002), arguing that the court misinterpreted the regulations, making important research impossible. See also Diane Hoffmann and Karen Rothenberg, Whose Duty is it Anyway? The Kennedy Krieger Opinion and Its Implications for Public Health Research, 6 J. Health Care L. & Pol'y 109 (2003), criticizing the case for misunderstanding the nature of public health research. How would the regulations on research with children have applied in the SUPPORT study in *Looney*, above?

3. *Clarifying "Any Risk"?* The Court denied the University's motion for reconsideration, but in doing so, it issued a *per curiam* opinion noting that "the only conclusion we reached as a matter of law was that . . . summary judgment was improperly granted." It went on to state:

In the Opinion, we said at one point that a parent "cannot consent to the participation of a child . . . in nontherapeutic research or studies in which there is any risk of injury or damage to the health of the subject." [W]e think is clear [that] by "any risk," we meant any articulable risk beyond the minimal kind of risk that is inherent in any endeavor. The context of the statement was a nontherapeutic study that promises no medical benefit to the child whatever, so that any balance between risk and benefit is necessarily negative. As we indicated, the determination of whether the study in question offered some benefit, and therefore could be regarded as therapeutic in nature, or involved more than that minimal risk is open for further factual development on remand.

Does this statement clarify or confuse the Court's position? Subsequent Maryland cases considered the application of *Grimes*. See, e.g., Kennedy Krieger Inst., Inc. v. Partlow, 460 Md. 607, 191 A.3d 425 (2018) (research institute owes a duty of care to a sibling of research participant who lived in the same house as the research participant); Partlow v. Kennedy Krieger Inst., 2017 WL 4772626 (Md. Ct. Spec. App. 2017) (duty of researcher to subject identified in *Grimes* extended to the sibling who lived in the same house as a child enrolled as a subject); Smith v. Kennedy Krieger Inst., 2017 WL 1076481 (Md. Ct. Spec. App. 2017) (upholding refusal to instruct jury that consent form was a contract and upholding admission of testimony of mayor as to seriousness of lead paint hazards in Baltimore); White v. Kennedy Krieger Inst., 110 A.3d 724 (Md. Ct. Spec. App. 2015) (distinguishing therapeutic trial of treatment for children with moderate lead blood levels from nontherapeutic research in *Grimes* but holding that a duty existed as between researcher and subject).

4. *Nuremberg Code as Evidence of Legal Standard.* The *Grimes* majority noted that the "breach of obligations imposed on researchers by the Nuremberg Code, might well support actions sounding in negligence in cases such as those at issue here." Compare Washington Univ. v. Catalona, 437 F. Supp. 2d 985 (E.D. Mo. 2006), aff'd on other grounds, 490 F.3d 667 (8th Cir. 2007), in which the trial court concluded that the Nuremberg Code and the Declaration of Helsinki (see Section V, below) were irrelevant to legal standards for research in the U.S.

5. *Protection or Exclusion?* Does the protection of particular groups always work for good, or can it have undesirable consequences? For example, pregnant women may participate as subjects in research only if the research is therapeutic and the risk to the fetus is minimized, or if the risk to the fetus is minimal. 45 C.F.R. § 46.204–207. Fetal protection practices which raise the cost or inconvenience of including women in research contribute to the exclusion of all women who could be pregnant or become pregnant during the course of a study presenting any fetal risk. As a result, nearly all women, from early adolescence through post-menopause, were excluded from most drug studies.

Fetal protection policies were not the only reason women have been excluded from research studies, but these practices contributed to the absence of research examining how life-threatening conditions behave in the female body as compared to the male body and on the effectiveness and safety of pharmaceuticals on other than the male body. See Cynthia Hathaway, A Patent Extension Proposal to End the Underrepresentation of Women in Clinical Trials and Secure Meaningful Drug Guidance for Women, 67 Food & Drug L. J. 143 (2012); Barbara Noah, The Inclusion of Pregnant Women in Clinical Research, 7 St. Louis U. J. Health L. & Pol'y 353 (2014).

In this context, can "equitable" selection of subjects be read to require inclusion of certain groups? Several federal statutes, for example, have encouraged and provided significant incentives for pharmaceutical research with children. See, e.g., Best Pharmaceuticals for Children Act and the Pediatric Research Equity Act.

People with disabilities have also been excluded from human subjects research. See Dianne Rios et al., Conducting Accessible Research: Including People with Disabilities in Public Health, Epidemiological, and Outcomes Studies, 106 Am. J. Publ. Health 2137–2144 (2016); A. S. Williams and S. M. Moore, Universal Design of Research: Inclusion of Persons with Disabilities in Mainstream Biomedical Studies, 3(82) Science Translational Medicine 82cm12 (2011).

E. COMMERCIAL INTERESTS IN RESEARCH

Several developments have worked a revolution in the financial infrastructure for research in the U.S. In 1980, Congress enacted the Bayh-Dole Act, 35 U.S.C. § 200, and unleashed the entrepreneurial spirit in academic research, transferring property rights in the products of federally funded research to the researchers and their institutions. The Act encouraged researchers to form for-profit companies for the development and marketing of the products of research for quicker translation into clinical use. Universities developed "tech transfer" offices to capitalize on this opportunity, and the number of patents issued to universities increased more than tenfold in the early years after Bayh-Dole.

Direct financial support for research also has shifted significantly, with much more research on pharmaceuticals, biologics, testing materials, and devices being supported directly by industry rather than public funds. In fact, industry support for pharmaceutical research dwarfs that available from government sources. See Marc Rodwin, Independent Clinical Trials to Test Drugs: The Neglected Reform, 6 St. Louis U. J. Health L. & Pol'y 113 (2012). The advent of entrepreneurial incentives and a profit-oriented infrastructure has raised serious concerns, however.

Critics of financial incentives in research argue that they can compromise research design; lead to overreaching in enrollment and retention of subjects; and lead to censoring of research results that are adverse to the sponsor. Because these financial interests are viewed as conflicting with goals of validity and credibility in research, as well as protection of human subjects, they are often called "conflicts of interest." To the extent that a conflict of interest generally requires that there be a fiduciary relationship between the parties—as between a treating physician and patient—using that term to embrace this situation may not be entirely accurate. See, e.g., William M. Sage, Some Principles Require Principals: Why Banning "Conflicts of Interest" Won't Solve Incentive Problems in Biomedical Research, 85 Tex. L. Rev. 1413 (2007). By short-handing "financial relationships" to "conflicts of interest," the latter term can be misunderstood to represent an individualized judgment about the personal integrity of the researcher. A financial conflict of interest, however, is identified at the point when the financial relationship is established and the research proposed, without an assessment of the researcher's personal character and before there is even opportunity for bad conduct. Robert Gatter, Walking the Talk of Trust in Human Subjects Research: The Challenge of Regulating Financial Conflicts of Interest, 52 Emory L.J. 327 (2003); Greg Koski, Research, Regulations, and Responsibility: Confronting the Compliance Myth: A Reaction to Professor Gatter, 52 Emory L.J. 403 (2003); Jesse Goldner, Regulating Conflicts of Interest in Research: The Paper Tiger Needs Real Teeth, 53 St. Louis U. L. J. 1211 (2009).

Financial payments and ownership interests of researchers and research institutions are not the only incentives in research, and so are not the only source of conflicts of interest. This narrow focus on financial interests alone fails to account for the broader range of incentives (reputational, competitive, etc.) that can influence researcher behavior in directions that may harm research subjects or compromise the validity of results. These incentives are often apparent in scientific misconduct cases where researchers have falsified results, rather than under the financial conflicts of interest regulations. See generally Gary Marx, An Overview of the Research Misconduct Process and an Analysis of the Appropriate Burden of Proof, 42 J.C. & U.L. 311 (2016).

The revised Common Rule addresses the issue of commercial exploitation of research results and products as a matter of consent. See § ___.116(c)(7) as to biospecimens. See also Section IV.F, below. Consider the following case:

GREENBERG V. MIAMI CHILDREN'S HOSPITAL RESEARCH INSTITUTE

United States District Court, S.D. Florida, 2003.
264 F. Supp. 2d 1064.

[Eds. Note: Canavan disease is a progressive, neurological genetic condition for which there is no cure. Symptoms usually appear in infancy, and most children with the disease do not live past the age of 10. It is genetic and primarily affects children of eastern and central European Jewish (Ashkenazi) or Saudi Arabian descent.]

* * *

The Complaint alleges a tale of a successful research collaboration gone sour. In 1987, Canavan disease still remained a mystery—there was no way to identify who was a carrier of the disease, nor was there a way to identify a fetus with Canavan disease. Plaintiff Greenberg approached Dr. Matalon, a research physician who was then affiliated with the University of Illinois at Chicago, for assistance. Greenberg requested Matalon's involvement in discovering the genes that were ostensibly responsible for this fatal disease, so that tests could be administered to determine carriers and allow for prenatal testing for the disease.

At the outset of the collaboration, Greenberg and the Chicago Chapter of the National Tay-Sachs and Allied Disease Association, Inc. ("NTSAD") located other Canavan families and convinced them to provide tissue (such as blood, urine, and autopsy samples), financial support, and aid in identifying the location of Canavan families internationally. The other individual Plaintiffs began supplying Matalon with the same types of information and samples beginning in the late 1980s. Greenberg and NTSAD also created a confidential database and compilation—the Canavan registry—with epidemiological, medical and other information about the families.

Defendant Matalon became associated in 1990 with Defendants Miami Children's Hospital Research Institute, Inc. and Variety Children's Hospital d/b/a Miami Children's Hospital. Defendant Matalon continued his relationship with the Plaintiffs after his move, accepting more tissue and blood samples as well as financial support.

The individual Plaintiffs allege that they provided Matalon with these samples and confidential information "with the understanding and expectations that such samples and information would be used for the specific purpose of researching Canavan disease and identifying mutations in the Canavan disease which could lead to carrier detection within their families and benefit the population at large." Plaintiffs further allege that it was their "understanding that any carrier and prenatal testing developed in connection with the research for which they were providing essential support would be provided on an affordable and accessible basis, and that Matalon's research would remain in the public domain to promote the discovery of more effective prevention techniques and treatments and, eventually, to effectuate a cure for Canavan disease." . . .

There was a breakthrough in the research in 1993. Using Plaintiffs' blood and tissue samples, familial pedigree information, contacts, and financial support, Matalon and his research team successfully isolated the gene responsible for Canavan disease. After this key advancement, Plaintiffs allege that they continued to provide Matalon with more tissue and blood in order to learn more about the disease and its precursor gene.

In September 1994, unbeknownst to Plaintiffs, a patent application was submitted for the genetic sequence that Defendants had identified. This application was granted in October 1997, and Dr. Matalon was listed as an inventor on the gene patent and related applications for the Canavan disease. Through patenting, Defendants acquired the ability to restrict any activity related to the Canavan disease gene, including without limitation: carrier and prenatal testing, gene therapy and other treatments for Canavan disease and research involving the gene and its mutations.

Although the Patent was issued in October 1997, Plaintiffs allege that they did not learn of it until November 1998, when MCH revealed their intention to limit Canavan disease testing through a campaign of restrictive licensing of the Patent. Specifically, on November 12, 1998, Plaintiffs allege that Defendants MCH and MCHRI began to "threaten" the centers that offered Canavan testing with possible enforcement actions regarding the recently-issued patent. Defendant MCH also began restricting public accessibility through negotiating exclusive licensing agreements and charging royalty fees.

Plaintiffs allege that at no time were they informed that Defendants intended to seek a patent on the research. Nor were they told of Defendants' intentions to commercialize the fruits of the research and to restrict access to Canavan disease testing.

... Plaintiffs generally seek a permanent injunction restraining Defendants from enforcing their patent rights, damages in the form of all royalties Defendants have received on the Patent as well as all financial contributions Plaintiffs made to benefit Defendants' research. Plaintiffs allege that Defendants have earned significant royalties from Canavan disease testing in excess of $75,000 through enforcement of their gene patent, and that Dr. Matalon has personally profited by receiving a recent substantial federal grant to undertake further research on the gene patent.

Informed Consent

... Plaintiffs, who served as research subjects, ... claim that Defendants owed a duty of informed consent. ... Defendants breached this duty, Plaintiffs claim, when they did not disclose the intent to patent and enforce for their own economic benefit the Canavan disease gene. ... Finally, the Plaintiffs allege that if they had known that the Defendants would "commercialize" the results of their contributions, they would not have made the contributions.

* * *

Defendants assert that extending a possible informed consent duty to disclosing economic interests ... would have pernicious effects over medical research, as it would give each donor complete control over how medical research is used and who benefits from that research. The Court agrees and declines to extend the duty of informed consent to cover a researcher's economic interests in this case.

Plaintiffs cite a variety of authorities in support of their contention that the duty of informed consent mandates that research subjects must be informed of the financial interests of the researcher. They first rely on *Moore v. Regents of the University of California,* where the court held that a physician/researcher had a duty of informed consent to disclose that he was both undertaking research and commercializing it. [ ] . . . Defendants correctly contest Plaintiffs' interpretation of *Moore.* . . . *Moore* involved a physician breaching his duty when he asked his patient to return for follow-up tests after the removal of the patient's spleen because he had research and economic interests. . . . The allegations in the Complaint are clearly distinguishable as Defendants here are solely medical researchers and there was no therapeutic relationship as in *Moore.*

In declining to extend the duty of informed consent to cover economic interests, the Court takes note of the practical implications of retroactively imposing a duty of this nature. First, imposing a duty of the character that Plaintiffs seek would be unworkable and would chill medical research as it would mandate that researchers constantly evaluate whether a discloseable event has occurred. Second, this extra duty would give rise to a type of dead-hand control that research subjects could hold because they would be able to dictate how medical research progresses. Finally, these Plaintiffs are more accurately portrayed as donors rather than objects of human experimentation, and thus the voluntary nature of their submissions warrants different treatment. . . .

* * *

Unjust Enrichment

. . . Plaintiffs allege that the [researchers' profit from the patent] violates the fundamental principles of justice, equity, and good conscience [as they would not have contributed to the research had they known that this would happen]. While Defendants claim that they have invested significant amounts of time and money in research, with no guarantee of success and are thus entitled to seek reimbursement, the same can be said of Plaintiffs. . . . The Complaint has alleged more than just a donor-donee relationship for the purposes of an unjust enrichment claim. Rather, the facts paint a picture of a continuing research collaboration that involved Plaintiffs also investing time and significant resources in the race to isolate the Canavan gene. Therefore, given the facts as alleged, the Court finds that Plaintiffs have sufficiently pled the requisite elements of an unjust enrichment claim and the motion to dismiss for failure to state a claim is DENIED as to this count.

* * *

NOTES AND QUESTIONS

1. *The* Gelsinger *Case.* The issue of financial interests in research rose to public attention with the death of Jesse Gelsinger and the subsequent lawsuit filed by his father against the University of Pennsylvania, the research physician, and the bioethicist who had consulted with the IRB in its approval of the protocol. Jesse was 18 years old and had a liver disorder that was under control. Under a research protocol and with a signed consent form, physicians inserted, into Jesse and other subjects, a virus based on the cold virus, to deliver genetic material with the hope that this genetic therapy would cure the disease. An FDA investigation after Jesse's death revealed a number of violations of federal regulations including continuing with the protocol after four patients had suffered serious reactions; failing to notify the FDA of the adverse events, contrary to a specific agreement between the University and the FDA; and altering consent forms to eliminate information that monkeys subjected to the intervention had died.

Because the protocol involved genetic therapy it had been subjected to heightened review at the FDA itself as a condition of approval. Financial conflicts of interest on the part of both the researcher and the University attracted the most public attention, however. Although reports vary in detail, it appears that the University gave Genovo, a private corporation, the exclusive rights to patent the results from the researcher's laboratory. The University also allowed the researcher to control up to 30% of Genovo stock, a company he founded, in an exception to the University's conflicts of interest guidelines. The University would receive $21 million to support research, and it also owned stock in Genovo. The University had received legal advice, covering research regulation issues, from an outside law firm in arranging the tripartite financial relationship. Gelsinger's lawsuit was resolved by a confidential settlement. Informal reports indicate that the settlement was about $5 to $10 million.

For a detailed history of the case, see Robin Fretwell Wilson, Estate of Gelsinger v. Trustees of University of Pennsylvania: Money, Prestige, and Conflicts of Interest in Human Subject Research, in Health Law & Bioethics: Cases in Context (Sandra Johnson et al., eds. 2009). See also Steven Raper, An Artless Tale: Challenges Faced in Clinical Research, 71 Food & Drug L. J. 59 (2016), written by the principal investigator in the *Gelsinger* protocol. For a description of similar cases, see Gabriela Weigel, Unresolved Conflicts of Interests in University Human Subject Research, 43 J. C. & U. L. 77 (2017).

2. *Moore v. Regents and Henrietta Lacks.* The *Moore* case involved a treating physician who removed tissue, under allegedly false pretenses, and used that tissue to develop a profitable cell line. The court in *Moore* held that the patient stated a claim for informed consent, subject to the elements required for that claim (see Section IV.C, above), because the actions occurred in a therapeutic relationship. The court rejected plaintiff's claim for conversion, however, which would have allowed the plaintiff to claim some of the profit

because the court believed that would significantly harm the research enterprise.

Henrietta Lacks was an African-American woman being treated for cancer when tissue removed in the course of treatment was used to produce one of the most widely used and lucrative cell lines in the U.S. without her knowledge or consent. Neither she nor her family or descendants received an acknowledgement, much less compensation, for decades. See Rebecca Skoot, The Immortal Life of Henrietta Lacks (2010), placing the episode in the larger context of race and human experimentation. In 2013 the surviving Lacks family reached an agreement with the National Institutes of Health to have some control over the use of the "HeLa" cell line in future research. See Carl Zimmer, A Family Consents to a Medical Gift, 62 Years Later, N.Y. Times, Aug. 7, 2013. More recently, the family hired a civil rights attorney to pursue damages from pharmaceutical companies that profited from research using the "HeLa" cell line, and possibly from Johns Hopkins University where Henrietta Lacks' cells were sampled by a physician nearly 70 years ago while treating her. See Tim Prudente, Family of Henrietta Lacks Hires Civil Rights Attorney to Seek Funds Over Famous Cells, Wash. Post, July 31, 2021.

Both the "Mo" and "HeLa" cell lines survive and continue to be used to this day. On the differences between property claims and informed consent liability, see Jorge Contreras, Genetic Property, 105 Geo. L. J. 1 (2016).

3.   *Greenberg Settlement.* The *Greenberg* parties ultimately settled by agreeing that the defendants would retain their patents but that they would allow the gene to be used without fee in any research relating to Canavan disease. Kevin L. J. Oberdorfer, The Lessons of Greenberg: Informed Consent and the Protection of Tissue Sources' Research Interests, 93 Geo. L.J. 365, 376 (2004).

4.   *Further Reading.* Conflicts of interest affect human subjects research conducted abroad, further exposing gaps in regulatory protections for human subjects. See Robert Gatter, Conflicts of Interest in International Human Drug Research and the Insufficiency of International Protections, 32 Am. J.L. & Med. 351 (2006); Marc A. Rodwin, Conflicts of Interest in Human Subject Research: The Insufficiency of U.S. and International Standards, 45 Am. J.L. & Med. 303, 320 (2019).

PROBLEM: MANAGING FINANCIAL INTERESTS

Several federal agencies have promulgated regulations or policies to address financial interests of researchers. These rules generally do not prohibit researchers from patenting and licensing their work, investing in research companies, or receiving financial support from industry. Instead, they place obligations on research institutions to manage these relationships.

The HHS regulation, for example, requires that organizations receiving Public Health Service grants have a written policy that defines what financial interests must be reported to the organization and identifies the procedures used by the organization to review and manage those interests. 42 C.F.R. § 50.601 et seq. Financial transactions or relationships that must be reported include intellectual property rights; sponsored travel; "any remuneration" in excess of $5,000 (except from the research institution itself); and equity interests in a publicly traded entity in excess of $5,000. The regulation requires the organization to maintain a publicly accessible website with its policies and procedures posted, as well as specifics on financial relationships that the organization has identified as a conflict of interest on the part of a researcher. The regulation provides many options, beyond terminating the financial relationship or prohibiting the researcher from conducting the study, for managing financial conflicts of interest. These include, for example, public disclosure of the financial relationship, disclosure to prospective subjects, modifying the research plan, or appointing an independent monitor for the study. See F. Lisa Murtha et al., Conflicts of Interest in Research: Assessing the Effectiveness of COI Disclosure Program, 18 J. Health Care Comp. 5 (July/August 2016).

1. If you decide to require disclosure of a conflict of interest to subjects participating in a study, is the following statement, used in the *Gelsinger* protocol (see Note 1 following *Greenberg*, above) adequate? If not, how would you change it?

> Please be aware that the University of Pennsylvania, Dr. James M. Wilson, (the Director of the Institute for Human Gene Therapy) and Genovo, Inc. (a gene therapy company in which Dr. Wilson holds an interest) have a financial interest in a successful outcome from the research involved in this study.

Would clearer language resolve the issue? See Kevin Weinfurt et al., Disclosure of Financial Relationships to Participants in Clinical Research, 361 NEJM 916 (2009), describing an empirical study of patients' decision making.

2. Should the following protocols be approved by your IRB or should your conflicts policy prohibit the research? If you believe it can go forward but with specific requirements, what would they be?

> a. A protocol covers a clinical trial of a new medication. The principal investigator for the protocol is on the speakers' bureau for the pharmaceutical firm that is funding the protocol and receives a fee of $1500 for every presentation she gives at any conference in relation to any research she does. The firm does not require that she submit any information regarding the content of her presentations. Last year, she received $24,000 for such presentations.

> b. A university is courting a biotech firm for a major donation for a new research building. The university's development office has asked researchers whose research is financially supported by the firm to make a presentation on the importance of the new building at the meeting at which the firm will be asked for a donation. The associate provost for

research is a member of the fundraising committee and a member of the university's IRB.

c.   An untenured faculty member is in his last year before tenure. He has had difficulty in getting publishable results from his research because he has had problems in enrolling and retaining subjects for his research on early signs of dementia. This research involves a "vulnerable population" and presents significant challenges, including assessment of the subject's capacity to consent. See Section IV.D and §§ ___.111(a)(3) and (b) and 109(g), above.

F. RESEARCH AND BIOBANKS

The use of human specimens and private health information in research is included within the definition of "human subject" (§ ___.102(e), above), whether the research protocol involves the actual removal or collection of tissue or data or whether the research merely uses material that is already stored in some sort of repository. The question of how this "secondary use" of stored specimens and data should be handled was one of the most controversial issues in the revision of the Common Rule and remains so.

There are over 600 biobanks (repositories of human specimens and related data) in the U.S. alone. The number of biobanks continues to grow explosively. The organizational structure of biobanks varies: some are associated with academic institutions or networks, others are industry created and controlled, and still others are held by federal and state governments. See Kara Swanson, Body Banking from the Bench to the Bedside (2014). 23andMe, a firm that sells popular DNA screening tests directly to consumers, has built a significant biobank with the information and specimens provided by its customers. The firm's website includes its policies on privacy and consent for research access to this material, which it sells to researchers.

Claims over ownership of biobanked material and resulting products have been litigated among researchers, research organizations, biobanks, and patients whose tissue has been used in research and product development. For example, approximately 6,000 former patients of a surgeon-researcher signed forms, at his request, directing that their tissue, stored by a university hospital, be released to the custody of the surgeon-researcher for his move to another university. Washington Univ. v. Catalona, 490 F.3d 667 (8th Cir. 2007), cert. denied, 552 U.S. 1166 (2008). The court ultimately concluded that the earlier consent forms that the patients had signed when their tissue was removed and stored governed and that their later request to remove them did not constitute withdrawal from a research protocol and would not be enforced. See § ___.116(b)(8), above. The court held that the university, and not the researcher, owned the specimens. See also Section IV.E, above, and Chapter 4 concerning property rights in tissue removed from cadavers.

Other legal protections for privacy and confidentiality of data have a significant effect on research using human biospecimens and health information. These restrictions emerge from a great number of statutory and common law provisions. See Heather Harrell and Mark Rothstein, Biobanking Research and Privacy Laws in the United States, 44 J.L. Med. & Ethics 106 (2016).

The critical issues facing regulators in the revision of the Common Rule were whether research on stored tissue and data ("secondary use") would continue to be embraced within the definition of "human subject;" and, if yes, whether use of such material requires the consent of the individual from whom the specimen or information was collected. The core value of biobanks is that they create a pool of raw material to be used in research many years down the line, much of which research cannot be anticipated when the tissue or data is collected. See Arti K. Rai, Risk Regulation and Innovation: The Case of Rights-Encumbered Biomedical Data Silos, 92 Notre Dame L. Rev. 1641 (2017).

Many argue that this material should be treated as a commons from which researchers can draw material without consent. Others argue that individual dignity and respect for persons are at risk when something as intimate as health records and body parts are used without the consent of the individual. See, e.g., Institute of Medicine, Beyond the HIPAA Privacy Rule: Enhancing Privacy, Improving Health through Research (2009), recommending that consent not be required for research on health information; Mark Rothstein, Improve Privacy in Research by Eliminating Informed Consent? IOM Report Misses the Mark, 37 J.L. Med. & Ethics 507 (2009); Sharona Hoffman & Andy Podgurski, Balancing Privacy, Autonomy, and Scientific Needs in Electronic Health Records Research, 65 S.M.U. L. Rev. 85 (2012); Ken Gatter, Biobanks as a Tissue and Information Semicommons: Balancing Interests for Personalized Medicine, Tissue Donors and the Public Health, 15 J. Health Care L. & Pol'y, 303 (2012); Chao-Tien Chang, Bank on We the People: Why and How Public Engagement Is Relevant to Biobanking, 25 Mich. Tech. L. Rev. 239, 301 (2019).

Before the revision of the Common Rule in 2017, guidance from OHRP provided researchers using repositories of data or biospecimens with only two options. They could secure consent for their specific research project from each individual whose health information or tissue they wanted to use, or they could follow a mandatory detailed procedure for "deidentifying" the material (which could result in removing relevant health data in an effort to assure that the individual could not be identified and which still presented the risk that the material could be "reidentified").

In its final form, the 2017 revision of the Common Rule appears to have landed on a compromise that retains this research within the definition of human subjects research (§ ___.102(e)(1)(ii), above) but places a good deal of research on stored tissue and data within the category of exempt research. Under the revised Rule, consent is not required for studies using identifiable private information or identifiable biospecimens that have been deidentified (as was the case before the revision) or are publicly available or are the subject of a study by or for federal agencies using information generated or collected by the government for nonresearch purposes, if the information is maintained in a fashion that meets other federal statutes regarding privacy and security. Research in these categories is considered exempt. § ___.104(d)(4), above.

For research that does not fit those categories, the Rule adopts a new concept of "broad consent" that applies to the "storage, maintenance, and secondary use of identifiable private information or identifiable biospecimens." The advantage of broad consent over the prior regime is that it accepts a one-time consent for any research that may be done in the future. For broad consent to be effective, the individual must be provided with the information listed in the regulation, and IRBs are prohibited from waiving those elements. §§ ___.116(d) and (f), above. Broad consent allows studies to use identifiable information and biospecimens, without deidentification, for many studies over many years and requires only a limited IRB review to assure merely that broad consent has been secured and documented and that, if there is a change in the way the material is stored, there have been adequate protections for privacy and confidentiality. §§ ___.104(d)(7) and (8) and ___.116(d), above.

The revised Common Rule adopted an additional significant change regarding use of medical records within organizations governed by HIPAA privacy rules for research consistent with HIPAA. Such studies are considered exempt. § ___.104(d)(4)(iii). For an analysis of "broad consent" and "secondary use" research, see Holly Fernandez Lynch, Leslie E. Wolf & Mark Barnes, Implementing Regulatory Broad Consent Under the Revised Common Rule: Clarifying Key Points and the Need for Evidence, 47 J.L. Med. & Ethics 213 (2019).

PROBLEM: DRAFTING STATE LEGISLATION

The Common Rule does not preempt state laws or tribal laws that provide additional protections for human subjects or that "require additional information to be disclosed in order for informed consent to be legally effective." §§ ___.101(f) and ___.116(i), above. States, for example, have regulated research uses of state-held biobanks such as those created by storing bloodspots collected as part of mandated screening of newborns for specific genetic conditions. See, e.g., Tufik Shayeb, Informed Consent for the Use and

Storage of Residual Dried Blood Samples from State-Mandated Newborn Genetic Screening Programs, 64 Buff. L. Rev. 1017 (2016).

Assume that you are legislative counsel to the state legislature charged with developing proposed legislation for your state. A state senator wants to propose legislation to require consent for research studies on biospecimens or health records that may offend an individual's religious or cultural values or that might stigmatize a particular group. She is also concerned that individuals might object to certain products that may be produced using their stored tissue.

The state senator has brought a particular controversy to your attention. A research team from Arizona State University collected blood samples from members of the Havasupai tribe with a consent form stating that the research would study "the causes of behavioral/medical disorders." After discussions with the researchers, those who provided blood samples thought that they would be used in diabetes and related public health research. The stored blood samples were actually used in subsequent research studies of schizophrenia, inbreeding, and migration genetics. Tribal members objected when the research turned toward areas that brought unwanted notice to the tribe; that might mark the tribe members as having a propensity toward mental illness; and that would undermine core tribal beliefs about the origins of the tribe. The tribe sued and the University settled the case. See Michelle Mello & Leslie Wolf, The Havasupai Indian Tribe Case: Lessons for Research Involving Stored Biologic Samples, 363 NEJM 204 (2010).

An earlier version of the revisions to the Common Rule proposed that consent to future unspecified research with biospecimens be allowed (as does the "broad consent" in the final Rule) but that individuals be offered the opportunity to refuse consent for use of their health records or tissue for specific categories of research that presented "unique concerns." 76 Fed. Reg. 44512 (July 26, 2011). That option was eliminated in the final version, however. Would you draw on this model for the state legislation the senator desires? Would you restrict certain uses without specific consent, or would you allow the individual to specify what uses would be prohibited? What public policy concerns arise in these proposals, and what stakeholders would be interested in the result? What enforcement tools would you use if you do adopt such a restriction?

V. INTERNATIONAL RESEARCH

The modern medical research enterprise is a global venture conducted through international academic collaborations and by multinational corporations. In 2010, an OIG report determined that over 80% of applications for FDA approval for drugs and biologics contained information generated abroad. OIG, Challenges to FDA's Ability to Monitor and Inspect Foreign Clinical Trials (June 2010). See also Darby Hall, Reining in the Commercialized Foreign Clinical Trial, 36 J. Leg. Med. 367 (2015).

International research in developing countries, in particular, presents several distinctive ethical challenges and has engendered considerable controversy. Clinical trials of pharmaceuticals in developing countries often test medications that the country will never be able to afford to provide to the very population from whom the research subjects are drawn. Ruqaiijah Yearby, Good Enough to Use for Research, But Not Good Enough to Benefit from the Results of that Research: Are the Clinical HIV Vaccine Trials in Africa Unjust?, 53 DePaul L. Rev. 1127 (2004). Limits on access to needed care may also influence research design in ethically challenging ways. In African trials of AZT, for example, U.S. agencies and pharmaceutical firms conducted a trial of low-dose AZT for pregnant women with HIV to prevent transmission of the virus to the newborn. The cost of the standard dose of AZT used for this purpose in the U.S. made its use in impoverished countries impossible without manufacturer discounts. The goal of the AIDS trials in Africa was to test whether a dose of AZT that was very significantly lower and cheaper than that used in the U.S. would be equally effective. As a control arm, however, the studies used a placebo rather than the higher, and more expensive dose used in the U.S. Jerry Menikoff and Edward Richards, What the Doctor Didn't Say: The Hidden Truth About Medical Research 67 (2006); Claire Wendland, Research, Therapy and Bioethical Hegemony, The Controversy over Perinatal AZT Trials in Africa, 51 African Studies Rev. 1 (2008); Sheryl Stolberg, U.S. AIDS Research Abroad Sets Off Outcry Over Ethics, N.Y. Times (Sept. 18, 1997).

In addition to the serious ethical issues involved in studies conducted in developing countries outside of the U.S., there are scientific concerns relating to whether specific populations (defined by race, genetics, gender, or age) and specific clinical environments might influence results in a way that would make outcomes in the U.S. less predictable. See, e.g., FDA, Draft Guidance, Acceptance of Medical Device Clinical Data from Studies Conducted Outside the United States (April 22, 2015), withdrawn as Guidance but still useful for illustrations. See also Blake Wilson, Clinical Studies Conducted Outside the United States and their Role in the FDA's Drug Marketing Approval Process, 34 U. Pa. J. Int'l L. 641 (2013).

Most nations have established standards and procedures governing medical research conducted within their own boundaries. See, e.g., Obiajulu Nnamuchi, Biobank/Genomic Research in Nigeria: Examining Relevant Privacy and Confidentiality Frameworks, 43 J.L. Med. & Ethics 776 (2015), part of a special issue on comparative approaches to research regulation; Jacob Kolman et al., Conflicts Among Multinational Ethical and Scientific Standards for Clinical Trials of Therapeutic Interventions, 40 J.L. Med. & Ethics 99 (2012).

The Common Rule governs research within the scope of the Rule even if it is performed abroad. The Rule requires that its criteria apply to that research unless the agency head determines that the country's policies provide protections for human subjects that are "at least equivalent" to those in the Common Rule. See § ___.101(h), above. This would include provisions related to vulnerable populations. See Section IV.D, above.

The FDA accepts certain investigational drug studies conducted abroad if the study was conducted "in accordance with good clinical practice (GCP)" 21 C.F.R. § 312.120. GCP is defined as a standard that assures that studies are designed, conducted, and reported in a manner that assures that the results are "credible and accurate" and that "the rights, safety and well-being of trial subjects are protected." The regulation specifies that GCP includes review and approval by an independent ethics committee and voluntary consent from the research subject as well as other protections and procedures. GCP is a concept proffered by the International Conference on Harmonisation of Technical Requirements for Registration of Pharmaceuticals for Human Use (ICH), an international organization of the pharmaceutical regulatory agencies and pharmaceutical industry in the U.S., Europe, and Japan. Earlier FDA regulations had accepted foreign clinical trials if they followed the laws of the host country or the principles of the Declaration of Helsinki, "whichever represents the greater protection of the individual." The FDA substituted compliance with GCP for drug studies in 2008 (73 Fed. Reg. 22800 (April 28, 2008)) and for device studies in 2018 (83 Fed. Reg. 7366 (Feb. 21, 2018)).

The Declaration of Helsinki, Ethical Principles for Medical Research Involving Human Subjects, is promulgated by the World Medical Association (WMA), an organization consisting of representatives from most of the world's national medical associations. Revisions of the Declaration over the past several years have struggled with issues relating to vulnerable populations. For example, the Declaration states that:

> Medical research involving a disadvantaged or vulnerable population or community is only justified if the research is responsive to the health needs or priorities of this population or community and if there is a reasonable likelihood that this population or community stands to benefit from the results of the research.

A 2000 revision of the Declaration included a statement that "every patient entered into the study should be assured of access to the best proven prophylactic, diagnostic and therapeutic methods identified by the study." This provision attracted strong opposition and may have influenced the action of the FDA to adopt GCP. The most recent revision of the Declaration (2013) modifies that article to provide:

> In advance of a clinical trial, sponsors, researchers and host country governments should make provisions for post-trial access for all participants who still need an intervention identified as beneficial in the trial.

In contrast, the Common Rule does not require that research subjects, in the U.S. or elsewhere, continue to receive medication after the clinical trial has ended. See Abney v. Amgen, 443 F.3d 540 (6th Cir. 2006), holding that the company had no fiduciary duty to subjects to provide the drug tested after a study was terminated. See also Michelle M. Mello et al., Compact v. Contract: Industry Sponsors' Obligations to their Research Subjects, 356 NEJM 2738 (2007).

International law may form a basis for claims in U.S. courts by persons injured in clinical trials conducted in foreign countries. In 1996, for example, Pfizer conducted clinical trials of the antibiotic Trovan on children in Kano, Nigeria, during an epidemic of bacterial meningitis. According to news reports, only one child had received Trovan prior to these trials, and animal testing had indicated significant potential for serious adverse effects on children. Parents consenting to administration of the drug reported that they had not been informed that the medication was experimental or that conventional treatment was available to them at no charge at the same site. Parents sued Pfizer in the U.S. under the Alien Tort Statute (ATS) alleging that the medication caused the deaths of 11 children and paralysis, deafness, and blindness in many others.

The Court of Appeals, reversing the trial court, held that nonconsensual medical research violated universal norms of customary international law and, therefore, supported a claim under the ATS. Abdullahi v. Pfizer, 562 F.3d 163 (2d Cir. 2009). Pfizer settled the case through an agreement with the Nigerian government to establish a trust fund of up to $35 million to compensate injured parties; to underwrite several health care initiatives ($30 million over two years) in Nigeria; and to reimburse Nigeria for $10 million in litigation costs. In 2011, Pfizer also settled with the private plaintiffs in an agreement that included them in the compensation trust fund. See Pfizer Settles Lawsuits Over Drug Trials on Children in Nigeria, Corp. Counsel (Feb. 23, 2011). See also Fazal Khan, The Human Factor: Globalizing Ethical Standards in Drug Trials through Market Exclusion, 57 DePaul L. Rev. 877 (2008).

A medical historian investigating archives relating to the Tuskegee syphilis study (See Section II.B., above) discovered documents revealing that the U.S. government had conducted similar research in Guatemala in the 1940s. See Kayte Spector-Bagdady & Paul Lombardo, "Something of an Adventure:" Postwar NIH Research Ethos and the Guatemala STD Experiments, 41 J. Law, Med. & Ethics 697 (2013). The Presidential Commission for the Study of Bioethical Issues conducted an investigation of the matter, and subsequently issued a report: Ethically Impossible: STD Research in Guatemala from 1946–1948. Over 5,000 prison inmates, sex workers, soldiers, children, and psychiatric patients were included in the study, and over 1,300 adults were intentionally infected with STDs. The report concluded that the study was unethical and would have been viewed as unethical even under the standards of the mid-1940s.

President Obama issued a formal apology for the research, but the federal government claimed sovereign immunity in the suit filed by those who were subjects in the study. The federal district court agreed and dismissed the suit. Garcia v. Sebelius, 867 F. Supp. 2d 125 (D.D.C., 2012), vacated in part 919 F. Supp. 2d 43 (D.D.C. 2013). In a related case (Estate of Alvarez v. Johns Hopkins Univ., 275 F. Supp. 3d 670 (D. Md. 2017)), a federal district court considered claims against a number of private parties, including the university and the pharmaceutical company, and refused to dismiss the suit as barred by the statute of limitations, stating that:

> The experimenters allegedly targeted the subjects of the Guatemala Experiments because of their inability to understand, discover, and control what was happening to them. Many of the subjects were young schoolchildren, patients in an asylum, or prisoners without access to proper medical care, education, or resources. These types of persons would have been unlikely to discover his or her exact injury and its cause.

CHAPTER 8

POPULATION HEALTH AND PUBLIC HEALTH LAW

■ ■ ■

I. INTRODUCTION

Between 1900 and 2000, life expectancy in the U.S. increased by 30 years. Five of those 30 years are attributable to the individualized health care that is the subject of most of this casebook and the basis of most of our health care industry. Twenty-five of those additional years of life, however, are a result of public health interventions. The primary goal of those interventions is population health. The Centers for Disease Control and Prevention (CDC) lists the following as the "Ten Great Public Health Achievements" of the 20th Century (48 MMWR 241):

- Vaccination

- Motor-vehicle safety

- Safer workplaces

- Control of infectious diseases

- Decline in deaths from coronary heart disease and stroke

- Safer and healthier foods

- Healthier mothers and babies

- Family planning

- Fluoridation of drinking water

- Recognition of tobacco use as a health hazard

Public health, then, is what a community, generally through its government, does to establish the conditions that allow members of that community to lead long and healthy lives. As the National Academy of Medicine has pointed out, public health generally involves collective action. Because the principle of autonomy and the value of individualism are so important in the United States, and because we tend to be suspect of government and collective action, we are called upon to weigh the benefit of public health interventions against the cost of government-imposed requirements more often than is the case in similar countries. While we do not often think about the way public health interventions protect us, we do when we need public health resources and help—when a pandemic sweeps across the nation, for example.

Because public health generally requires collective and sometimes coercive action, law is implicated in most public health interventions. The law may set standards for purity of food and water, for example, or at least designate those with appropriate expertise who will. The law may encourage or require immunization, as we will see, and it may require medical testing for diseases or conditions—either among the general population or subgroups such as teachers, students, pilots, food workers, or prisoners. Public health law may also protect the public by placing exceptions on limitations imposed by other laws, for example, by allowing needle exchanges or treatment with medication to avoid communicable diseases among individuals with substance use disorders even though drug laws may otherwise prohibit those actions. Additionally, the law may require the reporting of medical status to government agencies to allow those agencies to determine if there is a public health problem, and then how to protect the public health.

The application of public health law sometimes breaches some notions of individual privacy. Indeed, the collection of huge amounts of health information, some of it identifiable and some not, is one of the most important tools, and one of the greatest risks, of public health operations. When individually identifiable data is collected, particular follow-up measures—including, for example, tracing the contacts of those who could be transmitting communicable diseases—heighten privacy concerns. In rare circumstances, the law can require patients to receive treatment (or be placed in isolation or quarantine).

What makes an issue a "public health" issue? Is the quality of the gene pool a public health issue? The widespread use of alternative medicines? of tobacco? Gun violence? The availability and use of video games (or just violent video games)? Domestic violence? Restrictive licensure for midwives? What difference does it make whether an issue is characterized as a public health issue? Are we more likely to defer to governmental decisions, and more likely to accept governmental intrusions on individual liberties, where the public health is involved? Article 25 of the Universal Declaration of Human Rights says:

> Everyone has a right to a standard of living adequate for the health and well-being of [oneself] and [one's] family, including food, clothing, housing and medical care and necessary social services, and the right to security in the event of unemployment, sickness, disability, widowhood, old age or other lack of livelihood in circumstances beyond one's control.

Is this a principle of public health?

Public health can endeavor to encourage healthy individual behavior just as it can act to prohibit or discourage risky behavior. For example, while the Affordable Care Act (ACA) was designed to support a private health care system, it also recognizes the value of public health approaches to many issues. The Act provides subsidies for small employers who wish to establish employee wellness programs, and it allows employers of all sizes to give premium discounts to those employees who participate in some kinds of wellness programs. In addition, the ACA eliminates copays, deductibles, and other payments for some preventive care provided by private health plans, Medicare, or Medicaid, and it uses public relations strategies to increase national rates of immunization and other forms of preventive care. The ACA also requires the collection of data, including health disparity data, that will make the creation of public health approaches easier in the future.

In addition to supporting these traditional public health strategies, the ACA moves the health care system toward a greater concern for population health. The ACA, for example, encourages the formation of accountable care organizations (ACOs) as a system for delivering care. An ACO's performance is measured against its impact on outcomes relating to population health and wellness. See Susan DeVore and R. Wesley Champion, Driving Population Health through Accountable Care Organizations, 30 Health Aff. 41 (2011).

The ACA also institutes some more focused public health changes. It requires chain restaurants (those with more than 20 outlets serving essentially standardized products) to provide nutritional information on those products by posting calorie information on the menu or menu board and making additional nutritional information available to customers. It imposes the same obligations on those who sell products through vending machines, at least where customers cannot read the nutritional information already on the packaged food. The ACA also requires that employers with more than 50 employees (and others where it would not cause hardship) allow employees to express milk whenever that is necessary during the year following the birth of their children; employers must provide a clean and private space for this need and allow adequate break time, although it may be unpaid time. The law also provides for public education on a host of lifestyle and preventive care issues.

As we move toward government programs that encourage wellness, is public health entering a new era in the United States—one that depends more on incentives and price regulation than on statutory mandates? Are these new efforts—encouraging Americans to exercise, eat healthy food, stop smoking, and breastfeed their children—appropriately within the realm of government?

For a discussion of the ACA and public health, see Elizabeth Weeks Leonard, ed., Public Health Reform: Patient Protection and Affordable Care Act Implications for the Public's Health (Symposium), 39 J.L. Med. & Ethics 312 et seq. (2011). For a very thoughtful summary of the overlap of law and public health, see Barry Levy, Twenty-First Century Challenges for Law and Public Health, 32 Ind. L. Rev. 1149 (1999). For a comprehensive account of the law of public health, see Lawrence Gostin and Lindsay Wiley, Public Health Law: Power, Duty, Restraint (3d ed. 2016). See also Lawrence Gostin et al., The Law and the Public's Health: A Study of Infectious Disease Law in the United States, 99 Colum. L. Rev. 59 (1999). For an interesting historical and constitutional approach to these issues, see Wendy Parmet, Populations, Public Health, and the Law (2009). For a good account of the international public health issues, see Roger Detels et al., eds., Oxford Textbook of Public Health (4th ed. 2004).

II. POPULATION HEALTH AND (IN)EQUITY

The inequities of our society are laid bare by population health statistics, revealing that the U.S. does not have one population with a roughly uniform, shared health status. Instead, we are an amalgam of sub-populations—white and non-white, wealthy and impoverished, able-bodied and disabled, educated and uneducated, socially connected and unconnected—that are associated with significantly different levels of population health.

- Black Americans have an average life expectancy of 72 years as compared to 78 years for white Americans. See Elizabeth Arias et al., Provisional Life Expectancy Estimates for January through June, 2020, Vital Statistics Rapid Release (Feb. 2021).

- The wealthiest Americans live, on average, 14.6 years longer than the poorest Americans. Raj Chetty et al., The Association Between Income and Life Expectancy in the United States, 2001–2014, 315(16) JAMA 1750 (2016).

- In 2017 Americans with a 4-year college degree had a life expectancy 7–10 years longer than those with only a primary education and 5–6 years longer than those with only a secondary education. Isaac Sasson and Mark D. Hayward, Association Between Educational Attainment and Causes of Death Among White and Black U.S. Adults, 2010–2017, 322(8) JAMA 756 (2019).

These statistics highlight the fact that social determinants play a substantial role in health. According to the Healthy People 2030 project,

Social determinants of health (SDOH) are the conditions in the environments where people are born, live, learn, work, play, worship, and age that affect a wide range of health, functioning, and quality-of-life outcomes and risks.

* * *

. . . Examples of SDOH include:

- Safe housing, transportation, and neighborhoods
- Racism, discrimination, and violence
- Education, job opportunities, and income
- Access to nutritious foods and physical activity opportunities
- Polluted air and water
- Language and literacy skills

SDOH also contribute to wide health disparities and inequities. For example, people who don't have access to grocery stores with healthy foods are less likely to have good nutrition. That raises their risk of health conditions like heart disease, diabetes, and obesity—and even lowers life expectancy relative to people who do have access to healthy foods.

Just promoting healthy choices won't eliminate these and other health disparities. Instead, public health organizations and their partners in sectors like education, transportation, and housing need to take action to improve the conditions in people's environments.

HHS, Office of Disease Prevention and Health Promotion, Healthy People 2030.

While public health policy recommendations often reflect the association between key social factors and health, case law about public health issues rarely addresses the social determinants of health. Perhaps this is because cases focus closely on specific parties rather than on populations where relationships between social factors and health are clear.

The following opinion is an exception. In a lawsuit challenging the constitutionality of voting procedures implemented by Alabama for the November 2020 election, which took place during the COVID-19 pandemic, the court dove deeply into the health of the sub-populations represented by the plaintiffs. In so doing, the court accounted for many of the social determinants of health and demonstrated the link between those social factors, health, and the risk of disenfranchisement of plaintiffs through voting regulation.

PEOPLE FIRST OF ALABAMA V. MERRILL

United States District Court, N.D. Alabama, 2020.
491 F. Supp. 3d 1076.

ABDUL K. KALLON, JUDGE.

I.   FINDINGS OF FACT

A.   The COVID-19 Pandemic

1.   Our nation is now in the midst of a public health emergency due to the spread of COVID-19, the respiratory disease caused by the novel coronavirus SARS-CoV-2. Since April, the United States has led the world in the total number of COVID-19 cases. [ ]. As of September 29, 2020, the United States has confirmed 7,129,383 cases of COVID-19 and reported 204,598 deaths and counting due to COVID-19. [ ]. The COVID-19 pandemic has deeply affected Alabama, and as of September 29, Alabama reported 137,564 confirmed COVID-19 cases and 2,399 confirmed COVID-19 deaths, including family members and a friend of some of the individual plaintiffs.[ ].

2.   The most common symptoms of COVID-19 are fever, cough, and shortness of breath, and other symptoms include chest pain, muscle pain, vomiting, anorexia, confusion, and lack of senses of taste and smell. [ ]. Even mild cases of COVID-19 can be more severe than the flu and involve about two weeks of fevers and dry coughs. [ ]. Severe cases of COVID-19 cause acute respiratory distress syndrome ("ARDS") in which fluid displaces air in the lungs, leaving patients to "essentially drown[ ] in their own blood and fluids because their lungs are so full." [ ]. Due to the respiratory impacts of the disease, individuals with severe cases may need supplemental oxygen or to be intubated and put on a ventilator, or suffer a permanent loss of respiratory capacity. [ ]. Along with damaging lung tissue, severe cases of COVID-19 can damage the kidneys, or lead to strokes and heart attacks. [ ].

3.   In Alabama, as of September 3, approximately 11.5 percent of those infected by COVID-19 have been hospitalized. [ ]. Thus, surges of COVID-19 cases can strain healthcare systems, leading to critical shortages of doctors, nurses, hospital beds, medical equipment, and personal protective equipment. [ ]. For that reason and others, health officials recommend that all individuals take steps to prevent the spread of the disease, not just for their own health but for the health of the community. [ ]. As the CDC put it, "[e]veryone has a role to play in slowing the spread and protecting themselves, their family, and their community." [ ].

4.   COVID-19 is highly infectious and spreads through respiratory droplets from infected individuals. [ ]. Strong evidence now suggests that SARS-CoV-2 is aerosolized, such that tiny droplets containing the virus

remain in the air for a period of time, and the virus can be transmitted to others who inhale that air. [ ]. Further evidence indicates that aerosolized droplets with SARS-CoV-2 can linger in closed, stagnant air environments for up to fourteen minutes. [ ]. Consequently, transmission of the virus is more likely to occur indoors when people share space in a room for an extended period of time. [ ]. COVID-19 may also spread through contaminated surfaces, i.e., "when an infected individual touches a surface with a hand they have coughed into and then another person touches that same surface before it has been disinfected and then touches their face." [ ]. Finally, COVID-19 is particularly dangerous because it can be spread by individuals who are a- or pre-symptomatic, and who may not know they have the disease and can spread it to others. [ ].

5.   No vaccine for COVID-19 currently exists, and it is unlikely a vaccine will become widely available until well into 2021. [ ]. Accordingly, to slow the spread of COVID-19, public health officials have been left to urge the public to practice "social distancing," i.e., avoidance of close contact with others. [ ]. For purposes of social distancing, the CDC recommends that individuals stay at least six feet away from others who do not live in their households and avoid crowded places. [ ]. Following that recommendation, Governor Kay Ivey recognizes that "[m]aintaining a 6-foot distance between one another is paramount," and the Alabama Department of Public Health ("ADPH") has instructed the public to "spend as much time as possible at home to prevent an increase in new infections." [ ]. The CDC and ADPH also recommend that all individuals wear masks or face coverings, which decrease the spread of COVID-19 by helping to prevent an infected individual from spreading droplets containing the virus that can infect others. [ ]

6.   Though people of all ages have contracted COVID-19 and suffered severe manifestations of the disease, the illness poses special risks for older people, and the risk of COVID-19 increases steadily with age. [ ]. As of September 1, 77.3 percent of COVID-19 related deaths in Alabama have been people 65 or older, and 80 percent of deaths in the United States have been people over 65. [ ]. Thus, the CDC warns that older people should avoid interacting with people outside of their households and people who have been exposed to the virus. [ ].

7.   COVID-19 also presents special risks for people with disabilities. "The CDC [ ] warns that people with disabilities may be at increased risk of becoming infected with COVID-19 to the extent that they "(1) 'have limited mobility or cannot avoid coming into close contact with others who may be infected;' (2) 'have trouble understanding information or practicing preventative measures;' or (3) are 'not able to communicate symptoms of illness.'" [ ]. As Susan Ellis, the Executive Director of People First of Alabama, explained, people with intellectual disabilities are not able to identify risky behavior in the same way as other people can. [ ]. For

example, they may pick up and throw away a tissue on the floor without wearing gloves in an effort to be helpful to others without thinking of the germs on the tissue. [ ]. In addition, adults with disabilities are three times more likely than adults without disabilities to have underlying chronic medical conditions that put them at higher risk from COVID-19. [ ].

8.   People with certain preexisting medical conditions, including immunological conditions, hypertension, certain heart conditions, lung diseases (asthma and chronic obstructive pulmonary disease), diabetes mellitus, obesity, chronic kidney disease, and sickle cell anemia also have an elevated risk of severe complications or death from COVID-19. [ ]. In fact, statistics from the ADPH reveal that approximately 96.2 percent of Alabamians who have died from COVID-19 had underlying health conditions. [ ]. Due to their heightened risk from COVID-19, the CDC advises people with underlying conditions to limit interactions with people outside of their households as much as possible and to avoid others who are not wearing masks. [ ].

9.   Alabamians suffer from high rates of the underlying health conditions that increase their risk for COVID-19 complications: more than 14 percent have diabetes, 41.9 percent have high blood pressure, 10.4 percent have asthma, approximately 33 percent are obese, and about 1 in 11 have kidney disease. [ ]. As Alabama State Health Officer Dr. Scott Harris stated, "[c]hronic disease factors are a real risk for dying from this disease, and chronic diseases are found in about a third of our citizens." [ ]. The prevalence of underlying diseases is even more profound for older Alabamians, who are already at a higher risk from COVID-19. For example, the COVID-impact survey revealed that in the Birmingham metro region, 76.1 percent of White people and 80.3 percent of Black people over the age of 60 had at least one underlying condition that put them at an increased risk from COVID-19. [ ].

10. Racial minorities, particularly Black people, are more likely to suffer from underlying conditions that increase the risks from COVID-19, such as diabetes, heart disease, and lung disease. [ ]. That trend is reflected in Mobile, Alabama, where areas of the city with the highest percentage of Black residents have markedly higher prevalence of asthma, high blood pressure, kidney disease, COPD, and diabetes, than predominately White areas of the city. [ ]. Moreover, about 42.2 percent of Black people over the age of 65 have a disability, compared to 38.1 percent of their White cohorts. [ ].

11. People who are Black, Latinx, or Native American suffer higher rates of COVID-19 infections than non-Hispanic White individuals, and they are more likely to suffer severe outcomes and be hospitalized or die if they are infected with COVID-19. [ ]. Indeed, a CDC report published April 8, 2020, which included data from patients hospitalized across 14 states,

found that Black COVID-19 patients made up 33 percent of those for whom race or ethnicity information was available, despite representing only 18 percent of the states' populations. [ ]. This disparity has continued. As of September 21, the CDC reports that Black people represent 18.5 percent of COVID-19 cases and 21 percent of COVID-19 deaths in individuals for whom race or ethnicity is known, despite representing only 12.5 percent of the U.S. population. [ ]. And, Latinx people represent 29.5 percent of COVID-19 cases, though they represent only 18.4 percent of the population. [ ].

12. Like the rest of the country, Alabama has reported alarming racial disparities in serious illness and mortality due to COVID-19. [ ]. As of August 13, the ADPH reported that Black people in Alabama account for 41.1 percent of COVID-19 related deaths, despite making up just 27 percent of the population. [ ]. And, in Mobile County, as of August 13, Black people accounted for 51.1 percent of COVID-19 deaths, despite representing only 36 percent of the population. [ ]. In addition, counties in the Black Belt[28] in Alabama that are predominately African American, including Lowndes and Wilcox, have higher COVID-19 related death rates compared to predominantly White counties, and Lowndes County has the State's highest case fatality rate. [ ]. Dr. Latesha Elopre, the plaintiffs' expert on internal medicine, infectious diseases, and disparities in access to health care and health outcomes in Alabama, explained, "there is strong data that supports . . . looking [across] age groups, that Black people, regardless of health conditions, have higher rates of death compared to Whites." [ ].

13. The higher risk of COVID-19 infection for African Americans is tied to pre-existing and evolving inequities in structural systems and social conditions. [ ]. To begin, people who are Black, Latinx, or Native American are more likely to hold jobs that do not provide paid leave, cannot be performed remotely, and require more exposure to the public and, therefore, to COVID-19. [ ]. Also, due to patterns resulting from a history of housing discrimination, Black and Latinx individuals are more likely to live in areas impacted by environmental pollutants, or in densely populated areas, making it harder for those individuals to practice social distancing. [ ]. In addition, evidence reveals that testing resources for COVID-19 are scarcer in Black communities and that Black people are less likely to be referred for testing than White people when presenting comparable signs of infection. [ ]. Indeed, for these and other reasons, the CDC identified discrimination in housing, education, finance, and health care as a factor that puts racial and ethnic minorities at increased risk of getting sick and dying from COVID-19. [ ].

---

[28] The Black Belt is named for the region's fertile black soil. [ ].

14. The discrimination and systemic racism that contribute to elevated COVID-19 risk for Black people and other minorities nationally are evident in Alabama. [ ]. For example, in Alabama's Black Belt counties, residents face multiple stressors that lead to a relatively poor quality of life: high levels of poverty and unemployment, inadequate education, high rates of illiteracy, lack of jobs, many single-parent households, and food insecurity. [ ]. On top of that, Black Belt counties have fewer primary care physicians per resident than other counties, and, despite the high rates of infection, people living in these counties receive fewer COVID-19 tests than elsewhere in Alabama. [ ]. Disparities in access to transportation exacerbate those obstacles and make COVID-19 testing and health care even less accessible for Black people in those counties, which in turn makes it even more difficult for communities in the Black Belt to manage and slow the spread of the disease. [ ].

B. Alabama's Response to the Pandemic

[The court summarized a variety of orders by the Governor of Alabama and the state's public health agency, including orders that closed schools, prohibited gatherings of more than 25 persons, and prohibited bars and restaurants from allowing patrons to consume food or drink on site. Additionally, the court described later orders that required every person to stay at home except to perform "essential activities," that closed entertainment venues and businesses providing close-contact services, and that set standards for how "essential retailers" would operate so as to mitigate the risk of disease transmission. Finally, the court outlined the state's Safer-at-Home order, which required people to wear masks outside the home and to practice social distancing. Importantly, this order especially encouraged vulnerable populations to remain at home.]

C. Voting and Alabama's Elections in the Pandemic

* * *

21. With respect to voting in the pandemic, Dr. Anthony Fauci, head of the National Institute of Allergy and Infectious Diseases and a member of the White House's coronavirus taskforce, stated:

> I don't see any reason why, if people maintain [six-foot] distancing, wearing a mask and washing hands—why you cannot, at least where I vote, go to a place and vote.

> And you can do that, if you go and wear a mask, if you observe the physical distancing, and don't have a crowded situation, there's no reason why you shouldn't be able to do that. I mean, obviously if you're a person who is compromised physically or otherwise, you don't want to take the chance. There's the situation of mail-in voting that has been done for years in many places. So there's no reason why we shouldn't be able to vote in person or otherwise.

[ ]. Still, according to Dr. Reingold, traditional in-person voting exposes voters to a risk of contracting COVID-19 due to the proximity of a large number of individuals in a limited space and the large number of common surfaces that many people may touch. [ ]. And, the risk of infections increases for people who have to use public transportation or ride in a car with people outside of their household to get to the polls. [ ]. Moreover, Dr. Reingold indicated that, in Alabama, the risk of in-person voting is compounded because the State's mask order does not apply to "[a]ny person who is voting," which means that not everyone voting will wear a mask in polling places, leading to greater risk of COVID-19 transmission. [ ]. Similarly, in part because of the chance that people at polling sites will not be wearing masks, Dr. Elopre recommends that people at-risk from COVID-19 vote absentee rather than in-person. [ ].

22. Evidence exists of transmission of COVID-19 from in-person voting. Indeed, after the April 7 primary election in Wisconsin, that state's contact tracing efforts identified 71 individuals who had COVID-19 cases linked to working at the polls or voting in person at the polls. [ ]. The plaintiffs' expert, Dr. Chad Cotti, . . . concluded that counties that had higher numbers of in-person votes per polling location in the Wisconsin primary had a higher rate of COVID-19 spread at a community level after the election than counties with relatively fewer in-person votes. [ ]. In particular, he found that a 10 percent increase in in-person voter density, i.e., in-person votes per polling location, corresponds to about a 17–18 percent increase in the positive test rate. [ ]. Ultimately, Dr. Cotti concluded from his analysis that approximately 700 cases of COVID-19 in the five-week period after the April 7 Wisconsin primary were associated with in-person voting. [ ]. Based on his conclusion, Dr. Cotti opined that to reduce COVID-19 risk, "it would be prudent for election officials and state leaders to engage in as many practices to 'de-densify'[ ] polling locations as possible," including by offering more hours at the polls, early voting, and curbside voting, and by encouraging absentee voting. [ ].

23. In contrast to Dr. Cotti, the State defendants' expert, Dr. Quentin Kidd, . . . found that there was no increase in COVID-19 infection rates in the 14-day period following elections in Wisconsin, Virginia, West Virginia, Georgia, and Kentucky, based on his review of data from Johns Hopkins University. [ ]. Dr. Kidd's review, however, analyzed only the 7-day moving average for COVID-19 cases reported in databases for the 14 days after the elections. [ ]. Noticeably, he admitted he did not account for the acknowledged lag times between exposure, the onset of symptoms, testing, and results. [ ]. And, on cross examination, Dr. Kidd admitted that during the time-period between 15 and 30 days after the elections, which accounted for the lag time, a noticeable increase in COVID-19 cases could be seen in all of those jurisdictions, except Virginia. [ ]. Virginia, however, allows curbside voting for individuals over 65 and for individuals with a

disability, and only 57,500 people voted in person. [ ]. Thus, Dr. Kidd's analysis, even if the court were to accept it, does not refute Dr. Cotti's testimony that there is a link between higher numbers of in-person votes per polling location and increased COVID-19 spread in communities.

24. The CDC has issued specific guidelines to address concerns about voting during the COVID-19 pandemic. [ ]. Among other things, the CDC recommends that states "offer alternative voting methods that minimize direct contact and reduce crowd size at polling locations," including early voting, and it asked states to consider "drive-up voting for eligible voters if allowed in the jurisdiction" as a means of complying with social distancing rules and limiting personal contact during in-person voting. [ ]. The plaintiffs' experts, Dr. Reingold and Dr. Elopre, agree with the CDC's recommendations, including the recommendation for curbside or drive-up voting, though Dr. Reingold admits that curbside voting is not necessarily safer than absentee voting. [ ].

25. Alabama does not offer early, traditional in-person voting or curbside voting. [ ]. Still, Alabama has taken specific steps to make voting safer during the COVID-19 pandemic. To begin, on March 13, Secretary Merrill sent a letter to probate judges to suggest that they should sanitize polling machines and equipment frequently during voting, provide hand sanitizer to voters, and provide gloves to poll workers. [ ]. Then, on March 18, Governor Ivey rescheduled the March 31 primary runoff election to July 14. [ ]. That same day, Secretary Merrill promulgated an emergency rule titled "Absentee Voting During State of Emergency," providing in part as follows:

> [A]ny qualified voter who determines it is impossible or unreasonable to vote at their voting place for the Primary Runoff Election of 2020 due to the declared states of emergency, shall be eligible to check the box on the absentee ballot application which reads as follows: "I have a physical illness or infirmity which prevents my attendance at the polls. [ID REQUIRED]."

[ ]. In effect, this rule meant that any voter concerned about potential exposure to COVID-19 at a polling place could vote absentee in the July 14 primary runoff election. [ ]. Secretary Merrill issued a press release in March informing voters about the emergency absentee voting rule for the July election. [ ].

26. In April, Congress appropriated funds for Alabama to prepare for voting during the pandemic in the July and November elections, and the Secretary of State's office also dedicated increased funding to this effort. [ ]. The Secretary of State will use some of those funds to reimburse counties for masks, gloves, and cleaning supplies for polling places. [ ]. And, Secretary Merrill testified that his office will provide a mask to any voter at a polling place who would like one. [ ]. Still, Secretary Merrill's office

received letters from some probate judges expressing concern about their ability to comply with CDC recommendations and State and Federal election laws during the July runoff election. [ ]. And, the Secretary received a report that in Mobile County, the polling sites that would have the biggest issues complying with CDC guidelines serve predominately minority communities. [ ].

27. Alabama saw a record turnout of voters for a runoff election on July 14, with 17.6 percent of eligible voters voting. [ ]. During and after the July 14 election, Secretary Merrill received complaints that poll workers and voters did not wear masks at polling sites. [ ]. When he learned that voters in certain precincts were not wearing masks, Secretary Merrill took steps to ensure those voters were still allowed to vote in the runoff. [ ]. According to Secretary Merrill, "[n]o voter can be turned away [from a polling site] for any reason if they are a qualified elector[,]" even if they are visibly ill or have a known case of COVID-19. [ ].

28. Secretary Merrill extended the emergency absentee voting rule to apply to the November 3, 2020 general election. [ ]. Thus, any voter in Alabama who wishes to vote absentee in the November election due to concerns about COVID-19 may do so, subject to the witness and photo ID requirements. Secretary Merrill expects voter turnout in November to exceed the 2016 presidential election, and although most Alabama voters have traditionally voted in person on Election Day, he expects voters to cast a record number of absentee ballots in the November election. [ ]. On at least 14 occasions, Secretary Merrill's office issued a press release encouraging voters concerned about exposure to COVID-19 to vote absentee and providing information on how to apply for an absentee ballot and the applicable deadlines. [ ].

29. Absentee voting for the general election began on September 9, and as of September 11, more than 30,000 people had applied for absentee ballots across the State. [ ]. In addition, several hundred people have completed absentee ballots in person in their county AEM's office. [ ]. To cast an absentee ballot, a voter must complete an absentee ballot application, and return the application to the AEM by mail or in person with a copy of her photo ID at least five days before the election. [ ]. Then, after the voter receives the absentee ballot packet, she must seal the completed ballot inside an affidavit envelope, and sign the affidavit in the presence of two witnesses or a notary. [ ]. Finally, the voter must return the completed ballot and affidavit to the AEM in person up until the day before Election Day, or by mail in time to be received by noon on Election Day. [ ].

30. Voters may go in person to an AEM's office to vote before the election, and they can complete the absentee voting process in a single trip. [ ]. Alleen Barnett, the Absentee Election Coordinator for Mobile County,

estimates that her office has been serving between 40–50 voters a day since absentee voting began on September 9, and she anticipates that the number will increase. [ ], In addition to serving voters at the AEM's office, Secretary Merrill encourages AEMs and Circuit Clerks to hold events outside of their offices, such as outdoors on college campuses, in parking lots, or in nursing homes, to allow voters additional opportunities to cast absentee ballots in person before the election. [ ].

31. In the Mobile County AEM's office, Ms. Barnett and her staff wear masks, have hand sanitizer and rubber gloves available, and wipe their work areas and the voter areas with disinfectant spray. [ ]. They also practice social distancing in the office and have spaced chairs for voters at least six-feet apart. [ ]. But, Ms. Barnett's office cannot turn away a voter who does not wear a mask. [ ]. At least in part for that reason, Mr. Howard Porter, Jr., one of the individual plaintiffs from Mobile County, testified that he would not consider voting in person in the AEM's office. [ ]. Mr. Porter explained his decision for not voting in person: "I don't want any vote that I cast to be my last vote. And in Alabama, a person can vote even if they don't have on a face mask. And that's just too much for me. [ ] [A]s important as the right to vote is, I just can't endure that." [ ]. Mr. Porter, a Black man in his seventies, further explained: "[S]o many of my [ancestors] even died to vote. And while I don't mind dying to vote, I think we're past that—we're past that time. And that should not be a requirement . . . . Because voting is the only day that rich, the poor, sick, the healthy, all should be counted as one and just as easy. And any obstacle placed in the way of the opportunity to vote places an effect on the process itself." [ ].

D. Alabama's History of Disenfranchisement of Black Voters

32. For the individual plaintiffs who are Black and the organizations who advocate for the voting rights of Blacks, voting is fundamental and sacred in light of the many thousands who died to secure the right. They view any restrictions to voting as part of the centuries-old effort to deprive Blacks from accessing the ballot. As the plaintiffs testified, Black Alabamians have consistently overcome barriers to exercising their fundamental right to vote, only to later have that right curtailed. [ ]. The relevant historical facts are largely undisputed. . . .

[The Court summarized the history of changes to the law that ended slavery and promised the right to vote, and it described the many ways white Alabamians circumvented these laws to deny or otherwise undermine the ability of black Alabamians to exercise that right.]

45. Alabama's history of voter discrimination interacts with systemic disparities that currently affect Black residents' education, employment, and healthcare outcomes to negatively impact Black political participation. [ ].

46. For example, in addition to voting discrimination, racial discrimination in Alabama's public schools is also well-documented. [ ]. State and local officials fought to maintain segregation in Alabama schools long after the Supreme Court ruled the practice unconstitutional in Brown v. Board of Education, [ ]. Efforts to desegregate school systems in Alabama have continued, with dozens remaining under desegregation orders today. [ ].

47. To no surprise, consistent with CitiBank's study released last week that found that the U.S. Economy has lost $16 trillion since 2000 because of discrimination against Blacks,[38] educational discrimination has had a lasting effect in Alabama. [ ]. Of adults aged 25 years and over, 16.6 percent of Black Alabamians have not completed high school compared to 11.4 percent of their White counterparts. [ ]. For the same age range, only 17.3 percent of African Americans in the State earned a bachelor's degree or higher compared to 28.3 percent of White adults. [ ].

48. Complementing Alabama's historical racial disparities in education are entrenched disparities in employment and economic opportunities. [ ]. Over the past half-century, litigation has uncovered many examples of state entities adopting racially discriminatory employment practices. [ ]. And private employers too have faced countless allegations of racial discrimination. In 2019, for example Alabama had a higher percentage of racially-based employment discrimination claims than any other state. [ ]. Alabama was also overrepresented nationally in racially-based employment discrimination claims based on its population. [ ].

49. Relatedly, Black communities in Alabama suffer from higher concentrations of unemployment and under-employment. [ ]. The unemployment rate among Black Alabamians over age 16 is more than double the rate among White residents of the same age. [ ]. Those Black Alabamians who are employed are more likely to work lower paying jobs than White workers. Of employed Black adults in the State, 20.7 percent work in service occupations compared to just 14.8 percent of Whites. [ ]. Only 26.2 percent of employed Black Alabamians hold management or professional roles; 39.1 percent of White Alabamians hold such positions. [ ].

50. In Alabama, Black households also have fewer economic resources. The median household income for Black Alabamians is $33,503 compared to $58,257 for White households. [ ]. More than one quarter of Black Alabamians live in poverty compared to only 11.3 percent of White Alabamians. [ ]. This economic inequality has consequences. Roughly 13 percent of Black families in Alabama lack access to a vehicle compared to only 3.9 percent of White families. [ ]. Black households are eight percent

---

[38] NPR, *Cost of Racism*, [ ].

less likely than White households to own a computer, smartphone, or tablet. [ ]. Black households are also less likely to have broadband internet access—29.6 percent compared to 17.2 percent of White households. [ ].

51. These economic conditions intersect with other legacies of racial discrimination in Alabama to make the State's Black residents more likely to suffer negative health outcomes. [ ]. Historic racial residential segregation is a principal driver. [ ]. State and federal laws and practices "produced and maintained" racial residential segregation in Alabama by incentivizing discriminatory zoning, redlining, and predatory lending. [ ]. Racial residential segregation has devalued neighborhoods, reduced access to quality and affordable food, and produced disproportionately high concentrations of poverty in Black communities. [ ]. It is "one of the strongest indicators of chronic illness patterns in the Black community." [ ]. Because of racial residential segregation, Black individuals are more likely than White individuals to suffer from chronic conditions such as obesity, cancer, and asthma. [ ].

52. This is particularly true in Alabama's Black Belt counties, where "a shortage of infrastructure to support economic stability and growth compounds the inability of [residents] to move upward economically, contributing to generational poverty, especially among rural Black people who are less likely to be able to mobilize out of rural areas." [ ]. Health outcomes for Black Alabamians in rural counties are much worse than for White Alabamians in those counties. [ ]. Although impoverished people of all races face structural obstacles that adversely impact their health, a Black person living in poverty "typically [ ] would have a worse outcome compared to their counterparts." [ ].

53. These health disparities reflect, in part, callousness among elected officials in Alabama to the needs of Black residents. For example, Alabama's legislature has rejected requests to expand Medicaid under the Affordable Care Act notwithstanding the urging of Black leaders and a racial gap in insurance coverage. [ ]. Expanding Medicaid would have insured an additional 220,000 Alabamians, particularly benefiting Black residents. [ ]. This healthcare and insurance inequality helped to make Black Alabamians especially vulnerable when a novel coronavirus surfaced in early 2020. [ ].

E.   The Parties in this Lawsuit

(1)   Individual Plaintiffs

(a)   *Howard Porter, Jr.*

54. Mr. Porter is a Black man in his seventies, and a registered voter in Mobile County, Alabama who lives with his wife and adult son. [ ]. Mr. Porter has asthma and Parkinson's Disease, which causes him to have difficulty walking, and he uses a cane to walk. [ ]. Mr. Porter is at higher

risk for complications or death from COVID-19 because of his age and underlying medical conditions. [ ]. He takes the COVID-19 pandemic seriously and follows the CDC's guidance closely. [ ]. Sadly, his sister and uncle both died from COVID-19, and Mr. Porter is very fearful about becoming infected. [ ].

* * *

57. Mr. Porter has always voted in-person, but now he does not want to risk a COVID-19 infection. [ ]. Mr. Porter chose not to vote in the July 14 primary runoff election when he did not receive his absentee ballot. [ ]. He did not vote in-person in the election because he did not want to risk being infected with COVID-19. [ ]. When Mr. Porter votes in person at his polling place, he is required to walk 100 to 150 feet from his disabled parking space to the entrance. [ ]. Walking this distance is challenging for Mr. Porter due to his Parkinson's. [ ]. Mr. Porter would prefer to vote curbside as opposed to voting absentee or in-person. [ ]. He believes it would be "less strenuous" given his disability, provide less contact than going inside the polling place, and reduce the time he spends now to vote. [ ].

* * *

(c) Annie Carolyn Thompson

63. Ms. Thompson is a 68-year-old retiree who lives alone at her home in Mobile, Alabama. [ ]. She is Black, a U.S. citizen, has never lost her right to vote by reason of a felony conviction or court order, and is a lawfully registered voter in Alabama. [ ]. Ms. Thompson is at higher risk of contracting and having severe complications from COVID-19 because of her age and preexisting conditions, including diabetes and high blood pressure. [ ]. Ms. Thompson takes medication for both conditions. [ ].

* * *

66. Ms. Thompson testified that she would have difficulty meeting the photo ID requirement to vote absentee because she does not have the necessary technology or equipment to make a copy of her photo ID at home and would, therefore, have to venture out of her home to obtain the copy. [ ]. She also indicated that she would have difficulty meeting the witness requirement because she does not regularly interact with two people at the same time. [ ]. She further testified, however, that, having learned about the Mobile Public Library's curbside notary services in this litigation, she would use them to notarize her absentee ballot for the November election. [ ].

67. Ms. Thompson requested an absentee ballot for the July runoff election because of COVID-19, but her ballot did not arrive until the day of the election. [ ]. Accordingly, because Ms. Thompson could not vote absentee, she called a friend who works at her polling site to find out a time

when few people would be at the site. [ ]. She felt safe voting in person on that occasion because all of the poll workers wore masks and gloves, there were sanitizing stations all over, they had sneeze guards, and people practiced social distancing. [ ]. Ms. Thompson testified that she would not feel safe voting in person in November because her polling site, which is not a large facility, will be very crowded since it is a major election that will have a larger turnout. [ ]. Ms. Thompson also would not be comfortable voting in person in November because people are not required to wear masks when voting. [ ].

* * *

CONCLUSIONS OF LAW

[The court held that Alabama laws requiring that either a notary public or two adults witness an absentee ballot, requiring that a copy of the voter's photo ID accompany a mailed absentee ballot, and banning curbside voting violated the constitutionally protected liberty interest in voting when applied to the plaintiffs, who were vulnerable voters because they could not safely vote in person or satisfy the requirements for absentee voting during COVID-19 pandemic given their heightened risk of severe COVID-19 complications due to their ages, disabilities, pre-existing conditions, and race.]

NOTES AND QUESTIONS

1. *Social Determinants of Health.* What specific social determinants of health did the court recognize as playing a role in the inequitable incidence of COVID-19 infections, hospitalizations, and deaths? How, according to the court's findings, did that inequitable incidence contribute to the burden imposed by the challenged state voting laws on the plaintiffs' right to vote?

2. *Snowball Effect.* When two or more social factors combine to undermine the health of a sub-population, that combination can have a snowball effect by increasing the risk that members of that sub-population experience ever-worsening health outcomes. See Silvia Yee et al., Compounded Disparities: Health Equity at the Intersection of Disability, Race, and Ethnicity (2019). This phenomenon was evident during the COVID-19 pandemic:

> . . . The lower a person's socioeconomic status, the more limited their resources and ability to access essential goods and services, and the greater their chance of suffering from chronic disease, including conditions like heart disease, lung disease, and diabetes that may increase the mortality risk of COVID-19. Individuals and families in poverty have less control over their environment and few to no alternatives to substandard housing. These effects are exacerbated for people of color who are subjected to the consequences of discrimination and segregation in housing, on top of affordability challenges.

Many low-income individuals and families face significant challenges that prevent them from protecting themselves and others from COVID-19. Many lack the disposable income, flexible work schedules, and ability to do paid work from home. Nor do they have paid leave required to take care of children whose schools are closed and whose educational attainment and social development may be set back for months.

Emily A. Benfer and Lindsay F. Wiley, Health Justice Strategies to Combat COVID-19: Protecting Vulnerable Communities During A Pandemic, Health Aff. Blog, Mar. 19, 2020.

3.   *Perpetuating Inequity Through Law. People First of Alabama* does more than highlight certain social determinants of health. It also identifies how the law—in this case voting laws that unduly burden the rights of those already burdened inequitably by COVID-19—can perpetuate and exacerbate existing inequity by extending it into yet another activity, in this case, voting. Of course, laws that make voting during a pandemic more difficult or risky for those most heavily burdened by the pandemic undermine the ability of that sub-population to be heard at the polls and thus adequately represented in government in the future. In this way, *People First of Alabama* also provides an example of what Daniel E. Dawes calls the "political determinants of health." Daniel E. Dawes, The Political Determinants of Health (2020).

4.   *Additional Reading.* For additional reading on health equity, law and the social determinants of health during the pandemic, see Emily A. Benfer et al., Health Justice Strategies to Combat the Pandemic: Eliminating Discrimination, Poverty, and Health Disparities During and After COVID-19, 19 Yale J. Health Pol'y L. & Ethics 122 (2020); Ruqaiijah Yearby and Seema Mohapatra, Law, Structural Racism, and the COVID-19 Pandemic, 7 J.L. & Bioscien. 1 (2020).

III.   PUBLIC HEALTH LAW AND THE POLICE POWER

JACOBSON V. MASSACHUSETTS

Supreme Court of the United States, 1905.
197 U.S. 11.

[Eds. Note: The following two-paragraph description is reprinted from the syllabus of the Court.]

The Revised Laws of that Commonwealth [Massachusetts], c. 75, § 137, provide that "the board of health of a city or town if, in its opinion, it is necessary for the public health or safety shall require and enforce the vaccination and revaccination of all the inhabitants thereof and shall provide them with the means of free vaccination. Whoever, being over twenty-one years of age and not under guardianship, refuses or neglects to comply with such requirement shall forfeit five dollars."

* * *

Proceeding under the above statutes, the Board of Health of the city of Cambridge, Massachusetts, on the twenty-seventh day of February, 1902, adopted the following regulation: "Whereas, smallpox has been prevalent to some extent in the city of Cambridge and still continues to increase; and whereas, it is necessary for the speedy extermination of the disease, that all persons not protected by vaccination should be vaccinated; and whereas, in the opinion of the board, the public health and safety require the vaccination or revaccination of all the inhabitants of Cambridge; be it ordered, that all the inhabitants of the city who have not been successfully vaccinated since March, 1, 1897, be vaccinated or revaccinated."

MR. JUSTICE HARLAN, after making the foregoing statement, delivered the opinion of the court.

We pass without extended discussion the suggestion that the particular section of the statute of Massachusetts now in question [ ] is in derogation of rights secured by the Preamble of the Constitution of the United States. Although that Preamble indicates the general purposes for which the people ordained and established the Constitution, it has never been regarded as the source of any substantive power conferred on the Government of the United States or on any of its Departments. Such powers embrace only those expressly granted in the body of the Constitution and such as may be implied from those so granted. Although, therefore, one of the declared objects of the Constitution was to secure the blessings of liberty to all under the sovereign jurisdiction and authority of the United States, no power can be exerted to that end by the United States unless, apart from the Preamble, it be found in some express delegation of power or in some power to be properly implied therefrom. [ ]

We also pass without discussion the suggestion that the above section of the statute is opposed to the spirit of the Constitution. . . . We have no need in this case to go beyond the plain, obvious meaning of the words in those provisions of the Constitution which, it is contended, must control our decision.

What, according to the judgment of the state court, is the scope and effect of the statute? What results were intended to be accomplished by it? These questions must be answered.

* * *

The authority of the State to enact this statute is to be referred to what is commonly called the police power—a power which the State did not surrender when becoming a member of the Union under the Constitution. Although this court has refrained from any attempt to define the limits of that power, yet it has distinctly recognized the authority of a State to enact quarantine laws and "health laws of every description;" indeed, all laws that relate to matters completely within its territory and which do not by their necessary operation affect the people of other States. According to settled principles the police power of a State must be held to embrace, at least, such reasonable regulations established directly by legislative enactment as will protect the public health and the public safety. [ ]. It is equally true that the State may invest local bodies called into existence for purposes of local administration with authority in some appropriate way to safeguard the public health and the public safety. The mode or manner in which those results are to be accomplished is within the discretion of the State, subject, of course, so far as Federal power is concerned, only to the condition that no rule prescribed by a State, nor any regulation adopted by a local governmental agency acting under the sanction of state legislation, shall contravene the Constitution of the United States or infringe any right granted or secured by that instrument. . . .

We come, then, to inquire whether any right given, or secured by the Constitution, is invaded by the statute as interpreted by the state court. The defendant insists that his liberty is invaded when the State subjects him to fine or imprisonment for neglecting or refusing to submit to vaccination; that a compulsory vaccination law is unreasonable, arbitrary and oppressive, and, therefore, hostile to the inherent right of every freeman to care for his own body and health in such way as to him seems best; and that the execution of such a law against one who objects to vaccination, no matter for what reason, is nothing short of an assault upon his person. But the liberty secured by the Constitution of the United States to every person within its jurisdiction does not import an absolute right in each person to be, at all times and in all circumstances, wholly freed from restraint. There are manifold restraints to which every person is necessarily subject for the common good. On any other basis organized society could not exist with safety to its members. Society based on the rule that each one is a law unto himself would soon be confronted with disorder and anarchy. Real liberty for all could not exist under the operation of a principle which recognizes the right of each individual person to use his own, whether in respect of his person or his property, regardless of the injury that may be done to others. . . . The good and welfare of the Commonwealth, of which the legislature is primarily the judge, is the basis on which the police power rests in Massachusetts. [ ]

Applying these principles to the present case, it is to be observed that the legislature of Massachusetts required the inhabitants of a city or town to be vaccinated only when, in the opinion of the Board of Health, that was necessary for the public health or the public safety. The authority to determine for all what ought to be done in such an emergency must have been lodged somewhere or in some body; and surely it was appropriate for the legislature to refer that question, in the first instance, to a Board of Health, composed of persons residing in the locality affected and appointed, presumably, because of their fitness to determine such questions. To invest such a body with authority over such matters was not an unusual nor an unreasonable or arbitrary requirement. Upon the principle of self-defense, of paramount necessity, a community has the right to protect itself against an epidemic of disease which threatens the safety of its members. . . . Smallpox being prevalent and increasing at Cambridge, the court would usurp the functions of another branch of government if it adjudged, as matter of law, that the mode adopted under the sanction of the State, to protect the people at large, was arbitrary and not justified by the necessities of the case. We say necessities of the case, because it might be that an acknowledged power of a local community to protect itself against an epidemic threatening the safety of all, might be exercised in particular circumstances and in reference to particular persons in such an arbitrary, unreasonable manner, or might go so far beyond what was reasonably required for the safety of the public, as to authorize or compel the courts to interfere for the protection of such persons. . . . There is, of course, a sphere within which the individual may assert the supremacy of his own will and rightfully dispute the authority of any human government, especially of any free government existing under a written constitution, to interfere with the exercise of that will. But it is equally true that in every well-ordered society charged with the duty of conserving the safety of its members the rights of the individual in respect of his liberty may at times, under the pressure of great dangers, be subjected to such restraint, to be enforced by reasonable regulations, as the safety of the general public may demand.

* * *

Looking at the propositions embodied in the defendant's rejected offers of proof it is clear that they are more formidable by their number than by their inherent value. Those offers in the main seem to have had no purpose except to state the general theory of those of the medical profession who attach little or no value to vaccination as a means of preventing the spread of smallpox or who think that vaccination causes other diseases of the body. What everybody knows the court must know, and therefore the state court judicially knew, as this court knows, that an opposite theory accords with the common belief and is maintained by high medical authority. We must assume that when the statute in question was passed, the legislature of Massachusetts was not unaware of these opposing theories, and was compelled, of necessity, to choose between them. It was not compelled to commit a matter involving the public health and safety to the final decision of a court or jury. It is no part of the function of a court or a jury to determine which one of two modes was likely to be the most effective for the protection of the public against disease. That was for the legislative department to determine in the light of all the information it had or could obtain. . . . Upon what sound principles as to the relations existing between the different departments of government can the court review this action of the legislature? If there is any such power in the judiciary to review legislative action in respect of a matter affecting the general welfare, it can only be when that which the legislature has done comes within the rule that if a statute purporting to have been enacted to protect the public health, the public morals or the public safety, has no real or substantial relation to those objects, or is, beyond all question, a plain, palpable invasion of rights secured by the fundamental law, it is the duty of the courts to so adjudge, and thereby give effect to the Constitution." [ ]

Whatever may be thought of the expediency of this statute, it cannot be affirmed to be, beyond question, in palpable conflict with the Constitution. Nor, in view of the methods employed to stamp out the disease of smallpox, can anyone confidently assert that the means prescribed by the State to that end has no real or substantial relation to the protection of the public health and the public safety. Such an assertion would not be consistent with the experience of this and other countries whose authorities have dealt with the disease of smallpox. [The Court then summarizes the history of vaccination in Europe and America.]

* * *

Since then vaccination, as a means of protecting a community against smallpox, finds strong support in the experience of this and other countries, no court, much less a jury, is justified in disregarding the action of the legislature simply because in its or their opinion that particular method was—perhaps or possibly—not the best either for children or adults.

* * *

We are not prepared to hold that a minority, residing or remaining in any city or town where smallpox is prevalent, and enjoying the general protection afforded by an organized local government, may thus defy the will of its constituted authorities, acting in good faith for all, under the legislative sanction of the State [by refusing to be vaccinated]. If such be the privilege of a minority then a like privilege would belong to each individual of the community, and the spectacle would be presented of the welfare and safety of an entire population being subordinated to the notions of a single individual who chooses to remain a part of that population. . . . The safety and the health of the people of Massachusetts are, in the first instance, for that Commonwealth to guard and protect. They are matters that do not ordinarily concern the National Government. So far as they can be reached by any government, they depend, primarily, upon such action as the State in its wisdom may take; and we do not perceive that this legislation has invaded by right secured by the Federal Constitution.

Before closing this opinion we deem it appropriate, in order to prevent misapprehension as to our views, to observe . . . that the police power of a

State . . . may be exerted in such circumstances or by regulations so arbitrary and oppressive in particular cases as to justify the interference of the courts to prevent wrong and oppression. Extreme cases can be readily suggested. Ordinarily such cases are not safe guides in the administration of the law. It is easy, for instance, to suppose the case of an adult who is embraced by the mere words of the act, but yet to subject whom to vaccination in a particular condition of his health or body, would be cruel and inhuman in the last degree. We are not to be understood as holding that the statute was intended to be applied to such a case, or, if it was so intended, that the judiciary would not be competent to interfere and protect the health and life of the individual concerned.

MR. JUSTICE BREWER and MR. JUSTICE PECKHAM dissent [without opinion].

NOTES AND QUESTIONS

1.   *State Police Powers.* This oft-cited case established the foundation for the public health regulation that came during the ensuing century. Its holding plainly recognizes the authority of the state to exercise its police powers by imposing public health restrictions on willing and unwilling citizens (and others). Moreover, it clarifies that states may delegate their public health powers, whether to state agencies or to local governments. *Jacobson* also recognizes that a state's power to act in the name of public health is limited by the Constitution's protection of individuals against governmental interference. The following diagram depicts the police power, its delegation, and its constitutional limitations:

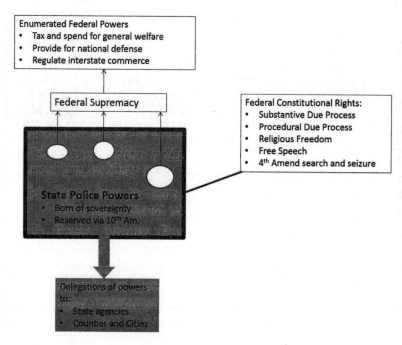

2.   *Federal vs. State Public Health Powers.* The federal government does not have the police power because it is a government of only limited powers, and the police power enjoyed by each of the states is not among the powers delegated to the federal government by the states under the federal Constitution. Moreover, the 10th Amendment clarifies that "powers not delegated to the United States by the Constitution, nor prohibited by it to the States, are reserved to the States respectively, or to the people." Thus, federal legislation designed to promote population health must find its constitutional authority under an enumerated power (e.g., power to regulate interstate commerce), while similar state legislation can be based on the state's police power.

3.   *"Arbitrary and Oppressive"?* *Jacobson* acknowledges that the Due Process Clause in the 14th Amendment to the federal Constitution restricts the state's power to act in the name of public health. Specifically, it held that the state may not act "by regulations so arbitrary and oppressive in particular cases as to justify the interference of the courts to prevent wrong and oppression." When might a state cross this line? The Court says that "[e]xtreme cases can readily be suggested," but it does not actually suggest them. Can you identify these cases?

4.   *Evolution of Substantive Due Process.* Substantive due process jurisprudence has developed considerably since *Jacobson*, and, as a result, several courts have imposed standards that are more rigorous than the "non-arbitrary" test articulated in *Jacobson*. For example, three courts reviewing state orders confining individuals with tuberculosis held that state officials must prove "substantial risk" or meet a strict scrutiny standard. See City of Newark v. J.S., 279 N.J. Super. 178, 652 A.2d 265 (1993) (state must prove by clear and convincing evidence that the individual poses a substantial risk of dangerous conduct); Greene v. Edwards, 164 W. Va. 326, 263 S.E.2d 661 (1980) (same); Best v. St. Vincents Hosp., 2003 WL 21518829 (S.D.N.Y.) (confinement must be necessary to serve a substantial governmental interest). During the 2014 Ebola scare, however, two federal courts reviewing quarantine orders employed a "reasonableness" test, citing to *Jacobson*. See Hickox v. Christie, 205 F. Supp. 3d 579 (D.N.J. 2016); Liberian Community Assoc. v. Malloy, 2017 WL 4897048 (D. Conn.), aff'd, Liberian Community Assoc. v. Lamont, 970 F.3d 174 (2d Cir. 2020). What, if any, substantive due process standard is required by *Jacobson* remains unsettled, and this uncertainty has resulted in inconsistent use of *Jacobson* in the many lawsuits brought to challenge the constitutionality of COVID-19 restrictions. See Part IV in this Chapter.

5.   *Wong Wai v. Williamson.* *Jacobson* was not the first vaccination case to be litigated in the federal courts. In Wong Wai v. Williamson, 103 F. 1 (N.D. Cal. 1900), the Circuit Court faced an action by Wong Wai, "a subject of the emperor of China, residing in the city and county of San Francisco," seeking to enjoin the city from enforcing a resolution of the board of health that prohibited Chinese residents from traveling outside of the city without proof that they had been inoculated with the "Haffkine Prophylactic," which was thought to provide immunization against bubonic plague. The resolution applied only to

those of Chinese extraction, although there was no evidence that they were any more likely than others to be subject to the plague. The Circuit Court recognized the public health authority of the city, but also recognized that there were limits to the exercise of this police power.

> The conditions of a great city frequently present unexpected emergencies affecting the public health, comfort, and convenience. Under such circumstances, officers charged with the duties pertaining to this department of the municipal government should be clothed with sufficient authority to deal with the conditions in a prompt and effective manner. Measures of this character, having a uniform operation, and reasonably adapted to the purpose of protecting the health and preserving the welfare of the inhabitants of a city, are constantly upheld by the courts as valid acts of legislation, however inconvenient they may prove to be, and a wide discretion has also been sanctioned in their execution. But when the municipal authority has neglected to provide suitable rules and regulations upon the subject, and the officers are left to adopt such methods as they may deem proper for the occasion, their acts are open to judicial review, and may be examined in every detail to determine whether individual rights have been respected in accordance with constitutional requirements.

* * *

> In the light of these well-established principles [that public health measures must have an appropriate relationship with the ends they seek to serve], the action of the defendants as described in the bill of complaint cannot be justified. They are not based upon any established distinction in the conditions that are supposed to attend this plague, or the persons exposed to its contagion, but they are boldly directed against the Asiatic or Mongolian race as a class, without regard to the previous condition, habits, exposure to disease, or residence of the individual; and the only justification offered for this discrimination was a suggestion made by counsel for the defendants in the course of the argument, that this particular race is more liable to the plague than any other. No evidence has, however, been offered to support this claim, and it is not known to be a fact. This explanation must therefore be dismissed as unsatisfactory.

103 F. at 12–15. The Court went on to find that the actions of the board of health violated not only "the express provisions of the constitution of the United States, in several particulars, but also of the express provisions of our several treaties with China and of the statutes of the United States." Id. at 23.

6. *Vaccine Mandate Medical Exception.* The Court in *Jacobson* suggests that it may be improper to apply the Massachusetts statute and the Cambridge order to an adult whose health would be placed at risk by vaccination, even though the only exception in the statute is for children whose health would be at risk. Could an adult with some legitimate health concerns about vaccination forgo it? Would this be a matter to be determined by the Board of Health, by the courts, or by an employer or school?

7.   *School Vaccination Mandates and Religious and Moral Exemptions.* In a long historical footnote, the *Jacobson* Court traces the history of vaccination from the "first compulsory act" in England in 1853. In fact, as early as 1827 the city of Boston required vaccination for school attendance, and in 1855 Massachusetts became the first state with required childhood vaccination laws associated with school attendance. Such laws were common by the end of the nineteenth century, but, for political and other reasons, they were not regularly enforced. Some were triggered only when there was a public health threat in the community, and some courts recognized religious defenses to the obligation to be vaccinated. See Rhea v. Board of Education, 171 N.W. 103 (N.D. 1919).

Today all states provide for exemptions when the vaccination would threaten the health of the child; almost all allow for religious exemption (although the nature of those exemptions varies from state to state); and many states allow for an exemption on "moral" or other grounds. Are states required to have religious exemptions under the First Amendment or other federal law designed to protect religious freedom? If a state has a religious exemption, should that legislation also permit a "moral" exemption? Is it constitutionally required to do so under the Establishment Clause of the First Amendment or the Equal Protection Clause of the Fourteenth Amendment?

8.   *Advisory Committee on Immunization Practices.* Requiring immunizations for children to be enrolled in school is a matter of state law, but, as a general matter, states adopt the CDC list of recommended immunization developed by the Advisory Committee on Immunization Practices. In order to keep some kinds of federal funding, the states now must make an effort to enforce their own immunization laws. This issue is not entirely a scientific one, though, and vaccine manufacturers do lobby state legislatures to have their own products included on the mandatory list.

9.   *Anti-Vax Fears and Misinformation.* While the health value of continued required immunization is now well established, those who fear government overreach have often had special concerns about vaccinations. It is unclear just why this issue has become a political one as well as a public health one, but it is not hard to find those who oppose vaccination with great zeal. The CDC, the World Health Organization (WHO), and other organizations have joined together to refute several myths about vaccination that sometimes scare parents into seeking exemptions.

For a full list of the most common arguments against vaccination and the refutation of each, see the WHO publication, 6 Common Misconceptions about Vaccinations and How to Respond to Them (2010). For an excellent account of this issue, and the basis for much of the material in these notes, see James Hodge, Jr., School Vaccination Requirements: Legal and Social Perspectives, 27 NCSL State Legislative Report, No. 14 (August 2002). See also S. Omer et al., Nonmedical Exemptions to School Immunization Requirements, 296 JAMA 1757 (2006); S. Omer, Correspondence: Vaccination Polices and Rates of Exemption from Immunization, 2005–2011, 367 NEJM 1170 (2012). For an excellent overview of the issue and a clever proposal to finesse the opposition to immunization while respecting individuals' liberties, see Ross Silverman, No More Kidding Around: Restructuring Childhood Immunization Exceptions, in Insure Public Health Protection, 12 Ann. Health L. 277 (2003).

The pattern of spreading false vaccine information and conspiracy theories has continued with respect to the COVID-19 vaccines. See I. Ullah et al., Myths and Conspiracy Theories on Vaccines and COVID-19: Potential Effect on Global Vaccine Refusals, 22(2) Vacunas 93 (2021); Ana Santos Rutschman, Social Media Self-Regulation and the Rise of Vaccine Misinformation, 4 U. Pa. J. Law & Innovation 25 (2021).

10. *Side Effects.* Of course, there are side effects to vaccination, and some are very serious, although a recent comprehensive study review found "no evidence of major safety concerns" associated with childhood vaccination, including current scheduling that requires delivery of multiple vaccines at one time. Institute of Medicine, Childhood Immunization Schedule and Safety: Stakeholder Concerns, Scientific Evidence, and Future Studies (2013).

11. *Vaccine Injury Compensation Program.* Compensation for injuries that arise out of immunization is specifically addressed in the National Vaccine Injury Compensation Program, 42 U.S.C. § 300aa–10 et seq. This statute provides scheduled (and thus limited) compensation for those who can prove that their injury has been caused by a covered vaccine. It does not require proof of negligence or products liability for compensation. The payment is made by a government agency and the compensation payments and administration of the program are financed by an excise tax on the vaccines. Thus, the manufacturers pay the cost of the program, but they can pass that cost on and spread it out among those who are vaccinated—virtually everyone. This statute may be unique in providing that attorneys must inform potential plaintiffs in vaccine injury cases of the existence of the law and the remedy it provides:

> It shall be the ethical obligation of any attorney who is consulted by an individual with respect to a vaccine-related injury or death to advise such individual that compensation may be available under the program for such injury or death.

42 U.S.C. § 300aa–10(b).

For an excellent history of vaccine shortages induced by manufacturers' fears of liability that gave rise to this no-fault compensation scheme and its predecessors, see Elizabeth Scott, The National Childhood Vaccine Injury Act Turns Fifteen, 56 Food & Drug L.J. 351 (2001). See also L. Rutkow et al., Balancing Consumer and Industry Interests in Public Health: The National Vaccine Injury Compensation Program and Its Influence During the Last Two Decades, 111 Penn. St. L. Rev. 681 (2007).

IV. INDIVIDUAL RIGHTS AND CONSTITUTIONAL STANDARDS OF REVIEW FOR PUBLIC HEALTH MEASURES

As noted in *Jacobson*, the police power is not absolute. Among other things, it is limited by individual rights recognized under the federal Constitution, including the rights to substantive and procedural due process and to the free exercise of religion. The COVID-19 pandemic has forced many courts to identify the relevant standard for reviewing the constitutionality of public health actions taken by state and local officials. As excerpted above, *Jacobson* asked only whether the City's smallpox vaccination requirement had some "real or substantial relation" to protecting population health and whether it was "beyond all question, a plain, palpable invasion of rights secured by the fundamental law." Of course, *Jacobson* was decided well before the development of today's standards for constitutional review. So, courts must now confront whether the standard applied in *Jacobson* remains good law, and if so, how to reconcile it with modern substantive due process jurisprudence. The following opinion addresses that very issue.

KLAASSEN V. TRUSTEES OF INDIANA UNIVERSITY

United States District Court, N.D. Indiana, 2021.
___ F. Supp.3d ___, 2021 WL 3073926.

DAMON R. LEICHTY, UNITED STATES DISTRICT JUDGE.

* * *

This case presents . . . [the following] question: whether Indiana University has acted constitutionally in mandating the COVID-19 vaccine for its students . . . . [T]his case does so only in the context of a preliminary injunction motion, not for a final decision on the merits.

* * *

Eight students sued Indiana University because of its vaccination mandate and because of the extra requirements of masking, testing, and social distancing that apply to those who receive an exemption. They ask the court to enter a preliminary injunction . . . .

* * *

FACTS

A. Parties.

Indiana University is a world-renowned public research university, . . . providing education to over 90,000 undergraduate and graduate students and employment for over 40,000 employees [ ]. . . .

The eight students here have varied backgrounds. [They include undergraduate, graduate and professional students, ranging in age from 18 to 39. They each object to the University's policy. Seven of the eight have received or are eligible to receive a religious exemption from the mandate, but each remains subject to the University's masking and testing requirements for unvaccinated, exempt individuals. One of the eight cannot qualify for an exemption. Most of the eight are unsure if they will remain students at the University if the policy remains in effect, but one will return regardless of the outcome of the case and one will not.]

* * *

B.   COVID-19.

COVID-19 is an infectious disease caused by the novel coronavirus. It primarily spreads through respiratory droplets, viral particles suspended in the air, and touching mucosal membranes with contaminated hands [ ]. The initial presentation of an infection ranges from no symptoms at all (asymptomatic) to severe illness and death; and even after recovery, various long-term health problems may linger [ ].

Individuals with longstanding systemic health inequities or preexisting or immunocompromising conditions, and elderly individuals prove at greater risk of severe illness or hospitalization following an infection [ ]. Children and young adults are less likely to experience serious illness or death from infection [ ]. Though data from the Centers for Disease Control and Prevention (CDC) suggest that more young adults are becoming infected with the virus than other age groups [ ], these individuals are less likely to require hospitalization or die [ ].

Worldwide COVID-19 has infected almost 189 million people and caused 4 million deaths, with these numbers still changing daily. In the United States, the novel coronavirus has infected over 33.5 million citizens, losing to death over 600,000 [ ]. Since March 6, 2020, Indiana has had over 750,000 confirmed COVID-19 cases and over 13,000 deaths [ ]. The COVID winter of 2020–2021 was particularly rough, until vaccines became options first in December 2020 and then in the early months of 2021.

As vaccination now increases, data gathered by the CDC point toward the waning of new COVID infections across the country—down from a peak of 312,325 new cases reported on January 8, 2021, with a seven-day average positive test rate of 13.85 percent, to 39,719 new cases reported on July 16, 2021, with a seven-day average positive test rate of 5.01 percent. The rate of new cases today is akin, if not greater, to the rate of new cases reported during the peak of the pandemic's first wave in the spring 2020, through the relative rate of positive tests thankfully remains much lower.

* * *

In Indiana, 561 new cases were reported on July 15, 2021; and the most recent data suggest a seven-day average positive test rate of 4.3 percent for unique individuals from July 3, 2021 to July 9, 2021, lower than the national average. Of all positive cases, 18.4 percent, the highest proportion of all age populations, comes from young adults aged 20–29. In Indiana, approximately 67.3 percent of all cases came from the Delta variant. Our country and our state have vastly improved, but challenges remain.

* * *

F.   Indiana University's Vaccine Mandate.

[The University created a "Restart Committee" to study and recommend to the University policies that would permit the resumption of face-to-face education. The committee consisted of experts from a variety of relevant fields, and it made a thorough study of relevant materials before making its recommendations.]

* * *

. . . The Board of Trustees adopted the restart committee's recommendations for the 2021 fall semester [ ].

The aim was short and strategic—vaccinate everyone, subject to certain exemptions [ ]. . . . The policy today requires all students, faculty, and staff to be fully vaccinated . . . before returning to campus between August 1 to August 15 for the fall 2021 semester [ ].

The choice of foregoing vaccination is not inconsequential. If not vaccinated, students are not permitted on campus, their emails and university accounts are suspended, and their access cards are deactivated [ ]. Although it seems from argument that the university will not create an informant culture, it reserves the right to pursue disciplinary action should a student deceive the process. Faculty and staff who refuse vaccination face termination. . . .

The university's COVID-19 vaccine policy has exemptions. A student may request an exemption for religious reasons [or for medical reasons]. Students who are enrolled in an online program, with no on-campus component, don't need to receive the vaccine [ ].

For those who receive exemption from vaccination, the policy imposes additional safety requirements. These requirements apply to six of the eight students here who have received exemptions and potentially a seventh who qualifies for an exemption. Such students must participate in more frequent mitigation testing, quarantine if exposed to someone who has tested positive for COVID-19, wear a mask in public spaces, and return to their permanent address or quarantine if there is a serious outbreak of COVID-19 [ ].

H.   Emergency Use Authorization of Vaccines.

* * *

Despite creating an expedited pathway to distribute new medical products during emergencies, products that receive [emergency use authorization (EUA)] . . . still must adhere to specified safety, efficacy, and manufacturing criteria, and HHS must ensure medical providers and individuals are informed of the product's EUA status, the "significant known and potential benefits and risks of such use, and of the extent to which such benefits and risks are unknown;" and for individuals, of the option to refuse and the consequences of such a decision. [ ]. An EUA generally allows a manufacturer to apply for EUA approval using interim clinical trial data, and the data need only demonstrate the product "may be effective" and that the known and potential benefits outweigh the known and potential risks. . . .

* * *

Not all EUAs are created equally. Because of the widespread use of a COVID-19 vaccine, the FDA informed manufacturers that it expected the same level of endpoint efficacy data as required for full approval, enough safety data to justify by clear and compelling evidence the vaccine's safety, and confirmation of the technical procedures and verification steps necessary to support full approval. In short, and as described in more detail below in this opinion's analysis, the FDA promulgated guidance that enhanced the basis on which any COVID-19 vaccine would meet EUA approval. In setting these more stringent standards, the FDA invited EUA applications only for vaccines positioned well to receive full approval.

I.   COVID-19 Vaccines.

[The court summarizes the development and testing of each of the three leading COVID vaccines, and the EUA each received from the FDA.]

With these vaccines, an emerging light appeared at the end of the tunnel. As of July 17, 2021, 337,239,448 doses of vaccine have been administered, and 161 million Americans, or 48.5 percent of the total population, is fully vaccinated. Of adults over the age of eighteen, 59.4 percent are fully vaccinated. In Indiana, 5,749,173 doses have been administered, and 2,888,239 Hoosiers, or 49.6 percent of those over the age of twelve, are fully vaccinated. Of ages 18–24, who account for 9.2 percent of the U.S. population, 11,720,847, or 42.2 percent, are fully vaccinated. In Indiana, 164,098 individuals aged 20–24, or 34.7 percent, are fully vaccinated.

J.   Risks of Vaccines.

Though the vaccines show remarkable effectiveness against infection and severe cases of COVID-19, and "have undergone and will continue to undergo the most intensive safety monitoring in U.S. history," they are not without risks, heretofore rare for serious risks [ ]. Many recipients experience mild local and systemic reactions, including fever, headache, muscle pain, chills, and tiredness. In very rare cases, more serious side effects seem to emerge such as allergic reactions or blood clots with low platelets [ ]. For young men specifically, experts are studying a temporal correlation between vaccines and myocarditis, an inflammation of the heart muscle, or pericarditis, inflammation of tissue around the heart [ ]. However, the risk of myocarditis appears to be exceptionally small [ ].

The medical community closely tracks adverse events from the vaccine in a national database called VAERS, or the Vaccine Adverse Event Reporting System. . . . Based on this surveillance, reports of anaphylaxis [appear] to be rare, blood clotting concerns are rare but higher in women under the age of 50, myocarditis is rare but more common in young people, and reports of death are rare.

The FDA has issued revisions to the patient and provider fact sheets about the risk of myocarditis and pericarditis acknowledging data about this risk. Furthermore, the FDA and CDC recommended a pause on the use of Johnson & Johnson's vaccine in light of reports of clotting in young women (a pause subsequently lifted). Recent changes last week occurred because of reported neurological impacts of the Johnson & Johnson vaccine, based on VAERS data. These refinements indicate that the ongoing safety of these vaccines are rigorously monitored by agency professionals.

* * *

PRELIMINARY INJUNCTION STANDARD

. . . To obtain an injunction, the students "must make a threshold showing that: (1) absent preliminary injunctive relief, [they] will suffer irreparable harm in the interim prior to a final resolution; (2) there is no adequate remedy at law; and (3) [they have] a reasonable likelihood of success on the merits." [ ] If they make these threshold showings, the court "consider[s] the balance of harms between the parties and the effect of granting or denying a preliminary injunction on the public interest." [ ]

ANALYSIS

A. *These Students Aren't Likely to Succeed on the Merits.*

* * *

1. *The Fourteenth Amendment.*

* * *

. . . [T]he Fourteenth Amendment says no "State [may] deprive any person of life, liberty, or property, without due process of law." [ ]. . . .

Indiana University is a state actor, [ ], so the Fourteenth Amendment also applies to it.

. . . The Fourteenth Amendment protects a person's substantive rights in life, liberty, and property. [ ] Certain rights or liberties have been deemed "fundamental," so they receive greater protection. [ ]

Bearing that in mind, the court initially approaches this case in a two-fold manner. First, the law requires a "careful description" of the asserted right or liberty. [ ] Second, the court must determine whether the so-defined right or liberty is fundamental under the Constitution. [ ] The Fourteenth Amendment's due process clause specially protects fundamental rights and liberties—those that objectively are "deeply rooted in this Nation's history and tradition" and so "implicit in the concept of ordered liberty" that "neither liberty nor justice would exist if they were sacrificed." . . .

* * *

The students and university disagree on the constitutional analysis. Declaring a right or liberty fundamental has important implications. Modern constitutional jurisprudence employs a different analysis when a person's fundamental right is at stake. If the government infringes on a fundamental right, the court often applies strict scrutiny. [ ] In such circumstances, the Fourteenth Amendment "forbids the government to infringe . . . fundamental liberty interests *at all*, no matter what process is provided, unless the infringement is narrowly tailored to serve a compelling state interest." [ ] This is the most rigorous form of constitutional scrutiny of government action.

Whereas infringements on other rights or liberties, though still constitutionally scrutinized, must meet what courts call rational basis review. [ ] . . . It is less stringent than strict scrutiny. Under rational basis review, "legislation is presumed to be valid and will be sustained if the classification drawn by the statute is rationally related to a legitimate state interest." [ ] The students argue for strict scrutiny, and the university argues for rational basis review.

2. *The Constitution in a Public Health Crisis.*

We live in the era of the COVID-19 virus—worldwide seeing to nearly 189 million cases and 4 million deaths, with these numbers changing daily. The United States hasn't been immune. Our citizens have recovered or struggled to recover from over 33 million cases of this novel coronavirus when over 606,000 tragically have passed. A public health crisis of this magnitude begs the question: how should the law respond to state action that infringes on the People's liberties during such times?

To be sure, the Constitution isn't put on the shelf. Indeed, in times of crisis, perhaps constitutional adherence proves the very anchor we all need against irrational and overweening government intrusion that would otherwise scuttle the ship. As the arbiters of the Constitution's checks and balances, [ ], the courts play an important role in ensuring that the government doesn't simply declare a never-ending public emergency and expand its powers *ad libitum* to the People's detriment.

Under our country's federalist system, state and federal governments share regulatory authority over public health matters. States traditionally exercise most authority under their inherent police power—and reasonably so when public health may flux and evolve by locale. States thus have the power, within constitutional limits, to pass laws that "provide for the public health, safety, and morals[.]" [ ]

To answer the question today, the court travels back in time to 1905: a time before the modern tiers of constitutional analysis (strict scrutiny and rational basis) and one rampaged by the smallpox epidemic. In that year, the United States Supreme Court issued a leading decision in answer to this question.

[Eds. Note: The court summarizes *Jacobson* and its relevant holdings. An excerpted version of *Jacobson* is presented in section III of this Chapter, above.]

* * *

The students want *Jacobson* confined to its time, whereas the university believes it applies with full force. In the years since, the high court has leaned on *Jacobson* to uphold government measures intended for the public welfare under effectively rational basis review, finding the measures reasonably advancing a legitimate state interest. For example, *Zucht [v. King]*, 260 U.S. [174 (1922)] . . . , relied on *Jacobson* to uphold a city ordinance excluding from its public schools children not having a certificate of vaccination, holding that it was within the state's police powers reasonably to so act. . . . This was not authorization of "arbitrary power," but only that broad discretion required for the protection of the public health." [ ] In doing so, "state and federal legislatures [enjoy] wide discretion to pass legislation in areas where there is medical and scientific uncertainty." [ ]

Based on this power, states and their authorized arms have historically adopted vaccination mandates. For instance, all fifty states and the District of Columbia have laws requiring students to receive certain vaccines before they may attend school. Many align their vaccine requirements with CDC's immunization recommendations, and all laws provide exemptions for medical reasons and nearly all religious exemptions. Adult vaccination mandates often have been limited to the private employment sector, though not always. For instance, the State of Indiana requires all public university students to receive vaccinations for diphtheria, tetanus, measles, mumps, rubella, and meningococcal disease, save for religious and medical exemptions. [ ]

* * *

. . . In this century . . . courts have returned again to [*Jacobson*'s] guidance during the COVID-19 pandemic. . . .

* * *

One such decision—and one heavily briefed by the parties—is *Roman Catholic Diocese of Brooklyn v. Cuomo*, ___ U.S. ___, 141 S. Ct. 63 [ ] (2020). In *Cuomo*, the State of New York adopted capacity restrictions on religious institutions that treated them less favorably than so-called "essential" businesses, [ ], including liquor and hardware stores, [ ]. *Cuomo* applied strict scrutiny because the law targeted religious practice contrary to the First Amendment, as incorporated against the states by the Fourteenth Amendment, and enjoined the limitations, saying they were not narrowly tailored to fulfill the state's compelling interest in controlling the spread of COVID-19. [ ]

Cuomo enhanced the law's focus under the First Amendment. [ ] So this begs another question: to what extent has *Cuomo*—if any—impacted the broad deference the court would seemingly afford a state during a pandemic under *Jacobson* to act in the interest of public health? *Cuomo*'s majority opinion never referenced *Jacobson*.

The students read *Cuomo* as implicitly overruling *Jacobson*, or at least as abrogating it. Though the Supreme Court may overrule a case without explicitly saying so, [ ], this is a tall task. Before a federal court concludes that the Supreme Court has implicitly overruled a prior decision, it must be "certain or almost certain that the decision or doctrine would be rejected by the higher court if a case presenting the issue came before it." [ ] This high bar is rarely met. [ ] It isn't met here. *Cuomo* and *Jacobson* involved entirely different modes of analysis, entirely different rights, and entirely different kinds of restriction. [ ] "*Jacobson* applied what would become the traditional legal test associated with the right at issue"—exactly what *Cuomo* did. [ ] The cases walk hand-in-hand.

* * *

Considering the modern tiers of constitutional scrutiny, the court reads *Jacobson* and *Cuomo* harmoniously, appreciating their respective spheres. Though *Jacobson* was decided before tiers of scrutiny, it effectively endorsed—as a considered precursor—rational basis review of a government's mandate during a health crisis. [ ] In its words, if a law purporting to be enacted to protect public health "has no real or substantial relation to [that legitimate aim]" or if the law proves "a plain, palpable invasion of rights secured by the fundamental law," the court's job is to give effect to the Constitution. [ ] Should the court have this melding of history and modernity wrong in faithfully adhering to the Fourteenth Amendment's plain original meaning of "life" and "liberty," comfort should come in knowing that *Jacobson*, whether rational basis review by any other name, leads to the same result today.

This view remains consistent with the right at stake in *Jacobson*: though a true "liberty" proved at stake—the right to refuse a vaccine during a smallpox epidemic—this interest in bodily autonomy, though protected by the Constitution, wasn't fundamental under the Constitution to require greater scrutiny than rational basis review. [ ] At the same time, *Jacobson* didn't hold that the government's authority in a pandemic balloons for it do whatever it wants in the name of public safety.

Jacobson instead counseled that federal courts should require a rational relation to a legitimate interest in public health. [ ] That *Cuomo* imposed heightened scrutiny of the government's interference with the free exercise of religion—a fundamental right under the First Amendment— was presciently contemplated a century beforehand by *Jacobson*: a court should intervene if a state imposes a regulation that is "beyond all question, a plain, palpable invasion of rights secured by the *fundamental law*." [ ] Because *Cuomo* involved a fundamental right, a "right[ ] secured by the fundamental law" under today's jurisprudence, the court intervened. [ ] The Constitution's original meaning should be so enduring.

The university seems to argue that *Jacobson* gave even more deference than rational basis review during a public health crisis, but not fairly so; and, even then, *Jacobson* cannot be taken once more too far. [ ]

Jacobson doesn't justify blind deference to the government when it acts in the name of public health or in a pandemic. For instance, the decision left the door open for people with legitimate medical concerns to challenge the vaccine mandate. [ ] And the deference owed to the States during a pandemic or public health crisis under *Jacobson* doesn't extend indefinitely. [ ]

> [A]t the outset of an emergency, it may be appropriate for courts to tolerate very blunt rules. . . . [B]ut a public health emergency does not give . . . public officials *carte blanche* to disregard the Constitution for as long as the medical problem persists. As more medical and scientific evidence becomes available, and as States have time to craft policies in light of that evidence, courts should expect policies that more carefully account for constitutional rights.

[ ]

In short, the Constitution doesn't permit the government to declare a never-ending public emergency and expand its powers arbitrarily. [ ] Instead, as our country and communities progress through a pandemic, the government must continually update its practices in light of the most recent medical and scientific developments. And a law or policy should be written with a mindset that medicine and science, and the circumstances that they create, will evolve, and so must the law or policy evolve or be revisited in amendment.

In sum, the law today recognizes *Jacobson* as a precursor to rational basis review. This is consistent with statements of many justices who continue to acknowledge *Jacobson* as good law, albeit with constitutional restraint. Government action that infringes on the liberty interest here, as in *Jacobson*, is subject to rational basis review. [ ]

3. *Defining the Right & Constitutional Analysis.*

The students assert a right to refuse the vaccine, saying the mandate infringes on their bodily autonomy and medical privacy. Indiana University throws a challenge flag here. To it, these students are merely saying they have a right to refuse a vaccine *so that* they may attend college. The university says the right being infringed then isn't the right to refuse a vaccine, but the right to attend college. Indeed, if they choose to forego college at Indiana University, there is no vaccine requirement. To the university, the students aren't being forced to take the vaccination against their will; they can go to college elsewhere or forego college altogether. If this case were merely that, merely the right to attend university, this state action wouldn't trample on their rights. There is no fundamental or constitutional right to a college education, [ ], much less one at a particular institution.

But that's not what this case concerns, and that's not the liberty at stake. The "unconstitutional conditions doctrine" forbids the university from pulling the rug out from under the students in a roundabout way. Under this doctrine, argued by the students as "coercion," "the government may not deny a benefit to a person because he exercises a constitutional right." [ ] This doctrine protects constitutional rights "by preventing the government from coercing people into giving them up." [ ] It "aims to prevent the government from achieving indirectly what the Constitution prevents it from achieving directly." [ ] The students say this state actor is denying a benefit—a public university education—because they are exercising a constitutional right to refuse a vaccine.

The first step in an unconstitutional condition claim "is to identify the nature and scope of the constitutional right arguably imperiled by the denial of a public benefit." [ ] Here, the Fourteenth Amendment "liberty" at stake is a college student's right to refuse a vaccine, today at this stage of the pandemic [ ]. The Supreme Court has assumed (using its word) and strongly suggested that individuals have a constitutional right to refuse unwanted medical treatment, [ ]. *Cruzan*[*v. Director, Missouri Dep't of Health*, 497 U.S. 261 (1990)] held that a competent individual had a constitutional right to refuse unwanted lifesaving hydration and nutrition; and [*Washington v.*] *Glucksberg*[, 521 U.S. 702 (1997),] recognized that an individual had a liberty interest in refusing unwanted lifesaving medical treatment, though not any fundamental right to assisted suicide. [ ]

* * *

The rights recognized (or assumed) in these cases weren't "simply deduced from abstract concepts of personal autonomy." [ ] They were rooted in longstanding common law rules or legal traditions consistent with this Nation's history. [ ] The students, quite skillfully represented in this emergency setting, offer no preliminary record of such historic rules, laws, or traditions that would facilitate the court's announcement, now in mere days from receiving this case, that a right to refuse a vaccine is anything more than a significant liberty under the Fourteenth Amendment.

The dearth of this record isn't a passing point. Indeed, both *Cruzan* and *Glucksberg* were limited to an individual's choice related to the refusal of lifesaving subsistence or medical treatment—with no ramifications to the physical health of others. Vaccines address a collective enemy, not just an individual one. Indeed, "the elimination of communicable diseases through vaccination [is] one of the greatest achievements of public health in the 20th century," [ ], and it continues to be so now in this century. A vaccine is implemented as a matter of public health, and historically hasn't been constitutionally deterred from state mandate. [ ]

In the backdrop of the Fourteenth Amendment's ratification in 1868, for instance, England had already passed its first compulsory vaccination act for smallpox in 1853 (and so had many countries). [ ] Science wasn't absolute or infallible at that time—nor is it today. But the "possibility that the belief may be wrong, and that science may yet show it to be wrong, is not conclusive; for the legislature has the right to pass laws which, according to the common belief of the people, are adapted to prevent the spread of contagious diseases." [ ] Appreciating the relative risks of vaccines, they nonetheless "are effective in preventing outbreaks of disease only if a large percentage of the population is vaccinated." [ ]

Added comfort comes from the consistent use of rational basis review to assess mandatory vaccination measures.[ ]

Given over a century's worth of rulings saying there is no greater right to refuse a vaccination than what the Constitution recognizes as a significant liberty, the court declines the students' invitation to extend substantive due process to recognize more than what already and historically exists. [ ]

* * *

The university is presenting the students with a difficult choice—get the vaccine or else apply for an exemption or deferral, transfer to a different school, or forego school for the semester or altogether. But this hard choice doesn't amount to coercion. The students taking the vaccine are choosing it among other options, and before the shot reaches their arms, they are made aware of the risks and the option to refuse.

* * *

4. *On This Preliminary Record, Non-Exempt Students Haven't Shown a Likelihood of Success on their Claim that Indiana University Lacks a Rational Basis for Its Vaccine Mandate.*

Determining that the students have a liberty interest under the Fourteenth Amendment's due process clause doesn't end the analysis. [ ] To decide whether the students' constitutional rights have been violated, . . . the court must balance their liberty against the relevant state interests in accord with the Constitution. [ ] . . .

"Stemming the spread of COVID-19 is unquestionably a compelling interest." [ ] According to the federal government and the State of Indiana, a state of emergency persists related to COVID-19, all the while restrictions are being scaled back gradually. Recognizing today's status of this pandemic, neither health professionals, government representatives, nor this court may say public health *vis-à-vis* COVID-19 has waned from being a legitimate state interest. Improved it undoubtedly has—today seems a world altogether different from last year—but public health remains a legitimate interest of the state to pursue. Indiana University too has a legitimate interest in promoting the health of its campus communities—students, and not least the faculty and staff who come daily in contact with them.

The students argue that the pandemic is basically over, but this goes against current proclamations from the Secretary of Health and Human Services, the Indiana State Department of Health, Governor Eric Holcomb, and the CDC, all then supported for institutions of higher learning by the U.S. Department of Education and the American College Health Association. . . . Vastly improved, yes; out of the woods we aren't, not on this preliminary record.

* * *

Let's not forget why we are here at this more promising stage of the pandemic, July 18, 2021. Antibody resistance developed naturally from prior cases has been a contributor to be sure; but, materially, improvement has come because of vaccinations—nationally over 161 million complete (over 337 million doses) and statewide nearly 3 million complete (and over 5.7 million doses). The vaccination campaign has markedly curbed the pandemic. In fact, certain age-stratified, agent-based modeling of COVID-19 has concluded that another 279,000 deaths and nearly 1.25 million more hospitalizations would have occurred by the end of June 2021 but for the vaccines. . . .

It isn't a foregone conclusion that this is overkill. This pandemic continues to evolve, and medicine and science with it. Science is a process in search of fact. One such moving target is the Delta variant (B.1.617.2). A mere four days ago Indiana reported 612 COVID cases—the highest count in more than six weeks (since May 27, 2021)—that health officials attributed largely to the Delta variant and the unvaccinated population. . . . The CDC labeled Delta a "variant of concern" in mid-June. Current science shows it more virulent and transmissible. A peer-reviewed study from scientists . . . found that the Delta variant has mutations that allow it to evade certain natural antibodies, with vaccination proving the best protection [ ]. Reports of surges of Delta cases among the 12–29 age group have occurred. . . .

* * *

For the impact of this vaccine mandate, the students focus only on the student body; and that is certainly an important part of the analysis here. The students argue that the fatality rate for healthy individuals age 20–49 is far less than older individuals [ ]. . . . The students point out that the university's student body had one death last year [ ]. Indiana University reasonably views a 0.15 percent death rate as unacceptable for its communities, particularly given the safe preventative measure of a vaccine [ ]. . . .

In addition, the student's position overlooks the larger Indiana University community. . . . The university analyzed the number of individuals within its campus population known to have increased risk factors for COVID-19 and determined that over 8,500 faculty and staff remained at increased risk of complications if they contracted the disease [ ], with the ongoing risk of asymptomatic spread that vaccines help address [ ]. Faculty and staff at Indiana University who have daily contact with students represent an even broader demographic than just the student body, and this policy was intended to protect them too. . . .

* * *

Focusing on just mortality risk from COVID-19 leaves out much of the debate. . . .

Without vaccination, college-aged students remain at risk for serious long-term complication from COVID-19, including prolonged debilitating symptoms that interfere with normal life such as myocarditis, reduced aerobic capacity, and brain damage [ ]. Long COVID remains a studied phenomenon. With Indiana reporting that individuals aged 20–29 have had more positive cases than any other age demographic [ ], and with more than 260,000 cases linked to American college and universities since January 1, 2021 [ ], this proves still a legitimate risk. . . .

The students say the risks of the vaccine, especially at this stage and to their age group, outweigh any benefits a vaccine might confer. These argued risks include myocarditis, clotting, death, and others [ ]. Some of these concerns are easier to assuage based on current science than others— and the court isn't the final arbiter of an evolving science, only of the law. The court must base today's decision on the snapshot of this preliminary record alone. It answers the question only whether the students have made a strong showing that Indiana University failed to act reasonably in achieving campus health to warrant the extraordinary remedy of a preliminary injunction.

* * *

. . . [R]eports have shown the risk of myocarditis (heart inflammation), while present and something worthy of continued investigation, to be seemingly rare—one study suggesting the risk is about eight in one million and the other study suggesting the risk is about twenty in one million. This issue has garnered increasing attention. . . . [A] CDC safety panel reported a "likely association" in young adults from mRNA COVID-19 vaccines and myocarditis and pericarditis, though it emphasized that it remained rare and typically mild, with the benefits of the vaccine still outweighing the risks [ ]. This assessment of heart inflammation's rarity and the overarching benefits of the vaccines has a bench of current medical support on this record, again giving Indiana University a rational connection between its mandate and its aim of campus health. . . .

* * *

The students return to VAERS to discuss the risk of death from the vaccines. . . .

The CDC has explored this issue as well and seems to have marshaled data, at this time, that any risk of death is rarer than the risk of death from a young adult COVID-19 infection. According to the CDC, 25,038,458 individuals aged 18–29 have been given their first dose of the vaccine as of July 18, 2021, with VAERS reporting a total of 68 deaths, or approximately 0.00027 percent, if in fact any are related. Alternatively, of the 6,174,415 cases of COVID-19 in this age group, 2,732 died, or approximately a 0.04 percent. In balancing the risks, Indiana University wasn't irrational in favoring the route that promoted greater safety for its students.

* * *

To be sure, EUA of the COVID-19 vaccines occurred on a tighter timetable and has existed only since December 2020 and February 2021. The students thus voice concerns about the experimental nature of the vaccines, though their counsel assures that their suit will persist even if the FDA grants the vaccine manufacturers full approval. Not all EUAs are equal, and the one required for COVID-19 vaccines was more robust than usual.

* * *

In October 2020, the FDA released industry guidance detailing the benchmark criteria for a COVID-19 vaccine to receive an EUA. . . . Because the virus would only be overcome through the sweeping immunity of the American public, the FDA informed manufacturers that approval would be given to those EUA applications that went beyond the safety and efficacy requirements prescribed by statute, and also expected manufacturers to consult with the FDA on the various non-clinical components of vaccine development and distribution as the clinical trial progressed. The FDA wanted the same level of efficacy data as for full approval, enough safety data to justify providing the vaccine to healthy individuals, and confirmation of the technical procedures and verification steps necessary to support full approval.

* * *

Indiana University closely considered the FDA's EUA requirements when adopting its policy. The specialists on the university restart committee appreciated that all three COVID-19 vaccines had been "studied in robust multi-centered, international, randomized-controlled trials and proven both effective and safe in millions of people" [ ]. These specialists explained that the EUA vaccines had been based on technology that has been studied for decades [ ]. Though even "small differences in chemical structure can sometimes make very large differences in the type of toxic response that is produced," [ ], much like the FDA, the university concluded that campus safety reasonably outweighed any lingering risks with the vaccines. This wasn't just any ordinary EUA process, but EUA on proverbial steroids. The university reasonably concluded that the "benefit dwarfs the potential rare risks" [ ].

* * *

Today, Indiana University has a rational basis to conclude that the COVID-19 vaccine is safe and efficacious for its students. The vaccine has been used on about 157 million Americans; and data now about eight months later, though it will grow more robust in years to come, is considerable and shows major side effects are rare. Much like over 500 universities and colleges in the United States that have done the same, Indiana University reasonably relies on the vaccine as a measure to return to normal school functioning. The students say the mandate is unreasonable because no other Indiana government agency mandates the vaccine. But just because it has gone above what others have done doesn't make it unreasonable. Indeed, universities are unique places, with lots of people gathered and living together in close quarters for months at a time. That Indiana University's mandate goes beyond what other public universities in Indiana have done doesn't compel a finding that this policy is unreasonable; indeed, other universities in the state have mandated the vaccine, and many others around the country have too.

Indiana University is following the recommendations of other well-established agencies, including the Centers for Disease Control, U.S. Department of Education, and the Indiana State Department of Health. These are reliable sources to assess the reasonableness of measures implemented, though the court must be cautious not to expand the guidance beyond what it says. [ ] To be sure, the CDC doesn't recommend that schools "mandate" the vaccine—a point the students make—but such a recommendation isn't consistent with the CDC's purview, which is to act as an informative agency. At the same time, the university's policy isn't inconsistent with the CDC's recommendations. The CDC says institutions of higher learning "can return to full capacity in-person learning, without requiring or recommending masking or physical distancing for people who are fully vaccinated" [ ]. The CDC's guidance to universities is that "[v]accination is the leading prevention strategy to protect individual from COVID-19 disease and end the COVID-19 pandemic." This will enhance the student body's opportunities, allowing them to have a more fulfilling college experience.

* * *

Overall, the students' arguments amount to disputes over the most reliable science. But when reasonable minds can differ as to the best course of action—for instance, addressing symptomatic versus asymptomatic virus spread or any number of issues here—the court doesn't intervene so long as the university's process is rational in trying to achieve public health. [ ] There is a rational basis for making distinctions here. No student, including those not yet exempt, have shown that Indiana University's vaccine mandate as applied to them violates rational basis review. The court thus denies their request to enjoin it preliminarily.

NOTES AND QUESTIONS

1.   *Subsequent History of* Klaassen. The Seventh Circuit affirmed the district court's decision with a very short opinion. See Klaassen v. Trustees of Indiana University, 7 F.4th 592 (7th Cir. 2021). The appellate court agreed that *Jacobson* had applied a rational-basis test to assess the constitutionality of the smallpox vaccine mandate at issue in that case. The court also held that Indiana University's vaccine mandate does not infringe on a fundamental liberty interest because our history does not reveal laws prohibiting such mandates. The court observed, "[t]o the contrary, vaccination requirements, like other public-health measures, have been common in this nation." Id. at

593. The Supreme Court summarily denied the plaintiffs' emergency petition for injunctive relief. See Klaassen v. Trustees of Indiana University, No. 21A15 (Aug. 13, 2021) (Barrett, J.).

2. *Supreme Court on* Jacobson. While the Supreme Court has not ruled directly on the precedential value of *Jacobson*, at least two Justices have signaled their views. In *Cuomo*—a decision analyzed by the district court in *Klaassen*—Justice Gorsuch, in a concurring opinion, expressed the interpretation of *Jacobson* that was adopted by the court in *Klaassen*. "*Jacobson* didn't seek to depart from normal legal rules during a pandemic, and it supplies no precedent for doing so. Instead, *Jacobson* applied what would become the traditional legal test associated with the right at issue— exactly what the Court does today." Cuomo, 141 S. Ct. at 70. Additionally, in *South Bay United Pentecostal Church v. Newsom*, 140 S. Ct. 1613 (2020) (mem.), the Court denied an emergency request to enjoin California's enforcement of an executive order limiting the number of worshippers at in- person religious services. Chief Justice Roberts, in a concurrence, relied on and quoted from *Jacobson*: "Our Constitution principally entrusts '[t]he safety and the health of the people' to the politically accountable officials of the States 'to guard and protect.'" Id. at 1613.

3. *Other Interpretations of* Jacobson. The Fifth Circuit gave *Jacobson* an entirely different interpretation, holding that it created a special constitutional standard to be used in the judicial review of all public health measures imposed by state actors during an emergency. In *In re Abbott*, 954 F.3d 772 (5th Cir. 2020), *vacated by* Planned Parenthood v. Abbott, 141 S. Ct. 1261 (2021), the appellate court granted a writ of mandamus vacating a district court's preliminary injunction that prevented Texas from enforcing an executive order prohibiting all medical procedures not necessary to save the life of or prevent serious harm to patients. The executive order was designed to preserve medical resources during the COVID-19 pandemic. Abortion providers in Texas sued, claiming that the executive order effectively bans abortions for the duration of the order. The district court granted a preliminary injunction banning enforcement of the order against abortion providers and did so relying on the "undue burden" test conventionally applied in constitutional challenges to abortion restrictions.

The Fifth Circuit held that the district court's analysis was so flawed as to justify mandamus because the lower court had not relied on *Jacobson*, which the appellate court interpreted as creating a lower standard of review applicable during public health emergencies.

> when faced with a society-threatening epidemic, a state may implement emergency measures that curtail constitutional rights so long as the measures have at least some "real or substantial relation" to the public health crisis and are not "beyond all question, a plain, palpable invasion of rights secured by the fundamental law." [quoting *Jacobson*] . . .

* * *

Jacobson remains good law. [ ] And, most importantly for the present case, nothing in the Supreme Court's abortion cases suggests that abortion rights are somehow exempt from the *Jacobson* framework.

954 F.3d at 784–85.

This approach has been called the "suspension" interpretation of *Jacobson* because it suspends during an emergency the application of conventional constitutional standards of review in favor of the "reasonableness" standard articulated in *Jacobson*. See Lindsay F. Wiley and Stephen I. Vladeck, Coronavirus, Civil Liberties, and the Courts: The Case Against "Suspending" Judicial Review, 133 Harv. L. Rev. Forum 179 (2020) (critiquing the doctrine). As noted above, the Supreme Court vacated *In re Abbott*. Since then, district courts in the Fifth Circuit have not deployed the suspension interpretation of *Jacobson*. See, e.g., New Orleans Catering, Inc. v. Cantrell, 523 F. Supp. 3d 902 (E.D. La. 2021) (applying conventional substantive due process standards to review constitutional challenges to executive orders restricting certain business operations in New Orleans, where the authority for such executive orders was derived from a statute granting emergency powers to the governor that are triggered during a declared state of emergency, and where the COVID-19 pandemic justified the governor in declaring a state of emergency).

In contrast to the "suspension" interpretation, some courts have applied *Jacobson* as if it creates a special constitutional standard for the review of public health measures, whether or not an emergency has been declared. For example, during the 2014 Ebola scare, federal courts reviewing quarantine orders employed a "reasonableness" test, citing to *Jacobson*. See Hickox v. Christie, 205 F. Supp. 3d 579 (D.N.J. 2016); Liberian Community Assoc. v. Malloy, 2017 WL 4897048 (D. Conn), aff'd, Liberian Community Assoc. v. Lamont, 970 F.3d 174 (2d Cir. 2020). To understand how to apply *Jacobson* in the case of a quarantine order, judges relied on federal quarantine cases regardless of whether those cases involved declared disease emergencies. Is this interpretation of *Jacobson* viable given later case law?

4.   *Should Strict Scrutiny Apply?* The district and appellate courts in *Klaassen* each held that individuals have a significant, but not a "fundamental," liberty interest. As a result, the university's vaccine mandate was subject to rational basis testing only, and not strict scrutiny. The students had argued that individuals have a fundamental liberty interest to avoid unwanted violations of their bodily integrity. They relied on Supreme Court precedent acknowledging that informed consent law widely and historically adopted by states is evidence of a fundamental liberty interest to avoid a medical battery. For presentation and discussion of *Glucksberg*, see Chapter 6. *Klaassen* acknowledged that there is a liberty interest in refusing unwanted medical treatment, but the opinions distinguish infectious disease vaccination from standard medical treatment because the decision to receive or refuse such a vaccine has implications for others, and not just the person making the decision. As the district court noted, "[v]accines address a collective enemy, not

just an individual one." Id. at *24. The opinions also cited to other vaccine cases to which only rational basis testing has been applied.

Is the liberty interest at stake properly understood as a non-consensual bodily invasion because the vaccine pierces the skin and introduces medicine to the body? If so, was the court right that the liberty interest is not fundamental? When reviewing the constitutionality of a vaccine mandate, should the fundamental/non-fundamental determination take into account whether population health is being actively threatened by an infectious disease? If so, does that mean a chickenpox vaccination requirement to attend school implicates a fundamental liberty interest in the absence of an epidemic and a non-fundamental liberty interest during a chickenpox epidemic? If the university's vaccine mandate had been found to implicate a fundamental liberty interest, would it have survived strict scrutiny because it is necessary to achieve a compelling state interest?

5.   *University Vaccine Mandates.* More than 1,000 of the nation's nearly 4,000 degree-granting public and private colleges and universities required at least some students or employees to receive a COVID vaccination prior to the fall 2021 semester. See Andy Thomason and Brian O'Leary, Here's a List of Colleges That Require Students or Employees to Be Vaccinated Against Covid-19, Chron. Higher Ed. (Sept. 2, 2021). The schools include the University of Massachusetts and Western Michigan University, each of which was subject to a lawsuit alleging that the universities' vaccine requirements violated the constitutional rights of students. In each suit, the students moved for a preliminary injunction. In one case, the court denied the motion, following the same analysis as in *Klaassen.* See Harris v. Univ. of Massachusetts, No. 21-CV-11244-DJC, 2021 WL 3848012 (D. Mass. Aug. 27, 2021) (students are unlikely to prevail on the merits of their substantive due process claim that the university's COVID vaccination mandate, together with the mandate's religious exemption, violates the students' protected liberty interests). In the other case, the court granted the motion, holding that strict scrutiny would apply if, as alleged, the university's vaccine mandate did not permit religious exemptions. See Dahl v. Bd. of Trustees of W. Michigan Univ., No. 1:21-CV-757, 2021 WL 3891620 (W.D. Mich. Aug. 31, 2021) (granting an *ex parte* motion for preliminary injunction to prevent the university's enforcement of its COVID vaccine mandate against members of the women's soccer team).

NOTE: PUBLIC HEALTH AND RELIGIOUS FREEDOM

The First Amendment to the federal Constitution, among other things, protects the right of individuals to exercise their religious beliefs free of governmental prohibitions (the "free exercise" clause). Public health measures occasionally burden an individual's religious freedom. For example, during the COVID-19 pandemic, orders limiting the size of gatherings or closing places where individuals gather disrupted conventional, in-person worship services. Worshippers challenged some of these orders as unconstitutional.

In "free exercise" cases, federal courts apply a rational-basis test to the allegedly unconstitutional law if that law is generally applicable to both secular and non-secular activity and if it is neutral toward religion. The burden is on the challenger, under that standard, to establish that the law in question lacks any rational basis. If, however, the challenged law is aimed at religious activity or is not neutral toward religion, then federal courts apply strict scrutiny. Under this much higher standard, the government bears the burden of establishing that the challenged law is necessary to serve a compelling state interest. See Roman Cath. Diocese of Brooklyn v. Cuomo, 141 S. Ct. 63 (2020). Thus, a court's determination of the applicable constitutional standard of review often decides the outcome of "free exercise" cases.

When assessing whether a law is neutral as to religion, courts will not only look to the law's language, but also to its application to comparable secular and non-secular activities. If, as a result of such a comparison, a court determines that the law burdens religious activities in a way it does not burden secular activities, the court will rule that the law is not neutral and that strict scrutiny applies. See id.

In *Tandon v. Newsom*, 141 S. Ct. 1294 (2021), plaintiffs were individuals seeking to gather for at-home religious services, and they challenged California's COVID-19 restriction that prohibited members of more than three different households from gathering together for any purpose in any one person's home. The plaintiffs argued that the restriction violated the Free Exercise clause. The trial court denied the plaintiffs' motion for preliminary injunction. Plaintiffs appealed and sought an emergency injunction from the Ninth Circuit pending the appeal, which emergency injunction was denied. The plaintiffs then sought from the Supreme Court—and were granted—an emergency injunction for the duration of the Ninth Circuit's appellate process.

A five-justice majority of the Supreme Court approved the application, ruling that California's uniform restriction on the size in-home gatherings for both secular and non-secular purposes was *not* a neutral law with respect to religious activities. Rather than compare secular and non-secular in-home activities under the challenged restriction, the majority opinion compared at-home religious activity with secular activities occurring outside of anyone's home. "California treats some comparable secular activities more favorably than at-home religious exercise, permitting hair salons, retail stores, personal care services, movie theaters, private suites at sporting events and concerts, and indoor restaurants to bring together more than three households at a time." Id. at 1297. Because of these differences, the majority determined that strict scrutiny applies and that the Ninth Circuit had failed to find that the restrictions were necessary to confront the comparable risk of COVID-19 transmission associated with secular activity outside of the home and religious activity inside a home.

PROBLEM: EXEMPTIONS FROM
IMMUNIZATION REQUIREMENTS

You are the legal counsel to the school district in Gadsden, Alabama. The Code of Alabama, § 16–30–1 et seq., requires certain immunizations as a condition of school attendance. Exemptions to this provision are found in § 16–30–3:

The provisions of this chapter shall not apply if:

(1) In the absence of an epidemic or immediate threat thereof, the parent or guardian of the child shall object thereto in writing on grounds that such immunization or testing conflicts with his religious tenets and practices; or

(2) Certification by a competent medical authority providing individual exemption from the required immunization or testing is presented the admissions officer of the school.

First grader Miranda Black's parents, who resent the heavy burden of government, have decided that they will not submit to any immunization requirement. They reason that Miranda should not have to shoulder the public health value of immunization by risking the side effects of the immunizations. Her doctor will not provide the "certification" that the school authorities want because there is no medical reason for her to avoid immunization, and thus her parents have submitted to the school authorities a statement that "compulsory immunization violates our Christian view that the government cannot make us do anything."

School health officials are concerned because there were four cases of whooping cough in Gadsden during the last year—up from two the year before, and one the year before that. They attribute the increase to the increasing number of people who have obtained exemptions. The school officials have asked you how they should react to the request that Miranda be exempt. How will you advise them to proceed and why?

NOTE: PUBLIC HEALTH EVIDENCE AND THE LAW

Regardless of the standard applied, courts hearing constitutional challenges to public health measures should account for public health evidence, as recognized by the Supreme Court in a 1987 disability discrimination case. In School Board of Nassau County, Fla. v. Arline, 480 U.S. 273 (1987), the Court was asked to apply Section 504 of the Rehabilitation Act, a federal disability discrimination statute, in the case of a teacher with latent tuberculosis who was discharged by a public school district. School officials were concerned that the teacher might infect students or others at school if her disease became active while she was employed. The Rehabilitation Act prohibits discrimination against any person who is disabled or perceived as disabled, which includes someone who has or is perceived to have a dangerous infectious disease. The statute protected the teacher from dismissal based on disability only if the teacher was "otherwise qualified" for her position. The

Court opined that, in the context of a person with or perceived to have an infectious disease, a trial court cannot determine whether the person is "otherwise qualified" unless the court makes:

> '[findings of] facts, based on reasonable medical judgments given the state of medical knowledge, about (a) the nature of the risk (how the disease is transmitted), (b) the duration of the risk (how long is the carrier infectious), (c) the severity of the risk (what is the potential harm to third parties) and (d) the probabilities the disease will be transmitted and will cause varying degrees of harm.' In making these findings, courts normally should defer to the reasonable medical judgments of public health officials.

Id. at 288 (internal citation omitted). See also Chapter 2 (discussing federal laws prohibiting discrimination based on disability). Thus, the Supreme Court has recognized the need to account for the scientific facts and public health risks associated with governmental action designed to safeguard population health.

Although *Arline* did not involve a constitutional challenge, its underlying principle of holding courts accountable to public health evidence should apply to any case in which the judiciary reviews measures taken in the name of public health. See Scott Burris, Rationality Review and the Politics of Public Health, 34 Vill. L. Rev. 933 (1989) (arguing for a medical rationality test that focuses judicial scrutiny on the relevant scientific facts concerning a dangerous infectious disease); Robert Gatter, Reviving Focused Scrutiny in the Constitutional Review of Public Health Measures, 64 Wash. U. J. L. & Pol'y 151 (2021).

V. SEPARATION OF POWERS AND ADMINISTRATIVE LAW AS LIMITATIONS ON PUBLIC HEALTH AUTHORITY

Constitutionally protected individual rights are not the only limitations on public health authority. Because executive agencies take action during a crisis to protect public health actions, they are also subject to federal or state administrative procedural codes and to constitutional doctrines deriving from the separation of powers. As exemplified in the materials in this section, public health officials who fail to comply with administrative procedure when implementing a public health measure may find that their measure is legally void. Likewise, public health actions taken by executive agencies or officials can be declared void if they are found to usurp policy-making authority assigned exclusively by a state constitution to the legislative branch. Opponents of ongoing business and social restrictions imposed by health officials during the COVID-19 pandemic have used both of these avenues to challenge those restrictions.

BESHEAR V. ACREE

Supreme Court of Kentucky, 2020.
615 S.W.3d 780.

* * *

FACTS AND PROCEDURAL HISTORY

... On January 31, 2020, the United States Department of Health and Human Services declared a national public health emergency, effective January 27, 2020, based on the rising number of confirmed COVID-19 cases in the United States. . . .

On March 6, 2020, Governor Andy Beshear, under the authority vested in him pursuant to KRS Chapter 39A, declared a state of emergency in Kentucky. [ ]. . . . After the statewide declaration, Kentucky's Cabinet for Health and Family Services (the Cabinet) began issuing orders designed to reduce and slow the spread of COVID-19 and thereby promote public health and safety. Those orders included directives such as prohibiting on-site consumption of food and drink at restaurants, closing businesses that encourage congregation, and prohibiting mass gatherings. As knowledge regarding the heretofore unknown novel coronavirus (COVID-19) grew, the Governor and the Cabinet modified their orders accordingly.

On March 17, 2020, the Cabinet issued an order requiring all public-facing businesses that encourage public congregation to close, including gyms, entertainment and recreational facilities, and theaters. These emergency measures worked to reduce COVID-19 cases by limiting gatherings where the virus could be transmitted. The Governor announced on April 21, 2020, the "Healthy at Work" initiative, a phased reopening plan based on criteria set by public health and industry experts to help Kentucky businesses reopen safely. On May 11, 2020, the Commonwealth began reopening its economy and the Cabinet issued minimum requirements that all public and private entities were required to follow, such as maintaining social distance between persons, requiring employees to wash hands regularly, and routinely cleaning and sanitizing commonly touched surfaces.

[The Court described legal actions taken by various business owners and the state Attorney General in two counties in the spring and summer of 2020. Each sought to enjoin both temporarily and permanently the enforcement of the Governor's COVID-19 orders. One trial court temporarily enjoined enforcement of some of the Governor's orders. The Governor filed for a writ of mandamus, and the Kentucky Supreme Court issued a stay of all lower court orders so as to consolidate the actions and hear them at once. Among the issues before the Court was whether (1) the Governor's orders were an exercise of legislative authority that had not been properly delegated to the executive branch and thus violated the separation of powers doctrine, and (2) the Governor was obligated to act through the promulgation of administrative regulations rather than through a series of blanket executive orders.]

* * *

ANALYSIS

Before turning to the specific issues presented, we briefly address the history of emergency powers legislation, which has existed in Kentucky since 1952. On March 5, 1952, the General Assembly enacted Chapter 39 of the Kentucky Revised Statutes, relating to civil defense. . . .

Recognizing that the Commonwealth is always subject to both contained and widespread threatening occurrences, in 1998 the General Assembly replaced KRS Chapter 39 with KRS Chapter 39A, which establishes a statewide comprehensive emergency management system. In enacting the Chapter, the General Assembly expressly noted that "response to these occurrences is a fundamental responsibility of elected government in the Commonwealth." [ ]. KRS Chapter 39A further expanded the scope of disasters and emergencies which necessitate the Governor's response and, notably, added biological and etiological hazards to the list of threats to public safety. The General Assembly recognized that the purpose of Kentucky's emergency management response had evolved from responding only to security and defense needs to responding to all types of natural and man-made hazards in order to address the contemporary needs of Kentucky citizens. [ ]. . . .KRS Chapter 39A powers have been invoked by every Governor who has served since the law's adoption in 1998. The emergencies have ranged from widespread events such as destructive storms to more localized concerns such as bridges and water supply. Since 1996, an emergency of some magnitude has been declared on approximately 115 occasions, leaving aside the accompanying orders in the face of those occurrences which prohibit price gouging or allow pharmacists to address prescription needs. As we address the issues in this case, we are cognizant of the Commonwealth's history and experience with emergency response.

* * *

II. During the Emergency, the Governor Has Exercised Executive Powers But to the Extent, If Any, KRS Chapter 39A Grants Him Legislative Authority, No Violation of the Separation of Powers Provisions of the Kentucky Constitution Has Occurred, the General Assembly Having Properly Delegated that Authority.

The Kentucky Constitution directs the separation of powers among the legislative, executive and judicial branches [ ] and prohibits any one branch from exercising "any power properly belonging to either of the others, except in the instances hereinafter expressly directed or permitted," [ ]. The

Governor maintains that in responding to the COVID-19 pandemic he has exercised executive powers derived from the Kentucky Constitution and that KRS Chapter 39A simply "recognizes, defines, and constrains" executive authority to direct an emergency response. To the extent any of his actions could be characterized as legislative, he notes that he is exercising authority lawfully delegated to him by the General Assembly in KRS Chapter 39A.

The Attorney General seemingly acknowledges some role for the Governor in the event of an emergency such as COVID-19 but generally insists that the Governor's response these last months via executive orders and emergency regulations is an unconstitutional encroachment on legislative authority. In advocating the striking of those portions of KRS Chapter 39A that permit the Governor to exercise legislative authority, . . . the Attorney General asks us to "use this case to restore the original meaning of the Constitution's separation of powers." To the extent we decline that invitation, he argues that the legislative authority in KRS Chapter 39A has been improperly delegated to the Governor. As we consider this argument, we do so guided by the presumption that the challenged statutes were enacted by the legislature in accordance with constitutional requirements. [ ]. "A constitutional infringement must be 'clear, complete and unmistakable' in order to render the statute unconstitutional." [ ]. Ultimately, we conclude that the Governor is largely exercising emergency executive power but to the extent legislative authority is involved it has been validly delegated by the General Assembly consistent with decades of Kentucky precedent, which we will not overturn.

The current Kentucky Constitution . . . does not address emergency occurrences or events directly except as to military matters which are firmly assigned to the Governor as the "commander-in-chief" of military affairs. [ ]. Generally, Section 69 vests the Governor with the "supreme executive power of the Commonwealth" and Section 81 mandates the Governor "take care that the laws be faithfully executed." Also instructive for the present case, Section 80 provides that the Governor "may, on extraordinary occasions, convene the General Assembly at the seat of government, or at a different place, if that should have become dangerous from an enemy or from contagious diseases. . . . When he shall convene the General Assembly it shall be by proclamation, stating the subjects to be considered, and no other shall be considered."

Although "extraordinary occasions" has been construed customarily to allow special legislative sessions for reasons of immediate import relating to funding and other matters, it plainly extends to those events or occurrences that qualify as a natural or man-made emergency, underscored by the "clue" regarding the convening of the legislature somewhere other than Frankfort in the event of an enemy or contagious diseases. Notably, Section 80 contains the permissive "may . . . convene" as opposed to the mandatory "shall . . . convene." Even in times when the Commonwealth is confronted with something extraordinary, to include enemies and contagious diseases, the decision to convene the General Assembly in a special session is solely the Governor's.

The implied tilt of the Kentucky Constitution toward executive powers in times of emergency is not surprising, given our government's tripartite structure with a legislature that is not in continuous session. . . .

* * *

Having a citizen legislature that meets part-time as opposed to a full-time legislative body that meets year-round, as some states have, generally leaves our General Assembly without the ability to legislate quickly in the event of emergency unless the emergency arises during a regular legislative session. The COVID-19 pandemic arose during the latter part of the 2020 legislative session, after the deadline for introducing a new bill, resulting in fourteen proposed COVID-19 related amendments to existing bills, five of which eventually passed. Most notably, Senate Bill 150, "AN ACT relating to the state of emergency in response to COVID-19 and declaring an emergency," acknowledged the Governor's declared emergency and provided:

> Notwithstanding any state law to the contrary, the Governor shall declare, in writing, the date upon which the state of emergency in response to COVID-19, declared on March 6, 2020, by Executive Order 2020-215, has ceased. In the event no such declaration is made by the Governor on or before the first day of the next regular session of the General Assembly, the General Assembly may make the determination.

. . . .

The Attorney General invites the Court to adopt a strict separation of powers stance by identifying the Governor's issuance of any rules, regulations or orders in an emergency as exercises of non-delegable legislative power (excepting only the Governor's initial declaration of an emergency perhaps) and then holding those emergency responses constitutionally invalid . . . . We decline. First, our reading of the Kentucky Constitution leaves us with no evidence that the powers at issue must be deemed legislative. The "extraordinary occasion," [ ], of a global pandemic gives rise to an obvious emergency and, as noted, the Constitution impliedly tilts to authority in the full-time executive branch to act in such circumstances. Indeed, the Governor's "commander-in-chief" status . . . reinforces the concept. Second, the structure of Kentucky government as discussed renders it impractical, if not impossible, for the legislature, in session for only a limited period each year, to have the primary role in steering the Commonwealth through an emergency.

On this latter point, the Attorney General argues that Section 80 [of the state Constitution] allows the Governor to call an extraordinary session and thus "envisions that the Governor will not go it alone during a crisis, but instead will work hand in hand with the People's representatives." Again, the language of the section is permissive not mandatory, leaving it to the Governor—also duly elected by the People—whether the General Assembly should be convened. Moreover, the view advocated by the Attorney General creates an obvious dilemma: if the Governor is not empowered to adopt emergency measures because that constitutes "legislation," the Commonwealth is left with no means for an immediate, comprehensive response because either the General Assembly is not in session and cannot convene itself or even if in session it will have limited time to deal with the matter under constitutionally mandated constraints on the length of the session. So, our examination of the Kentucky Constitution causes us to conclude the emergency powers the Governor has exercised are executive in nature, never raising a separation of powers issue in the first instance.

Fortunately, the need to definitively label the powers necessary to steer the Commonwealth through an emergency as either solely executive or solely legislative is largely obviated by KRS Chapter 39A, "Statewide Emergency Management Programs," which reflects a cooperative approach between the two branches. Plaintiffs and the Attorney General insist that the statute is in large part unconstitutional, however, because it grants the Governor legislative authority in violation of the nondelegation doctrine. We disagree.

We acknowledge, of course, that making laws for the Commonwealth is the prerogative of the legislature. Addressing a statute that authorizes the Governor to reorganize governmental bodies during the period between annual legislative sessions, we recently observed, "[t]he legislative power we understand to be the authority under the constitution to make the laws, and to alter and repeal them." Beshear v. Bevin, 575 S.W.3d 673, 682 (Ky. 2019) [ ]. "The nondelegation doctrine recognizes that the Constitution vests the powers of government in three separate branches and, under the doctrine of separation of powers, each branch must exercise its own power rather than delegating it to another branch." Id. at 681 [ ]. Nevertheless, we found KRS 12.028, at issue in that case, to be a valid delegation of legislative power, recognizing that legislative power can be delegated "if the law delegating that authority provides 'safeguards, procedural and otherwise, which prevent an abuse of discretion'" thereby " 'protecting against unnecessary and uncontrolled discretionary power.'" Id. at 683 (citations omitted). Our holding was but one in a series of Kentucky cases over several decades addressing the proper delegation of legislative power.

The United States Supreme Court [ ] held that "[i]f Congress shall lay down by legislative act an intelligible principle to which the person or body authorized to [act] . . . is directed to conform, such legislative action is not a forbidden delegation of legislative power." [ ]. Recognition of the delegation of legislative powers in Kentucky largely began with Commonwealth v. Associated Industries of Kentucky, 370 S.W.2d 584, 586 (Ky. 1963): "We find nothing in our State Constitution that declares explicitly: 'Legislative power may not be delegated.' " . . . More recently, [ ], we recognized "given the realities of modern rule-making" a legislative body "has neither the time nor the expertise to do it all; it must have help." [ ]. Examining the nondelegation doctrine generally and finding the "intelligible-principle rule" instructive if somewhat "toothless" in application by the federal courts, [ ], the Court reviewed several Kentucky cases wherein a delegation of legislative authority was deemed unlawful because the "powers were granted without 'legislative criteria,' " [ ], or the delegation lacked "standards controlling the exercise of administrative discretion," [ ]. The "unintelligible" legislative pension statute at issue in [that case] failed for those reasons—lack of "an intelligible principle" and the absence of any "standards controlling the exercise of administrative discretion." [ ].

In the case before us, the intelligible principle enunciated by the General Assembly and the legislative criteria pertinent to the use of emergency powers are set forth in KRS 39A.010 quoted above. In the event of any of those multitude of threats, the Governor (and the Division of Emergency Management and local emergency agencies) are authorized to take action "to protect life and property of the people of the Commonwealth, and to protect public peace, health, safety and welfare . . . and in order to ensure the continuity and effectiveness of government in time of emergency, disaster or catastrophe. . . ." In KRS 39A.100(1), the Governor is granted twelve enumerated "emergency powers" including in subsection (j) the following: "Except as prohibited by this section or other law, to perform and exercise other functions, powers, and duties deemed necessary to promote and secure the safety and protection of the civilian population." Given the wide variance of occurrences that can constitute an emergency, disaster or catastrophe, the criteria are necessarily broad and result-oriented, "protect life and property . . . and . . . public . . . health," KRS 39A.010, allowing the Governor working with the executive branch and emergency management agencies to determine what is necessary for the specific crisis at hand. Floods, tornadoes and ice storms require different responses than threats from nuclear, chemical or biological agents or biological, etiological, or radiological hazards but the emergency powers are always limited by the legislative criteria, i.e., they must be exercised in the context of a declared state of emergency, KRS 39A.100(1); designed to protect life, property, health and safety and to secure the continuity and effectiveness of government, KRS 39A.010; and exercised "to promote and secure the safety and protection of the civilian population." KRS 39A.100(1)(j).

In addition, KRS Chapter 39A contains procedural safeguards to prevent abuses. All written orders and administrative regulations promulgated by the Governor "shall have the full force of law" upon the filing of a copy with the Legislative Research Commission. [ ]. This provides the requisite public notice. The duration of the state of emergency, at least the one at issue in this case, is also limited by the aforementioned 2020 Senate Bill 150, Section 3, which requires the Governor to state when the emergency has ceased but, in any event, allows the General Assembly to make the determination itself if the Governor has not declared an end to the emergency "before the first day of the next regular session of the General Assembly." The enunciation of criteria for use of the emergency powers, the timely, public notice provided for all orders and regulations promulgated by the Governor and the time limit on the duration of the emergency and accompanying powers all combine to render KRS Chapter 39A constitutional to the extent legislative powers are delegated.

Recently the Michigan Supreme Court, in a sharply divided opinion, addressed two certified questions posed by the federal district court regarding the Michigan Governor's exercise of emergency powers under that state's Emergency Management Act of 1976 (EMA) and Emergency Powers of the Governor Act of 1945 (EPGA). [ ]. The EPGA gave the Governor power, indefinite in duration, to declare an emergency and issue "reasonable orders, rules, and regulations as he or she considers necessary to protect life and property or to bring the emergency situation within the affected area under control." [ ]. The majority concluded the "reasonable" and "necessary" standard failed to provide sufficient guidance to the Governor regarding the exercise of her powers and failed to constrain her actions "in any meaningful manner." [ ].

Finding the power delegated to be "of immense breadth and . . . devoid of all temporal limitations," [ ], the majority struck the statute as an unlawful delegation of legislative power to the executive branch violative of the Michigan Constitution's separation of powers provision. Chief Justice McCormack, writing for the three-justice minority, observed that the majority departed from one part of their longstanding test for delegation of legislative power, namely that "the standard must be as reasonably precise as the subject matter requires or permits." [ ]. Citing Gundy v. United States, ___ U.S. ___, 139 S. Ct. 2116, 2130, 204 L.Ed.2d 522 (2019), for the proposition that delegations of such authority must give the delegee "the flexibility to deal with real-world constraints," she noted that "given the unpredictability and range of emergencies the Legislature identified in the statute, it is difficult to see how it could have been more specific." [ ].

Our case differs from the Michigan case in several important ways but most notably our Governor does not have emergency powers of indefinite duration, 2020 S.B. 150, § 3, and our legislature is not continuously in session, ready to accept the handoff of responsibility for providing the government's response to an emergency such as the current global pandemic. Moreover, with the breadth of potential emergencies identified in KRS 39A.010, the standards of protection of life, property, peace, health, safety and welfare (along with the "necessary" qualifier in KRS 39A.100(j)) are sufficiently specific to guide discretion while appropriately flexible to address a myriad of real-world events. While the authority exercised by the Governor in accordance with KRS Chapter 39A is necessarily broad, the checks on that authority are the same as those identified in Chief Justice McCormack's dissenting opinion: judicial challenges to the existence of an emergency or to the content of a particular order or regulation; legislative amendment or revocation of the emergency powers granted the Governor; and finally the "ultimate check" of citizens holding the Governor accountable at the ballot box. [ ].

Whatever import the principle of properly delegated legislative authority has in the ordinary workings of government, its import increases dramatically in the event of a statewide emergency in our Commonwealth. A legislature that is not in continuous session and without constitutional authority to convene itself cannot realistically manage a crisis on a day-to-day basis by the adoption and amendment of laws. In any event, we decline to abandon approximately sixty years of precedent that appropriately channels and limits the delegation of legislative power in Kentucky. Applying that delegation precedent, KRS Chapter 39A passes muster as a constitutional delegation of power to the extent any of the powers accorded to and exercised by the Governor are in fact legislative.

In sum, the powers exercised by a Kentucky Governor in an emergency are likely executive powers in the first instance given provisions of our Kentucky Constitution, but to the extent those powers are seen as impinging on the legislative domain, our General Assembly has wisely addressed the situation in KRS Chapter 39A. That vital and often-used statutory scheme validly delegates any legislative authority at issue to the Governor with safeguards and criteria sufficient to pass constitutional muster.

III. KRS Chapter 13A Does Not Limit the Governor's Authority to Act Under the Constitution and KRS Chapter 39A in the Event of an Emergency.

KRS Chapter 13A, "Administrative Regulations," provides for the promulgation of administrative regulations—defined in relevant part as a "statement of general applicability . . . that implements, interprets, or prescribes law or policy," [ ]—both in the ordinary course of state government . . . and in the event of an emergency, [ ]. Plaintiffs and the Attorney General challenge the executive orders and regulations issued by the Governor as violative of KRS Chapter 13A. The Governor maintains that KRS Chapter 39A by its plain terms controls in a declared emergency, granting him the authority he has exercised but that if any conflict is perceived then the more specific statutory enactment pertaining to emergencies prevails. The plain language of the statutes supports the Governor's position. [ ].

KRS 39A.100 recognizes the authority of the Governor to declare an emergency and exercise the enumerated emergency powers. In furtherance of that authority, KRS 39A.090 provides that "[t]he Governor may make, amend, and rescind any executive orders as deemed necessary to carry out the provisions of KRS Chapters 39A to 39F." Nothing in the plain words used requires consideration of KRS Chapter 13A or even requires promulgation of regulations; the Governor can choose to act solely through executive orders. The Governor may also promulgate regulations, however, as authorized by KRS 39A.180:

> (2) All written orders and administrative regulations promulgated by the Governor, the director, or by any political subdivision or other agency authorized by KRS Chapters 39A to 39F to make orders and promulgate administrative regulations, shall have the full force of law, when, if issued by the Governor, the director, or any state agency, a copy is filed with the Legislative Research Commission .... All existing laws, ordinances, and administrative regulations inconsistent with the provisions of KRS Chapters 39A to 39F, or of any order or administrative regulation issued under the authority of KRS Chapters 39A to 39F, shall be suspended during the period of time and to the extent that the conflict exists.

[ ]. This statute plainly provides that the orders and regulations issued pursuant to the emergency authority granted the Governor in KRS Chapter 39A "shall have full force of law" upon filing with the Legislative Research Commission (LRC), the same entity that compiles, publishes and distributes administrative regulations generally. [ ]. To the extent KRS Chapter 13A contains anything "inconsistent" with either Chapter 39A or an order or regulation issued under the authority of that chapter then the General Assembly has expressly directed that it "shall be suspended during the period of time and to the extent that the conflict exists." [ ]. In short, while a state of emergency prevails the Governor can issue executive orders he or she "deem[s] necessary," ...

* * *

In insisting that the Governor must use the regulatory process to affect private rights, the Attorney General emphasizes our statement in Bowling v. Department of Corrections, [ ]: "Regulation is ... mandated by KRS 13A.100, which requires regulation if, as here, the regulation will prescribe statements of general applicability which implement laws ... or affect private rights." Missing from this argument is any recognition that no state of emergency existed in Bowling, but, more importantly, any acknowledgement of the General Assembly's specific directive in KRS 39A.180(2) that inconsistent laws, ordinances and regulations are suspended by KRS Chapter 39A and the executive orders and regulations issued pursuant thereto. We find nothing strange about the legislature giving the Governor flexibility in the event of an emergency to act through either executive orders or regulations, the former being more suited to immediate response in the acute state of an emergency. . . .

In short, the General Assembly has answered this argument for us. KRS 39A.180(2) suspends any inconsistent laws. To the extent KRS Chapter 13A requires more than KRS Chapter 39A, the regular process applicable to administrative regulations has been displaced.

[Temporary injunction issued by the trial court is reversed and the matter is remanded.]

NOTES AND QUESTIONS

1. *Judicial Review of Administrative Actions.* Several lawsuits have challenged COVID-19 restrictions issued through executive branch orders on the same grounds as those used by the plaintiffs in *Beshear*, and—while most of the public health measures at issue in those lawsuits survived—some did not. For example, in *Wisconsin Legislature v. Palm*, 942 N.W.2d 900 (Wis. 2020), the Wisconsin legislature challenged a statewide "Safer at Home" order issued by the state's chief health officer in response to the population health risks posed by the COVID-19 pandemic. The legislature pursued two claims. First, the legislature alleged that the order was, in fact, a "rule" under the state's administrative procedures act and was void because it was issued without following "rulemaking" procedures. A majority of the Wisconsin Supreme Court was persuaded. The Court held that, because the executive order required individuals statewide to stay at home except to conduct certain essential tasks and close all non-essential businesses statewide, it was an agency action of "general applicability" and thus was a rule subject to standard or emergency rulemaking procedures. A dissenting opinion argued that the majority failed to account adequately for language in the public health agency's authorizing statute that empowers the agency to promulgate rules and issue "orders" necessary to protect population health. Additionally, the dissent highlighted the absurdity of relying on emergency rulemaking to respond to a fast-moving infectious disease threat because even emergency rules require between 18 and 49 days to be implemented.

Second, the legislature argued that the statute authorizing the Wisconsin Department of Health Services to all actions "necessary" to protect public health against an infectious disease threat lacked the substantive specificity and procedural safeguards required to delegate legislative authority to an executive agency. Again, a majority of the Court agreed. Rather than rely on a constitutional non-delegation doctrine, the Court based its holding on a state statute that required all agency actions to be based on an explicit, statutory delegation of authority. The majority ruled that this statute eliminated the state's public health agency's ability to rely on its broadly worded authority in its enabling statute (to take all actions necessary to prevent the spread of communicable diseases) as the basis for its Safer at Home order.

2. *Non-Delegation Doctrine.* The non-delegation doctrine prohibits the legislative branch from generally delegating its legislative authority to any other branch of government, but it permits such delegations where the legislature has provided an "intelligible principle" that will guide and constrain those exercising the delegated powers. A key issue is whether statutes broadly authorizing executive agencies and officials to take any and all actions "necessary to protect the public health" from disease threats constitute an "intelligible principle."

In *Beshear*, the Kentucky Supreme Court held that such broad language provided an intelligible principle because the statute attempted to account for a wide variety of emergencies that could arise, which justified the use of broadly worded language designed to direct its use, and because the state legislature is often not in session. Several other courts have addressed the same issue. In Casey v. Lamont, 258 A.3d 647 (Conn. 2021), for example, a state statute granted emergency power to the governor to achieve "the efficient and expeditious execution of civil preparedness functions or the protection of the public health" and to take "such other steps as are reasonably necessary in the light of the emergency to protect the health, safety and welfare of the people of the state." The court held that the statute provides an "intelligible principle" limiting that power and thus was not an unconstitutional delegation of legislative authority to the executive branch. Similarly, in Newsom v. Superior Ct. of Sutter Cty., 278 Cal. Rptr. 3d 397 (Cal. App. 2021), the statement in an emergency management statute that its purpose is "to ensure that 'all emergency services functions' of the State and local governments, the federal government, and 'private agencies of every type,' 'be coordinated . . . to the end that the most effective use be made of all manpower, resources, and facilities for dealing with any emergency that may occur' " was held sufficient "to direct implementation of legislative policy" and thus was not an unconstitutional delegation of legislative authority to the executive branch. But see, e.g., In re Certified Questions From United States Dist. Ct., W. Dist. of Michigan, S. Div., 958 N.W.2d 1 (2020) (delegation of power to governor under expansive emergency powers statute to "promulgate reasonable orders, rules, and regulations as he or she considers necessary to protect life and property" lacked sufficient legislative guidance and thus was an unconstitutional delegation of legislative authority to the executive branch).

3. *Rules vs. Orders.* During the COVID-19 pandemic, several courts in addition to those in *Beshear* and *Wisconsin Legislature* have addressed the distinction between administrative "rules" and "orders" as those are defined in a state administrative procedure act (APA). See, e.g., Grisham v. Romero, 483 P.3d 545 (N. Mex. 1021) (suggesting that emergency public health orders are not rules because they do not create standards for the future but, instead, respond temporarily to current emergency circumstances, and holding that the

state statute authorizing health officials to close businesses so as to protect population health creates a power to issue orders in addition to the power to promulgate rules). The Model State APA defines an order as "an agency decision that determines or declares the rights, duties, privileges, immunities, or other interests of a specific person," and it defines a rule as "an agency statement of general applicability that implements, interprets, or prescribes law or policy . . . and has the force of law." Because state and local mask orders, stay-at-home orders, and other social distancing orders issued during the pandemic apply to populations rather than specifically named individuals, they arguably fit the definition of being an administrative "rule." Rules—unlike orders—are subject to time-consuming procedures such as publishing notice of the proposed rule and providing time for public comment.

The Model State APA recognizes an exception for emergency rules designed to protect, among other things, public health. Yet such emergency rules expire after six months and, in some jurisdictions, cannot be renewed. See, e.g., Grisham v. Romero, cited above, which describes such limitations to emergency rulemaking. How does the distinction between administrative rules and orders affect legal preparedness for infectious disease threats?

4. *Legislative Claw-Back of Executive Public Health Powers.* State legislators unhappy with executive orders designed to prevent the spread of COVID-19 and unsatisfied with judicial application of doctrines related to the separation of powers have used their legislative powers to claw back public health powers from governors and executive branch public health agencies and officials. For example, new legislation in Ohio authorizes the state legislature to rescind any order or rule issued by the state's department of health or its director. Ohio Rev. Code Ann. § 101.36. Montana has enacted legislation that authorizes local legislative bodies to override emergency public health orders of local public health officials and that prohibits local public health officials from taking any action during an emergency that would interfere in any way with a person's attending a religious service in person or with the operation of a place of worship. Mont. Code Ann. § 50–2–118. For additional examples and an analysis of the risk associated with such legislative claw-backs, see NACCHO & Network for Public Health Law, Proposed Limits on Public Health Authority: Dangerous for Public Health (May 2021).

PROBLEM: EBOLA AND RETURNING AID WORKER

Chris is a nurse who volunteers with the organization "Nurses Without Borders" (NWB) to provide international medical aid. He spent the last 4 weeks treating Ebola patients in Liberia. While there, Chris wore head-to-toe protective gear (including gloves and enclosed headgear) whenever he was treating patients. He never experienced a needle prick, and neither his skin nor any other part of his body ever came into direct contact with the bodily fluids of, or any object that had come into contact with, any patient. The CDC's Ebola factsheet states:

Ebola . . . is a rare and deadly disease caused by infection with one of the Ebola virus species. . . .

Ebola is spread through direct contact (through broken skin or unprotected mucous membranes in, for example, the eyes, nose, or mouth) with:

- blood or body fluids (including but not limited to feces, saliva, sweat, urine, vomit, breast milk, and semen) of a person who is sick with Ebola,

- objects (like needles and syringes) that have been contaminated with the virus, . . .

- and . . . possibly from contact with semen from a man who has recovered from Ebola (for example, by having oral, vaginal, or anal sex).

Ebola is not spread through the air or by water, or in general, by food. . . . There is no evidence that mosquitos or other insects can transmit Ebola virus. . . .

A person infected with Ebola virus is not contagious until symptoms appear. Signs and symptoms of Ebola include: fever, severe headache, fatigue, muscle pain, weakness, diarrhea, vomiting, stomach pain, unexplained bleeding or bruising. Symptoms may appear anywhere from 2 to 21 days after exposure to the virus, but the average is 8 to 10 days.

There is a 24- to 48-hour time period after a person infected with Ebola first experiences a fever, headache or fatigue before he exhibits the "wet symptoms" (diarrhea, vomiting, bleeding) that pose a significant risk of transmission to others. Thus, of the roughly 30,000 Ebola cases recorded over 42 years, all have resulted from an uninfected person coming into contact with the wet symptoms of an infected person. There are no recorded cases of an infected person transmitting Ebola to another while experiencing only fever, headache or fatigue.

Throughout his deployment in Liberia, Chris felt fine. A day after his last shift, he took a series of flights until he re-entered the U.S. at Newark International Airport in New Jersey. Again, Chris felt fine throughout his return trip, and he did not display any symptoms.

When Chris de-planed in New Jersey, a U.S. Department of Homeland Security (DHS) officer asked him whether he had traveled to any west African countries, and Chris replied, "Yes, I was stationed with NWB in Liberia." The DHS officer asked if Chris had had any contact with any person sick with Ebola, and Chris said he had been treating Ebola patients as part of his deployment. The officer asked Chris additional questions about his exposure to Ebola patients and his health. Chris explained that he had worn protective gear whenever he treated a patient, that he had never come into direct contact with any patient, and that he had no other exposure to Ebola other than while treating his patients. Chris also reported that his health was fine. The officer noted that Chris did not report or appear to be suffering from any symptoms of any illness. Additionally, the officer took Chris's temperature three times, and each time it was normal.

The DHS officer asked Chris to wait while he informed New Jersey public health officials about Chris's status and the fact that DHS was about to permit him to enter New Jersey. An hour later, a New Jersey public health official visited Chris and told him that, "in an abundance of caution," the state would quarantine him at a nearby hospital until the incubation period for Ebola lapsed in about 19 days.

Is the proposed quarantine of Chris constitutional? If so, why? If not, what, if any, public health action may the state of New Jersey take without violating the federal Constitution?

VI.   NON-COMMUNICABLE DISEASE AND PUBLIC HEALTH

Public health has focused primarily on infectious disease. Yet, several non-communicable diseases pose greater threats. The CDC has identified heart disease, cancer, chronic lower respiratory disease, stroke, and diabetes as among the ten leading causes of death in the U.S. Actions by public health agencies designed to reduce the incidence of these diseases— restrictions on tobacco use, warnings on cigarette boxes, disclosure of fast- food nutritional information—often face legal challenge because those actions seek to change individual lifestyles and because those actions are opposed by entrenched business interests.

BOREALI V. AXELROD
Court of Appeals of New York, 1987.
517 N.E.2d 1350.

TITONE, J.

We hold that the Public Health Council overstepped the boundaries of its lawfully delegated authority when it promulgated a comprehensive code to govern tobacco smoking in areas that are open to the public. While the Legislature has given the Council broad authority to promulgate regulations on matters concerning the public health, the scope of the Council's authority under its enabling statute must be deemed limited by its role as an administrative, rather than a legislative, body. In this instance, the Council usurped the latter role and thereby exceeded its legislative mandate, when, following the Legislature's inability to reach an acceptable balance, the Council weighed the concerns of nonsmokers, smokers, affected businesses and the general public and, without any legislative guidance, reached its own conclusions about the proper accommodation among those competing interests. In view of the political, social and economic, rather than technical, focus of the resulting regulatory scheme, we conclude that the Council's actions were ultra vires and that the order and judgment of the courts below, which declared the Council's regulations invalid, should be affirmed.

I.   LEGISLATIVE BACKGROUND
AND REGULATORY SCHEME

More than two decades ago, the Surgeon General of the United States began warning the American public that tobacco smoking poses a serious health hazard. Within the past five years, there has been mounting evidence that even nonsmokers face a risk of lung cancer as a result of their exposure to tobacco smoke in the environment. As a consequence, smoking in the workplace and other indoor settings has become a cause for serious concern among health professionals [ ].

This growing concern about the deleterious effects of tobacco smoking led our State Legislature to enact a bill in 1975 restricting smoking in certain designated areas, specifically, libraries, museums, theaters and public transportation facilities [ ]. Efforts during the same year to adopt more expansive restrictions on smoking in public places were, however, unsuccessful. . . .

In late 1986 the Public Health Council (PHC) took action of its own. Purportedly acting pursuant to the broad grant of authority contained in its enabling statute [ ], the PHC published proposed rules, held public hearings and, in February of 1987, promulgated the final set of regulations prohibiting smoking in a wide variety of indoor areas that are open to the public, including schools, hospitals, auditoriums, food markets, stores, banks, taxicabs and limousines. Under these rules, restaurants with seating capacities of more than 50 people are required to provide contiguous nonsmoking areas sufficient to meet customer demand. Further, employers are required to provide smoke-free work areas for nonsmoking employees and to keep common areas free of smoke, with certain limited exceptions for cafeterias and lounges. Affected businesses are permitted to prohibit all smoking on the premises if they so choose. Expressly excluded from the regulations' coverage are restaurants with seating capacities of less than 50, conventions, trade shows, bars, private homes, private automobiles, private social functions, hotel and motel rooms and retail tobacco stores. . . .

II.   PROCEDURAL HISTORY

[Several parties affected by the PHC's antismoking regulations sued, and a trial court set aside the regulations, holding that they exceeded PHC's statutory authority. An intermediate appellate court affirmed, and this appeal followed.]

III. ANALYSIS

Preliminarily, we stress that this case presents no question concerning the wisdom of the challenged regulations, the propriety of the procedures by which they were adopted or the right of government in general to promulgate restrictions on the use of tobacco in public places. The degree of scientific support for the regulations and their unquestionable value in protecting those who choose not to smoke are, likewise, not pertinent except as background information. Finally, there has been no argument made concerning the personal freedoms of smokers or their "right" to pursue in public a habit that may inflict serious harm on others who must breathe the same air. The only dispute is whether the challenged restrictions were properly adopted by an administrative agency acting under a general grant of authority and in the face of the Legislature's apparent inability to establish its own broad policy on the controversial problem of passive smoking. Accordingly, we address no other issue in this appeal.

A.   The Delegation/Separation of Powers Issue

Section 225(5)(a) of the Public Health Law authorizes the PHC to "deal with any matters affecting the . . . public health". At the heart of the present case is the question whether this broad grant of authority contravened the oft-recited principle that the legislative branch of government cannot cede its fundamental policymaking responsibility to an administrative agency. As a related matter, we must also inquire whether, assuming the propriety of the Legislature's grant of authority, the agency exceeded the permissible scope of its mandate by using it as a basis for engaging in inherently legislative activities. . . .

However facially broad, a legislative grant of authority must be construed, whenever possible, so that it is no broader than that which the separation of powers doctrine permits [ ]. Even under the broadest and most open-ended of statutory mandates, an administrative agency may not use its authority as a license to correct whatever societal evils it perceives [ ]. Here, we cannot say that the broad enabling statute in issue is itself an unconstitutional delegation of legislative authority. However, we do conclude that the agency stretched that statute beyond its constitutionally valid reach when it used the statute as a basis for drafting a code embodying its own assessment of what public policy ought to be. Our reasons follow.

Derived from the separation of powers doctrine, the principle that the legislative branch may not delegate all of its law-making powers to the executive branch has been applied with the utmost reluctance—even in the early case law. . . .

The modern view is reflected in this court's statement in *Matter of Levine v. Whalen* [ ]: "Because of the constitutional provision that '[t]he legislative power of this State shall be vested in the Senate and the Assembly' (NY Const, art III, § 1), the Legislature cannot pass on its lawmaking functions to other bodies ... but there is no constitutional prohibition against the delegation of power, with reasonable safeguards and standards, to an agency or commission to administer the law as enacted by the Legislature"....

* * *

A number of coalescing circumstances that are present in this case persuade us that the difficult-to-define line between administrative rulemaking and legislative policymaking has been transgressed. While none of these circumstances, standing alone, is sufficient to warrant the conclusion that the PHC has usurped the Legislature's prerogative, all of these circumstances, when viewed in combination, paint a portrait of an agency that has improperly assumed for itself "[t]he open-ended discretion to choose ends" [ ], which characterizes the elected Legislature's role in our system of government.

First, while generally acting to further the laudable goal of protecting nonsmokers from the harmful effects of "passive smoking," the PHC has, in reality, constructed a regulatory scheme laden with exceptions based solely upon economic and social concerns. The exemptions the PHC has carved out for bars, convention centers, small restaurants, and the like, as well as the provision it has made for "waivers" based on financial hardship, have no foundation in considerations of public health. Rather, they demonstrate the agency's own effort to weigh the goal of promoting health against its social cost and to reach a suitable compromise....

Striking the proper balance among health concerns, cost and privacy interests, however, is a uniquely legislative function. While it is true that many regulatory decisions involve weighing economic and social concerns against the specific values that the regulatory agency is mandated to promote, the agency in this case has not been authorized to structure its decision making in a " cost-benefit" model [ ] and, in fact, has not been given any legislative guidelines at all for determining how the competing concerns of public health and economic cost are to be weighed. Thus, to the extent that the agency has built a regulatory scheme on its own conclusions about the appropriate balance of trade-offs between health and cost to particular industries in the private sector, it was "acting solely on [its] own ideas of sound public policy" and was therefore operating outside of its proper sphere of authority [ ]. This conclusion is particularly compelling here, where the focus is on administratively created exemptions rather than on rules that promote the legislatively expressed goals, since exemptions ordinarily run counter to such goals and, consequently, cannot be justified as simple implementations of legislative values [ ]....

The second, and related, consideration is that in adopting the antismoking regulations challenged here the PHC did not merely fill in the details of broad legislation describing the over-all policies to be implemented. Instead, the PHC wrote on a clean slate, creating its own comprehensive set of rules without benefit of legislative guidance. Viewed in that light, the agency's actions were a far cry from the "interstitial" rule making that typifies administrative regulatory activity [ ].

A third indicator that the PHC exceeded the scope of the authority properly delegated to it by the Legislature is the fact that the agency acted in an area in which the Legislature had repeatedly tried—and failed—to reach agreement in the face of substantial public debate and vigorous lobbying by a variety of interested factions. While we have often been reluctant to ascribe persuasive significance to legislative inaction [ ], our usual hesitancy in this area has no place here. Unlike the cases in which we have been asked to consider the Legislature's failure to act as some indirect proof of its actual intentions [ ], in this case it is appropriate for us to consider the significance of legislative inaction as evidence that the Legislature has so far been unable to reach agreement on the goals and methods that should govern in resolving a society-wide health problem. Here, the repeated failures by the Legislature to arrive at such an agreement do not automatically entitle an administrative agency to take it upon itself to fill the vacuum and impose a solution of its own. Manifestly, it is the province of the people's elected representatives, rather than appointed administrators, to resolve difficult social problems by making choices among competing ends.

Finally, although indoor smoking is unquestionably a health issue, no special expertise or technical competence in the field of health was involved in the development of the antismoking regulations challenged here. Faced with mounting evidence about the hazards to bystanders of indoor smoking, the PHC drafted a simple code describing the locales in which smoking would be prohibited and providing exemptions for various special interest groups. . . .

In summary, we conclude that while Public Health Law § 225 (5) (a) is a valid delegation of regulatory authority, it cannot be construed to encompass the policy-making activity at issue here without running afoul of the constitutional separation of powers doctrine. . . .

* * *

Accordingly, the order of the Appellate Division should be affirmed.

NOTES AND QUESTIONS

1.   *Administrative Law. Boreali* is representative of constitutional and administrative law challenges to public health actions aimed at non-communicable diseases. Because public health measures are commonly taken

by local, state, and federal agencies, doctrines such as separation-of-powers, non-delegation, and *ultra vires* can curtail public health powers. For example, the same court that decided *Boreali* later set aside a "soda ban" that prohibited many businesses from selling sugary drinks in sizes larger than 16 ounces—a measure designed to help fight obesity. See New York Statewide Coalition of Hispanic Chambers of Commerce v. New York City Dept. of Health & Mental Hygiene, 16 N.E.3d 538, 992 N.Y.S.2d 480, 23 N.Y.3d 681 (2014); c.f. NYC C.L.A.S.H., Inc. v. New York State Office of Parks, Recreation and Historic Preservation, 27 N.Y.3d 174, 51 N.E.3d 512, 32 N.Y.S.3d 1 (2016) (using the *Boreali* factors to uphold agency's outdoor smoking ban in parts of New York City's parks, historic places, and recreational facilities).

2. *Free Speech and Public Health.* The First Amendment, through the commercial speech doctrine, has also been interpreted so as to limit public health authority and expand the rights of tobacco, food, and drug manufacturers. For example, in Sorrell v. IMS Health, 564 U.S. 552 (2011), the Supreme Court struck down a Vermont statute restricting the collection and disclosure of pharmacy prescribing records of individual physicians. The Court held that the restriction suppressed protected speech under the First Amendment. It then applied heightened scrutiny, requiring the State to show that the statute "directly advanced" a "substantial governmental interest" and that it was "drawn to achieve" that interest, meaning that there must be "a fit between the legislature's ends and the means chosen to accomplish those ends." Under this standard, the Court held that Vermont's interests in protecting physician privacy, preventing harassing sales behavior by manufacturers, and preserving the doctor-patient relationship were not sufficient justifications for the restriction. Additionally, it ruled that the State failed to establish a fit between the restrictions and the State's goals of promoting public health and lowering medical costs.

By giving little weight to the State's interest in protecting public health, *Sorrel* undermines the stability of laws restricting language used to market food, tobacco, and prescription drugs, as well as laws mandating certain disclosures. For example, the Second Circuit subsequently ruled that off-label promotion of a pharmaceutical was protected commercial speech subject to at least heightened scrutiny. See U.S. v. Caronia, 703 F.3d 149 (2d Cir. 2012). Meanwhile, the D.C. Circuit flip-flopped on whether regulators may justify disclosure mandates based on interests other than correcting prior deception. In R.J. Reynolds Tobacco Co. v. FDA, 696 F.3d 1205 (D.C. Cir. 2012), the court set aside FDA regulations requiring cigarette manufacturers to place graphic warnings on packaging, holding that the First Amendment requires the government to first establish that existing cigarette packaging is misleading to consumers, which it did not show. But two years later, in American Meat Inst. v. U.S. Dept. of Agriculture, 760 F.3d 18 (D.C. Cir. 2014), the same court overruled its decision in *R.J. Reynolds*.

By broadening commercial speech protection and narrowing the discretion of public health officials, First Amendment doctrine poses a significant challenge to improving population health. See Joshua Sharfstein, Public Health and the First Amendment, 93 Milbank Q. 459 (2015).

PROBLEM: OBESITY AS A PUBLIC HEALTH PROBLEM

As a legislator in your state, you have become very concerned about obesity in the population, and you are worried about the long-range consequences of this condition on the vitality of the public. In particular, you have become concerned about the proliferation of fast-food sites that offer primarily economical but very high fat foods. You are concerned about "food deserts" in your state—areas where there are no supermarkets and limited public transportation. You are also disturbed by the relatively unhealthy foods available at the public schools throughout the state, many of which have contracts with beverage companies to run their vending machines and snack bars, and by the disappearance of mandatory physical education requirements at the high school level. There is also evidence that a comparatively sedentary lifestyle that includes watching television and playing video games contributes to what some have called the "epidemic" of obesity. Furthermore, studies have repeatedly shown that obesity alone is not a predictor for shortened lifespans or significantly greater health risks. Rather, a lack of exercise or fitness is more closely associated with health risks than is obesity. A person can be obese and healthy.

Is the problem of obesity a "public health" problem, or simply a problem for some individuals? What qualifies as "obesity" for public health purposes? Is it a matter of personal responsibility that should be unrelated to the obligations of the law? Would it make any difference if you classified obesity as a "public health" problem? What legal responses might be available to the state legislature to address this problem? Are there ways to use school law, tax law, criminal law, tort law, food and drug law, insurance rate-setting, public program eligibility standards or general public health regulatory law to address this problem? What evidence is there that pulling any of these legal levers would be effective at lowering the incidence of obesity? What are the legal, political and social consequences of doing so? Are there civil liberty interests of persons who are obese (or others) that are at stake in using the law in this way? Do you feel the same way about tobacco, gun violence, bullying, or "extreme sports" as public health problems?

INDEX

References are to Pages
